MW01526645

Food & Nutrition for You

Suzanne Weixel

Faithe Wempen

Prentice Hall

Boston • Columbus • Indianapolis • New York • San Francisco • Upper Saddle River
Amsterdam • Cape Town • Dubai • London • Madrid • Milan • Munich • Paris • Montreal • Toronto
Delhi • Mexico City • Sao Paulo • Sydney • Hong Kong • Seoul • Singapore • Taipei • Tokyo

Editor in Chief: Vernon Anthony
Acquisitions Editor: William Lawrensen
Editorial Assistant: Lara Dimmick
Director of Marketing: David Gesell
Campaign Marketing Manager: Leigh Ann Sims
Curriculum Marketing Manager: Thomas Hayward
Marketing Assistant: Les Roberts
Associate Managing Editor: Alexandrina Benedicto Wolf
Project Manager: Emergent Learning, LLC
Senior Operations Supervisor: Pat Tonneman
Operations Specialist: Deidra Schwartz
Text Designer: Vanessa Moore
Cover Designer: Keithley & Associates, Inc.
Cover Images: Shutterstock
Manager, Rights and Permissions: Zina Arabia
Image Permission Coordinator: Craig Jones
Full-Service Project Management: Emergent Learning, LLC
Composition: Vanessa Moore
Printer/Binder: Courier Kendallville
Cover Printer: Moore Langen, a Courier Company
Text Font: 11/14 Minion Pro

Credits and acknowledgments borrowed from other sources and reproduced, with permission, in this textbook appear on appropriate page within text.

---

10 9 8 7 6 5 4 3 2 1

Prentice Hall
an imprint of

PearsonSchool.com/careertech

ISBN-10: 0-13-508728-7
ISBN-13: 978-0-13-508728-2

Table of Contents

PART 1 ■ THE FOOD WE EAT

1 ■ The Food You Eat

2 ■ Making Smart Eating Decisions

3 ■ Planning a Meal

4 ■ The Physiology of Food

5 ■ Special Diets for Weight Loss and Fitness

6 ■ Eating and Nutrition-Related Disorders

PART 2 ■ KITCHEN BASICS

7 ■ Following and Modifying Recipes

8 ■ Sanitary Food Handling and Food Safety

9 ■ Using Basic Kitchen Equipment

10 ■ Kitchen and Dining Plans and Etiquette

PART 3 ■ NUTRITION AND COOKING

11 ■ Grains, Pasta, and Legumes

12 ■ Meats

13 ■ Poultry

14 ■ Fish

15 ■ Spices, Herbs, and Garnishes

16 ■ Fats and Oils

17 ■ Dairy and Eggs

18 ■ Fruits and Nuts

19 ■ Vegetables

20 ■ Soups and Casseroles

21 ■ Breads

22 ▪ Desserts

Introduction

Food & Nutrition for You is a comprehensive text that prepares students for a healthy lifestyle. It provides essential information about how food supplies the nutrients every body needs, and how to make healthy food choices. In addition, it introduces students to safe and sanitary food handling, proper use of kitchen equipment, culinary arts, and basic cooking techniques.

This clear, concise text is designed to engage students by focusing on subjects that are relevant to their lives today. It puts facts into context by utilizing real-life examples, and presents information in short, colorful segments that capture the attention and interest of students.

The book is organized into three sections:

In Part 1, The Food We Eat, students are introduced to basic information about nutrition, and how eating provides the fuel our bodies need. They learn about the factors that influence our decisions about what to eat, and how to make healthy choices using dietary guidelines and MyPyramid. Students investigate the role food plays in families and societies, and they learn how to plan meals for a variety of people and occasions, including budgeting and shopping.

Part 2, Kitchen Basics, prepares students for entering the kitchen and learning how to cook. This section covers all aspects of recipes, including how to measure, use fractions, and make conversions. Students learn how to identify and use kitchen equipment, basic cooking methods, and how to present food and serve meals. In depth coverage of sanitation and food safety includes information about how to recognize and eliminate potential risks and dangers, government regulations, and industry standard methods and procedures.

In Part 3, Nutrition and Cooking, students learn about the different types of foods, including nutritional value, purchasing and storage procedures, and basic preparation skills. Each chapter includes three or more recipes that highlight the selected type of food.

A careers appendix covers how to prepare for and obtain a career in the food service industry; and a comprehensive glossary includes food and nutrition terms and definitions.

Integrated throughout the book are features that provide tips, descriptions, and discussion points about relevant topics, such as assuring a safe food supply, the impact of technology on food, sustainable agriculture, food allergies, organic farming, and international cuisine. These features are visually designed to stand out from the main text using color and graphics. They include:

- *Hot Topics*, which present timely and sometimes controversial trends in food and nutrition
- *Cool Tips*, offering suggestions, fun facts, and tips
- *Safe Eats*, which provide food safety and sanitation information
- *What's Cooking?*, which include information about specific foods, nutrition, and making healthy food choices
- *Basic Culinary Skills* provide step-by-step, illustrated instructions for tasks
- *Career Counsel* for tips and facts about working in the food service industry
- *Check the Label*, which lists descriptive terms and definitions
- *Utility Drawer*, containing descriptions of specialized tools and equipment
- *Figures*, which are photos that include captions and questions to encourage critical thinking and discussion

Numerous activities in every chapter integrate the core subjects of math, science, language arts, and social studies with the food and nutrition content. These features are designed to teach important skills such as collaboration, global awareness, critical thinking, and community involvement while reinforcing important messages about health, wellness, nutrition, and smart choices.

Chapter activities include:

- *Why You Need to Know This*, a "bellringer" or opening activity designed to focus the class on the current subject while encouraging discussion and collaboration
- *Science Study* activities provide an opportunity for students to use the scientific process to test the food and nutrition concepts presented in the text
- *By the Numbers* activities integrate basic math skills with food and nutrition concepts by measuring, calculating, and converting
- *Case Studies*, which present real life scenarios with questions to prompt critical thinking and problem solving
- *Put It to Use* activities that provide students with the opportunity to put the information they studied in the chapter into context in their own lives

- *Write Now* activities challenge students to use language arts skills to produce essays, stories, and other types of written documents relating to the chapter content
- *Tech Connect* activities provide the opportunity for students to use technology in self-directed exploration of relevant topics, and to communicate the information to their peers
- *Team Players* are group activities where students can work collaboratively on projects ranging from research presentations to cooking competitions
- *Put It Together* is a matching activity that reinforces vocabulary and terminology
- *Try It!* are recipes in a range of difficulty and complexity that can be adapted to any cooking environment. They include nutritional information, safety guidelines, and chef's tips and suggestions.

Acknowledgments

We'd like to express our thanks to the following reviewers for their insightful feedback, support, and help throughout developing this book:

Alicia H. Benton

Linda Brothers

Layne Coyne, MS, Family and Consumer Sciences Education

Donna G. Donaldson, MS

Dorothy Dundas, BS, Vocational Homemaking;
MA, Child Development & Family

Mary Funk

Traci Gibson, BS, Family and Consumer Sciences

Paula Wright Long

Rose L. Martin, MS, RD

Cynthia Theiss, CFCS (Certified Family and Consumer Scientist)

Elizabeth Quintana, EdD, RD, LD, CDE

Kristine M. Westover, MS, RD, LD

Amanda Ziaer, BS

Part 1

The Food We Eat

1 The Food You Eat

In This Chapter, You Will . . .

- Understand the science of nutrition
- Learn how proteins, carbohydrates, and fats comprise the foods you eat
- Learn about the benefits of vitamins, minerals, and water
- Discover what calories are and how they are calculated
- Learn about food additives such as flavor enhancers, preservatives, and coloring

Why YOU Need to Know This

Have you ever wondered how the food you eat fuels your body? Why some foods make your body feel and perform better than others? Nutrition has the answers. The more you know about nutrition, the better you will be able to make food choices that will keep you healthy and strong and keep your body functioning optimally. Adequate nutrition can help prevent certain diseases, maintain an optimal body weight, and give you more energy to participate in the activities you enjoy. There are six essential nutrients in the human diet. Take a look at what's on the menu in your school cafeteria today, and see if you can identify items that contain these nutrients.

What Is Nutrition?

Nutrition is the science of food and how it affects our health and well-being. Human bodies need fuel in order to function, and food is that fuel. Each food provides a different balance of the valuable components that your body needs, such as carbohydrates, proteins, fats, vitamins, minerals, and water. You'll learn more about these components later in the chapter. Studying nutrition helps you understand these building blocks so that you can choose the foods and food combinations that deliver what your body needs.

FIGURE 1-1
At many fast food restaurants, healthier choices are becoming available, in response to consumer demand. The last time you were at a fast food restaurant, what were some of the lower-calorie, lower-fat options?

Nutrition scientists have developed various sets of guidelines over the years to build strong human bodies using optimal combinations of these building blocks. Part of studying nutrition involves understanding these guidelines and comparing them to our diets—and the diets of anyone for whom we are responsible. If you pursue a career in the culinary arts, you will probably be planning meals for others to enjoy, and you will want to make each meal nutritious as well as tasty.

The study of nutrition also covers factors that influence a person's eating patterns, including food availability, convenience, and safety. The "best" meal choices can vary, depending on where you live, how much money you can spend on food, how much time you have to prepare and eat meals, and the refrigeration and cooking techniques available to you. For example, your choices of what to eat differ when you have 15 minutes to grab a bite before class, rather than more time to grocery-shop and cook a family dinner.

Dietitians are experts in the field of nutrition. They help people develop diet plans that achieve specific goals, such as weight loss, recovery from surgery or illness, or lower cholesterol. Many dietitians are employed by hospitals or clinics, where they design custom meal plans for patients who need special diets such as low-sugar, low-sodium, bland, or liquid. Dietitians also work for schools and nursing homes to develop balanced meal plans.

A registered dietitian must, at the minimum, earn a Bachelors degree in nutrition, complete a 9- to 12-month internship, and pass a certification examination. In some states, dietitians must also be licensed.

A nutritionist performs some of the same duties as a dietitian, but is not registered or licensed and typically has less education/training. A nutritionist might provide nutrition counseling at a weight loss center, for example, or to families applying for government assistance programs involving food.

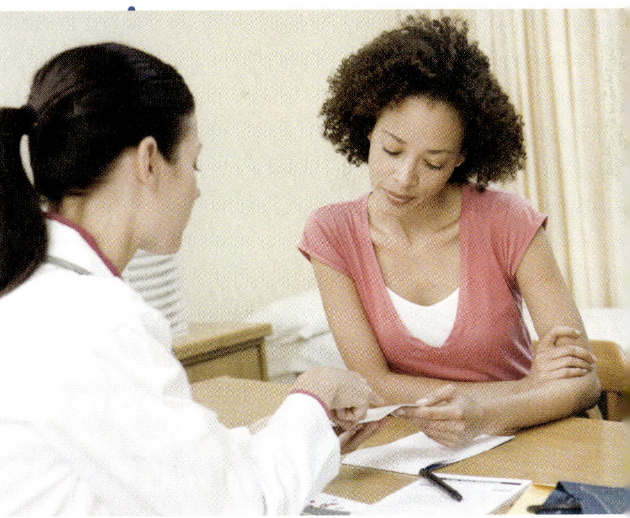

FIGURE 1-2
Hospitals typically hire registered dietitians to counsel patients on their diets, whereas weight loss centers and government agencies typically hire nutritionists, **who are then supervised by dietitians.** Why do you think hospitals insist on the higher level of training and education?

FIGURE 1-3
Organic chemistry is a required class for anyone pursuing a career as a dietitian. Why do you think it is important for a dietitian to understand chemistry?

Nutrition is a health science, a branch of **physiology**—the study of how bodies work. When scientists study nutrition, they analyze the chemical composition of food and how the body digests and processes it. Based on a breakdown of what goes into the body, what comes out of it, and what stays in it, they develop recommendations about what foods provide the body with the best fuel.

Nutrition As Chemistry

Technically, your whole digestive system is a long hollow tube, with openings at both ends. Food goes in one end; waste comes out the other. (Chapter 4 explains the digestive system in a lot more detail.) In between these two points, your body breaks down the food and extracts various elements and compounds it can use. In that way, nutrition is all about the chemistry.

The major organ of chemical digestion is the small intestine; that's where almost 80% of nutrient absorption happens. The small intestine produces a number of digestive enzymes, each acting upon a different type of nutrient. For example, let's say you just ate a big slice of pepperoni pizza. To digest the sugars found in starchy items like pizza crust, your small intestine secretes amylase, maltase, and sucrase. Peptidases are needed to digest certain proteins, like the meat on the pizza, and lipase is needed for digestion of certain fats, like the fat in the cheese. The enzymes mix with the food to break it down into substances that can be absorbed into the body via the walls of the small intestine. From there the digested food enters the bloodstream, which distributes the nutrients to all of the body's cells. Anything that isn't absorbed passes on through. That doesn't mean that if you eat perfectly, you won't have any body waste, though! A certain amount of waste passing through your body is necessary, and some waste comes from your own body processes. That's the primary function of fiber, for example; it passes through undigested, helping move other waste along with it. You'll learn more about the science of nutrition and digestion as you progress through this book.

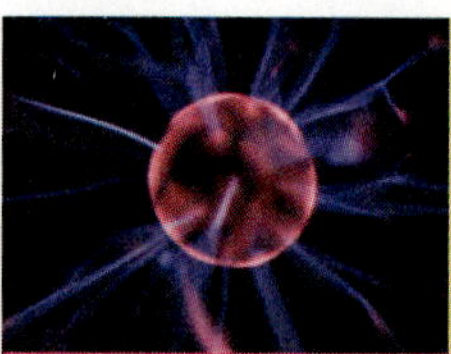

SCIENCE STUDY

Every time you eat something, you are performing a chemistry experiment.

For example, have you noticed that some foods give you a stomach ache, cause gas and bloating, or even make it difficult to breathe? Many people have food allergies that trigger symptoms like these. People with food allergies know about them via real-life experiences—usually unpleasant—in eating those foods.

Even people who don't have food allergies can experience unpleasant side effects from eating too much of certain foods. Remember the last time you ate too much candy—maybe at Halloween time? Your stomach was upset and you may have felt anxious or sleepy from the influx of refined sugar.

As an experiment, write down everything you eat for 24 hours (including portion sizes), and record how you feel—physically, mentally, and emotionally—before and 30 minutes after each meal or snack.

Nutrient Basics: The Building Blocks

Your body requires several classes of nutrients for optimal health; that's why it's important to eat a variety of foods each day. Table 1-1 summarizes them. Let's look at these basic building blocks and how each one contributes to your body's healthy functioning.

The first three of these—proteins, carbohydrates, and fats—are all energy sources for the body. Carbohydrates and fats are the major sources of fuel. Minerals and vitamins are like fuel additives that help with metabolism. They improve the way food is processed, but they don't actually deliver any energy. That's why you can't survive on a diet of vitamin/mineral pills.

Proteins

Protein is required for every structure, function, and repair of the body. Muscles, hormones, clotting, and antibody generation all depend on proteins. Animal products are the most common source of protein—meat, eggs, and dairy products. Protein is also found in certain grains and legumes, such as peas or beans.

Amino acids are the building blocks of proteins. Only dietary protein delivers amino acids to your system; carbohydrates and fats can't do it, and neither can vitamins and minerals. Your body can use 20 different amino acids. Some of them are made by your body. Others must come from the foods you eat, and are called **essential amino acids**.

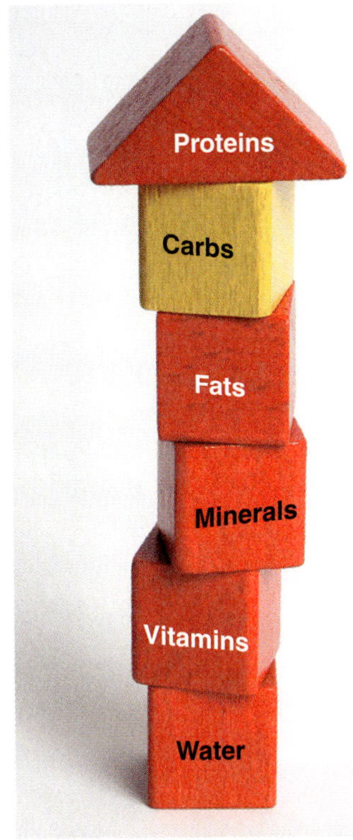

FIGURE 1-4
The six nutrients are all necessary, but in different quantities. Which three of these do you think you need the most of every day?

TABLE 1-1: ESSENTIAL NUTRIENTS

Nutrient	Purpose	Examples
Proteins	Build and renew body tissues Generate heat and energy Provide amino acids that aid in body functions Excess is converted to fat for storage	Meat Fish Legumes
Carbohydrates	Generate heat and energy Excess is converted to fat for storage Provides fiber	Pasta Bread Sugar
Fats (Lipids)	Provide fatty acids for growth and development Generate heat and energy Carry fat-soluble vitamins to cells Excess is converted to fat for storage	Butter Oil Nuts
Minerals	Regulate the activity of the heart, nerves, and muscles Build and renew teeth, bones, and other tissues	Iron Calcium, Potassium
Vitamins	Assist in metabolism, growth, and body development Prevent certain diseases	Vitamins A through K
Water	Carries other nutrients Regulates body temperature Helps eliminate wastes	

Some amino acids are essential for everyone:

- Isoleucine
- Leucine
- Lysine
- Methionine
- Phenylalanine
- Threonine
- Tryptophan
- Valine

Others are essential for infants and children:

- Arginine
- Cysteine
- Histidine
- Tyrosine

FIGURE 1-5
Although meat is one of the richest sources of protein, it is far from the only source. Vegetarians get their protein needs met from non-animal sources. What non-animal protein sources can you name?

Proteins can be divided into two types: complete and incomplete. **Complete proteins** contain all the amino acids in the proportion you need; **incomplete proteins** are deficient in one or more essential amino acids. Complete proteins are found in animal and soy products; other plant-based sources of protein are incomplete. That's why many people who choose to eat vegetarian carefully select combinations of foods containing complementary incomplete proteins. For example, combining rice (which is limited in lysine) with red beans (which are high in lysine but deficient in tryptophan) results in a complete protein. Vegetarians often include soy in their diet because soy beans have all the essential amino acids; most other plant sources do not.

One of the most influential proponents of protein combining, author Frances Moore Lappe, wrote a very popular book called *Diet for a Small Planet*

Fiction	Fact
Eating a very high-protein diet will help an athlete gain muscle mass quickly.	Eating extra protein does not build extra muscle. Once the body meets its protein needs, any extra protein is burned for energy or stored as fat. Muscle is gained by exercise, not by eating certain foods.

FIGURE 1-6
Rice and beans is a classic combination of incomplete protein sources that add up to a complete one. What other combinations are commonly used? Use the Internet to research this if needed.

in 1971 that strongly advocated for protein combining. In an updated edition, she recanted her original insistence on that technique and said:

> "With three important exceptions, there is little danger of protein deficiency in a plant food diet. The exceptions are diets very heavily dependent on [1] fruit or on [2] some tubers, such as sweet potatoes or cassava, or on [3] junk food (refined flours, sugars, and fat). Fortunately, relatively few people in the world try to survive on diets in which these foods are virtually the sole source of calories. In all other diets, if people are getting enough calories, they are virtually certain of getting enough protein."

One way to evaluate protein sources in terms of completeness is the Protein Digestibility Corrected Amino Acid Score (PDCAAS). A PDCAAS value of 1 is the highest and most desirable, and a 0 is the lowest. Generally speaking, the closer the value is to 1, the more complete the set of amino acids in the food. Table 1-2 lists some of the common PDCAAS scores. Notice that soy protein isolate, even though it is plant-based, is a very complete protein, and regular soybeans are not far behind. That's why so many vegetarians eat tofu and other soy-based food products.

Hot Topics

Even experts disagree about whether most vegetarians are actually in danger of not getting the right proteins if they don't pay attention to the food combinations they eat.

TABLE 1-2: PROTEIN DIGESTIBILITY CORRECTED AMINO ACID SCORES (PDCAAS)

Food	Score (0 to 1)
Whey	1
Egg whites	1
Milk	1
Soy protein isolate	1
Beef	0.92
Soybeans	0.91
Kidney beans	0.68
Rye	0.68
Whole wheat	0.54
Lentils	0.52
Peanuts	0.52

FIGURE 1-7
Carbohydrates are easily digested and provide energy to the body very efficiently. That's why marathon runners eat high-carbohydrate meals before the race. If you were getting ready to run a marathon, what would you eat?

Carbohydrates

Carbohydrates are the most common source of energy for the body. Foods high in carbohydrates include bread, pasta, legumes, potatoes, bran, rice, cereal, and table sugar.

Simple carbohydrates are molecules made when one or two smaller sugar molecules (called **saccharides**) combine. **Glucose** is a simple sugar, and is found in fruit and other foods. **Maltose** is another simple sugar, made of two glucose molecules bonded together. Table sugar, the familiar white crystals that most people call "sugar," is a simple sugar called **sucrose**, which is a pairing of one glucose molecule and one **fructose** molecule bonded together. Simple sugars are found in raw and cooked fruits, some cooked vegetables, and refined sugar and its products (like soda pop or candy).

Complex carbohydrates are found in grains, grain products (like bread or pasta), legumes, and some cooked vegetables and fruits. Complex carbohydrates are long chains of glucose molecules bonded together. The chains may be straight, but are usually branched, which connects more than one chain together. The body can use each of the glucose molecules in the chain. In this way they provide a steady supply of the sugar needed by the body for energy in each individual cell.

Dietary fiber is not digestible, but it's an important ingredient of a healthy diet. Fiber used to be called "roughage." Fiber is found in many foods that are also complex carbohydrates, like oatmeal or whole wheat bread, fruit and vegetable skins or pulp, and seeds. **Cellulose** is one type, and can carry water through the intestinal tract and provide bulk. **Bran** and **lignin** are other types that are actually "woody," and provide a healthful, gentle abrasion to the inside of the intestines. **Pectin** (in fruits) and **soluble fiber** (in oats) form a gel-like goo in the intestines that helps the body rid excess cholesterol or other wastes.

Dietary fiber helps prevent bowel and colon diseases. And as it passes through your body, it absorbs water and acts as a bulking agent. This keeps you from getting constipated. Fiber may, under certain circumstances, contribute up to 2 calories of energy per gram. However, in most cases fiber is not considered a source of energy. Eating adequate fiber may also lower the risk of heart disease and some cancers.

FIGURE 1-8
Simple versus complex carbohydrates.

COMPLEX CARBOHYDRATES

SIMPLE CARBOHYDRATES

FIGURE 1-9
Fat is a carrier of flavor, which is why high-fat foods taste so good. Given that fact, if you were going to make a low-fat snack food, how would you compensate for the lack of fat-delivered flavor in it?

Fats (Lipids)

Fats (lipids) are an important part of the diet—although not getting enough fat is very seldom a problem in the diet of the average U.S. citizen! Fats are a source of energy, containing more than twice as many calories as the same quantity of carbohydrates or protein. They also are needed for the absorption and transportation of fat-soluble vitamins and deliver essential fatty acids. Stored body fat cushions internal organs, insulates your body against cold, and stores fat-soluble vitamins until needed. Many meats provide animal fat in addition to lean protein, and fats are also found in nuts and in dairy products like cheese and whole milk in addition to lean protein.

Fats can be categorized as saturated or unsaturated, based on the hydrogen atoms bonded to the long chains of carbon atoms that make a fat molecule (fatty acid).

A **saturated fat** has a hydrogen atom bonded to every carbon bond site. When a hydrogen atom is missing, a double bond forms on the chain. When a fatty acid has one or more double bonds, it is considered **unsaturated**. The more unsaturated a fat is, the more vulnerable it is to spoilage. Therefore a lot of snack food uses saturated fat to extend its shelf life.

Unsaturated fat can be further defined as **monounsaturated** (containing exactly one double bond) and **polyunsaturated** (containing more than one double bond). Generally speaking, unsaturated fats are healthier for your body than saturated ones.

However, not all unsaturated fats are better for you than saturated fats. Fatty acids can be rigid or flexible. The arrangement of hydrogen atoms next to a double bond in a chain is called *cis* (when the hydrogen is arranged on the same side of the chain) or *trans* (when the hydrogen is on the opposite side of the chain), and that determines the flexibility of the fat molecule. Trans fat molecules are rather rigid, which is why **trans fats** can increase your risk of coronary heart disease—the rigid fat can collect in clumps in the bloodstream. The body performs better with the more flexible **cis fat** molecules.

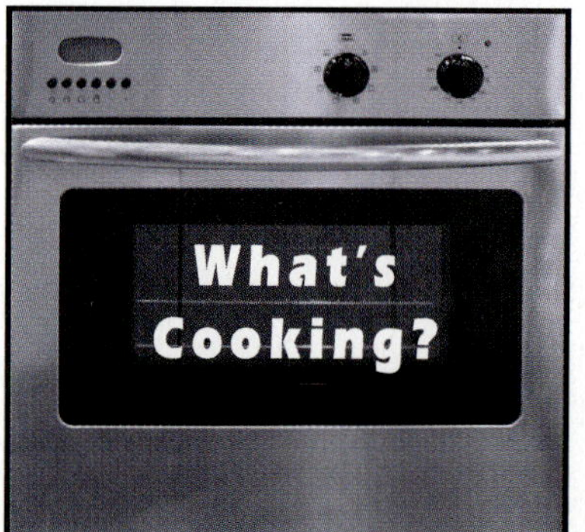

Fat is also used in some cooking techniques (like frying), so foods that start out being relatively low-fat can become very fatty as a result of cooking. Potato chips are a great example. The potato in its raw form is a high-carbohydrate, low-fat food.

Check the Label

The nutrition labels on prepared foods typically break down the types of fats they contain into, at a minimum, Total Fat, Saturated Fat, and Trans Fat. Some labels also include the amounts of polyunsaturated and monounsaturated fats as well. For example, reduced-fat wheat crackers might list their fat content as:

Total Fat 3.5 g
  Saturated Fat 0.5 g
  Trans Fat 0 g
  Polyunsaturated Fat 2 g
  Monounsaturated Fat 0.5 g

Notice that the parts don't add up to the whole in this example. That sometimes happens, especially when there are multiple values of less than 1 g involved, due to rounding. The FDA has some fairly complex rules for rounding numbers on labels; see http://www.cfsan.fda.gov/~dms/nutrguid.html for details.

Minerals

Minerals are inorganic compounds that human bodies need for maintenance and growth. They are required in varying quantities, sometimes in tiny trace amounts. Most of the necessary minerals are present in the average adult diet, but some people choose to supplement their intake with a multivitamin that includes minerals. Table 1-3 lists and describes the major minerals you need; Table 1-4 lists the trace minerals.

TABLE 1-3: MAJOR DIETARY MINERALS

Mineral	Role	Intake	Sources	Too Little	Too Much
Calcium (Ca)	■ Strong bones and teeth ■ Muscle contractions ■ Nerve function ■ Blood clots and other functions	■ Adults: 1,000–1,200 mg/day ■ Children/Teens: 1,300 mg/day	■ Dairy foods ■ Tofu ■ Green leafy vegetables ■ Canned sardines ■ Fortified products	■ Risk of osteoporosis (brittle bones)	■ Over 2,500 mg/day over time may cause kidney stones or calcium deposits in soft tissue
Chloride (Cl-)	■ Nerve impulses ■ Stomach acid ■ Water balance ■ Electrolyte	■ 2,300 mg/day	■ Table salt ■ Vegetables ■ Salted foods	■ Infant convulsions (deficiencies are rare)	■ Over 3,600 mg/day ■ High blood pressure
Magnesium (Mg)	■ Proper enzyme function ■ Nerve function ■ Heart function ■ Muscle function ■ Bone growth	■ Men: 400–420 mg/day ■ Women: 310–320 mg/day	■ Dark green leafy vegetables ■ Legumes ■ Nuts ■ Soybeans ■ Halibut	■ Muscle pain and weakness ■ Decreased heart function	■ No upper limit from diet. The upper limit for a supplement is 350 mg/day from a pill or from milk of magnesia, which acts as a laxative and causes diarrhea
Phosphorus (P)	■ Strong bones and teeth ■ pH balance (acid/base) ■ Important fluids in cells ■ Energy (ATP)	■ Adults: 700 mg/day ■ Children/Teens: 1,250 mg/day	■ Dairy products (good ratio with Ca) ■ Soda pop ■ Meat ■ Baking powder	■ Decreased bone health	■ Over 3–4 gm/day from a pill ■ Poor bone health
Potassium (K)	■ Nerve function ■ Water balance ■ Electrolyte	■ 4,700 mg/day	■ Orange juice ■ Bananas ■ Spinach ■ Melon	■ Muscle cramps ■ Irregular heart beat ■ Poor appetite	■ Decreased heart rate
Sodium (Na)	■ Nerve function ■ Water balance ■ Electrolyte	■ Adults: 1,500 mg/day ■ Over 50: 1,200 mg/day	■ Table salt ■ Salted foods ■ Canned soups and sauces	■ Muscle cramps	■ Over 2,300 mg/day ■ High blood pressure in some individuals
Sulfur (S)	■ Amino acid function ■ pH balance (acid/base)	■ Not specified	■ Meat & fish ■ Soy & legumes ■ Dairy	■ None	■ Not observed

TABLE 1-4: TRACE DIETARY MINERALS

Mineral	Role	Intake	Sources	Too Little	Too Much
Chromium (Cr)	■ Insulin function	■ 25–35 µg/day	■ Vegetables ■ Grape juice ■ Whole grains	■ High blood sugar	■ No upper level for diet
Copper (Cu)	■ Enzyme function ■ Iron function	■ 900 µg/day	■ Clams, oysters ■ Veal, beef	■ Blood problems ■ Poor growth	■ Over 8–10 mg/day ■ Nervous system disorders
Fluoride (F-)	■ Hardens tooth enamel	■ Starts for infants at 0.01 mg/day ■ Increases over life to adult level of 3–4 mg/day	■ Fluoridated drinking water ■ Toothpaste (for tooth brushing, not eating)	■ Tooth cavities	■ Over 10 mg/day for adults ■ In children exposed to too much: mottled teeth and bones
Iodine (I-)	■ Thyroid function	■ 150 µg/day	■ Ocean fish ■ Iodized salt	■ Goiter on thyroid gland ■ Mental retardation	■ Over 1.1 mg/day ■ Poor thyroid function
Iron (Fe)	■ Oxygen in blood	■ Men: 8 mg/day ■ Women: 18 mg/day	■ Red meat ■ Enriched bread and cereal	■ Anemia ■ Fatigue	■ Over 45 mg/day ■ Nausea ■ Heart problems
Manganese (Mn)	■ Enzymes ■ Antioxidants	■ 1.8–2.3 mg/day	■ Whole grains ■ Leafy vegetables	■ None observed	■ Over 11 mg/day ■ Not observed from foods
Molybdenum (Mo)	■ Enzymes	■ 45 µg/day	■ Legumes ■ Whole grains	■ None observed	■ Over 2 mg/day ■ Not observed
Selenium (Se)	■ Antioxidant	■ 55 µg/day	■ Organ meat ■ Seafood ■ Meat	■ Vulnerable to other physical problems	■ Over 400 µg/day ■ Nausea ■ Hair loss ■ Rash
Zinc (Zn)	■ Enzymes ■ Reproduction ■ Immunity ■ Antioxidant ■ Growth ■ Taste	■ Men: 11 mg/day ■ Women: 8 mg/day	■ Shellfish ■ Meat ■ Nuts ■ Legumes	■ Poor development ■ Rash ■ Decreased taste ■ Hair loss ■ Decreased immunity	■ Over 40 mg/day ■ Diarrhea ■ Decreased copper usage ■ Depressed immunity ■ Cramps

Vitamins

Vitamins are active organic compounds, in the sense that they are carbon-based chemicals. Even if they are man-made, they are organic, and in this way they differ from minerals. Vitamins function as coenzymes (that is, they help enzymes) for normal health and growth. Some vitamins also behave like hormones.

Vitamins are named with letters of the alphabet. Vitamins A, D, E, and K are fat soluble and can be stored in the body for later use. Vitamins B complex and C are water soluble; they can't be stored. The excesses are excreted and must be replaced every day, either in the diet or as pill supplements.

A **deficiency** is a disease caused by a lack of a nutrient; **toxicity** is a disease caused by too much of a nutrient. There is a range between the two that is optimal for most humans. Your vitamin intake must fall within this range for optimal good health. Table 1-5 lists the vitamins people need and the Recommended Dietary Allowances (RDAs) for each one, along with the diseases associated with getting too much or too little of a specific vitamin.

Cool Tips

Multivitamin/mineral pills do not usually deliver calcium, the mineral that is most commonly lacking in adult diets (especially in women). Drinking milk is one of the best ways of getting more calcium in your diet. In addition, calcium-based antacids such as Tums can serve as effective dietary supplements to provide extra calcium.

BY THE NUMBERS

The symbol μg stands for microgram. A **microgram** is 1 millionth of a gram, or one thousandth of a milligram.

TABLE 1-5: VITAMIN REQUIREMENTS

Vitamin	Chemical Name	RDA	Sources	Diseases
A	Retinoids (retinol, retinoids and carotenoids)	700 to 900 μg; Not more than 3,000 μg	Liver, milk, cheese, sweet potatoes	■ Too little: night-blindness and and dry corneas ■ Too much: birth defects, liver problems, reduced bone density, skin discoloration, hair loss, dry skin
B1	Thiamine	1.1 to 1.2 mg	Yeast, pork, whole grains, asparagus	■ To little: Beriberi (weakened heart, wasting, partial paralysis)
B2	Riboflavin	1.1 to 1.3 mg	Liver, wheat bran, eggs, meat, milk, cheese	■ Too little: Ariboflavinosis (sore throat, cracked lips, moist scaly skin on genitals, magenta-colored tongue, decreased red blood cell count)
B3	Niacin	14 to 16 mg; Not more than 35 mg	Meat, wheat germ, dairy products, yeast	■ Too little: Pellagra (sensitivity to sunlight, aggression, dry and reddened skin with sores, insomnia, weakness, mental confusion, diarrhea) ■ Too much: Flushing red of the face
B5	Pantothenic acid	5 mg	Meat, whole grains broccoli, avocados	■ Too little: Nausea, insomnia

(continued)

TABLE 1-5: VITAMIN REQUIREMENTS *(CONT)*

Vitamin	Chemical Name	RDA	Sources	Diseases
B6	Pyridoxine, pyridoxamine, pyridoxal	1.2 to 1.7 mg; Not more than 100 mg	Potatoes, bananas, garbanzo beans, chicken breast	■ Too little: Anemia (low red blood cell count) ■ Too much: Impairment of proprioreception (awareness of your own movement), nerve damage, skin lesions
B7	Biotin	30 µg	Liver, legumes, soybeans, tomatoes, carrots	■ Too little: Dry, scaly skin, inflammation of the stomach and intestines, hair loss
B9	Folic acid, folate	400 µg; Not more than 1,000 µg	Leafy vegetables, dried beans and peas, liver	■ Too little: During pregnancy, birth defects, megaloblastic anemia (low red blood cell count) ■ Too much: Masks the symptoms of vitamin B12 deficiency
B12	Cyanocobalamin, hydroxycobalamin, methylcobalamin	2.4 µg	Mollusks (clams, oysters), liver, trout, salmon	■ Too little: pernicious anemia (low red blood cell count)
C	Ascorbic acid	75 to 90 mg; Not more than 2,000 mg	Guava, kiwi, sweet red pepper, orange juice	■ Too little: Scurvy (paleness, depression, spongy gums, bleeding from mucous membranes) ■ Too much: Indigestion, diarrhea
D	Ergocalciferol, cholecalciferol	5 to 10 µg ; Not more than 50 µg	Cod liver oil, salmon, mackerel, sardines, tuna, milk	■ Too little: Rickets, osteomalacia (softening of the bones) ■ Too much: Dehydration, vomiting, decreased appetite, irritability, constipation, fatigue, kidney stones
E	Tocopherols, tocotrienols	15 mg; Not more than 1,000 mg	Sunflower seeds, almonds, turnip greens	■ Too little: mild anemia (low red blood cell count) in newborn infants
K	Phylloquinone, menaquinones	90 to 120 µg	Leafy green vegetables, avocados, kiwi	■ Too little: susceptibility to bleeding because blood does not clot properly

The vitamin and mineral supplement industry has sales of billions of dollars a year. According to Cornell University Cooperative Extension's Yates Association, nearly 40% of adults in the United States take vitamin and mineral supplements. But are they necessary? Are people who take supplements really healthier than other people?

Most dietary experts agree that a diet of fruits and vegetables, dairy products, enriched or whole grain products, along with some meat or meat substitute should provide all the nutrients that a healthy person needs without having to take supplements. If your diet doesn't measure up to that standard, however, a supplement might be of benefit.

There are also conditions where vitamin supplements may actually be needed. A doctor might suggest supplements for people with irregular eating habits or for people eating a diet low in calories, since the reduced quantity may not contain adequate amounts of some nutrients. Pregnancy is another condition for which supplementation may be advised; in fact, most doctors recommend folic acid supplements for all women of childbearing age.

A daily multivitamin may not be necessary for most people, but neither is it usually harmful. However, megadoses (doses many times the RDA) of vitamins taken without a doctor's advice can pose serious health risks. As you saw in Table 1-5, some vitamins, when taken in excess, can cause diseases or disorders, especially vitamin A.

What about the hype in the fitness and alternative medicine communities over megadoses of vitamins? Scientific research has explored the results of taking vitamins in excess of the recommended levels, but to date no credible studies have supported vitamin megadoses for improving athletic performance, cold or flu resistance, memory, or disease prevention.

Water and Electrolytes

Water is essential to the body. It carries nutrients to the body cells and carries waste products away from the body cells. It also lubricates the joints and helps regulate body temperature and body processes. Water makes up 55 to 65% of your body's total weight, including 74% of your brain and 22% of your bones. Because water is lost through evaporation, urination, and respiration, it must be replaced every day.

Some beverages, such as sports drinks like Gatorade, contain **electrolytes**—sodium and potassium salts that help the body replace its mineral loss after vigorous exercise, diarrhea, vomiting, intoxication, overhydration, or starvation. In normal bodies that are not unduly stressed, hormones maintain the electrolyte balance automatically. It is usually not necessary to replace losses of sodium, potassium, and other electrolytes that you might lose during exercise, unless you exercise for over 5 or 6 hours at a time. (A marathon runner might qualify, for example.)

FIGURE 1-10
Sports drinks can replenish electrolytes after vigorous exercise, but most people don't exercise intensely enough to require electrolyte rebalancing. Besides marathon runners, who else do you think might benefit from a sports drink, rather than plain water?

Plain tap water is a great beverage all by itself. Drinking an adequate amount of water each day can:

- Improve your mental and physical performance
- Remove toxins and wastes from your body
- Help you lose weight
- Avoid or cure headaches and dizziness caused by dehydration
- Make your skin look more clear and healthy

Some people prefer to add flavoring to their water, such as sugar free drink mix or a teabag or two; as long as the flavoring does not contain sugar, or any ingredients to which you are allergic, this should not be a significant nutrition issue.

Cool Tips

Most electrolyte-containing sports drinks contain sugar to help the body absorb those electrolytes and meet energy needs during intense exercise. But if you are exercising for weight loss, be aware that that sports drink you enjoy after your workout can negate all the calorie loss of your exercise!

Calories

Proteins, carbohydrates, and fats are all energy-producing nutrients, which means that they have **kilocalories**. A kilocalorie (kcal) is a unit of measurement that describes how much energy the food delivers to the body when it is digested and metabolized.

A kilocalorie is the amount of heat (energy) required to raise the temperature of one kilogram of water by one degree Celsius. Technically, since "kilo-" means thousand in the metric system, a kilocalorie is 1,000 calories. However, in popular usage, such as on nutrition labels, the terms kilocalorie and Calorie (uppercase C) refer to the same thing. So, when a food label states that the product has 400 Calories per serving, it's the same as saying it as 400 kilocalories. (To make things even more confusing, some labels lowercase the word *calories,* but it is still equivalent to kilocalories!)

Proteins and carbohydrates both deliver the same amount of energy per gram: 4 kcal. In other words, for every gram of those nutrients you digest, your body receives 4 kcal of fuel. Fats, on the other hand, deliver much more of a punch—9 kcal per gram. Fat is an efficient source of body fuel. That's why people who are trying to lose weight are often counseled to eat less fat. It's not that fat is necessarily off-limits, but you can eat much more food (in total ounces) and feel more satisfied, if you stick with proteins and carbohydrates. Table 1-6 summarizes the kcal content of each nutrient type.

FIGURE 1-11
For the same amount of energy you could have this filling apple or this little square of chocolate.

Calculating the kilocalories in foods is not always as straightforward as multiplying the grams by the energy per gram, because few foods are pure sources of a single nutrient class. You have to figure out what portion of the food—in grams—is fat and what portion is protein or carbohydrate. (For calorie-counting purposes, you can lump the latter two together, since they

TABLE 1-6: KILOCALORIES PER GRAM

Nutrient	Energy Delivered per Gram
Protein	4 kcal
Carbohydrate	4 kcal
Fat	9 kcal
Vitamins	None
Minerals	None
Water	None

have the same value.) For foods where the fat is infused, such as a marbled piece of meat or a French fry that has soaked up the oil in which it was prepared, weighing out the fat separately is nearly impossible outside of a lab setting. In addition, most foods contain a certain amount of water, which contains no calories.

So what do we do, then? Well, mostly we rely on the food labeling and nutrition information that the food producers make available. Many restaurants publish nutrition information for their most popular dishes on their Web sites, and there are several great all-purpose food databases online that provide nutrition information—including calorie counts—for both home-cooked dishes and restaurant fare.

BY THE NUMBERS

For instant calorie and nutrition data on the food found at most fast food restaurants, visit http://www.foodfacts.info. To get nutrition information for unprocessed foods and dishes you prepare from scratch at home, an excellent source is http://www.nutri-facts.com.

Food Additives

Besides the nutrients you've learned about so far in this chapter, the food you eat may also contain other ingredients. These additives serve a variety of purposes, from prolonging the product's shelf life to making the food look or taste more appealing. Most additives are harmless unless you happen to be allergic to them. The following sections look at some of the additives you might see when you read the ingredients label for a food item.

Cool Tips

For many years, it was common knowledge that the human tongue could taste four attributes: sweet, salty, sour, and bitter. A fifth taste has recently been identified, however, called *umami*, which refers to the taste of glutamates. The word umami means "yumminess" in Japanese.

Flavor Enhancers

One of the best and easiest ways to improve the taste of food is to use herbs and spices. These all-natural flavor enhancers give many dishes their distinctive taste, from the garlic in shrimp scampi to the cumin seeds in Indian bread. You'll learn about the types of spices and their uses in Chapter 15, "Spices, Herbs, and Garnishes."

In addition to these natural herbs and spices, the food you eat may contain a variety of natural or artificial flavor enhancing chemicals. These chemicals have little or no taste of their own, but they bring out the flavor in foods to which they are added. The most common of these are the **glutamates**, which are derived from glutamic acid in dried seaweed or fish. There are many types of glutamates used for flavor enhancement, each one based on a different mineral (sodium, potassium, calcium, magnesium, and so on).

Hot Topics

Food that contains MSG may not say so on the ingredient list specifically. There are many food additives and ingredients that contain or create MSG that are called by different names. If you see any of the following ingredients on a label, the food might not be good for someone to consume who has a sensitivity to glutamates:

- Hydrolyzed vegetable protein
- Yeast extract
- Autolyzed yeast extract
- Natrium glutamate

Monosodium glutamate (MSG) is used to enhance the flavor of bouillon cubes, barbecue sauce, salad dressings, and seasoning mixes. It is also used in many of the cooked dishes in some Chinese restaurants. However, because some people have sensitivity to MSG, many restaurants have stopped using MSG; some even make a point to advertise the "No-MSG" status of their food.

Some people have a sensitivity to glutamates, especially MSG, that causes them to feel a burning sensation, tingling, numbness, headache, nausea, drowsiness, weakness, and/or sweating after consuming food flavored with MSG. The American Medical Association has concluded that glutamate in any form is not a significant health hazard for adults; however, it was removed from baby food in the 1960s due to concerns about causing over-stimulation in children.

Another common flavor enhancer is **maltol** (or ethyl maltol), a natural organic compound derived from the bark of the larch tree, pine needles, and roasted malt. It smells like caramelized sugar and tastes like freshly baked bread. Commercial bakeries often put it in bread and cakes.

Food Coloring

Food coloring is used in most processed foods. This coloring can be either natural or artificial. The most common natural food colorings are beet juice (red), beta-carotene (yellow), and caramel (brown). A variety of artificial colors are also used, but some people are allergic to them (especially yellow and red dyes), so product packaging must include the specific dye numbers that are used.

Bleaching agents are the opposite of food coloring; they remove color. They are used mostly in baked goods to make yellow flour ("young" flour) look whiter, which gradually happens naturally as it ages. Benzoyl peroxide is the most common bleaching agent used in flour.

FIGURE 1-12
Some foods are so strongly associated with their color that it is hard to think of them in some other color, like cola being brown. What commercially prepared foods can you think of that have a strong identity based on their color?

Texture and Balance Stabilizers

Texturizers are additives that improve the texture of food. For example, calcium chloride is commonly added to canned potatoes and tomatoes so they don't fall apart. Similarly, a **stabilizer** helps a food product maintain its texture or color over time. An **emulsifier** helps keep fat evenly dispersed within food. Have you ever seen "natural" peanut butter, where the oil separates from the rest floats to the top? Commercially processed peanut butter doesn't do this because of the emulsifiers. Humectants and desiccants maintain the correct moisture levels in food—**humectants** keep soft foods soft, like marshmallows and chewing gum; **desiccants** prevent food (table salt, for example) from absorbing moisture.

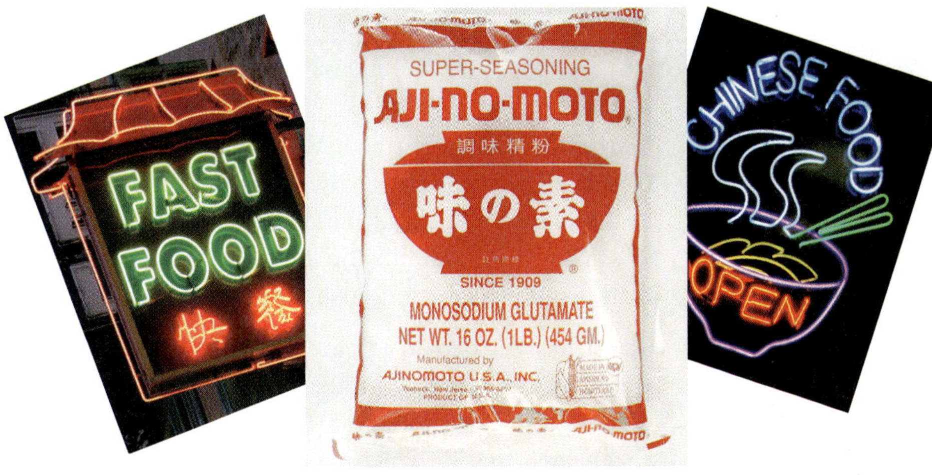

FIGURE 1-13
MSG is a very common flavor enhancer, especially in Asian food. However, some people have a sensitivity to it, and it makes them sick. If you owned a restaurant, would you use MSG to improve the taste of your food? Why or why not?

Extra Vitamins and Minerals

Yes, vitamins and minerals themselves can be food additives! In addition to the naturally occurring vitamins and minerals in food, some manufacturers enrich certain products with extra vitamins. Have you ever wondered why sugar-coated corn breakfast cereal provides essential vitamins and is "part of a nutritious breakfast"? It's because vitamins were added. Milk is nearly always enhanced with vitamin D, and iodine is added to salt.

Vitamins can also be used as natural preservatives. For example, vitamin E can be added to fat-based products to keep them from spoiling, and ascorbic acid (vitamin C) is often added to frozen fruit, dry milk, apple juice, soft drinks, and candy. Sodium ascorbate, a combination of salt and vitamin C, is used as an antioxidant in concentrated milk products, cereals, and cured meats.

Fiction	Fact
Food that contains artificial flavoring is not as good for you as food that contains only natural flavoring.	According to Gary Reineccius, a professor of food science at the University of Minnesota, artificial and natural flavorings are equal in health value. "It's like saying that an apple sold in a gas station is artificial and one sold from a fruit stand is natural." He also adds that artificial flavorings may actually be safer than natural ones, because the laws governing them are more stringent.

Case Study

Maria and her friend Tina have decided to try eating vegetarian. They've heard that a vegetarian diet can be healthier than a meat-filled diet, and can help with weight loss. They also both like the idea of avoiding eating animals for philosophical reasons.

Maria has decided to try an **ovo-lacto** vegetarian diet, which includes eggs and milk products. (*Ova* is Latin for egg; *lacto* is Latin for milk.) Tina has decided to try a **vegan** diet, one in which only plant products are consumed.

What do you think the differences will be in their experiences?

- Do you think Tina will be more likely to lose weight than Maria? Why?
- What lifestyle challenges will Tina face that Maria might not? For example, will Tina have trouble finding something she can eat when dining out with friends?
- Given the price of food in your area, which diet will be less expensive? Are eggs and milk more or less expensive per serving than fruits, vegetables, and grains?

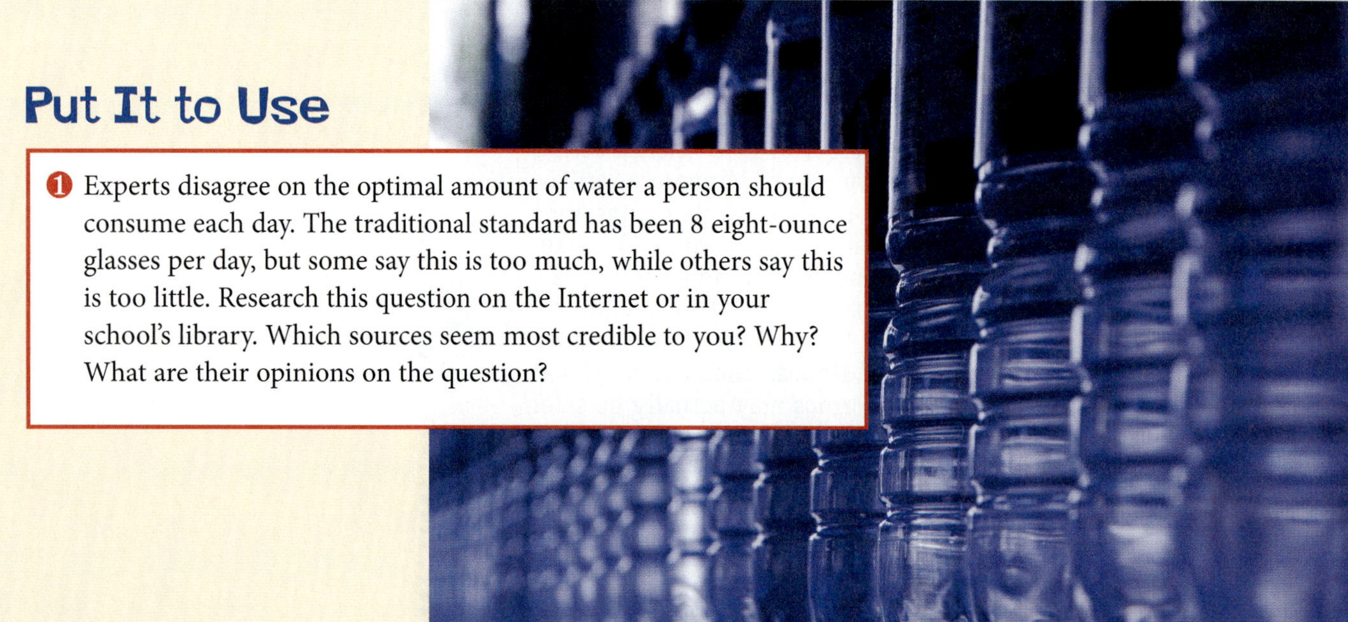

Put It to Use

1. Experts disagree on the optimal amount of water a person should consume each day. The traditional standard has been 8 eight-ounce glasses per day, but some say this is too much, while others say this is too little. Research this question on the Internet or in your school's library. Which sources seem most credible to you? Why? What are their opinions on the question?

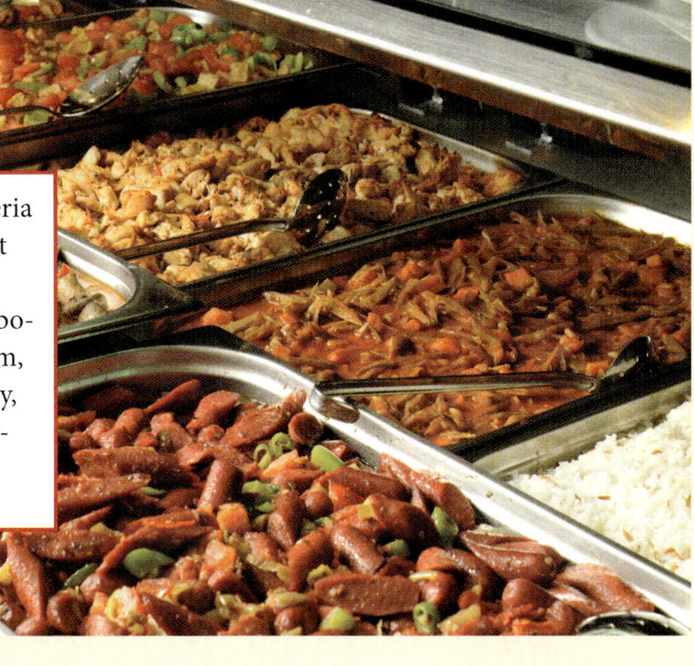

Put It to Use

❷ From among the food choices in your school cafeteria today, plan a lunch for someone who is on a low-fat diet. Use labeling information from any prepared foods available to determine the food's protein, carbohydrate, and fat content; use www.nutritiondata.com, a book of nutrition values from your school's library, or some other source to look up the nutrition information on any foods that do not have labeling.

Write Now

When someone says they are a "vegetarian," it is sometimes hard to know exactly what they mean because there several types of vegetarian diets. What are the different types of diets, and what are their restrictions? Research this topic on the Internet or in your school library, and write an essay explaining the differences. Provide a sample menu of foods that a person following each type of diet might eat in a typical day.

Tech Connect 

Most people think meat when they hear "protein," but protein is found in a variety of ingredients, including bread and dairy products. On the Web, go to http://www.mcdonalds.com/usa/eat/nutrition_info.html and look up the nutrition information for a Quarter Pounder with Cheese. How much protein does this sandwich provide? Of that amount, how much comes from the beef patty and how much from the other ingredients? If you ordered it without the cheese, what would be the effect on the protein you received?

Team Players

Most vitamins and minerals can be taken in supplement (pill) form. These supplements can be purchased in grocery stores, drugstores, at specialty stores, and online. Some sources say that for certain vitamins and minerals, the processing of the supplements and/or the form of delivery makes a big difference in how the nutrient is absorbed in the body, so it's worth it to pay more at a specialty store. Others say vitamins are vitamins, and there is no real difference.

Break into small groups, with each group choosing a different vitamin or mineral. Do some research to find out whether there is any controversy or expert disagreement regarding the optimal daily dosage, the form in which it is delivered, or the way the supplements are processed. Check several sources for supplements, including a local grocery store and an online store, and create a summary spreadsheet of the prices per recommended daily dose. Share your findings with the class.

Put It Together

Match the explanation in column 1 with the term in column 2.

Column 1

- **a.** the science of food and how it affects our health and well-being
- **b.** an expert in the field of nutrition
- **c.** ingredient plentiful in complex carbohydrates, but not in simple carbohydrates
- **d.** nutrient needed for the absorption of vitamins A and D
- **e.** an inorganic nutrient
- **f.** nutrient where examples include folic acid, niacin, and riboflavin
- **g.** the building blocks of proteins
- **h.** a type of lipid that can cause high cholesterol
- **i.** the chemical name of vitamin B1
- **j.** mineral salts found in sports drinks, like Gatorade

Column 2

1. amino acids
2. dietary fiber
3. dietitian
4. electrolytes
5. fat
6. minerals
7. nutrition
8. thiamine
9. trans fat
10. vitamins

2 Making Smart Eating Decisions

In This Chapter, You Will . . .

- Think about the factors that influence your food choices
- Learn about the MyPyramid and American food pyramid models
- Find out the recommended daily allowances of the nutrients you need
- Learn how to read a nutrition label
- Develop a personal eating plan
- Analyze how nutrition needs are different for people of various ages and activity levels

Why YOU Need to Know This

The choices people make about what to eat for each meal might seem straightforward, but there are many factors that influence them. These factors change depending on where you live, too. As a class, make a list of the factors that go into your decision about what to have for lunch or dinner on a particular day. In addition, make a list of things that you assume are true about *all* the food choices available to you.

The Choices You Make

What's for lunch today? The school cafeteria's special, or a brown bag lunch from home? Maybe a burger from an off-campus fast food joint, or just a handful of trail mix from a bag in your locker? For most people, every day presents a variety of eating choices and opportunities, each with a different cost and benefit. In this chapter, we'll look at some of those decision points—perhaps including some that you might not have thought of before.

Hot Topics

Lots of people eat when they're stressed out or depressed. Are there any foods that you crave when you're under a lot pressure? Do your emotions or other psychological factors influence the food choices that you make?

- **Availability:** You can't eat what's not there, of course. That might seem like a no-brainer, but in some cases, you can influence the availability if you plan ahead. You might bring your own food or lobby the school cafeteria, a local restaurant, or a supermarket to provide what you want.
- **Taste:** How a food tastes is, for a lot of people, one of the most important factors in whether they will eat it or not. If it's not pleasing to you, you probably won't eat it more than a few times, no matter how nutritious it may be. Taste is such an individual thing, though, that you can't count on other people liking the same foods that you like, especially people who have grown up in different cultures.
- **Hunger:** How hungry are you? Are you eating only when you are hungry, or do you eat at specific times of day regardless of hunger level? There are proponents of both ways of eating. Many weight loss books and articles say that people can lose weight simply by eating only when hungry and stopping eating when they get full—and that most "naturally thin" people do that anyway. Others, like Mireille Guiliano, author of *French Women Don't Get Fat,* maintain that eating only at mealtimes—and never snacking between meals—is important.
- **Nutritional Value:** To ensure that you are eating a balanced, healthy diet, it's important to consider the nutritive value of the food you consume. This includes not only finding the right balance of protein, carbohydrate, and fat, but also choosing products that have the vitamins and minerals you need. Chapter 1 covered these building blocks, and later in this chapter, you'll learn how to use nutrition labeling to evaluate a food.
- **Calories:** Do you want to lose weight, gain weight, or maintain your current weight? The number of calories you consume each day is the primary determinant of whether you meet that goal, along with the amount of exercise you get. In Chapter 5, "Special Diets for Weight Loss and Fitness," you'll learn how to calculate the number of kilocalories your body burns each day at various activity levels, which will then tell you what number of calories you should aim to consume.

FIGURE 2-1
In some schools, menus are planned by a nutritionist or a registered dietitian; in others, someone with nutrition knowledge but without any specific credentials does the planning, such as a head cook. How is it done in your school? Research this to find out.

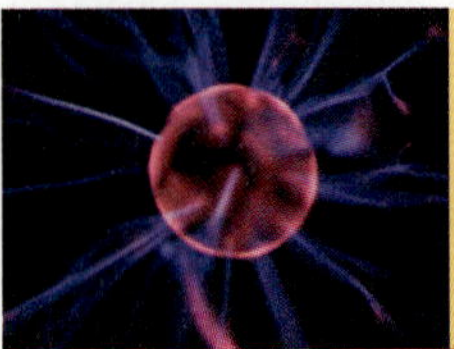

SCIENCE STUDY

Different people can actually taste the same food very differently, depending on their "taster" status.

One-third of all Americans are "super-tasters," people who are extra-sensitive to the taste of food, especially bitterness. Another one-third are in an average range, dubbed "tasters," and the final third, the "non-tasters," have less-than-average taste sensitivity.

Super-tasters seem to have a particularly strong ability to taste bitterness in certain foods such as broccoli, coffee, grapefruit juice, green tea, spinach, and alcoholic beverages. They have more food dislikes, and are more likely to ask for sauce and dressings on the side, according to research done by Virginia Utermohlen, MD, of Cornell University. Moderately sensitive tasters tend to think about food in the most positive way of all three groups; chefs are most likely to be moderate tasters. Non-tasters are the most likely to prefer food that is intensely sweet. Women are more likely to be supertasters, as are Asians and Africans.

Test strips are available to determine a person's taster status, but you don't have to run a lab test to find out what kind of taster you are. Just stick out your tongue and look in the mirror! The more taste buds you see, the more sensitive you're likely to be. You may have to compare your taste buds with those of a few friends for perspective. (Blue food dye makes it easier to see your taste buds, so you might want to drink a blue-colored beverage beforehand.)

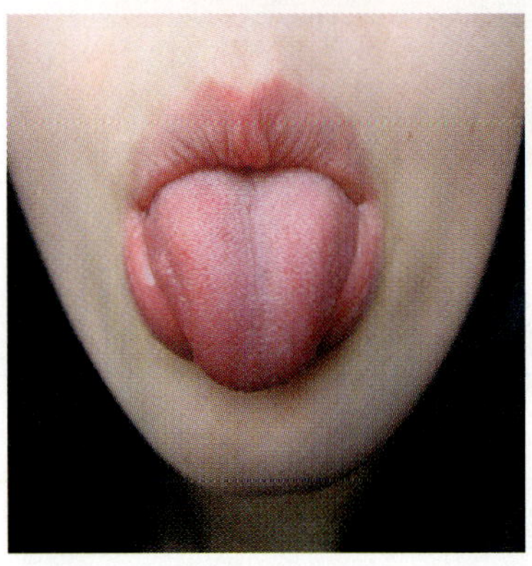

FIGURE 2-2
The density of taste buds on your tongue can determine how acutely you taste certain bitter foods, especially at a young age. If you were a parent with a child who was a "super-taster," would you insist that he or she eat everything on the plate? Why or why not?

- **Cost:** Depending on your budget and finances, food cost may or may not be an issue. Staples such as beans, rice, eggs, milk, and pasta are typically cheaper than luxury foods such as steak, seafood, and some desserts. Meat is usually more expensive than vegetables and breads because it usually cost more to raise animals than grow plants. In Chapter 3, "Planning a Meal," we'll look at the issue of food cost in planning both a specific type of meal and the actual menu.
- **Convenience:** Are you a busy person, constantly on the go, who finds it hard to find time to prepare and serve a sit-down meal? Today, many people are in that position. The convenience of a food is an important factor when time is short. It is often easier and faster to pop a prepared item into the microwave or go through a fast-food drive-through than it is to cook a meal from scratch.
- **Culture:** The culture in which we live determines a lot about the foods that we eat. For example, in American culture, sandwiches made with wheat-based bread are a common lunchtime meal. In Asian cultures, on the other hand, rice is much more common than wheat. Your family's ethnic background, and those of your friends, can also make a difference in the foods you eat, as your parents or grandparents may prefer foods that they ate when they were growing up in their country of origin.
- **Religion:** Some religions have special dietary restrictions. For example, many Jewish people follow a Kosher diet, and many Hindus are vegetarians. If you grew up in such a religion, you probably already know the rules fairly well, but if you are going to work in the culinary arts field, you will need to know about restrictions for many different religions, not just your own. Table 2-1 summarizes some of the dietary restrictions of popular religions.

- **Science and Technology:** Scientific advances in food growth, preparation, transportation, and preservation have made a big difference in the American diet over the last 100 years. For example, food that was previously available for only a short season each year, like fresh lettuce or tomatoes, is now available year-round, thanks to refrigeration. Trucks, trains, and planes with refrigerated compartments allow food to be transported across long distances without spoiling. Gamma irradiation of fruits and vegetables at the source maintains their freshness and color during transport across town or across the world. In addition, genetic research has allowed the development of new varieties of certain fruits, vegetables, and grains, much more resistant to disease or spoilage. Some people object to the fact that the Food and Drug Administration (FDA) does not require that genetically modified food be labeled as such for consumers; therefore, consumers cannot make informed decisions about whether or not to buy this type of product.

FIGURE 2-3
Many religions and cultures have dietary restrictions. What foods do you eat at home on a daily basis? Are there any foods never served in your home for a cultural or religious reason?

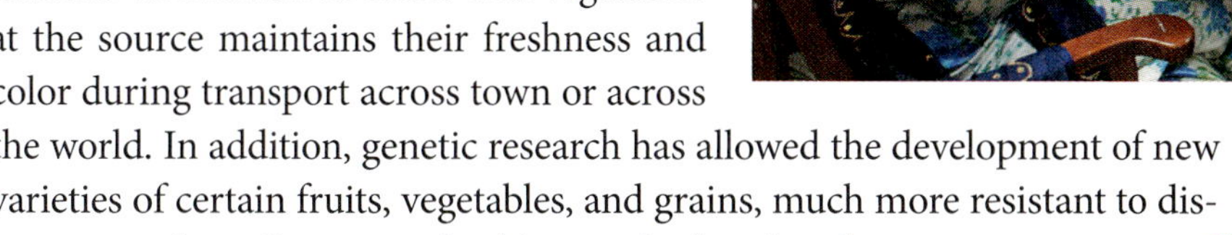

Hot Topics

Genetically engineered foods are controversial because some people believe these foods have not been sufficiently proven safe for human consumption. Other people feel the risks are minimal compared to the many benefits of genetic engineering, including creating crops that will grow well in poor soils, making it possible to feed hungry people who would otherwise starve.

TABLE 2-1: RELIGIOUS DIETARY RESTRICTIONS

Religion	Restrictions
Buddhist	Generally vegetarian.
Conservative Protestant (some denominations)	Avoid alcohol.
Church of Jesus Christ of Latter-day Saints (Mormon)	Avoid alcohol and caffeinated coffee/tea; some also abstain from caffeinated soft drinks. Abstain from eating on fast days (once per month).
Greek Orthodox	No meat or dairy products on fast days.
Hindu	Generally vegetarian.
Muslim	No alcohol, pork, or pork products.
Orthodox Jewish	Kosher diet: No shellfish, pork, or non-kosher meats; no serving milk products with meat; no eating leavened bread during Passover. Abstain from eating on specific fast days.
Roman Catholic	No food one hour before communion, and no meat on Ash Wednesday, Good Friday, and all Fridays during Lent.
Seventh-Day Adventist	Generally vegetarian.

FIGURE 2-4
Think of an ad for food that you recently have seen. What three adjectives come to mind about that food? Was your thinking of any of those words a result of something you saw in the ad?

- **Media:** Have you ever bought and eaten something solely because you saw a commercial for it on TV? Or read a magazine ad about it, or clipped a coupon? If you answered yes, you are not alone; advertising really works, and that's why it's a multi-billion dollar industry. Advertising can be helpful in informing people about new products, but it can also influence public thinking in negative ways. For example, many companies that make children's breakfast cereals advertise on the Saturday morning cartoon shows on TV. This teaches a whole generation of children that sugary cereals are "nutritious," when they are really just simple sugary carbohydrates, enriched with a few vitamins and minerals.
- **Friends:** Some food choices are made because of peer pressure—in other words, because "everyone else is eating it." That's not good or bad per se, but it's something to be aware of. If everyone else is eating healthy food, then peer pressure is not necessarily a bad thing. But if it suddenly becomes popular to eat nothing but gummi bears and potato chips for lunch, that's a problem.
- **Age:** People of different ages tend to make different food choices. For example, young children tend to be picky eaters, preferring bland foods. As we get older, we become more adventurous, trying out foods that combine several flavors, that have unusual spices or textures, or that have a certain amount of bitterness (like coffee). Senior citizens may enjoy traditional foods that remind them of their younger days.
- **Health Issues:** Illnesses, diseases, allergies, and other sensitivities may also play a role in people's food choices. For example, diabetics need to control the amount of carbohydrates they consume, and someone with an allergy to wheat would avoid bread and pasta made from wheat.

Cool Tips

The senses of taste and smell tend to diminish as we age. Children are more sensitive tasters than adults, which may explain why children—especially those who are super-tasters—are so reluctant to eat their vegetables, which may have a subtle bitter taste that most adults don't notice. By the time they reach adulthood, they have lost enough sensitivity to taste that the vegetables actually taste different to them. This gradual loss of ability to taste continues our whole lives. Senior citizens often request extra salt or other seasonings because their taste sense is not as strong as that of a younger person.

How Food Choices Influence Health

The food you eat can actually determine how healthy you will be, both now and as your body ages. Science has found a very strong correlation between certain types of diets and the likelihood of developing or avoiding certain diseases. In fact, the National Cancer Institute estimates that *three out of four* deaths in the United States each year are caused by diseases that are linked to what we eat: heart disease, high blood pressure, stroke, cancer, and diabetes.

Cool Tips

Vitamin C, vitamin E, and beta carotene, which forms vitamin A, are **antioxidants**. They protect body cells from oxidation, a process that can lead to cell damage and may play a role in cancer. Plant foods contain **phytochemicals**, chemicals that may affect human health. A growing body of evidence indicates that phytochemicals may also help protect against cancer. To get these benefits, eat more fruits and vegetables that contain vitamins A and C and beta carotene—dark green leafy vegetables such as spinach, kale, collards, and turnip greens. Citrus fruits, such as oranges, grapefruit, and tangerines, are also high in antioxidants. Other red, yellow, and orange fruits and vegetables, or their juices, are also healthful choices.

Diet and weight have a strong correlation to many types of **cancer**. According to the M. D. Anderson Cancer Center and the American Cancer Society, experts estimate that 30% to 40% of all cancers could be avoided if people would maintain a healthy body weight, eat a proper diet, and exercise regularly. They recommend these dietary guidelines to avoid cancer:

- Eat a plant-based diet high in fruits, vegetables, whole grains, and legumes.
- Eat at least five servings of fruits and vegetables every day.
- Limit your intake of red meat.
- Limit your intake of fat, especially saturated (animal-based) fats.
- Limit your consumption of alcohol, if you drink at all.
- Eat a variety of fruits, vegetables, and starchy plant foods (such as rice and pasta).
- Eat a diet that is high in fiber.

Certain **heart diseases** may be preventable, and are influenced by what we eat. You can drastically cut your risk of heart disease by eating a diet that reduces saturated fats and emphasizes fruits and vegetables. Regular exercise is also important.

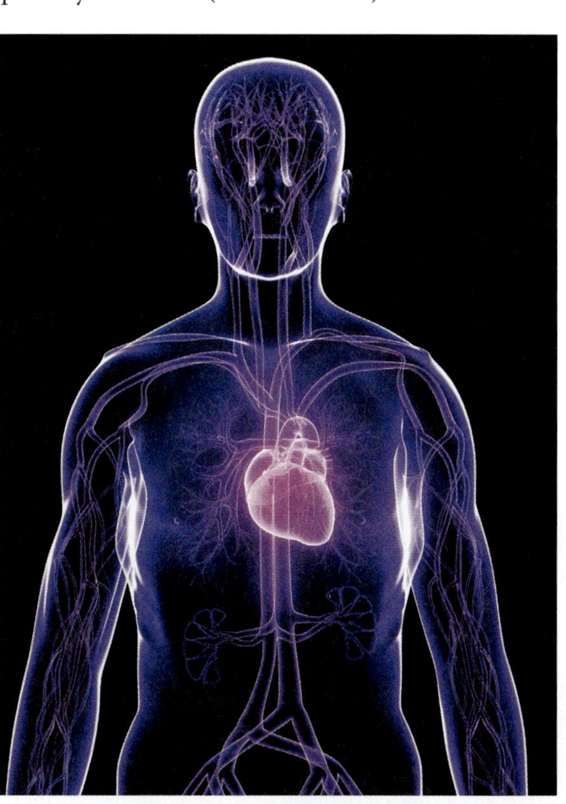

FIGURE 2-5
Your heart pumps the blood through your body, so if it isn't working right, your entire body suffers. What are some symptoms of heart disease? If you don't know, research it online.

Vascular diseases like **high blood pressure** may also be related to diet. To lower the risk of developing high blood pressure, or to help control it if you already have it, minimize consumption of animal and hydrogenated vegetable fats. This will reduce saturated fats, trans fats, and dietary cholesterol. You should also control the intake of sodium and exercise regularly to maintain normal body weight.

Dietary **cholesterol** has often been blamed for heart disease, and it may increase the risk in some individuals. However, cholesterol is made by your body in the liver and is necessary for your health and survival. It is involved in vitamin D production, hormone production, transport of nutrients, and digestion. Individuals with high blood cholesterol may have a genetic problem that is not closely related to their diet.

The risk of developing **Type 2 diabetes** (adult onset diabetes) can also be minimized with proper diet. Studies have found that overweight or obese people are much more likely to develop diabetes, and that if overweight or obese people lose weight in the early stages of having diabetes, they can greatly reduce or eliminate their diabetic symptoms. People with diabetes are advised to reduce and control the amount of carbohydrates they eat and to exercise regularly.

Osteoporosis, a disease that affects mainly people over 50, is characterized by a loss of bone density that makes bones brittle and, therefore, easily broken. By the time a person is diagnosed with this disease, there is little that can be done. If you start when you are young, however, and get enough calcium all through your life, you can minimize the risk of developing osteoporosis later in life. Studies have shown that people who get plenty of calcium as children and young adults are less likely to get osteoporosis as senior citizens. Exercise that requires your legs to bear the weight of your body, such as walking or playing sports, improves the strength of the bones.

FIGURE 2-6
The more calcium a person takes in as a child or young adult, the less likely they are to be affected by osteoporosis in later years. What foods should you be eating every day to minimize your osteoporosis risk?

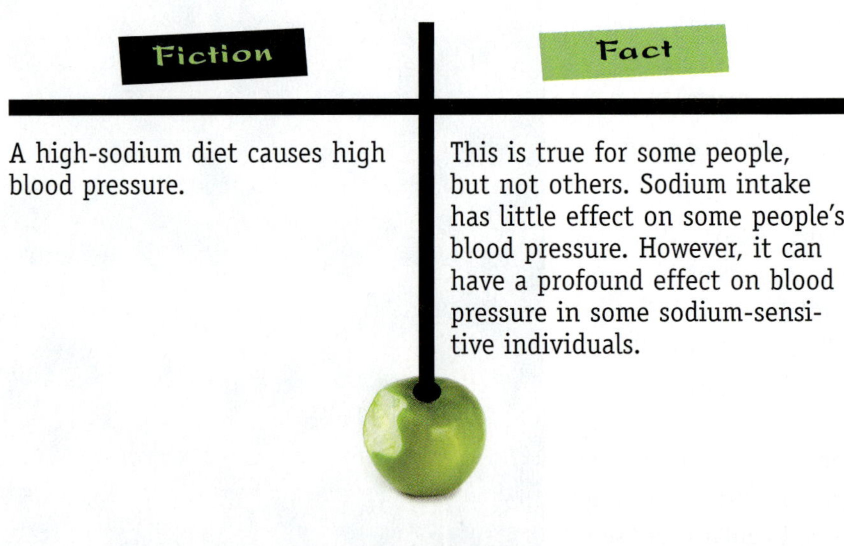

Fiction	Fact
A high-sodium diet causes high blood pressure.	This is true for some people, but not others. Sodium intake has little effect on some people's blood pressure. However, it can have a profound effect on blood pressure in some sodium-sensitive individuals.

What Influences Our Food Supply?

In the U.S., we have a vast array of foods to choose from in any supermarket. It might seem like we are making those choices completely autonomously, but many influences are constantly in effect, and constantly changing. You might not even realize that your choices are being affected by some of the following.

- **Weather:** Have you noticed that some years, certain fruits are less expensive and more plentiful than in other years? This is because weather fluctuations affect the crops. For example, if there are freezing conditions in Florida during a citrus growing season, the supply of citrus may be affected. As a result, citrus may be more expensive (because it has to be imported from some other country), and the fruit may be smaller and not as high in quality as usual.
- **Natural Disasters:** Hurricanes, floods, earthquakes, and other extreme weather conditions can not only affect the growing seasons of certain crops, but they can also disrupt the supply chain. For example, if the roads out of a village that supplies a certain type of food are destroyed by an earthquake, the price of that food may go up.
- **Wars and Conflicts:** During wartime, certain foods may be less readily available. For example, if the U.S. were at war with a country that supplied most of the bananas we consume, the price of bananas would increase and bananas might become difficult to get. In some countries, war also has destroyed the food production system, from farmlands to factories, making it hard for an entire population to get adequate nutrition.
- **Politics and Trade Agreements:** Governments often make trade agreements to encourage the flow of goods between the nations. If the U.S. has an agreement with some other country to import a certain food item at a low price, or with low tariffs (taxes), the cost of that food will be lowered in U.S. grocery stores. Conversely, if the U.S. is trying to discourage consumers from eating a particular imported commodity (to encourage U.S. producers to step up production, for example), the cost to the importers might go up, and with it the cost to consumers.
- **Federal, State, and Local Laws:** Government at any level can issue regulations or laws that change the way food is distributed. For example, a state might require stringent inspection for a certain type of food, making that food safer but also more expensive. And a city or town might pass laws to encourage or discourage farmers markets, hot dog vendors on the street, or food vendors at local fairs.
- **Distribution Systems:** Food does not just magically show up at a grocery store; there are very sophisticated distribution systems all over the world for moving food from place to place, including air, rail, and truck transport. Changes in any of those transport methods, such as higher or lower fuel prices, union strikes, or even steel shortages (because the transports are typically made out of steel) can affect prices.

- **Government Subsidies and Programs:** To ensure a steady and reliable flow of the food distribution system, governments sometimes offer subsidies (payments) to food producers to encourage a certain type of production behavior. For example, some farmers receive subsidy payments to leave their land fallow (unfarmed) for one or more years, so that the price of grain does not drop too low.
- **Trends in Health and Wellness:** As we learn more about health and wellness as a society, the food production and distribution industries respond to changing consumer demand for certain products. For example, organ meats (liver, kidneys, hearts, and so on) used to be much more popular than they are today, because of new information we have now that shows they are not as good for you as was once thought. Some foods become more popular over time, even to the point of being fads, as new research shows their benefits. For example, blueberries and pomegranates have increased in popularity in the last decade because of their antioxidant benefits.

What You Eat Can Change the World

The choice of whether to eat an apple or a piece of cheese as a snack might seem inconsequential, a simple matter of personal preference. But each choice you make has a ripple effect that can be felt all over the world. In the choice between an apple and cheese, for example, you are choosing to support either the apple industry or the dairy industry. You are contributing to the income of a worker in one part of the world versus another. You are informing the suppliers and distributors of the type of food you want them to produce more of. You are even making one country more economically powerful than another!

Here are some ways that your food choices impact the world:

- By eating meat instead of vegetables, you are encouraging farmers to produce more animals for human consumption. For this reason, many animal rights activists are vegetarian, and do not wear leather.
- By eating foods that are grown locally, you are supporting your local economy. This can include buying produce at a local farmers market, planning events and festivals that celebrate a particular food item that is grown in your area, and sending gifts of locally grown foods to friends in other areas.
- Some foods consume more natural resources than others in their production and transport. For example, food that is grown in a remote area of the world requires more fuel to get to you.
- Some foods are more efficiently produced than others. For example, it is much less efficient to feed grain to cattle—and then eat the beef—than it is to simply eat the grain yourself.
- By buying food produced in the U.S., you are directly supporting U.S. farming. However, buying food produced in other countries can also indirectly benefit the U.S. by supporting strong and beneficial trade agreements.

Dietary References

So far in this book, you've heard a lot about the benefits of eating a healthy diet and the risks of eating an unhealthy one. But what does that actually mean? What should you be eating on a daily basis, and what should you be avoiding? There's no need for guesswork. Several government and private organizations have published clear, easy-to-follow guidelines that can help you construct the diet that is right for you. The various guidelines do not agree with each other on every point, but there is enough similarity that following any of them—or your own combination of them—will set you on the path to healthy eating.

For many years, the dietary standards in the United States were developed by the Food and Nutrition Board of the National Research Council, National Academy of Sciences. They created a model called Recommended Dietary Allowances (RDA). This model tells people what nutrients they should consume in order to prevent nutrient-deficiency diseases. For example, the RDA specifies a certain amount of vitamin C per day to prevent the vitamin C deficiency disease called scurvy.

However, in developed countries like the United States, most deficiency-related diseases are now extremely rare, and nutrition experts have become more concerned with designing dietary plans that help people achieve optimal health and well-being, manage their weight, and proactively prevent chronic diseases, such as diabetes and heart disease. Because of this shift, new dietary guidelines that expanded upon the RDA were developed in 1993. The new model is called **Dietary Reference Intakes (DRI)**. DRI encompasses RDA, but it also provides new ways of looking at the question of "What constitutes a good diet?"

FIGURE 2-7
What constitutes a good diet? Are there any foods or food groups that you think you eat too little of? What about too much of?

The DRI for most nutrients is made up of four values:

- Estimated Average Requirement (EAR)
- Recommended Dietary Allowances (RDA)
- Adequate Intake (AI)
- Tolerable Upper Intake Level (UL)

There are separate guidelines for energy-producing macronutrient consumption (protein, carbohydrate, and fat):

- Estimated Energy Requirement (EER)
- Acceptable Macronutrient Distribution Range (AMDR)

Each of these is a different way of answering the question: *What is the optimal amount of each nutrient to consume?* Let's look briefly at each of these.

Estimated Average Requirement (EAR)

EAR is the average daily nutrient intake level estimated to meet the requirements of at least 50% of the healthy individuals in a particular life stage or gender group (in other words, the mean value). It's a very lax standard in that what it prescribes is adequate for half of the people in the group and inadequate for the other half. For example, suppose that you tested a statistically significant number of women between the ages of 19 and 30, and determined how much phosphorous they need per day. Then you would line up the people in order from the most phosphorous needed to the least. The amount needed by the middle person on the list would be the EAR for phosphorous for that group.

Hot Topics

The *Dietary Guidelines for Americans* is a set of guidelines for how good dietary habits can help keep you healthy and reduce risk for major chronic diseases. These guidelines, published by the Department of Health and Human Services (HHS) and the Department of Agriculture (USDA) every five years, serve as the basis for federal food and nutrition education programs. For more information, go to: http://www.health.gov/DietaryGuidelines.

Recommended Dietary Allowance (RDA)

RDA used to be the name by which experts referred to all nutrient requirements in the United States, but now it is just one part of the larger model of DRI. The RDA is the average daily nutrient intake level that meets the requirements of 97% to 98% of healthy individuals in a particular life stage and gender group. For example, the RDA for phosphorous is 700 mg per day for women between the ages of 19 and 30 because that value satisfies the needs of all but 2% to 3% of the people studied. Most food products, when they list dietary information, compare their product to the RDA values for each nutrient.

Adequate Intake (AI)

Some nutrients cannot be adequately measured scientifically to create an RDA or EAR value, so scientists have to make an educated guess. For such nutrients, an AI is provided. An AI is like an EAR (that is, it is based on 50%), but it is less certain. Nutrients that have an AI value include calcium, vitamin D, vitamin K, and fluoride.

Tolerable Upper Intake Level (UL)

The UL is the highest level of a nutrient that you can consume without posing a risk of adverse health effects for almost all individuals in a particular life stage and gender group. It's kind of like a reverse RDA. Whereas the RDA specifies the amount most people need to prevent a deficiency, the UL specifies the amount most people should *not* exceed in order to prevent problems that could be caused by an excess of the nutrient. For example, in Table 1-5 (Chapter 1), you learned that there are specified upper limits for certain vitamins. The upper limit for vitamin C is 2,000 mg; if you consume more than that per day, you may experience indigestion or diarrhea.

Estimated Energy Requirement (EER)

The EER is the average dietary energy intake that is predicted to maintain energy balance in a healthy adult. In other words, if the individual is currently a healthy weight, this amount of intake will cause him or her to maintain that healthy weight. EER deals with overall calorie intake; it does not specify which nutrients should comprise that total. The overall number of calories a person needs depends on their age, gender, weight, height, and physical activity level and health status. For many generic dietary calculations, such as the percentages of RDA met by a particular food as noted on the nutrition label, 2,000 calories is the usual "assumed" amount. However, in actual practice, this varies greatly from person to person.

Acceptable Macronutrient Distribution Ranges (AMDR)

The AMDR works along with EER to specify the split of nutrients among the three macronutrient categories: protein, carbohydrate, and fat. The AMDR recommends the following split:

Nutrient	Percentage
Carbohydrate	45% to 65%
Protein	10% to 35%
Fat	20% to 35%

These all must add up to 100%, so if you consume carbohydrates in the upper percentage range, say, 65%, the percentage of protein and/or fat for that day would decrease. This gives you some flexibility. One day, you could consume 65%/10%/25%, and then next day 45%/35%/20%, for example. Both of these distributions are fine, as long as you stay within the ranges for each nutrient class on average, over time.

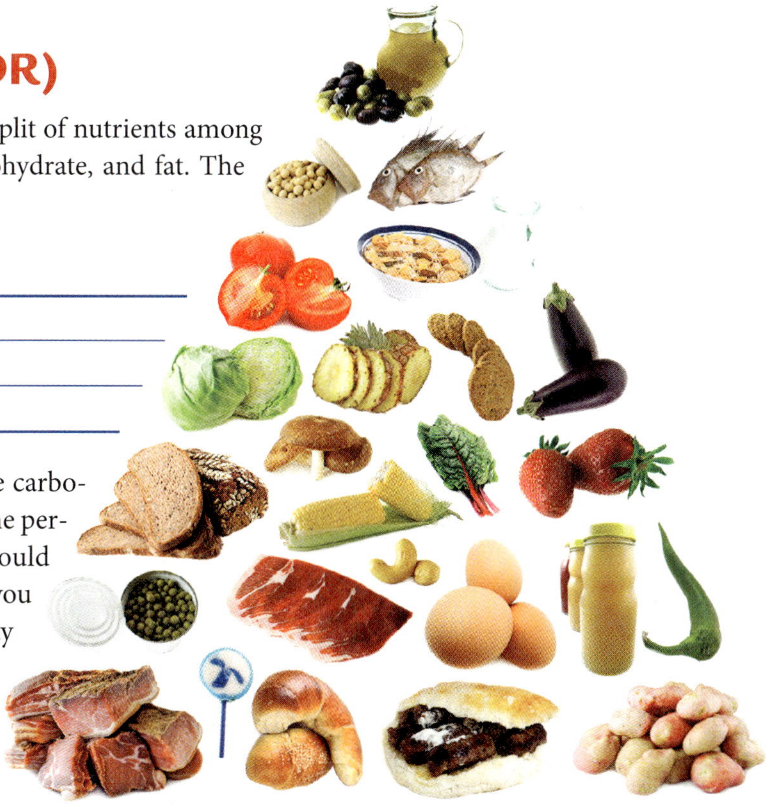

FIGURE 2-8
Focusing on the percentages of nutrients rather than the quantities (serving sizes and number of servings) is one way of thinking about dietary needs. What are the advantages and disadvantages of looking at nutrition in this way?

Standardizing the Requirements for Food Labeling

In the United States, most pre-packaged food is required to provide a Nutrition Facts label that lists the serving size, servings per container, and nutrition information per serving, including what percentage of your daily requirements are being met by the product.

But wait—that label is generic, and your situation is specific, so how can it possibly predict what you need as an individual? Because all of the other standards you've learned about so far in this chapter are unique for a particular group of

Nutrition Facts
Serving Size: 32g
Servings Per Container: 4.5

Amount Per Serving

Calories 110	Calories from Fat 5
	% Daily Value*
Total Fat 0.5g	1%
Saturated Fat 0g	0%
Cholesterol 0mg	0%
Sodium 420mg	18%
Total Carbohydrate 22g	7%
Dietary Fiber 2g	10%
Sugars 2g	
Protein 4g	
Vitamin A 25%	Vitamin C 70%
Calcium 4%	Iron 8%

*Percent Daily Values are based on a 2,000 calorie diet. Your daily values may be higher or lower depending on your calorie needs.

	Calories:	2,000	2,500
Total Fat	Less than	65g	80g
Sat Fat	Less than	20g	25g
Cholesterol	Les than	300mg	300mg
Sodium	Less than	2,400mg	2,400mg
Total Carbohydrate		300g	375g
Dietary Fiber		25g	30g

Calories per gram:
Fat 9 Carbohydrate 4 Protein 4

FIGURE 2-9
Do you usually look at the labels on the foods that you eat?

people (age, sex, activity level, and so on), they cannot be used as a standardized benchmark to apply on food labels. In other words, the fruit smoothie you eat might give *you* 50% of the calcium you need for the day; but if *someone else* drank the same smoothie, it might only give them 40% of the calcium they need because their needs are higher than yours. Therefore, there has to be a standard that can be applied uniformly to all food nutrition labels, and that standard is called Daily Values (DV).

It is based on two standardized values: Reference Daily Intakes (RDI) and Daily Reference Values (DRV). The RDIs provide standardized values for nutrients that have RDAs, including protein and vitamins. The DRVs provide standards for foods that do not have an RDA, such as fiber, cholesterol, and saturated fats. Table 2-2 provides the DRVs used for food labeling, and Table 2-3 provides the RDIs. All of these are based on a 2,000-calorie diet.

TABLE 2-2: DRVs FOR FOOD LABELING

Food	DRV
Fat	65 g
Saturated fat	20 g
Cholesterol	300 mg
Total carbohydrate	300 g
Fiber	25 g
Sodium	2,400 mg
Potassium	3,500 mg

TABLE 2-3: RDIs FOR FOOD LABELING

Food	RDI	Food	RDI
Protein	50 grams (g)	Vitamin A	1,000 Retinol Equivalents (RE)
Vitamin D	400 International Units (IU)	Vitamin E	30 IU
Vitamin K	80 micrograms (μg)	Vitamin C	60 milligrams (mg)
Folate	400 μg	Thiamin	1.5 mg
Riboflavin	1.7 mg	Niacin	20 mg
Vitamin B6	2 mg	Biotin	0.3 mg
Panthothenic acid	10 mg	Calcium	1,000 mg
Phosphorous	1,000 mg	Iodide	150 μg
Iron	18 mg	Magnesium	400 mg
Copper	2 mg	Zinc	15 mg
Chloride	3,400 mg	Manganese	2 mg
Selenium	70 μg	Chromium	120 μg
Molybdenum	70 μg		

Pyramid Models for Nutrition

In 1992, the U.S. Department of Agriculture (USDA) released a now-famous nutrition guide known as the Food Guide Pyramid (or American Food Pyramid), shown in Figure 2-10, as a guideline for selecting the right proportions of foods. The pyramid illustrates the recommended daily portions of each food group.

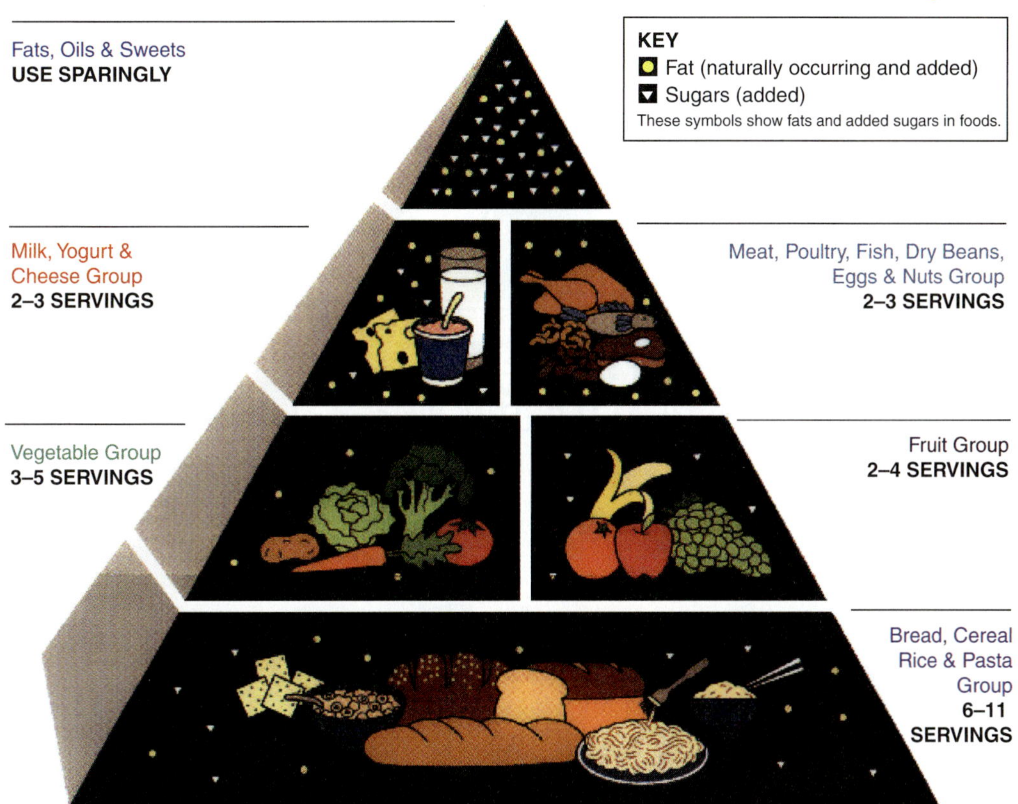

FIGURE 2-10
The Food Guide Pyramid. How does your own daily diet match up with its recommendations?
Source: U.S. Department of Agriculture

Many dietary experts, however, complained that the 1992 pyramid was not accurate based on the newest research on nutrition. They protested that it placed too much emphasis on beef and dairy and not enough on fruits and vegetables. For example, the food pyramid allowed certain dietary choices that may or may not be good, depending on the individual. Some had been linked to heart disease, such as three cups of whole milk and an eight-ounce serving of hamburger every day. In addition, the pyramid lumped all members of the protein-rich group together (meat, poultry, fish, dry beans, eggs, and nuts) and made no distinction between whole grains and refined products. Many people complained that the USDA was influenced by corporate interests, such as the dairy, meat, and sugar industries, and had allowed lobbyists to change the wording. Another argument made was that it did not take individual differences into account such as sex, age, and health.

As a result, in 2005 the USDA released a new nutrition guide known as MyPyramid. There are no foods pictured on the MyPyramid image, and no text.

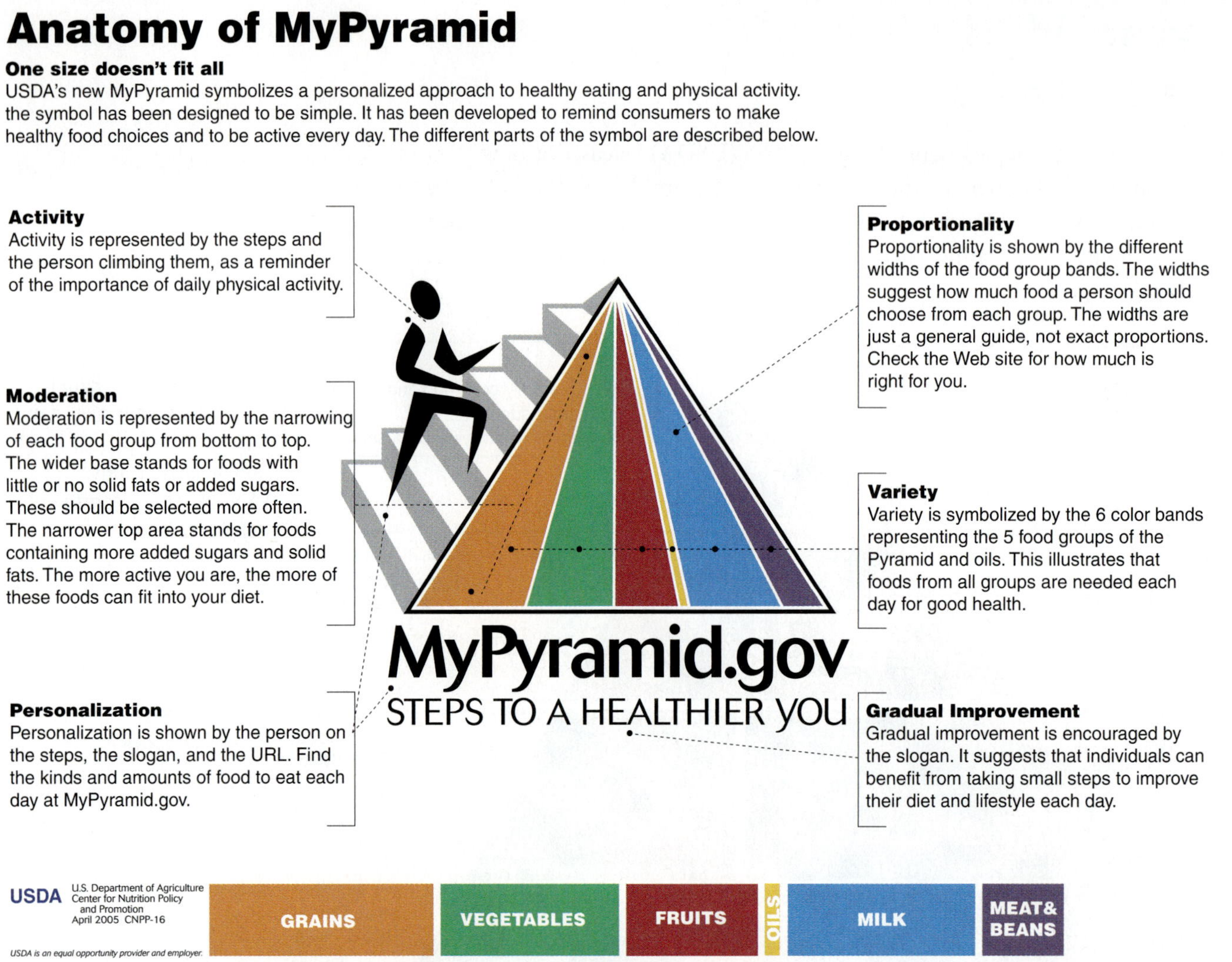

Anatomy of MyPyramid
One size doesn't fit all
USDA's new MyPyramid symbolizes a personalized approach to healthy eating and physical activity. the symbol has been designed to be simple. It has been developed to remind consumers to make healthy food choices and to be active every day. The different parts of the symbol are described below.
Activity
Activity is represented by the steps and the person climbing them, as a reminder of the importance of daily physical activity.
Moderation
Moderation is represented by the narrowing of each food group from bottom to top. The wider base stands for foods with little or no solid fats or added sugars. These should be selected more often. The narrower top area stands for foods containing more added sugars and solid fats. The more active you are, the more of these foods can fit into your diet.
Personalization
Personalization is shown by the person on the steps, the slogan, and the URL. Find the kinds and amounts of food to eat each day at MyPyramid.gov.
Proportionality
Proportionality is shown by the different widths of the food group bands. The widths suggest how much food a person should choose from each group. The widths are just a general guide, not exact proportions. Check the Web site for how much is right for you.
Variety
Variety is symbolized by the 6 color bands representing the 5 food groups of the Pyramid and oils. This illustrates that foods from all groups are needed each day for good health.
MyPyramid.gov
STEPS TO A HEALTHIER YOU
Gradual Improvement
Gradual improvement is encouraged by the slogan. It suggests that individuals can benefit from taking small steps to improve their diet and lifestyle each day.
USDA
U.S. Department of Agriculture
Center for Nutrition Policy and Promotion
April 2005 CNPP-16
USDA is an equal opportunity provider and employer.
GRAINS
VEGETABLES
FRUITS
OILS
MILK
MEAT& BEANS

FIGURE 2-11
The MyPyramid model. How does it compare to the American Pyramid shown in Figure 2-10? *Source: U.S. Department of Agriculture*

Instead, the new logo emphasizes the importance of physical activity by showing a sort of stick figure climbing the stairs of the pyramid. MyPyramid was designed to be simple; colored vertical bands represent the different food groups. Six bands of color run from the top to the base: orange for grains, green for vegetables, red for fruits, a thin band of yellow for oils, blue for milk, and purple for meat and beans. The widths of each colored segment recommend how much food a person should choose from each group. See Figure 2-11.

One confusing aspect of MyPyramid is the absence of recommended serving sizes. The USDA has stated that these were not specified because they differ depending on the person's age, weight, gender, and activity level. The agency advises that you visit www.mypyramid.gov to create personalized pyramids. However, this means that millions of people without access to the Internet have trouble getting essential information.

To complicate matters further, the Harvard School of Public Health has issued their own "Healthy Eating" pyramid as an alternative to the one provided by the USDA. The Harvard pyramid can be accessed at http://www.hsph.harvard.edu/nutritionsource/pyramids.html. Harvard's pyramid includes calcium and multi-

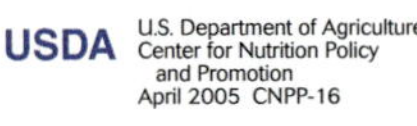

vitamin supplements as well as moderate amounts of alcohol (for adults). Many observers believe that the Harvard pyramid follows the results of scientific nutrition studies more closely than the USDA pyramid, which they still believe to be influenced by political pressure exerted by food production lobbyists. For example, note that if you combine MyPyramid's fruit and vegetable categories into one band, that band would have totally dominated the pyramid and sent a very different message to consumers about nutritious choices. The fact that the pyramid recommends a diet dominated by fruits and vegetables, therefore, is somewhat camouflaged.

Harvard's nutrition pyramid is far from the only alternative to the USDA recommendations. There are also special nutrition guides/pyramids based on Asian, Latin, Mediterranean, and vegetarian diets. The "pyramid wars" will likely continue in the nutrition community for years to come.

Hot Topics

Healthy People 2010 is a program that provides dietary and fitness guidelines for overall health, created as a collaborative effort between scientists, government agencies, and the public. The 2010 part of the name is a reference to the fact that the organizers hope to fulfill their goals by the year 2010. This program is much wider in scope than just nutrition; it encompasses substance abuse issues, sexual behavior, environmental quality, immunization, access to health care, tobacco use, and more. For more information, visit www.healthypeople.gov.

Developing Your Personal Eating Plan

The tools and techniques you've learned about in this chapter should be enough to get you started in developing a healthy eating plan for yourself and for those who depend on you. These basic nutrition facts and guidelines are just the tip of the iceberg as far as what's available to you. The Internet offers a wealth of resources, from nutrition label graphics to calorie counters, and your school library and local public library can provide additional resources as well.

Evaluating Sources of Dietary Information

Not all sources of information are equally reliable; you probably know that already from surfing the Internet! When evaluating nutrition claims, consider the following:

- **Expertise:** Is the person a known expert in the field of nutrition, or someone with a degree in the subject? A student might post a Web site about nutrition for a class project, for example, that is full of errors. On the other hand, a Web site sponsored by a large, well-known organization such as the American Dietetic Association is likely to present only that content approved by experts. Trustworthy experts in nutrition may have a degree or certification such as Registered Dietitian, Licensed Nutritionist, Master of Science, PhD, or Medical Doctor.

- **Objectivity:** Is this source seeking to sell you something, or to profit somehow if you voice a certain opinion? Nutrition Web sites sponsored by companies that sell vitamins might overstate the value of supplements; books about fad diets are mostly designed to sell books, not to provide balanced help for weight loss. Recall the controversy about the food pyramids discussed earlier in the chapter, where some people thought the information was influenced by lobbyists. Government agencies are usually trustworthy sources of information, both because they are not selling anything and because they are usually monitored by professionals in the field.

Tracking Your Diet

Thanks to the government guidelines you've learned about so far in this chapter, it doesn't take a credentialed expert to develop a healthy diet plan. What it *does* take to succeed at implementing the plan, though, is someone who is motivated and who has the time and energy to follow through. Healthy eating is not always effortless, especially when you are first making changes to your eating patterns. It takes time to read labels, to calculate nutrient amounts, and to make informed choices between foods.

Cool Tips

Some of the government agencies providing nutrition information that is generally considered to be accurate and reliable include:

- Centers for Disease Control (CDC) www.cdc.gov
- National Health and Nutrition Examination Survey (NHANES) www.cdc.gov/nchs/express.htm
- The National Institutes of Health (NIH) www.nih.gov

Some non-profit and professional organizations providing reliable information include:

- American Dietetic Association (ADA) www.eatright.org
- American Society for Clinical Nutrition (ASCN) www.faseb.org/ascn
- The Society for Nutrition Education (SNE) www.sne.org
- The American College of Sports Medicine (ACSM) www.acsm.org

Fortunately, there are many Web sites—some for free, some for pay—that make it easy to find nutrition facts for the foods you eat and to record what you eat in journal entries. These sites typically have large databases of information about fresh and prepared foods, along with a private-login area where you can record what you eat each day.

For example, FitDay (www.fitday.com) provides a free diet journal where you can track not only the macronutrients you consume (protein, carbohydrates, and fats), but also your daily intake of essential vitamins and minerals. Figure 2-12 on the next page shows a typical day of food entered into the journal; Figures 2-13 and 2-14 on the following pages show how this list can be easily evaluated to point out the strengths and weaknesses of the day's intake.

An online journal and calculator makes it easy to track your daily nutrients, but you can also do this manually just by reading the Nutrition Facts label for each food you eat, and looking up any foods that do not have a label in a reference guide (like fresh fruits and vegetables, for example).

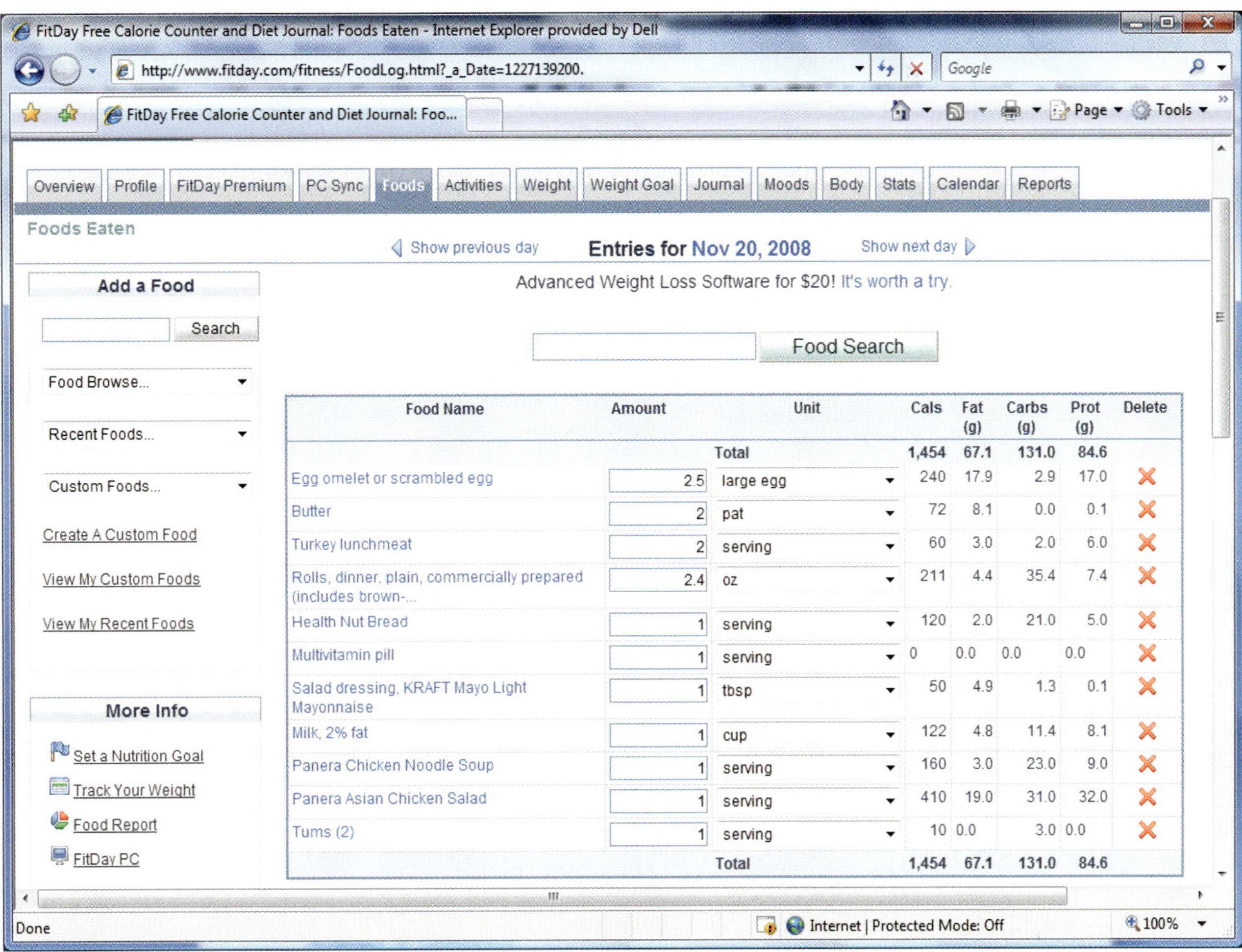

FIGURE 2-12
A sample day of food journaling at fitday.com.

BY THE NUMBERS

If you want to precisely track the amount of food you are eating, an inexpensive food scale can be a very handy tool. That way when you record in your food journal that you had, for example, 3 ounces of mashed potatoes, you can be sure that it was really 3 ounces and not 4 or 5!

Don't have a scale and don't want to bother with one? Here are tips for estimating portions, from the National Center on Physical Activity and Disability.

- One ounce is the size of a small matchbox, an average human thumb, or four dice.
- Three ounces is about the size of a deck of playing cards, or the palm of your hand.
- Two tablespoons is the size of a golf ball.
- A half-cup of cooked beans is the size of a light bulb.
- A cup of cooked rice, cereal, or pasta is the size of a tennis ball.
- A cup of chopped fruits or vegetables is the size of a baseball.
- A medium apple or orange is the size of an average woman's fist.
- A medium potato is the size of a computer mouse.

FIGURE 2-13
A quick analysis of the food journal shows that even though only 1,454 calories were consumed, the diet was too heavy in fat.

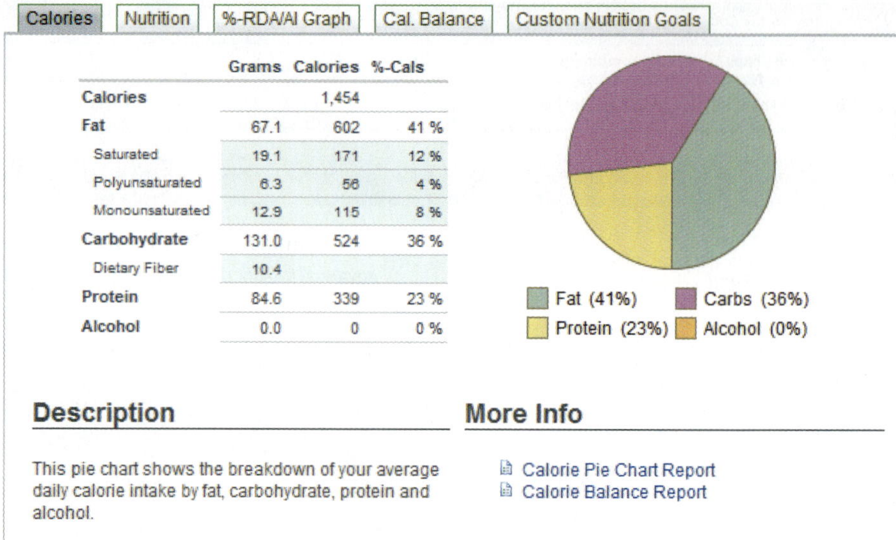

	Grams	Calories	%-Cals
Calories		1,454	
Fat	67.1	602	41 %
Saturated	19.1	171	12 %
Polyunsaturated	6.3	56	4 %
Monounsaturated	12.9	115	8 %
Carbohydrate	131.0	524	36 %
Dietary Fiber	10.4		
Protein	84.6	339	23 %
Alcohol	0.0	0	0 %

FIGURE 2-14
The food consumed satisfied most of the RDA requirements, but was lacking in magnesium and potassium.

Calories | Nutrition | %-RDA/AI Graph | Cal. Balance | Custom Nutrition Goals

			RDA	% RDA
Vitamin A	751.3	mcg	700.0	107
Vitamin A	5,123.9	IU	–	–
Vitamin B6	2.3	mg	1.3	180
Vitamin B12	8.7	mcg	2.4	365
Vitamin C	90.7	mg	75.0	121
Vitamin D	14.3	mcg	5.0	285
Vitamin D	570.7	IU	–	–
Vitamin E	22.2	mg	15.0	148
Vitamin E	33.1	IU	–	–

			RDA	% RDA
Calcium	1,421.4	mg	1,000.0	142
Cholesterol	686.6	mg	–	–
Copper	1.2	mg	0.9	128
Iron	24.4	mg	18.0	136
Magnesium	163.9	mg	320.0	51
Manganese	2.8	mg	1.8	153
Niacin	25.2	mg	14.0	180

			RDA	% RDA
Pant. Acid	3.1	mg	5.0	63
Phosphorus	709.0	mg	700.0	101
Potassium	776.4	mg	4,700.0	17
Riboflav	3.1	mg	1.1	283
Selenium	120.6	mcg	55.0	219
Sodium	4,499.8	mg	1,500.0	300
Thiamin	2.2	mg	1.1	196
Water	362.0	g	–	–
Zinc	14.2	mg	8.0	178

Description

This table lists your average daily intake for all nutrients. If a nutrient has an RDA (recommended dietary allowance) or AI (adequate intake) then the intake as a percentage of RDA/AI is displayed in the %RDA column.

More Info

Set a Nutrient Goal
Long-Term Nutrition Report

Nutrition for All Ages

Most of the guidelines you've learned about so far in this chapter have caveats with them, such as "a healthy person" or "for a certain age group or sex." That's because one size *doesn't* fit all when it comes to nutrition! The following sections look at how dietary needs differ for children, expectant mothers, and the elderly.

Nutrition for Children and Adolescents

Nutrition guidelines are available for all ages from newborn infants through the oldest senior citizens. Starting around puberty, nutrition guidelines begin to differentiate between male and female, with females requiring slightly fewer calories. Table 2-4 summarizes some age and sex-based differences outlined by NetWellness, a service of The Ohio State University College of Medicine. Other sources may differ slightly from these, and may also provide specific vitamin and mineral recommendations.

FIGURE 2-15
Suppose you were planning a picnic lunch for a group of children. How might your menu choices be different than if you were planning for adults?

Nutrition Concerns for Pregnant and Soon-to-Be-Pregnant Women

The food that a woman eats during pregnancy can affect her unborn child's growth and development, so it's important to choose food carefully for someone who is "eating for two." There's no one magical formula for pregnancy nutrition, but a few nutrients do deserve special attention.

Folate (a.k.a. folic acid) is at the top of the list. Folate is a B vitamin that helps prevent serious abnormalities of the brain and spinal cord before a woman may even realize that she is pregnant. Lack of folate may increase the risk of problems with the formation of the spinal cord and brain, premature birth, low birth weight, and poor fetal growth. Pregnant women—and women who are planning to soon become pregnant—should have at least 600 micrograms (μg) per day. One of the best sources is vitamin-fortified breakfast cereal. Naturally occurring folate comes from leafy green vegetables, citrus fruits, and legumes. It is also found in vitamin-fortified cereal.

FIGURE 2-16
Our culture is now more aware of the special nutrition needs during pregnancy than it was 100, 50, or even 20 years ago. Was your mother or grandmother concerned about nutrition during pregnancy? What was the prevailing wisdom about what you should eat during pregnancy at that time? Ask and find out.

TABLE 2-4: AGE AND SEX DIFFERENCES IN NUTRIENT RECOMMENDATIONS

<table>
<tr><th>Age</th><th colspan="2">Recommendations</th></tr>
<tr><td>0–24 months</td><td colspan="2">900 calories per day:
350–350 from fat
550–600 from protein and carbohydrates, including the following:
■ 16 oz of whole milk or 2% milk
■ 1.5 oz of meat/proteins
■ 1 cup fruit
■ ¾ cup vegetables
■ At least 2 oz of grain
■ 19 grams of fiber
■ Additional calories should be from carbohydrates</td></tr>
<tr><td colspan="3">Notes:
■ Human milk if possible; breastfeed any time the baby is hungry for first 12 months.
■ Delay the introduction of 100% juice until at least 6 months and limit to no more than 4–5 oz per day.
■ Start other sources of nutrition such as rice and pureed baby foods between 4–6 months.
■ Limit salt/sodium intake to less than 1,500 mg per day.</td></tr>
<tr><td>2–3 years</td><td colspan="2">1,000 calories per day:
300–350 from fat
650–700 from protein and carbohydrates, including the following:
■ 16 oz fat-free milk
■ 2 oz meat/proteins
■ 1 cup fruits
■ 1 cut vegetables
■ At least 3 oz of grain
■ 19 grams of fiber
■ Additional calories should be from carbohydrates</td></tr>
<tr><td colspan="3">Notes:
■ Use fat-free or low-fat dairy products.
■ Encourage 2 servings of fish per week, but avoid shark, swordfish, mackerel, and tilefish.
■ Limit salt/sodium intake to less than 1,900 mg per day.
■ Limit intake of sweet beverages, such as soda and juice drinks, to a maximum of 4–6 oz per day.</td></tr>
<tr><td>4–8 years</td><td>Boys
1,400 calories:
350–500 from fat
900–1,050 from protein and carbohydrates, including:
■ 16 oz fat-free milk
■ 4 oz meat/proteins
■ 1.5 cup fruits
■ 1.5 cup vegetables
■ At least 5 oz grains
■ 25 grams of fiber
■ Additional calories should be from carbs</td><td>Girls
1,200 calories:
350–450 from fat
750–900 from protein and carbohydrates, including:
■ 16 oz fat-free milk
■ 3 oz meat/proteins
■ 1.5 cup fruits
■ 1 cup vegetables
■ At least 4 oz grains
■ 25 grams of fiber
■ Additional calories should be from carbs</td></tr>
<tr><td colspan="3">Notes:
■ Use fat-free or low-fat dairy products.
■ Encourage 2 servings of fish per week, but avoid shark, swordfish, mackerel, and tilefish.
■ Limit salt/sodium intake to less than 1,900 mg per day.
■ Limit intake of sweet beverages, such as soda and juice drinks, to a maximum of 4–6 oz per day if between 4 and 6 years old, or 8–12 oz per day if between 6 and 8 years old.</td></tr>
</table>

(continued)

TABLE 2-4: AGE AND SEX DIFFERENCES IN NUTRIENT RECOMMENDATIONS *(CONT)*

Age	Recommendations	
9–13 years	**Boys** 1,800 calories: 450–650 from fat 1,150–1,350 from protein and carbohydrates, including: ■ 24 oz fat-free milk ■ 5 oz meat/proteins ■ 1.5 cup fruits ■ 2.5 cup vegetables ■ At least 6 oz grains ■ 31 grams of fiber ■ Additional calories should be from carbs	**Girls** 1,600 calories: 400–550 from fat 1,050–1,200 from protein and carbohydrates, including: ■ 24 oz fat-free milk ■ 5 oz meat/proteins ■ 1.5 cup fruits ■ 2 cup vegetables ■ At least 5 oz grains ■ 26 grams of fiber ■ Additional calories should be from carbs
Notes: ■ Use fat-free or low-fat dairy products at home. ■ Encourage 2 servings of fish per week, but avoid shark, swordfish, mackerel, and tilefish. ■ Limit salt/sodium intake to less than 2,200 mg per day. ■ Limit intake of sweet beverages, such as soda and juice drinks, to a maximum of 8–12 oz per day.		
14–18 years	**Boys** 2,200 calories: 550–750 from fat 1,450–1,650 from protein and carbohydrates, including: ■ 24 oz fat-free milk ■ 6 oz meat/proteins ■ 2 cups fruit ■ 3 cups vegetables ■ At least 7 oz grains ■ 38 grams of fiber ■ Additional calories should be from carbs	**Girls** 1,800 calories: 450–650 from fat 1,150–1,350 from protein and carbohydrates, including: ■ 24 oz fat-free milk ■ 5 oz meat/proteins ■ 1.5 cups fruit ■ 2.5 cups vegetables ■ At least 6 oz grains ■ 29 grams of fiber ■ Additional calories should be from carbs
Notes: ■ Use fat-free or low-fat dairy products at home. ■ Encourage 2 servings of fish per week, but avoid shark, swordfish, mackerel, and tilefish. ■ Limit salt/sodium intake to less than 2,200 mg per day. ■ Limit intake of sweet beverages, such as soda and juice drinks, to a maximum of 8–12 oz per day.		

Calcium is also especially important for expecting mothers. If a pregnant woman doesn't get enough calcium, the calcium needed to form the baby's bones and teeth will be drawn from her own bones, increasing her risk of osteoporosis. Dairy products are the richest source of calcium. Many fruit juices and breakfast cereals are also fortified with it.

Iron is an important part of hemoglobin, the protein in the red blood cells that carries oxygen to the tissues. During pregnancy, when a woman's body is creating the blood supply for her baby, her need for iron almost doubles, to 27 milligrams per day. Good sources of iron are red meat, poultry, and fish. Iron also comes from iron-fortified breakfast cereal, nuts, and dried fruit.

Not only should a pregnant woman actively choose healthy foods, but she should also stay away from certain foods. The Mayo Clinic offers these recommendations:

- **Seafood:** Up to 12 ounces of fish per week is excellent for the health of both mother and baby. The FDA recommends that swordfish, shark, mackerel, and tilefish be avoided during pregnancy, as these varieties might contain low levels of mercury that could harm the baby. It is best to also avoid raw fish during pregnancy, including raw oysters and sashimi.
- **Meats:** Avoid rare meats because of the risk of bacterial food poisoning. Cook meats thoroughly or avoid them completely.
- **Dairy:** Use only pasteurized dairy products, including cheeses. Check the labels to make sure. Unpasteurized milk products may carry live bacteria.
- **Caffeine:** Caffeine enters the baby's bloodstream and can affect growth. Avoid it altogether, or ask your health care provider to recommend limits for intake.
- **Herbal tea:** Herbal tea ingredients vary widely, and some medicinal herbs can be harmful to the baby. Check specific types of teas with your health care provider before consuming them.
- **Alcohol:** Avoid alcohol completely. Even moderate drinking can impact the baby's brain development.

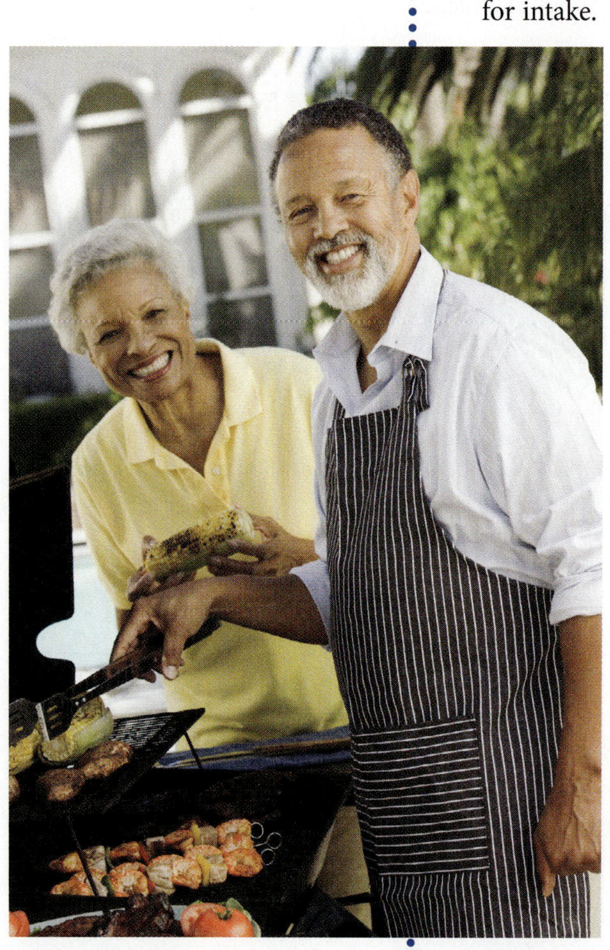

FIGURE 2-17
Look over some of the challenges in Table 2-5 on page 48 that seniors face in getting good nutrition. What challenges might a senior citizen have in eating in your school cafeteria?

Nutrition Concerns for Senior Citizens

There has not been enough research to date to know the exact nutritional requirements of elderly people. However, it is especially important to get adequate amounts of certain nutrients in old age, including water, calcium, fiber, iron, protein, and vitamins A and C, because each of these is associated with the prevention of some health problems that affect seniors in higher proportion than the general population. For example, calcium helps keep bones strong, fiber helps avoid constipation, and iron helps with red blood cell count.

During the aging process, changes that occur in the body can affect nutrition. There may be less absorption and greater excretion of nutrients. That means seniors must take in more nutrients to absorb the same amount as a younger adult. At the same time, calorie needs decrease with age because the metabolism slows down. Therefore seniors need nutrient-dense foods so they can meet their needs without gaining weight. Fruits, vegetables, plain breads and cereals, low-fat dairy products, lean meats, and low-fat meat substitutes, such as beans, peas, lentils, fish, and eggs, are all great choices.

Healthy eating can be a challenge for many seniors, due to lifestyle issues such as:

- Both appetite and the sense of taste diminishes as people age, so food does not taste as good as it once did, resulting in less enthusiasm for eating.
- Widowed people may feel socially isolated or depressed and may lose interest in preparing meals or eating.
- If the widowed person was not the primary cook in the family, they may not have and the necessary skills or may be unsure how to cook for themselves.
- People with limited mobility or energy may find it too tiring to cook full meals.
- Elderly people experiencing gradual mental deterioration may forget to eat, or forget how to cook; conversely, they may forget they have already eaten, and eat too much.
- People on a fixed income with high medical bills may scrimp on food in order to make ends meet.
- Some people may eat an overly narrow selection of foods because of age-related health problems, such as missing or painful teeth, constipation, diarrhea, heartburn, or upset stomach.
- People who have special dietary needs due to other health problems may be unable or unwilling to go to the trouble of conforming to that diet, such as low-sodium for high blood pressure, low-sugar for diabetes, or low-fat for heart disease. The elderly may also become confused by conflicting dietary instructions from multiple sources.

Table 2-5 offers some tips for addressing common food-related difficulties that seniors face, based on advice from the U.S. Food and Drug Administration.

Optimal nutrition is more available now than it has ever been. The U.S. has the safest and most nutritious food supply available in the history of the world. We have problems, but we have the longest lifespan and the tallest physical stature (growth) in history. We also have the least food-borne illness of any civilization in history. Our problems are real, but so are our successes.

Hot Topics

Proper nutrition and fitness are key to a healthy body and mind. Many senior centers employ nutritionists to make sure that the seniors are eating properly. Some communities have programs such as Meals On Wheels that deliver nutritious food to home-bound elderly. Are there any programs like this in your community? Are there any other resources in your community, such as food programs for students?

TABLE 2-5: NUTRITION TIPS FOR SENIORS

Problem	Instead of this...	Try this...
Difficulty chewing	Fresh fruit	Fruit juices and soft canned fruits.
	Raw vegetables	Vegetable juices, canned vegetables, creamed, mashed, or cooked vegetables.
	Meat	Ground meat, eggs, milk, cheese, yogurt, pudding.
	Sliced bread	Cooked cereal, rice, bread pudding, soft cookies.
Upset stomach	Milk	Milk products designed for lactose-intolerant people, fermented milk products such as cheese, buttermilk, and yogurt.
		Drinking small amounts of milk multiple times a day.
		Drinking milk with meals.
	Cabbage, broccoli	Vegetable juices and other vegetables, such as green beans, potatoes, and carrots.
	Fresh fruit	Fruit juices, canned fruits.
Can't shop	Shopping for groceries	Find a local food store that delivers.
		Ask someone at your church or synagogue for volunteer help.
		Ask a family member or neighbor to shop for you.
		Hire a home health worker for a few hours a week to run errands.
Can't cook	Cooking full meals	Use a microwave to cook frozen dinners.
		Take part in group meal programs offered through senior citizen programs.
		Move to a place where someone else will cook.
No appetite	Not eating	Eat with family and friends.
		Take part in group meal programs.
		Ask your doctor if your medicines could be causing appetite problems.
		Increase the flavor of food by adding more spices and herbs.
Short on money	Skimping on food budget	Buy low-cost food, such as dried beans and peas, rice, and pasta.
		Use coupons.
		Buy foods on sale.
		Buy store brands.
		Find out if your local church or synagogue offers free or low-cost food programs.
		Visit food pantries.
		Get food stamps.

Case Study

Michael is having an argument with his mother over breakfast cereal. He claims that his favorite high-sugar cereal (see figure A on the next page) is actually just as nutritious as the oatmeal-based "adult" cereal (see figure B on the next page) his mother prefers. He plans to compare the Nutrition Facts from the two cereals to prove his point.

When identifying the "better" cereal, which factors should Michael and his mother consider? Here are some ideas to get you started:

Compare how much of each vitamin and mineral you will receive from each cereal alone, not including the milk. Complete the following chart:

Nutrient	Cereal A	Cereal B
Vitamin A	______	______
Vitamin C	______	______
Calcium	______	______
Iron	______	______
Vitamin E	______	______
Thiamin	______	______
Riboflavin	______	______
Niacin	______	______
Vitamin B6	______	______
Folic Acid	______	______
Phosphorous	______	______
Magnesium	______	______
Zinc	______	______

Compare how much protein, carbohydrates, and fat you will receive from each cereal alone. Again, do not include the milk. Complete the following chart:

Nutrient	Cereal A	Cereal B
Protein	______	______
Total Fat	______	______
Saturated Fat	______	______
Trans Fat	______	______
Total Carbohydrate	______	______
Dietary Fiber	______	______
Sugars	______	______
Other Carbohydrates	______	______

Compare how much sodium, potassium, and cholesterol each cereal provides. Complete the following chart:

Nutrient	Cereal A	Cereal B
Sodium	______	______
Potassium	______	______
Cholesterol	______	______

After this initial analysis, take a look at the serving sizes. Notice that Cereal A's serving size is 27 g, whereas Cereal B's serving size is 56 g. That's almost two-for-one! Repeat the above analysis, this time doubling the values for Cereal A to more fairly compare the two products by weight. Now which cereal provides better nutrition? And what about calories? Will doubling the serving size on Cereal A result in consuming many more calories than eating a single serving of Cereal B?

A.

Nutrition Facts

Serving Size: 3/4 cup (27g)

Servings Per Container: about 22

Amount Per Serving	**Cereal Alone**	**with 1/2 Cup Vitamin A&D Fortified Skim Milk**
Calories	110	150
Calories from Fat	25	25
	% Daily Value**	
Total Fat 2.5g*	**4%**	**4%**
Saturated Fat 1g	**5%**	**6%**
Trans Fat 0g		
Polyunsaturated Fat 0.5g		
Monounsaturated Fat 1g		
Cholesterol 0mg	**0%**	**1%**
Sodium 200mg	**8%**	**10%**
Total Carbohydrate 21g	**2%**	**7%**
Dietary Fiber 1g	**7%**	**9%**
Sugars 9g	**3%**	**3%**
Other Carbohydrate 11g		
Protein 2g		
Vitamin A	**0%**	**4%**
Vitamin C	**0%**	**0%**
Calcium	**0%**	**15%**
Iron	**25%**	**25%**
Thiamin	**25%**	**30%**
Riboflavin	**25%**	**40%**
Niacin	**25%**	**25%**
Vitamin B6	**25%**	**25%**
Folic Acid	**100%**	**100%**
Zinc	**25%**	**30%**

* Amount in Cereal. One half cup skim milk contributes an additional 65mg Sodium, 200mg Potassium, 6g Total Carbohydrate (5g Sugars), and 4g Protein.

** Percent Daily Values are based on a 2,000 calorie diet. Your daily values may be higher or lower depending on your calorie needs.

	Calories:	2,000	2,500
Total Fat	Less than	65g	80g
Sat Fat	Less than	20g	25g
Cholesterol	Les than	300mg	300mg
Sodium	Less than	2,400mg	2,400mg
Total Carbohydrate	300g	375g	
Dietary Fiber		25g	30g

Ingredients: Corn flour, rice flour, sugar, peanut butter (peanuts, dextrose, hydrogenated vegetable oil, salt), oat flour, salt, caramel color, flaxseed oil, niacinamide, reduced iron, zinc oxide, EDTA (a preservative), riboflavin*, folic acid*.

* One of the B vitamins.

B.

Nutrition Facts

Serving Size: 1 cup (56g)

Servings Per Container: about 12

Amount Per Serving	**Cereal Alone**	**with 1/2 Cup Vitamin A&D Fortified Skim Milk**
Calories	210	250
Calories from Fat	25	25
	% Daily Value**	
Total Fat 2.5g*	**4%**	**4%**
Saturated Fat 0.5g	**2%**	**3%**
Trans Fat 0g		
Polyunsaturated Fat 1g		
Monounsaturated Fat 1g		
Cholesterol 0mg	**0%**	**1%**
Sodium 250mg	**10%**	**13%**
Potassium 210mg	**6%**	**9%**
Total Carbohydrate 44g	**15%**	**17%**
Dietary Fiber 5g	**18%**	**18%**
Soluble Fiber 2g		
Sugars 10g		
Other Carbohydrate 29g		
Protein 6g		
Vitamin A	**10%**	**15%**
Vitamin C	**10%**	**10%**
Calcium	**10%**	**20%**
Iron	**90%**	**90%**
Vitamin E	**10%**	**10%**
Thiamin	**25%**	**30%**
Riboflavin	**25%**	**40%**
Niacin	**25%**	**25%**
Vitamin B6	**25%**	**30%**
Folic Acid	**100%**	**110%**
Phosphorus	**20%**	**25%**
Magnesium	**15%**	**15%**
Zinc	**25%**	**30%**

* Amount in Cereal. One half cup skim milk contributes an additional 65mg Sodium, 200mg Potassium, 6g Total Carbohydrate (5g Sugars), and 4g Protein.

** Percent Daily Values are based on a 2,000 calorie diet. Your daily values may be higher or lower depending on your calorie needs.

	Calories:	2,000	2,500
Total Fat	Less than	65g	80g
Sat Fat	Less than	20g	25g
Cholesterol	Les than	300mg	300mg
Sodium	Less than	2,400mg	2,400mg
Total Carbohydrate	300g	375g	
Dietary Fiber		25g	30g

Put It to Use

❶ Suppose you had to develop an awards dinner menu at which there will be several attendees who are vegetarian. Develop a nutritious and appealing menu including a salad, entrée, and dessert that everyone will be able to eat.

Put It to Use

❷ Interview an older person in your life, such as a grandparent, about their eating habits, and record their answers on paper or on audio or video tape. Suggested questions to ask:

- How has your taste in food changed over the years? Are there any foods that you used to dislike that you now like? Or vice versa?
- Do you prefer foods to be more or less heavily seasoned now than you did 20 years ago?
- Do you think your daily diet is more or less healthy than it was 20 years ago, on the average? Why?
- What, if anything, stands in the way of you eating a healthy and balanced diet every day?

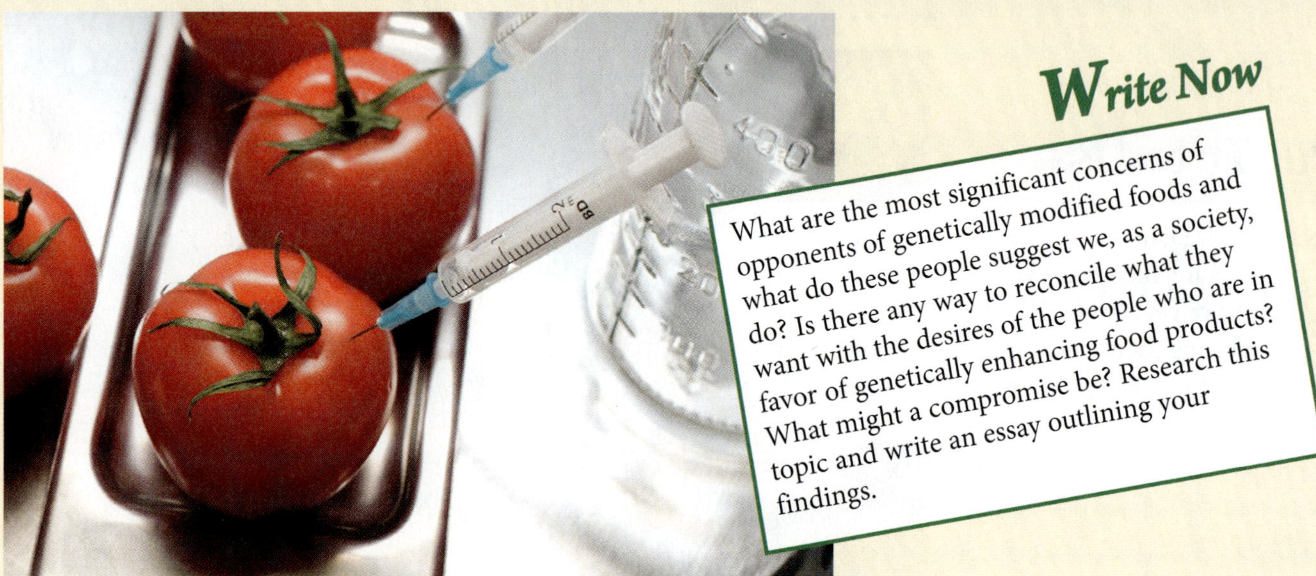

Write Now

What are the most significant concerns of opponents of genetically modified foods and what do these people suggest we, as a society, do? Is there any way to reconcile what they want with the desires of the people who are in favor of genetically enhancing food products? What might a compromise be? Research this topic and write an essay outlining your findings.

Tech Connect

There are many dietary guidelines and recommendations published by various sources, including many major universities, food manufacturing and distribution companies, industry associations for particular types of food, and government agencies. Not all of them are equally reliable. Using the Internet, find a source of dietary advice that is provided by a source that has an agenda or product to promote. Explain why you think this is not a reliable source of accurate information.

Team Players

Pair up into teams of two or more people. Create a 2,000-calorie meal plan (breakfast, lunch, and dinner) for a person who is trying to lower their dietary sodium level. For each meal, write a paragraph explaining the choices you made and provide the reference sources you used to look up the sodium in each food item.

Put It Together

Match the explanation in column 1 with the term in column 2.

Column 1

a. a waxy substance that builds up in the arteries, making them harden and narrow
b. a disease that makes bones brittle and easily broken
c. a dietary model encompassing RDA and several other measurements
d. a dietary model that proposes a number of calories to consume per day
e. the standardized values on which Nutrition Facts percentages are based on food labels
f. a 2005 USDA nutrition guide that represents food groups with vertical color bands
g. dietary restriction common to both Buddhists and Hindus
h. DNA modification that makes some food products less prone to disease or spoilage
i. mineral that helps prevent osteoporosis
j. the highest level of a nutrient that you can consume without posing a risk of adverse health effects

Column 2

1. calcium
2. cholesterol
3. DRI
4. EER
5. genetic engineering
6. myPyramid
7. osteoporosis
8. RDI
9. UL
10. vegetarian

3 Planning a Meal

In This Chapter, You Will . . .

- Examine social trends in food selection and meal types
- Plan a menu for a meal
- Find out how to cut costs when buying ingredients for a dish
- Learn how to shop more efficiently and save money

Why YOU Need to Know This

Meals are not just opportunities for feeding our bodies, but in many cases, social times as well. Families and friends share meals together to enjoy each other's company and catch up on what's new in one another's lives. Yet as schedules become increasingly busy, some do not make time for a family dining experience around a table at home. What are some of the different ways in which your family and friends eat meals together? How have your eating styles changed over the last five years? How do you expect them to change in the next several years?

Societal Trends in Eating

Sociologists study people's eating habits as a reflection of the culture in which the people live. Watching the way in which a group of people cook and eat meals tells you about what they think is important and what their lives are like. For example, is it typical in a particular culture for someone to spend a long time preparing a meal? Is it typical for multiple family members to participate in meal preparation? Is it typical for everyone to eat at the same time, at the same table? Is snacking the norm, or do people eat only at specific meal times?

Hot Topics

Eating food on the go, rather than enjoying it in a more leisurely way, may be one contributing factor to the problem of obesity in our society. Meals eaten just to relieve hunger are likely to be consumed more quickly, so the stomach does not have time to signal the brain that it is full before we overeat. Fast food is also likely to be higher in calories, sodium, and fat than food prepared at home.

Family Eating Patterns

Fifty years ago in the United States, it was assumed that most families would eat meals at home, together around the family dinner table. The family dinner was an important part of the culture of that time, and provided a consistent opportunity for the family members to interact each day. One or more family members (frequently females) would prepare and serve a full meal, and everyone would eat together.

While many families today still eat meals together, busy lifestyles often make it difficult for the whole family to sit down to eat at the same time each night. A meal might consist of home-cooked dishes and freshly purchased local produce, a frozen entrée that you cook at home, carry-out food from a restaurant, a snack from a vending machine, or any combination of these. The entire family might sit down to the meal together, or just one or two members, or everyone might eat separately, depending on their schedules.

How have eating patterns become so diverse in such a short span of time? It all has to do with choices. People have more choices now for every meal than they did 50 years ago. When faced with meal choices, you must balance the cost and benefit of each option in terms of expense, convenience, nutrition, and time.

The benefits and drawbacks of the various options have changed over time as well. For example, more restaurants are now offering low-fat cuisine, making it possible to eat as healthy a meal in a restaurant as you might fix for yourself at home. However, restaurant food is usually more expensive, and sometimes there is no time even for a quick sit-down meal at a restaurant.

FIGURE 3-1
Carry-out food mixes the convenience of a restaurant with the more personal atmosphere of dining at home. What are some restaurants in your area that offer healthy carry-out cuisine?

Food Customs and Cultures

The culture or country in which you were raised helps to determine what foods you think of first when you are hungry. For example, families of Asian descent may get more of their carbohydrates through rice; families of Mexican descent may eat more beans; and Europeans may eat more potatoes. These preferences originated based on availability; rice is plentiful in Asia, just as beans are in Central America. Societies develop cultures and families create traditions around the food that is most readily available, so a family's favorite "special occasion" dish is likely to include ingredients common in their country of origin.

FIGURE 3-2
A family's country of origin may dictate some of the daily foods served. What foods are the most common in your home? Which of those are a result of the culture or nationality of your parents and grandparents?

Culture does not only determine the choices of ingredients, but also the rituals, traditions, and habits surrounding meals. These rituals could include saying a prayer before a meal, waiting for an elder to take the first bite of food, pouring drinks for one another, eating from a common bowl, and even belching at the end of the meal. (In China, belching after a meal is considered a compliment to the cook!)

Here are some notable differences in meal habits worldwide:

- In Pakistan and Japan, it is polite to "clean your plate" to show how delicious the food was, whereas in Russia, it is polite to leave some morsels on the plate to show that the host served ample portions.
- In some Asian countries, it is good manners to leave a friend's house immediately after a meal; if you linger, it indicates you did not get enough to eat. In India, Europe, and the Americas, a quick departure is considered "eat and run" and is rude because it indicates you were just there for the food, not the company.
- In some European countries, sausages are held between the fingers while being eaten, rather than put in a bun like in the U.S.
- In France, both hands should be visible above the table at all times. In contrast, in the U.S., it is considered polite to keep the left hand in the lap when eating with the right hand.
- In India, maintain silence while eating; do not make conversation. In the U.S., it is polite to make small talk with those around you.
- In Japan, it is acceptable, and even encouraged, to make a slurping noise when eating noodles. There's no slurping allowed in the U.S.

Cool Tips

What if you had a Japanese foreign-exchange student living with your family? It's your first dinner together and your mom serves spaghetti. Your new houseguest loves the spaghetti, which he has never had, and makes hearty slurping noises with every bite. How would you politely explain that this is not the normal way to eat "noodles" in the U.S.?

- In Mediterranean European countries, Latin America, and Sub-Saharan Africa, it is normal, or at least widely tolerated, to arrive half an hour late for a dinner invitation, whereas in Germany and the United States, this is considered very rude.

Food on Special Occasions and Holidays

Besides the nutritional and social value, food is also symbolic of family and cultural traditions and occasions. For example, in the United States, it is customary to serve cake to the guests after a wedding. For Thanksgiving dinner, many families eat a traditional menu that includes roasted turkey, mashed potatoes, and pumpkin pie. For those of Christian faith in the United States, an Easter dinner often includes ham; for others, St. Patrick's Day or New Year's Day may be celebrated with corned beef and cabbage.

FIGURE 3-3
Many families have holiday traditions involving food. Make a list of the foods typically served at a holiday dinner in your household, and compare the list to that of your classmates.

Religious ceremonies also may involve food. For example, some Christian churches use bread and either wine or grape juice for Communion ceremonies. Many people of the Jewish faith celebrate Passover with a ceremonial Seder meal that includes wine, bitter herbs, a hard-boiled egg, bread, matzo, and parsley or celery.

FIGURE 3-4
Some religions mix food with ceremony or ritual, such as a Jewish Seder. Each ingredient shown here has a special significance. Can you name one of the ingredients here and explain its significance? Use the Internet to research if you do not know.

Types of Meals

Suppose you have decided you want to organize a meal for family or friends to share. What kind of meal would best fit the occasion, your budget, and your guests' tastes?

- **Home-cooked meals:** A meal you cook yourself can be a gift to your family and friends, showing that you don't mind putting forth the effort involved in planning and preparing the meal. Home-cooked food may be less expensive than ready-to-serve food, and you can more precisely control the flavors, variety, and nutrition content of the meal. A home-cooked meal is usually the most work of all the meal types, however, and takes the most time.
- **Prepared food served at home:** If you want to offer the comfortable atmosphere of your own home without all the work involved in a home-cooked meal, serving a meal consisting mostly of ready-to-serve food can be an attractive alternative. Frozen or fresh entrees, side dishes, and desserts are all readily available at your local grocery store or as carry-out from restaurants, and can dramatically cut down on your preparation time. However, ready-to-serve food tends to be more expensive, and the quality and nutritional value may or may not be up to the same standards that you could produce yourself with home-cooked food.
- **Restaurant meals:** Hosting a meal in a restaurant is easy—just make a reservation! But it's also expensive, and you don't have control over what's on the menu.

There are certainly many situations in which prepared food or restaurant food is most appropriate. However, in this book, we'll focus on meals that you cook yourself. These meals are usually most economical and offer the best opportunity to practice your skills in planning, cooking, and serving nutritious and delicious meals.

FIGURE 3-5
Special occasions are often celebrated with food, such as birthdays and anniversaries. Think of a special occasion you recently celebrated with family or friends. What was on the menu?

Planning a Menu

Let's say you want to plan a meal for your family, or for a group of friends, or for an even larger group, such as a class or club. Where do you begin? What kinds of things do you need to think about? How do you decide what to serve, and how to serve it?

Occasion

First of all, why are you serving the meal? Is it "just because," or is there an occasion you are celebrating? Is this a birthday or an anniversary party? A celebration of a team victory, a job promotion, or a new baby? A holiday meal? In some cases, the occasion may dictate one or more menu items. The classic example, of course, is that at a birthday party, guests will expect birthday cake for dessert. Guests may expect certain dishes at holidays as well, such as turkey at Thanksgiving.

Courses

How many courses will you serve? Simple at-home meals are typically only one or two courses—perhaps a green salad and an entrée. Depending on your budget, the time you have available, and the occasion and theme, you may want to add other courses such as appetizers, soups, or desserts.

FIGURE 3-6
Next time you are at your favorite local restaurant, see how many sections of the menu include soups. When would you consider a soup to be an entrée rather than a first course?

Food Theme

Do your plans include serving a menu around a theme? For example, you might serve an entire menu of food from a certain country or region, such as Mexico or Asia. Themes aren't limited to just nationalities. You could serve foods that are commonly available at a football stadium when hosting a Super Bowl party (like hot dogs, curly fries, and popcorn).

Cost

Most people don't have unlimited funds to spend on a meal; there is usually a budget for it. The menu you will prepare will depend, in part, on the ingredients you can afford. We'll look at food prices in detail later in the chapter, in the "Food Costs and Budgeting" section.

Nutrition

Unless you are serving food every meal, every day to the same people, you don't have to worry about everyone's nutrient needs. However, given the other goals you wish to accomplish, you should do your best to provide the most nutrient-rich food you can. In many cases, it is just as easy, inexpensive, and tasty to use an ingredient with a lot of nutritive value as to use one that is mostly empty calories.

Cool Tips

There are many variations on a multi-course meal, and courses are served in different orders in different parts of the world. Here is one possibility, modified from the model provided by Fordham University Hospitality Services:

- Hors d'oeuvres: Pre-dinner small appetizers, typically served before guests are seated.
- Appetizer: May be either hot or cold.
- Soup: Usually hot, but a few gourmet types are served cold.
- Sorbet: A light fruit sherbet that cleanses the palate before the entrée.
- Entrée: The main dish of the meal, usually meat or pasta based. Sometimes a separate shellfish or pasta course is served earlier in the meal as well. The entrée may optionally be broken down into two separate courses, Fish and Meat.
- Salad: Usually of leafy greens. In the U.S., the salad is sometimes served before the entrée, but at a formal gourmet meal it is served afterwards, as is European tradition.
- Dessert: Usually something sweet, although cheese is sometimes served as a dessert.
- Coffee: At a formal dinner, coffee is served after the dessert, and not in conjunction with it.

Balance of Foods

A satisfying meal should contain a mixture of textures, flavors, temperatures, and colors. For example, you might balance a heavy spaghetti-and-meatball dish with a crisp green salad, and top that off with a light fruit sherbet for dessert.

Cool Tips

If you pursue a career in food services, you may be asked to develop a menu for a restaurant at some point. There are many types of restaurant menus:

- A la carte: Each food or beverage is priced and served separately. For example, you would order a steak, salad, and baked potato separately.
- California menu: Breakfast, lunch, and dinner are all available at any time of the day or night.
- Du jour menu: Food are served only on that particular day. Du jour is French for "of the day."
- Table d'hôte or prix fixe menu: A complete meal, including appetizer, dessert, and beverage, for a fixed price. A banquet meal is an example.
- Fixed menu: Offers the same items every day.
- Cyclical or cycle menu: A series of rotating menus planned for a specific time period, such as a week. The menu is different for each day during the cycle and then repeats. Cycle menus are often used in school and hospital cafeterias.

Equipment Required

As you are perusing recipes, keep in mind the kitchen in which you will work, and its equipment, including appliances and hand tools such as knives and bowls. Crème brulée is a wonderful dessert, but you can't make it properly without a butane kitchen torch to caramelize the sugar on top and ramekins (a special type of bowl) in which to bake the individual servings.

Choices and Preferences

Will you offer your guests a choice of dishes, or are you making one fixed menu for everyone? If you're preparing a meal for family and friends, the expectation is that you will prepare only one entrée, but if you are catering food for a large group, you will be expected to provide alternatives. For example, a large group is likely to include some vegetarians, so consider having a vegetarian entrée available.

Besides all the standard considerations, you must address an important subjective factor: will your guests *like* the food? Think about what you know about them—their likes and dislikes, the types of food with which they are familiar, the kinds of food they may have never had before. For example, if your guests are all small children, you probably will achieve more success with a simple menu of hamburgers and fruit salad than a fancy tray of sushi.

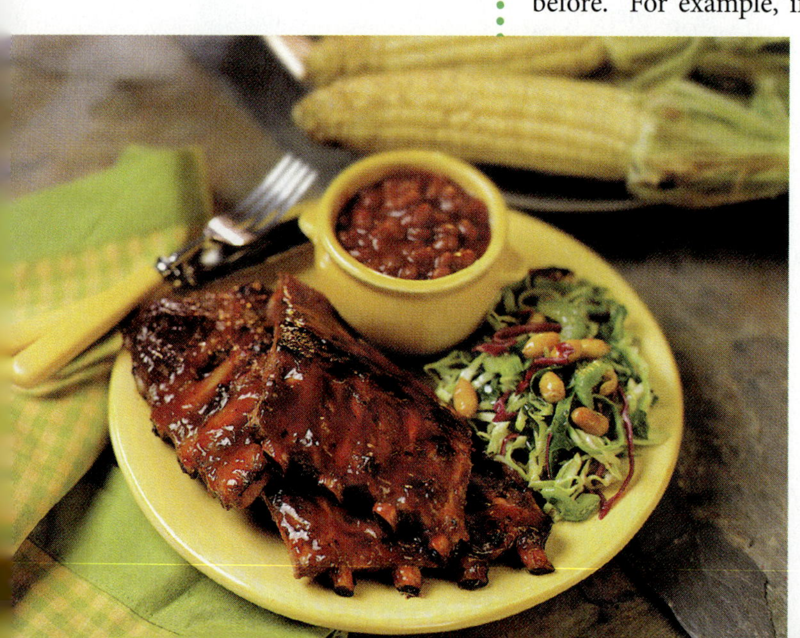

FIGURE 3-7
A plate of food containing many colors and textures is more appealing to eat than one that contains only one color and texture. Think about the last meal you had: How many different colors of food did it include?

Planning Meals Using Patterns

How do you know what combinations of dishes to include in a meal? As you become more experienced at cooking and meal planning, it will become second nature to put together meals that include a variety of healthful ingredients. As you are getting started, though, it may be helpful to follow meal pattern guidelines.

The U.S. Department of Agriculture (USDA) has published a set of meal pattern guidelines for infants, children, and adults through their Child and Adult Care Food Program (CACFP). These guidelines were developed to specify what food components must be present in a meal in order for the provider to get reimbursed by the government for providing meals to participants in government-sponsored food/nutrition programs. These guidelines can be applied to general planning as well to build balanced meals for almost anyone.

The CACFP meal pattern guidelines specify foods from four groups: milk, fruits/vegetables, grain/bread, and meat or meat alternative. The version of the guidelines that organizations follow for government reimbursement specifies minimum serving sizes for each food. For full information about the program, including detailed guidelines on servings, see http://www.fns.usda.gov.

However, because calorie intake requirements vary from person to person, it is more useful for most of us (that is, those of us who are not administering government programs) to use the meal pattern guidelines as general templates for the types of food to include in each meal, rather than the specific quantities or ingredients. Table 3-1 shows meal patterns for adults; the Web site provides patterns for infants and children.

TABLE 3-1: USDA MEAL PATTERN GUIDELINES FOR ADULTS

Meal	Food Group		Example
Breakfast	Milk		1 cup milk
	Fruit/Vegetable		½ cup fruit juice or fruit
	Grain/Bread		1 cup hot cooked cereal
Lunch	Milk		1 cup milk
	Fruit/Vegetable		1 cup raw vegetables
	Grain/Bread		2 slices whole wheat bread
	Meat/Meat Alternative		2 ounces lean meat
Dinner	Fruit/Vegetable		1 cup cooked vegetables
	Grain/Bread		2 slices whole wheat bread
	Meat/Meat Alternative		2 ounces lean meat
Snack	Select two:	Milk	1 cup yogurt and ½ cup fruit
		Fruit/Vegetable	OR
		Grain/Bread	Sandwich with 2 slices whole wheat bread and 2 ounces cheese
		Meat/meat Alternative	

Food Costs and Budgeting

How much will it cost to prepare this wonderful meal? That all depends on the ingredients in the recipes you have chosen to prepare and the quantities you need to purchase. For example, an entrée that is primarily pasta and vegetables is likely to be much less expensive to prepare than an entrée that contains lobster or crab; and a dinner at which you serve everyone an 8-ounce steak will cost more than a dinner where everyone gets a 6-ounce portion. Table 3-2 categorizes some common ingredients according to price, to help you choose recipes that fit your budget.

Portion sizes can be tricky. As you learned in Chapter 2, dietary guidelines provide recommended serving sizes for most foods, such as a 3-ounce portion of meat or fish or a 1-cup portion of pasta. Those servings can seem pretty small, however, when placed on a big plate.

When preparing and serving food at home, one way to circumvent the problem of judging portion size is to serve most of the dishes family-style, where the food is placed into serving bowls and passed among the diners. See Chapter 10, "Kitchen and Dining Plans and Etiquette," to learn more about methods of meal service.

Calculating the Cost of a Recipe

Have you ever wanted to be on *The Price is Right* to show off your knowledge of how much items cost? People who cook on a budget increase their knowledge every time they go to the grocery store. You can do it too! It just takes practice. When you know how much a certain item *should* cost, you can better distinguish a so-so discount price from a really good one. The average cost of an item frequently changes, of course, because food costs are always fluctuating. However, the more shopping you do, the more of a feel you will get for the price range.

TABLE 3-2: INGREDIENTS AND COST ESTIMATION

Inexpensive	Moderate	Expensive
■ **Vegetables:** Vegetable soup, vegetarian casseroles	■ **Chicken:** Baked chicken, chicken and pasta casserole	■ **Shrimp:** Fried shrimp, shrimp scampi
■ **Pasta:** Spaghetti with marinara sauce, macaroni and cheese	■ **Beef or pork roasts:** Chuck roast, arm roast, rump roast	■ **Lobster and crab:** Crab legs, broiled lobster
■ **Bread:** Sandwiches, breadsticks, rolls	■ **Pork chops:** Pork steak (cheaper, more fatty), mixed cuts of chops (moderate)	■ **Steak:** Filet mignon, New York strip steak, ribeye steak
■ **Rice:** Fried rice, rice pilaf	■ **Turkey:** Whole turkey, turkey breast, turkey legs	■ **Nuts:** Cashews, macadamia nuts, pecans
■ **Beans:** Red beans and rice, black beans, refried beans	■ **Dairy (butter, cheese, milk):** Grilled cheese sandwich, cream-based soups	■ **Exotic mushrooms:** Morels, chanterelles, truffles
■ **Potatoes:** Baked potatoes, mashed potatoes	■ **Ground beef:** Meatloaf, hamburgers	■ **Veal:** Veal cutlets, veal scaloppini
■ **Lentils:** Lentil soup, curried lentils		■ **Fresh fish:** salmon, halibut, orange roughy

To calculate how much it is going to cost to prepare the meal, add up the cost of all the ingredients for each recipe. You may not always have to buy every ingredient, because a well-stocked kitchen is likely to already have basic ingredients like salt, flour, sugar, cooking oil, and certain spices. So check your pantry as you are planning. If you can find a recipe that uses ingredients you already have on-hand, you can save money.

FIGURE 3-8
Pesto is a sauce made from chopped up leaves (usually basil), olive oil, some kind of nuts (usually pine nuts or walnuts), and parmesan cheese. What nutrient does pesto sauce consist of primarily: carbohydrates, fat, or protein?

Ways to Cut the Cost of a Recipe

Let's calculate the cost of a recipe, and learn how to reduce that cost by making minor substitutions based on ingredients you already have on hand.

Suppose you want to make a pesto pasta dish for four people that calls for the following ingredients:

1 pound bow-tie pasta
½ cup pine nuts
1 teaspoon salt
½ teaspoon black pepper
3 cups loosely packed fresh basil
⅔ cup olive oil
¼ cup grated parmesan cheese

We'll assume that the kitchen already has salt and pepper on hand. At your local grocery store, you price the ingredients as follows:

3 cups fresh basil	$ 2.79
Bow-tie pasta (1 pound)	$ 2.19
Pine nuts (at least ½ cup, or 4 ounces)	$ 5.98
Olive oil (smallest bottle available)	$ 6.49
Grated parmesan cheese (smallest quantity available)	$ 2.99
Total	**$20.44**

Your total cost for this recipe is $20.44, which comes out to about $5 per person. That's cheaper than hosting the dinner at a restaurant, but it's still pricey for a meal that does not even include meat. And that's just one dish—we haven't even gotten to the salad, bread, and dessert yet!

As you can see, preparing a recipe in a basically "empty" kitchen can be expensive because you must buy all the ingredients, including staples like olive oil and parmesan cheese. Those costs can add up quickly. In a well-stocked kitchen, however, many basic items will already be available, so you need only to buy the main ingredients for each recipe. Therefore, the first few times you cook, you may spend more money on groceries than you will after you have acquired many of the basics.

FIGURE 3-9
Different types of pasta. Would you consider substituting one similar ingredient for another to save money?

For this example, let's assume that you don't have any of these ingredients on hand. There are several ways you could economize:

- Look for generic or store-brand versions of products. These are typically a little cheaper than the brand names.
- Look for items on sale or for which you have coupons.
- Substitute similar ingredients, those that you may have on hand, are on sale, or are less expensive.

Looking at this recipe, what are the most significant expenses? The olive oil and pine nuts are all over $5 apiece. The dish wouldn't be the same without either of these, so there is no obvious substitution opportunity. However, you could possibly substitute a cheaper type of pasta, such as plain spaghetti, for the bow-tie pasta. A 1-pound package of the store-brand spaghetti costs only $1.09, which would save you $1.10. You could also get store-brand parmesan cheese ($1.99, a savings of $1), and you could cut the quantity of pine nuts to 2 ounces (saving $2.99). Final cost:

3 cups fresh basil	$ 2.79
Spaghetti (1 pound)	$ 1.09
Pine nuts (at least ¼ cup, or 2 ounces)	$ 2.99
Olive oil (smallest bottle available)	$ 6.49
Grated parmesan cheese (smallest quantity available)	$ 1.99
Total	**$15.35**

Evaluating the Cost of Ready-to-Serve Alternatives

But wait—there may still be another way to fix a pesto pasta dish without spending all that money.

Generally speaking, it is less expensive to make dishes from scratch than it is to buy ready-to-serve food. However, that savings often happens over the long run, given the quantities that you buy and the number of dishes you can make with them. For example, with the olive oil, we only need 3 tablespoons of it, and it comes in a 17-ounce bottle. You can make this dish dozens of times using the same bottle of oil, so every time you re-make the dish, you save money. If you are going to make this dish only once, it's not such a good value. So let's look into some pre-prepared pesto.

Suppose your local store has a 7-ounce container of basil pesto for $5.50. The ingredients are listed as follows:

> Extra Virgin Olive Oil, Canola Oil, Parmesan Cheese (Cultured Milk, Salt, Enzymes), Reduced Lactose Whey, Parsley, Walnuts, Romano Cheese [(Made from Cow's Milk), Cultured Milk, Enzymes, Salt], Water, Whey Protein Concentrate, Pine Nuts, Salt, Garlic Puree, Spices.

It seems to have most of the same ingredients as the pesto you had planned to make, although it does use walnuts (a cheaper nut) as the main nut source. Pine nuts are included, but lower down on the ingredient list. By using the pre-prepared pesto instead of buying the ingredients (except the pasta of course, for which we'll stick with the cheaper spaghetti), you can reduce the total cost of the dish to $6.59 (just over $1.60 per serving). That's a pretty significant savings! And, your preparation time would also be greatly reduced.

Spaghetti (1 pound) $1.09
Prepared pesto sauce $5.50
Total $6.59

Exploring Alternative Lower-Cost Recipes

As we just explained, sometimes prepared food is actually cheaper than cooking from scratch, depending on the ingredients you must buy. But, we could have also have used another "from-scratch" recipe that called for either cheaper ingredients or ingredients that we had on hand. For example, here's a recipe for a pesto that uses parsley instead of basil, and walnuts instead of pine nuts:

1 pound pasta (any type)
¼ cup walnuts (¾ oz)
½ garlic clove
1⅓ cups packed fresh flat-leaf parsley
3 tablespoons extra-virgin olive oil
2½ tablespoons finely grated parmesan cheese
1 tablespoon water
1 teaspoon salt
¼ teaspoon black pepper

Here's the shopping list for the above recipe:

Spaghetti (1 pound). . . . . $ 1.09
¼ cup walnuts . . . . $ 2.70
Garlic, 1 bulb . . . . $ 0.50
1⅓ cups fresh parsley . . . . $ 0.69
Olive oil (smallest bottle available) . . . . $ 6.49
Grated parmesan cheese (smallest quantity available). . . . . $ 1.99
Total $13.46

That's pretty good, but it could be even better. Since this recipe calls for only 3 tablespoons of olive oil, you could try substituting a less expensive vegetable oil at a dramatic savings: a small bottle of vegetable oil is only $1.79, bringing down the total cost to $8.76. Yes, it's a different recipe with a totally different taste, but your guests won't know that it's not what you had originally planned to make!

Buying in Bulk: A Good Deal?

As you saw in the previous section, ingredients like olive oil can be a big expense upfront, but then you'll have it for other recipes that require it. The larger the quantity you buy, the better deal you may be able to get.

How do you tell if the larger quantity is really a better deal? Calculate the cost per ounce. This is easy if you carry a calculator with you to the store; just type in the price, and then divide it by the number of ounces in the container. For example, suppose you're shopping for olive oil, and you have the following choices:

17 ounces for $6.49

25 ounces for $9.25

34 ounces for $11.99

44 ounces for $14.99

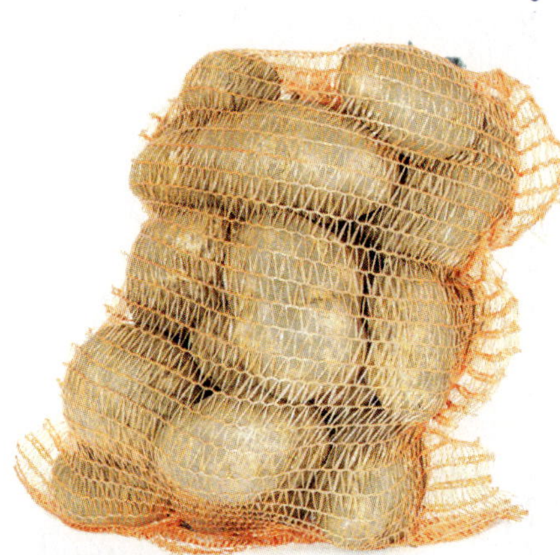

Divide each of those prices by the number of ounces:

$6.49 ÷ 17 = $0.38 per ounce

$9.25 ÷ 25 = $0.37 per ounce

$11.99 ÷ 34 = $0.35 per ounce

$14.99 ÷ 44 = $0.34 per ounce

FIGURE 3-10
Buying in bulk can potentially save a lot of money, but it's not always the best choice. What potential drawbacks can you think of for buying in bulk?

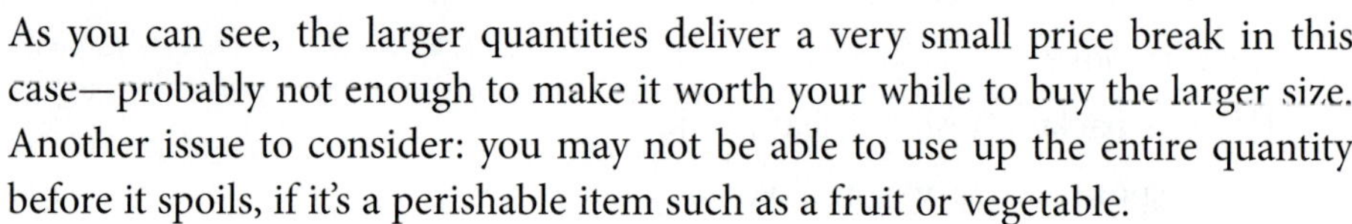

As you can see, the larger quantities deliver a very small price break in this case—probably not enough to make it worth your while to buy the larger size. Another issue to consider: you may not be able to use up the entire quantity before it spoils, if it's a perishable item such as a fruit or vegetable.

Some price differences are much more dramatic, though. Let's say you need to buy some basmati rice (a type of white rice). You can buy:

1 pound (16 ounces) for $2.07

2 pounds (32 ounces) for $3.39

25 pounds (400 ounces) for $35.00

Fiction	Fact
Large quantities always cost less per ounce than smaller quantities.	Large quantities often cost less per ounce, but not always. Sometimes when small quantities are on sale, they are a better deal than the larger ones, so shop carefully.

In this case, since all the measurements are in pounds, you can calculate the price per pound, rather than per ounce:

$2.07 ÷ 1 = $2.07 per pound

$3.39 ÷ 2 = $1.69 per pound

$35.00 ÷ 25 = $1.40 per pound

No calculator? That may not be a problem, because many grocery stores show a price per ounce on the shelf price labels, making it easy to compare products without doing the math.

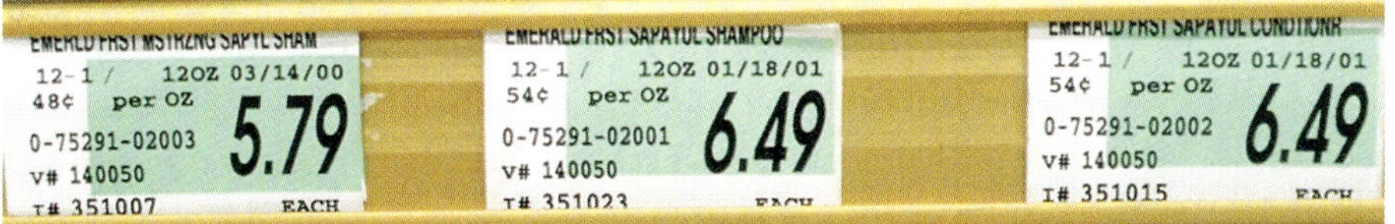

FIGURE 3-11
Shelf labels in a grocery store often show the price per ounce of the item. This provides a quick way to compare one size to another. What other comparisons could you do using the prices per ounce?

Buying Store-Brand and Generic Products

When you buy a product that you see advertised on TV, like a certain brand of frozen vegetables or breakfast cereal, part of the money you spend goes toward paying for that TV spot. Is the product really any better than a similar one that isn't advertised on TV, or is the extra money just wasted? If you said that the unadvertised product was just as good, then you are the type of consumer that store-brand and generic products target.

A **store-brand** product bears the brand of the grocery store where you bought it, rather than that of an outside manufacturer. Whether store brands are as good as the more heavily advertised ones is a matter of debate, and often of personal preference. **Generic** products are a step below the store-brand ones. Rather than carrying the store's name, they carry no name at all. The packaging is no-frills, and the cost is often rock-bottom. The quality may or may not be the same as that of a store brand—but usually, it is of somewhat lower (though acceptable) quality.

Fiction	Fact
Store-brand products are inferior to name brand.	Store-brand products are packaged and sold by the individual store chains, so their quality depends on that store's suppliers and contracts with packagers. In some cases, the store brands may be exactly the same as name-brand products, and even packaged in the same plant.

Grocery Shopping Tips

Ready to go shopping for the ingredients you will need for your recipes? Here are some tips.

Where to Buy Groceries

There are many different types of stores that sell groceries, each with their own strengths and weaknesses. You're probably familiar with many of these already.

For the lowest prices, consider a **deep-discount grocery store**. These stores don't offer much in the way of service—for example, you might need to bag your own purchases or even bring your own bags—but the prices are good, and many generic and store-brand items are available. For staples such as flour, sugar, cooking oil, salt, and milk, generic products are not much different from the higher-priced versions, so shopping for such items at a discount grocery store makes good sense. These stores typically don't have much in the way of fresh produce, and most of the meat is frozen. If you do buy fresh meat or dairy here, check the expiration dates carefully.

Cool Tips

Grocery store owners use a variety of layout tricks to entice customers to buy more products. Perishable items like milk, meat, cheese, and eggs are usually near the back of the store, along the back wall. Grocers put them there because those are the most popular items, and they want you to walk by as much other merchandise as possible to tempt you to make additional purchases.

If you need large quantities, try a **warehouse club**. These stores offer good values on many grocery items available in large quantities. They may not have many brands from which to choose, and their selection may vary from week to week. If they happen to have what you are looking for, though, and you need a lot of it, you're likely to get a good deal. You may have to sign up for a membership card to shop there, which can cost $25 or more per year.

FIGURE 3-12
What types of grocery stores does your family shop at? Does your family have a membership to a warehouse club such as BJ's Wholesale Club or Sam's Club? What items might you find at a local store that you might not find at a warehouse club?

Low prices can also be found at **superstores**—that is, large national retail stores that have grocery departments, such as Target, Wal-Mart, and Meijer. These stores typically have a good selection of products, including most brand names, at very good prices. Their secret is that they buy in bulk; these retail chains can negotiate good deals with suppliers by buying huge quantities at a time.

Almost every town has its own **local grocery stores**, some independent and some part of regional or national chains. It's difficult to generalize about these stores; check them out yourself to compare price, selection, and quality.

For fresh produce, look for **farmers markets**. These are usually regularly occurring open-air marketplaces where local farmers can sell their wares. Here you may find not only better prices than in a grocery store, but a higher quality of produce, often grown without pesticides or other chemicals.

In some areas, **food co-ops** are available. These are somewhat like members-only warehouse clubs, in that you must join in order to buy there, but they are generally small stores that focus on organic foods and/or foods that were grown or prepared in a certain way, such as free-range chicken or shade-brown coffee.

Online grocery shopping may be available in some cities. You place your order via a Web interface, and then it is delivered to you by a local grocery store. It is convenient, but a delivery charge may apply, and you may not get exactly what you want because of a limited catalog selection online compared to the selection available in-store.

Convenience stores—like the mini-marts attached to gas stations—are the least desirable place to shop for groceries. Prices are high, and the selection is poor.

Evaluating Coupon Deals

Coupons can dramatically cut the cost of your grocery shopping. Many coupons are available in newspapers and magazines; others can be printed from online sources. Some grocery stores even offer double or triple coupon promotions, increasing the value of your coupons. Not every coupon may be eligible for doubling or tripling; usually only coupons under a certain amount qualify. Check the store's policies.

Remember, if the price difference between the brand-name item on the coupon and the store brand is more than the coupon amount, you aren't really saving any money by going with the brand name. For example, suppose the brand name olive oil is $7.50 a bottle and the store brand is $6.50. Even with a $0.75 coupon, the store brand is still the better value. You should also look around to see whether another brand or size of the product may be on sale at a price low enough to meet or beat the coupon price.

Have you ever seen "Buy one, get one free" on a savings sign or coupon? Keep in mind that you might not need to get both items to reap the savings. You might be able to simply buy one item at half off. This policy varies from store to store, so be sure to ask about the policy where you are shopping.

Another "gotcha" with coupons is that they expire. Always check your coupons before you get to the cashier to make sure they haven't expired.

When creating your grocery shopping list, try to arrange the items on the list in the order you will come to them as you move through the grocery store. For example, you might want to start with the aisles that contain non-perishable foods such as canned and boxed goods. Next work your way around the outside of the store, where the produce and refrigerated items usually are, and finish up in the frozen foods, so they have less time to thaw.

Reading and Interpreting Food Labels

When comparing two products, you can find a lot of useful information on their labels. Here are some things to watch out for:

- Compare the weights of each package, as reported on the label. Quantities are usually measured in **net weight**, which is the total weight minus the packaging. "Packaging" can also include the liquid in a canned product, such as oil or water in tuna. When you buy a can of tuna that is marked as 6 ounces, you get 6 ounces of tuna plus a few ounces of water or oil.
- Look at the **ingredient list**. Ingredients are listed in order of largest to smallest. So, for example, a can of pork and beans that lists pork as the second ingredient (right after beans) has more pork in it than a can that lists pork as the fifth ingredient.
- Read the product description carefully, and understand what you are getting. For example, there's a big difference between real *cheese* (like Colby or cheddar) and *processed cheese food* (like American cheese slices), and an even bigger difference between those and *natural cheese flavor* (like the cheese powder covering fried cheese-flavored snacks, such as Cheetos).

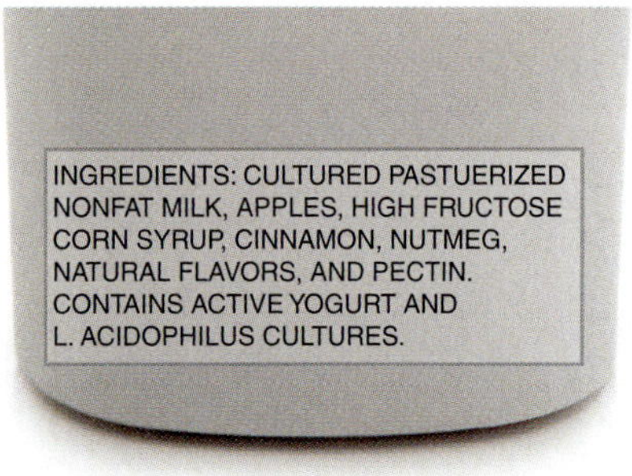

FIGURE 3-13
Products ingredients are listed in descending order, largest to smallest. What product do you think the above ingredient list came from?

Checking Expiration Dates

Many products have a date printed on them to give you an estimate of their age. This labeling is voluntary; there are no federal laws governing it, except on baby food and baby formula.

Grocery stores are supposed to pull expired food off the shelves and not sell it, but some stores are more vigilant about doing this than others. As you are shopping, it's a good idea to check the date on each of the items you put in your cart to make sure it doesn't expire before you are going to be using it.

The tricky part is that not all the dates on foods refer to the same thing. You may see any of these:

- **Use By or Expiration Date:** This is the last date on which the product should be eaten. Food is no longer safe to eat after this date. Throw it out.
- **Best If Used By:** This is a date recommended for best flavor, texture, or quality. It does not mean the food is no longer safe once the date has passed. This type of date is common on bread, snack chips, and soda pop, for example.

- **Sell By or Pull By:** This date tells the grocers when to stop selling the food. It doesn't mean the food is bad once it reaches that date. For example, milk can be safe and taste fine as much as 10 days after the sell-by date if kept refrigerated. Chicken can be good for one or two days after its sell-by date.
- **Closed or coded dates:** A string of mysterious numbers that do not directly resemble a date may be found on products that have a long shelf life, such as canned or boxed foods. If you could decipher the code, you could determine when the product was packaged. This could help in the case of a recall. These are not expiration dates and are generally of little use to the consumer.

Monitoring Freshness of Food After Purchase

Once you have opened a product, you can't use the date on the package to determine its freshness, because oxygen gets into the package. This causes food to oxidize, which makes it "go bad" faster.

Because most grocery stores rotate their stock when new items are shelved, the freshest products are usually at the *back* of the shelf or display case. If you won't be using the product within a short time of purchase, buy yourself some extra time by pulling from as far back as you can reach. This is especially important when buying highly perishable items, like milk and bread.

Canned goods should be stored at a temperature no higher than 65 degrees Fahrenheit. At higher temperatures, the products don't last as long. Generally speaking, canned goods can be stored for about one year; some can last much longer. Mark the purchase date on each can as you put it in your pantry to make sure you are using the product before it expires. Certain types of canned products don't last that long; fruit juices, peppers, sauerkraut, green beans, and tomato products should be used within six months. After that time, the food is probably still safe to consume, but may not taste as good; for example, fruit juice may have a metallic taste absorbed from its can.

FIGURE 3-14
What do you think this date stamped on a canned good means?

Case Study

Adam, a high school senior, has made a commitment to himself this year to change the way he eats. In the past, he has regularly gone to nearby fast food restaurants at lunchtime, but he often feels sluggish and sleepy after those high-fat lunches. He also noticed that, according to some height/weight charts, he is slightly overweight. Finally, he doesn't like all the money he has been spending on fast food; there has to be a cheaper way to eat without missing out on being with his friends at lunchtime.

Suppose Adam was a student at *your* school. What would his alternatives be, and what are the advantages and drawbacks of each one?

Put It to Use

1. Think about the last 9 meals you ate, and create a spreadsheet that categorizes them in the following ways:
   - Where were you? (At home? At school? At a restaurant?)
   - Who were you with? (Family? Friends? Strangers? Alone?)
   - What type of meal was it? (Home-cooked? Pre-packaged? Restaurant-prepared?)
   - How was it served? (On a plate? In a wrapper?)
   - Generally, how nutritious was it? (Rank 1 to 5, with 5 being the highest.)
   - Generally, how much fat did it contain? (Rank 1 to 5, with 5 being the most fatty.)

   If you don't remember some of your meals, do this exercise going forward, and make notes on the food you eat for the next several days instead.

   Based on your notes, what are your typical eating patterns for each meal? Compare your patterns with those of your classmates. Ask your parent or an adult friend about their eating patterns, and compare them to your own. Write a brief report that summarizes your findings.

Put It to Use

❷ Suppose you are going to purchase or bake a cake for a friend's birthday party. Go to a grocery store that has a bakery, and get the price on a cake that would serve 12 people. Then go to the baking aisle and get a price on a cake mix and icing. Which is the better value? Now revisit the cake mix as if you had none of the additional ingredients it requires, such as eggs and cooking oil; and, add in the cost of buying those items. Now which is a better value? Finally, compare the prices again as if you had to buy a pan in which to bake the cake. (Most grocery stores sell cake pans.) Is the cake mix still the better value? Write a brief report that summarizes your findings.

Write Now

Suppose you want to host a party for a grandparent's birthday (or for another elderly person). Your guests will number 7 people plus the guest of honor, all aged 70 and up. Think about the best location in which to hold the party. Would you have it in a restaurant? In your home? In their home? What type of food would you serve? Write a paper explaining your choices.

Tech Connect

Different recipes can have very different ingredients for making the "same" dish (or, at least, two dishes with the same name). You saw this earlier in the chapter; pesto is usually made with basil and pine nuts, but a sauce made with parsley and walnuts is also considered pesto.

Suppose you want to serve hummus to your guests. Use the Web to find several recipes for hummus to find out what ingredients you will need to make it. Write a brief report listing the main ingredients on which most hummus recipes agree, and the ingredients that are present in some, but not all recipes.

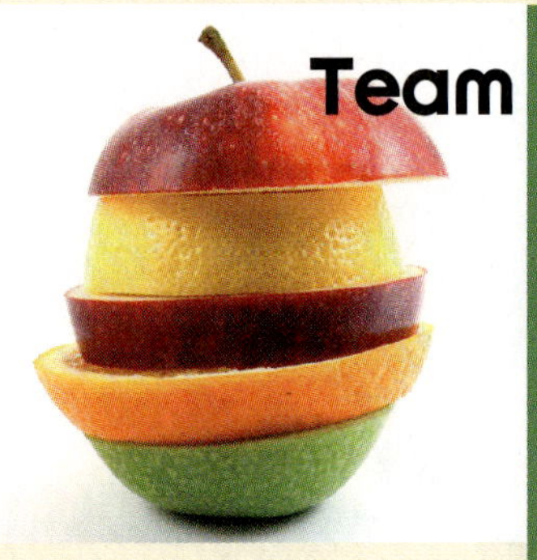

Team Players

Break into teams of two or three people. Each team is responsible for developing a menu for a dinner party for 4 people that includes the following items:

- A non-alcoholic beverage
- A hot appetizer made with artichokes
- A green salad including mandarin oranges and pecans
- A main entrée including chicken breasts
- A dessert of chocolate cake

Using cookbooks or Internet recipe databases, find recipes and collect ingredient lists. Go to a local grocery store and get prices on all the ingredients you will need. You can assume that the kitchen you are working in has the following items: salt, pepper, ketchup, yellow mustard, onions, garlic powder, butter, sugar, flour, and cooking oil. You can go to more than one store if needed to get the best prices.

Each team should present their menus and their total cost to the rest of the class. Which team had the lowest food cost? As a class, vote on which team had the most appetizing menu.

Put *It* Together

Match the explanation in column 1 with the term in column 2.

Column 1

- **a.** a complete meal for a fixed price
- **b.** a menu on which each item is priced and served separately
- **c.** a product packaged by the grocery store chain at which you are purchasing it
- **d.** a product with no company identified as its packager or manufacturer
- **e.** a large national retail store that has a grocery department
- **f.** a small store with a small selection of groceries and prepared foods
- **g.** the measure of a product minus its packaging
- **h.** the last date on which a product is safe to eat
- **i.** the date after which a product may not have optimal flavor, texture, or quality
- **j.** the date after which a grocer should pull a product from its shelves

Column 2

1. a la carte
2. best if used by
3. convenience store
4. expiration
5. generic
6. net weight
7. sell by
8. store brand
9. superstore
10. table d'hôte

4 The Physiology of Food

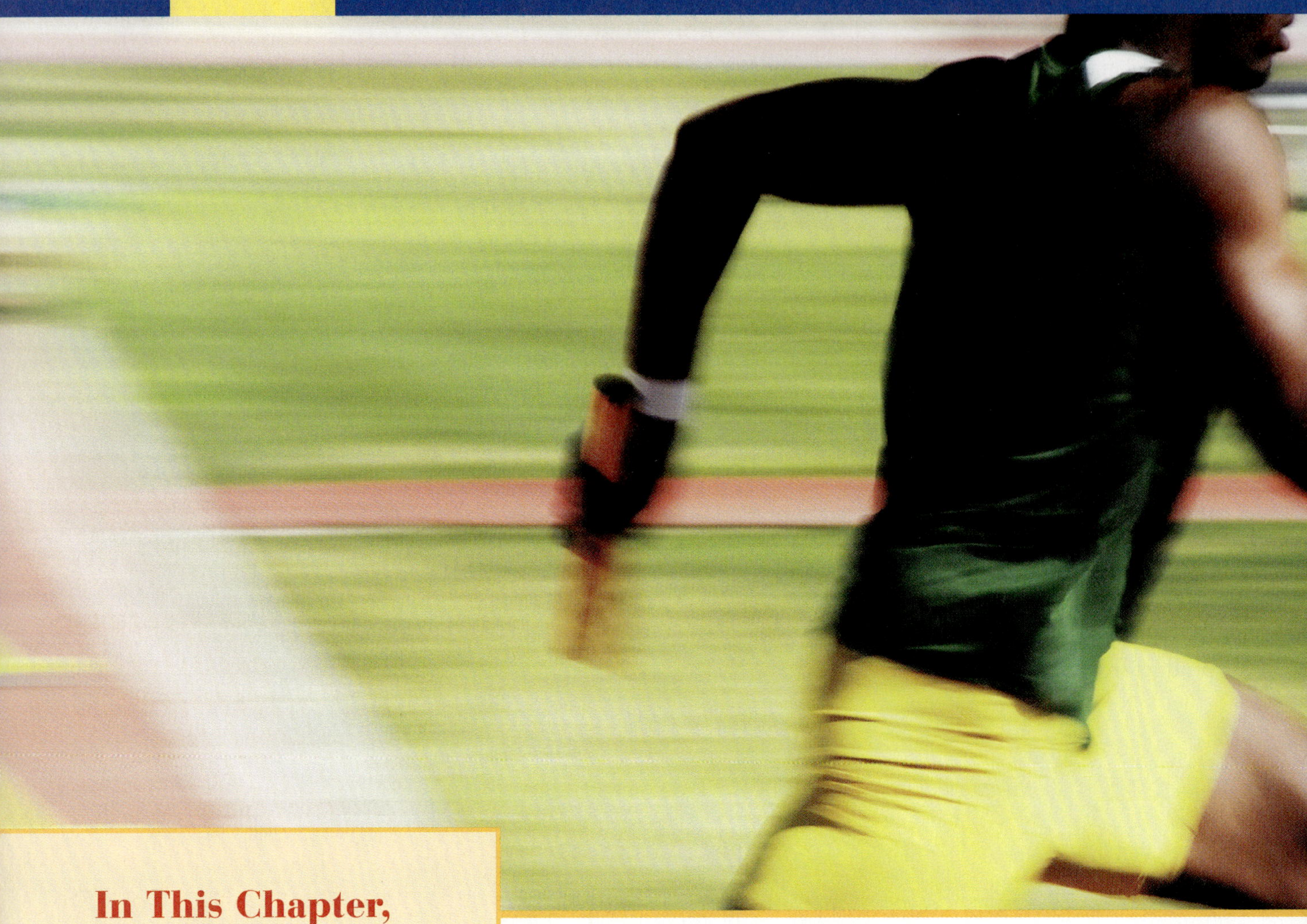

In This Chapter, You Will . . .

- Discover how food is converted into energy
- Trace the path of food through the human body
- Learn about each of the organs of the gastrointestinal system
- Understand common physical problems related to digestion

Why YOU Need to Know This

Have you ever noticed a relationship between what you eat and how you feel later, or how your body looks, performs, or smells? As a class, brainstorm combinations of foods and body effects that you have observed; for example, some people have found that eating a lot of garlic makes their skin smell like garlic, or that eating asparagus makes their urine smell unpleasant.

How Food Is Converted to Energy

Food is the fuel your body requires to work—to move, to think, and even to stay alive. The reason we study nutrition—and the reason it's so important to do so—is that the quality of the food you eat has a direct result on the way your body performs. Think of your body as a sports car that requires premium gas. If you give it the high-octane gas it needs, it will perform well. If you give it lesser-quality gas, it will still perform, but not as smoothly, and over time, it may develop mechanical problems.

FIGURE 4-1
Food = Fuel. What fuel did you have for breakfast this morning? Will it last you until lunch, or will you run out of gas before then?

In this chapter, you'll learn about the physical processes by which food is **metabolized**—that is, converted to energy. Understanding these processes, and how your choices of foods affect them, can help you understand at a gut level (pun intended!) the importance of good nutrition. Then, the next time you reach for junk food, you'll think about how it will affect your body.

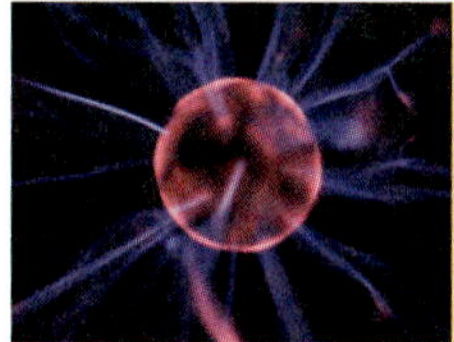

SCIENCE STUDY

Metabolism is the term used to describe all the processes that keep the body fueled. The fuel comes from outside the body in the form of food.

Digestion is just one aspect of metabolism, but it's an important one. In the context of digestion, metabolism refers to the way your body breaks down the raw materials of the food you eat into molecules that the body can use as fuel. A body's metabolism is the set of rules that determines what it finds nutritious and how much it needs to function. If you hear someone talk about having a "fast metabolism," they probably mean that their body needs an above-average amount of food.

Have you ever wondered why it seems that men can eat more than women without gaining weight? And why there are more obese women than men? It's all about metabolism. The more muscle mass you have, the faster your metabolism. Because men tend to have more muscle than women, they require more food than women—about 500 calories a day more, on the average.

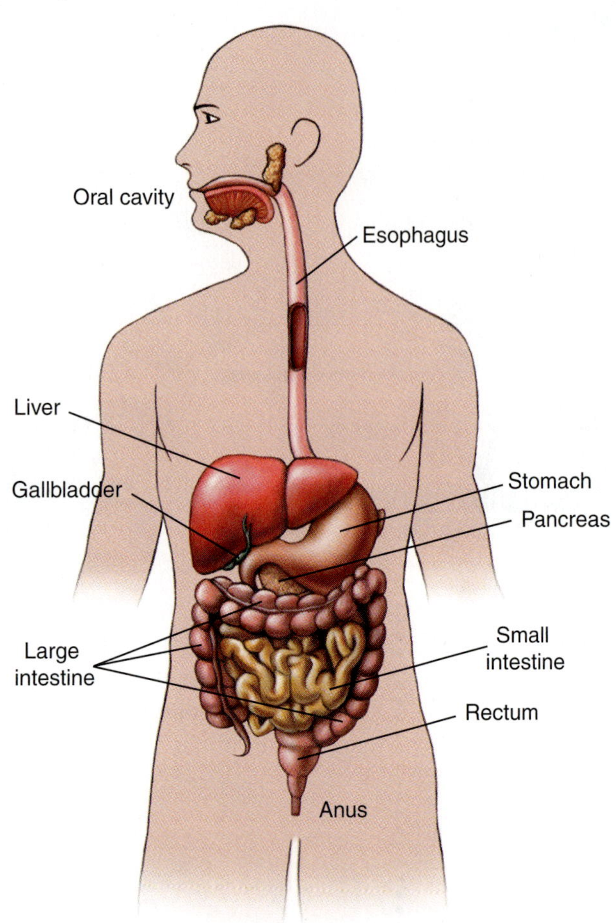

FIGURE 4-2
The organs of the gastrointestinal system.

Introducing the Digestive System

The parts of your body that process food are known as the **digestive system,** or the **gastrointestinal (GI) system**. GI breaks down to *gastro*, which is Greek for stomach, and *intestinal*, which refers to your intestines. This system includes the oral cavity (mouth), esophagus, stomach, small and large intestines, rectum, and anus. There are also accessory organs that help with digestion, like the pancreas, liver, and gallbladder. Figure 4-2 shows an overview of the GI system. We'll look at each of these components in more detail later in this chapter.

The Food-Processing Process

The components of the digestive system work together to perform the following functions:

1. **Ingestion:** Food enters your mouth.
2. **Mastication:** Food is chewed up and swallowed.
3. **Digestion:** The stomach breaks down the food.
4. **Secretion:** The body produces various chemicals that mix with the food.
5. **Absorption:** The intestines absorb nutrients from the food.
6. **Excretion:** Unusable portions of the food are expelled from the body.

These steps roughly correspond with the organs in Figure 4-2 from top to bottom, but it's not an exact match-up because some of the steps involve more than one organ. For example, secretion occurs in multiple places, including the oral cavity (saliva), the stomach (gastric juices), and the intestines (enzymes).

The Oral Cavity

During ingestion, the first phase of the food-processing process, food enters your body via your mouth, which is also called the oral cavity. Your **lips** are the front door and your **tongue** is the floor. The **palate** is the "roof" of your mouth. The **cheeks** form the side walls.

Cool Tips

The word *palate* is also sometimes used to refer to sense of taste, as in "The dinner was a delight to the palate." The word *palatable*, which means agreeable or pleasant, comes from that root.

Just behind the palate is the **uvula**, the little dangly thing you can see in the mirror when you open your mouth very wide. It may look like wasted skin, but there is actually a purpose to it. The uvula aids in swallowing because it helps to direct the food toward the pharynx and block food from coming out your nose!

The tongue performs many functions. It provides taste stimulus to your brain via the taste buds on its surface. It senses the temperature and texture of food. The tongue manipulates the food during chewing and aids in swallowing. As the tongue moves the food around in the oral cavity, **saliva** is added from the **salivary glands** to moisten and soften the food. Saliva also has digestive enzymes in it, so the chemical processing of food begins in the mouth. As the food is crushed up by the teeth, it turns the food into a ball-like mass called a **bolus**.

Cool Tips

If you can push a bolus into the pharynx with your tongue, why don't you swallow your tongue, too? A membrane under your tongue, called the *lingual frenulum*, prevents it. You can see it when you lift up your tongue. Not only is the frenulum important for swallowing, but it also aids in proper speaking. An abnormally short frenulum prevents clear speech—which is where the term "tongue-tied" comes from.

The Pharynx and Esophagus

The **pharynx** is a passageway between the oral cavity and the esophagus. It serves double duty by carrying both food and air, so it's also part of the respiratory system. A flap of tissue called the **epiglottis** covers the airway to the lungs when you swallow, so the food passes into the esophagus. See Figure 4-3.

The **esophagus** is a tube approximately 10 inches long, responsible for transporting food from the pharynx to the stomach. The esophagus is soft-sided, and normally it is collapsed, like a deflated balloon. When you swallow a bolus of food, a **sphincter** (a muscular ring) at the top of the esophagus relaxes, letting the bolus in. From there, muscles in the esophagus contract rhythmically to work the food down into the stomach, in a process called **peristalsis**. (That's why you can still swallow food when you are upside-down!) When the food gets to the bottom of the esophagus, a second sphincter opens to let it into the stomach. That sphincter then closes again to prevent the stomach's acidic gastric juices from squirting up into the esophagus.

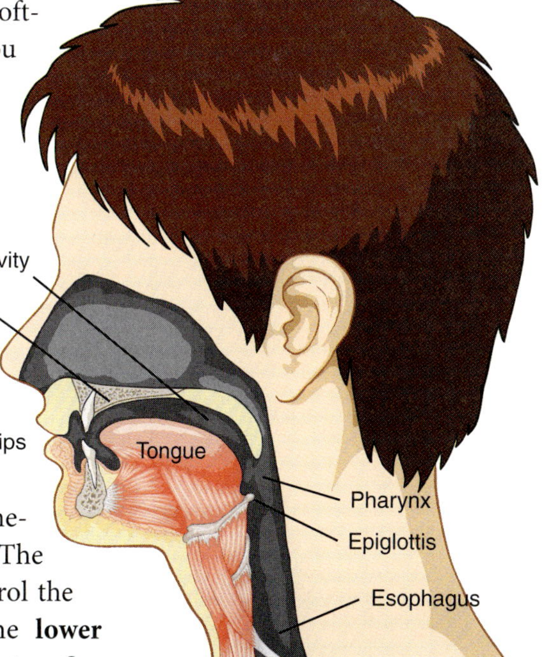

FIGURE 4-3
The upper section of the digestive system.

The Stomach

The **stomach** is a pouch, approximately 10 inches long, with a diameter that varies depending on how much you eat at any given time. The stomach, like the esophagus, has sphincters at both ends that control the food's movement in and out. At the stomach's entrance is the **lower esophageal sphincter**. At the stomach's exit is the **pyloric sphincter**. See Figure 4-4.

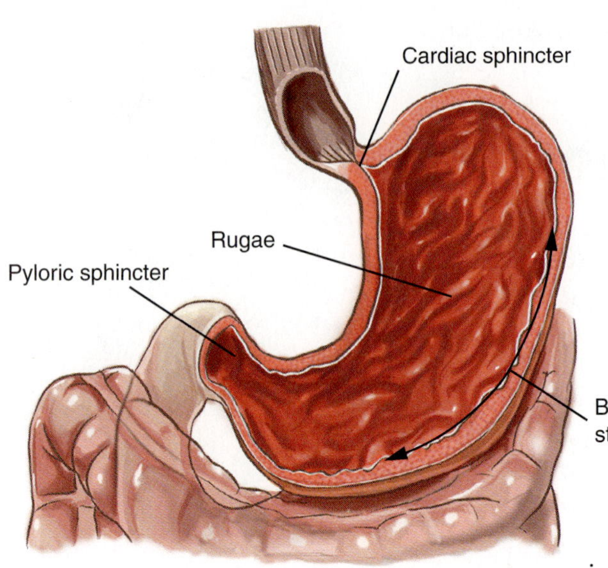

FIGURE 4-4
The parts of the stomach.

The stomach's walls have deep folds called **rugae** that allow it to expand like an accordion when you fill it with food, and then contract back to a smaller size as it empties. The maximum capacity of most people's stomachs is about 1 quart or liter. One of the functions of the stomach is to serve as a holding tank for food, releasing it into the intestines in a controlled manner. Therefore, the more you eat in one sitting, the longer it takes the stomach to empty afterwards. It takes about four hours for the stomach to empty after an average-size meal. Liquids and carbohydrates pass through fairly quickly. Protein takes a little more time, and fats take even longer, usually from 4 to 6 hours.

A second function of the stomach is to break down the food. The expansion of the rugae triggers the body to secrete gastric juices to process the incoming food. Gastric juice is a general term describing a combination of hydrochloric acid (HCl), pepsinogen, and mucus. HCl is the main digestive enzyme in the stomach. Pepsinogen combines with HCI to form **pepsin**, which breaks down proteins, and mucus protects the lining of the stomach from acid damage.

Gastric juices mix with the food, causing the food to disintegrate into a thick liquid called **chyme**. The muscular action (**gastric motility**) of the stomach works like a cement mixer to physically churn up the food and mix it with the gastric juices that the walls of the stomach secrete. The stomach also absorbs alcohol and small amounts of water, but it does not actually do much nutrient absorption.

The Small Intestine

The small intestine is the major organ of nutrient absorption; almost 80% of the absorption of usable nutrients takes place here. From here, amino acids, fatty acids, ions, simple sugars, vitamins, and water are all absorbed into the bloodstream, and then on to the body's cells. The remaining 20% is absorbed in the stomach or in the large intestine. Anything that is indigestible passes on to the large intestine for removal from the body.

The small intestine is a long thin coil of tubing, what most people think of when they hear "guts." There are three main sections. The first section, the **duodenum,** is about 10 inches long and connects to the stomach. The middle section, the **jejunum**, is approximately 8 feet in length. The last section, the **ileum**, attaches to the large intestine. See Figure 4-5.

Cool Tips

The name duodenum comes from Latin—*duo* means two, and *denum* means ten. Together they add up to 12, the number of finger-widths long that this organ is.

The **pyloric sphincter** at the bottom of the stomach opens to allow chyme to enter the duodenum, where it mixes with additional digestive enzymes from the gallbladder and pancreas. See Figure 4-6.

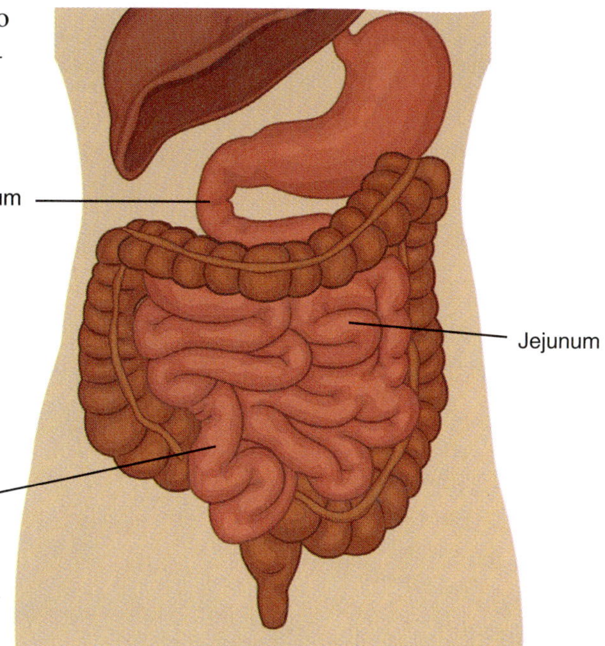

FIGURE 4-5
The parts of the small intestine.

The **gallbladder** is a small green organ adjacent to the liver. It stores and delivers bile to the duodenum. The liver produces the bile; the gallbladder is a holding tank for it. **Bile** is a digestive liquid that contains bile salts, pigments, cholesterol, electrolytes, and water. The salts in bile **emulsify** (make chemically soluble in water) fat into tiny droplets that can disperse in water, making the fat in the chyme easier to absorb.

The **pancreas** delivers pancreatic juice to the chyme. Pancreatic juice contains enzymes and **sodium bicarbonate** (like baking soda!), which neutralize the acid in the chyme. The pancreas is also responsible for producing insulin, which helps break down sugars.

After receiving enzymes from the gallbladder and pancreas, chyme passes into the jejunum and ileum. Two types of muscular action occur in the small intestines; **segmentation** mixes the chyme and digestive juices like a cement mixer, and **peristalsis** moves undigested food through the tube, toward the large intestine. As the small intestines absorb nutrients, more enzymes are excreted to further break down the food as peristalsis moves it along.

The walls of the small intestine contain tightly packed fingerlike protrusions called **villi**. Each villus (singular form of villi) contains a network of capillaries that absorb and transport sugars (from carbohydrates) and amino acids (from proteins) to the liver for further processing, before they are sent out into the body as fuel. They also absorb fatty acids (from fat) directly into the lymphatic system for distribution throughout the body.

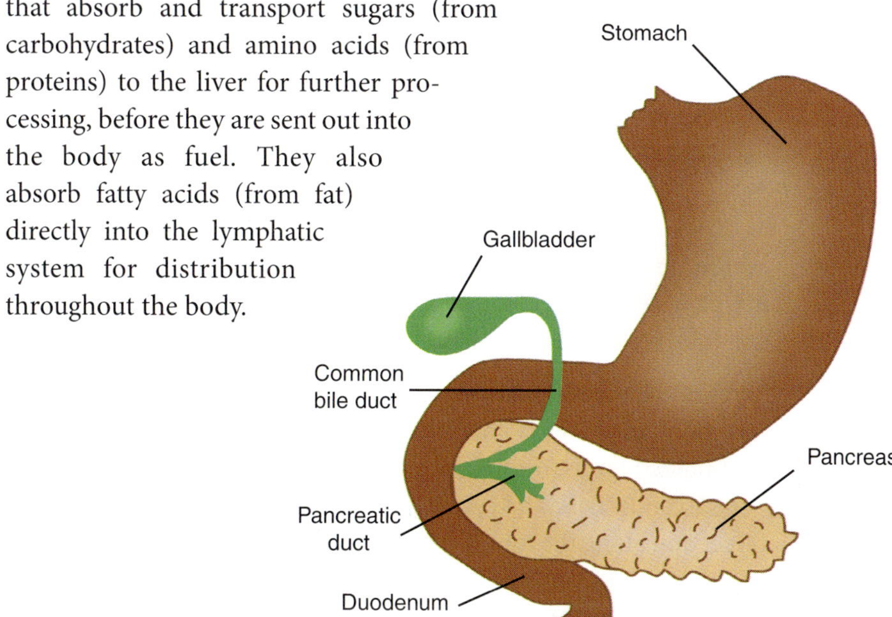

FIGURE 4-6
The gall bladder and pancreas feed enzymes into the duodenum.

The Liver

The liver plays an essential role in digestion. Nutrient-rich blood from the small intestine flows into the **liver**, where it is filtered so that any toxins are removed. The liver filters about 1.5 liters of blood a minute; all the blood in the body circulates through the liver many times per hour, constantly being purified and regenerated.

The liver also metabolizes nutrients and acts as a temporary storage area for some types of nutrients. For example, when you eat carbohydrates, which are converted into sugar glucose, not all of the glucose is immediately dumped into the bloodstream as you eat; the liver converts it temporarily to glycogen and holds it until the body signals that your blood sugar is low.

Cool Tips

The appendix has no real function; it's considered a **vestigial organ**—one that originally had a function in the human body, but no longer does. Because it contains lymphatic tissue, researchers think it might have once helped fight infection. If the opening to the appendix becomes blocked, it becomes inflamed and you get **appendicitis**. Treatment for appendicitis is either antibiotics or surgical removal.

The liver packages fats for transport in the bloodstream. Fat is packaged with proteins as **lipoproteins**, which are called HDL, LDL, and VLDL. HDL is sometimes called "good cholesterol" and both LDL and VLDL are sometimes called "bad cholesterol." The liver also processes amino acids for protein synthesis in the body, or for storage, and it makes several non-essential amino acids. (They are called non-essential because they do not come from the diet, but they are very important to proper functioning of the body!)

The Large Intestine

After the usable nutrients have been stripped from the food, there is still some remaining matter that the body cannot digest. This matter passes through to the large intestine, where it is compacted, dehydrated, and prepared for excretion as feces. The large intestine (Figure 4-7) is approximately 5 feet long and 2½ inches in diameter; it is shorter than the small intestine, but also wider. It has three main sections: the cecum, the colon, and the rectum.

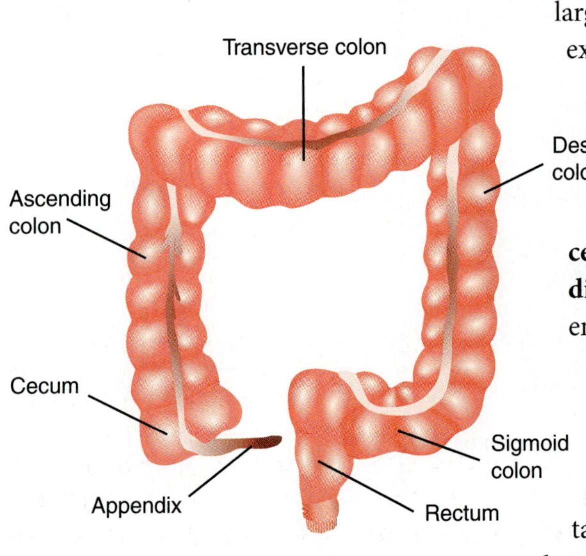

FIGURE 4-7
The parts of the large intestine.

Matter coming in from the small intestine first enters the **cecum**, in the lower-right corner of the large intestine. The **appendix** hangs off the bottom of the cecum. It's a slender hollow dead-end tube with some lymphatic tissue in it.

From the cecum, matter passes into the **colon**, which is the main body of the large intestine. The colon has four sections, each named for its location: the **ascending colon** (the area that rises, or ascends, from the cecum), the **transverse colon** (the horizontal section along the top), the **descending colon** (the vertical section along the left side), and the **sigmoid colon** (the S-shaped bottom section that connects to the rectum).

Peristalsis occurs in the large intestine, but at a slower rate. Intermittent waves move the fecal matter through the colon, and along the way, water is removed, turning the feces from a watery soup to a semi-solid mass.

The large intestine is rich in bacteria that play two important roles. They help to further break down some indigestible materials, and they produce much of the vitamin K that we need for proper blood clotting. A by-product of this bacterial action is **flatulence**, or the production of gas.

Hot Topics

Consuming alcohol along with chewing tobacco appears to raise the chances of developing cancer, not only in the mouth, but also in the throat, stomach, and bladder. Experts believe that alcohol acts as a solvent for the cancer-causing substances in tobacco, making them more harmful.

After fecal matter moves through the colon, it arrives at the **rectum**, a pathway leading to the anal canal. As the rectum fills, a reflex is triggered that causes the rectal muscles to contract and the **anal sphincter** to relax so the waste is expelled.

Common Physiological Problems Related to Digestion

When a person's digestive system is working properly, nutritious food goes in, the body processes it efficiently and without discomfort, and feces comes out regularly and painlessly. A variety of things can go wrong with that process, though, causing everything from occasional heartburn/gas pain to malnutrition. Let's look at some of the common digestive problems people have and how they are routinely treated.

Oral Disorders

We'll start at the entrance to the GI system—your mouth. The most common problem in the mouth is **dental caries** (cavities), the result of microorganisms attacking tooth enamel. Bacteria, diets rich in simple carbohydrates, poor dental hygiene, and the lack of regular visits to the dentist can result in the creation of a soft sticky substance called **plaque**. Plaque provides a great hideout for the bacteria that create the acids that attack the teeth. Once a cavity forms, it must be cleaned out and filled to protect the tooth. If a cavity is not filled, further destruction of the tooth will occur, and you might even need to have the tooth pulled. You can get a systemic infection, or even suffer heart valve damage from untreated cavities.

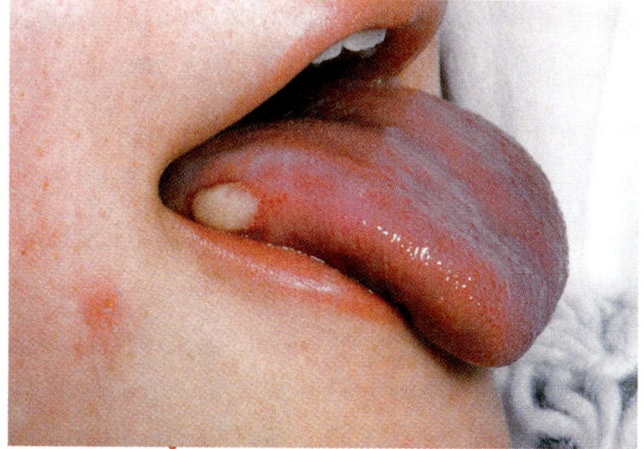

FIGURE 4-8
Tongue cancer.

The most common cause of tooth loss is not cavities, but **periodontal disease** (gum infection). As plaque builds up along the gum line, the gums become inflamed, causing a condition called **gingivitis**. The inflamed and sometimes bleeding gums can get infected; the result is that teeth may not be held securely in place. To avoid the plaque buildup that leads to both cavities and periodontal disease, brush your teeth frequently, floss, and have your teeth cleaned regularly by a dental assistant.

The lips, cheeks, and tongue can also be susceptible to **cancer**. The main cause of oral cancer is tobacco chewing. **Leukoplakia** is a precancerous condition usually attributed to chewing tobacco or snuff. The affected areas exhibit a white patch of tissue. Chewing tobacco can also increase the incident of cavities because of the sugar normally used in its processing.

Not all sores or ulcers of the mouth are cancerous. **Canker sores** are small temporary ulcers in the mouth; they are sometimes caused by eating certain foods, although their cause is often unknown. **Cheilosis**, which is the cracking and inflammation of the lips and corners of the mouth, can be caused by infections, allergies, or nutritional deficiency.

Esophagus, Stomach, and Duodenum Disorders

Problems with the esophagus, stomach, and duodenum are usually due to stomach acid (hydrochloric acid, HCl) that is burning body tissues. Stomach acid is very strong; it can cause real harm if the structures that are supposed to protect tissues do not function well.

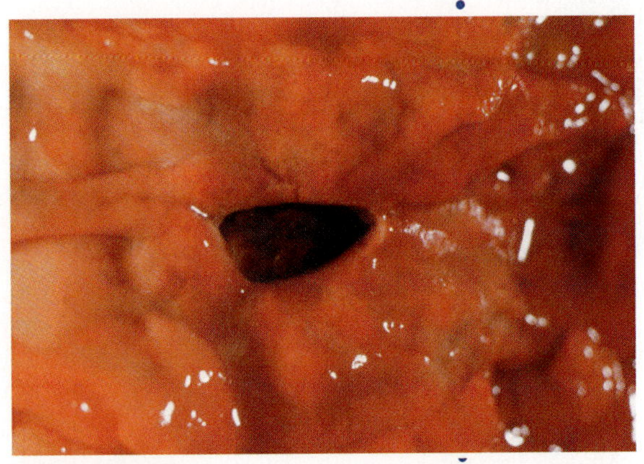

FIGURE 4-9
An ulcer in the stomach.

Heartburn, also called **gastroesophageal reflux disease** (GERD), occurs when the acidic content of the stomach squirts back into the esophagus. This happens because the lower esophageal sphincter does not close properly. Since the esophagus does not have the thick protective mucus barrier that the stomach has, it gets "burned" by the acid. GERD has become a major health concern in the United States because of our society's diet and exercise habits. As a nation, we overeat constantly, and instead of walking after a meal, we sit or recline. Antacids, which offer limited relief, are a quick fix for heartburn. Lifestyle changes are more effective long-term. These include eating less, reducing body weight, limiting fats, alcohol, caffeine, and chocolate, and walking or standing after eating.

A healthy stomach is protected from stomach acid by a thick layer of mucus, but sometimes people develop sores in the digestive organs anyway; these are called **ulcers**. We used to think that stress caused most ulcers, but recent research has shown that a major offender is a bacterium called *Helicobacter pylori*, which erodes the mucus lining and creates inflammation. Other contributing causes include smoking, heavy drinking, aspirin, caffeine, and certain medications.

A generalized inflammation of the stomach lining, causing a "sore stomach," is called **gastritis.** It can be caused by spicy foods, excess acid production, stress, alcohol, aspirin consumption, or heavy smoking. Both ulcers and gastritis are usually cleared up with antacids/antibiotics or other medications.

Disorders of the Small Intestine and Accessory Organs

Now let's move down to the small intestine, and take a look at some of the common problems that can occur there. A disease called **celiac sprue** is a chronic condition in which wheat gluten causes damage to the membranes of the small intestine, so absorption does not occur properly. Symptoms can include diarrhea, weight loss, flatulence, distended abdomen, anemia, and a variety of nutritional deficiency symptoms. The best treatment is a gluten-free diet and anti-inflammatory medication.

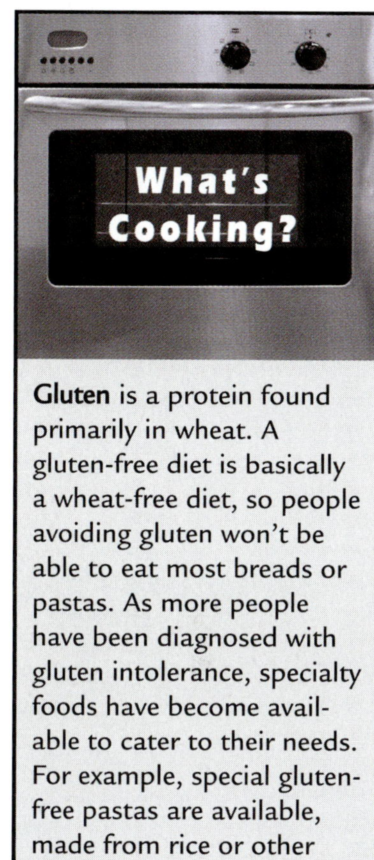

Gluten is a protein found primarily in wheat. A gluten-free diet is basically a wheat-free diet, so people avoiding gluten won't be able to eat most breads or pastas. As more people have been diagnosed with gluten intolerance, specialty foods have become available to cater to their needs. For example, special gluten-free pastas are available, made from rice or other grains.

Accessory organs can face problems as well. Recall from our earlier discussion that the gallbladder and pancreas feed enzymes into the duodenum. The pancreas can become inflamed (**pancreatitis**), causing pain, nausea, vomiting, and bloating. The gallbladder can develop **gall stones** (**cholecystitis**), which are chunks of cholesterol that block the exit tube and prevent bile from flowing. The liver can also become inflamed and/or infected (**hepatitis**), enlarged (**hepatomegaly**), or scarred (**cirrhosis**) so that its tissues thicken and no longer function.

Disorders of the Large Intestine, Rectum, and Anus

The speed at which the matter moves through the colon (large intestine) is important; if it moves too rapidly, not enough water is removed and **diarrhea** occurs. If it moves too slowly, too much water is removed and **constipation** occurs. Assuming there is no disease-related underlying cause, you can control these conditions on a temporary basis with over-the-counter or prescription drugs, and with diet and activity level changes.

Diverticulitis is an infection and inflammation of the pockets that may form in the intestinal tract, especially in the colon. It can cause bleeding, abdominal pain, and fever. Attacks are short-lived (24 hours or less in most cases), but may periodically recur. The best treatment is to address the short-term pain of the attack with pain medications, and then make dietary changes (more cellulose fiber, more water, avoiding foods with seeds or nuts in them) to prevent future attacks. Recurring acute attacks or complications may require surgery.

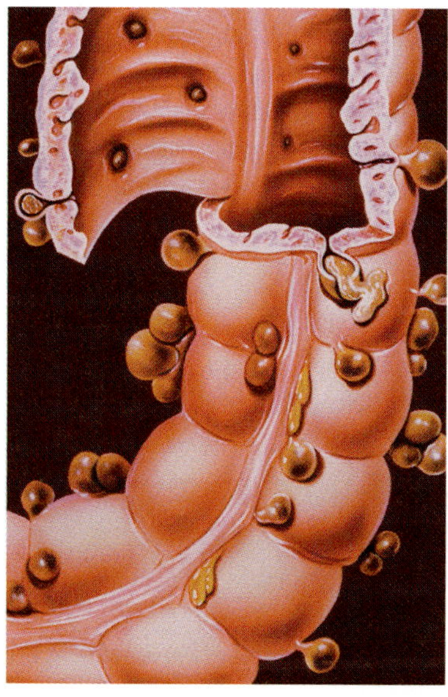

FIGURE 4-10
Diverticulitis in the colon.

Sometimes, the intestines become irritated for no apparent reason, causing pain, bloating, and a change in bowel activity (diarrhea or constipation). This is called **irritable bowel syndrome** (IBS); it is often seen in young to middle-aged adults and is due to stress, laxative abuse, or irritating foods. Some medications are available for chronic occurrences; lifestyle and dietary changes may also help.

Hemorrhoids are varicose veins in the rectum that cause the rectum and anus to become sore and swollen. They cause pain, itching, burning, and bleeding, but are usually not serious. Dietary changes are typically prescribed (more fiber and water), as are stool softeners and topical medications to reduce the discomfort.

Food Poisoning

Sometimes food itself can cause digestive disorders and diseases. More than 250 known diseases can be transmitted through food, causing stomach and intestinal distress commonly known as **food poisoning**. Not all food poisoning cases are the same, of course, because of the many different infectious and toxic agents that can be involved. Food poisoning can be caused by viruses, bacteria, parasites, poisons (like poisonous mushrooms), improperly prepared exotic foods, or pesticides on fruits and vegetables. Typical symptoms include nausea, vomiting, abdominal cramping, and diarrhea that occur within 48 hours of consuming a contaminated food or drink. Depending on the contaminant, you may experience bloody stools, dehydration, fever and chills, and even nervous system damage. Severe cases can even result in death.

Hot Topics

"Employees Must Wash Hands." We've all seen these signs in resaurant bathrooms here in the U.S., but have you ever thought about how important this really is? Some other countries do not have the same strict standards for their food-service employees. In fact, travelers to these countries often contract food poisoning for this very reason. Common symptoms of this type of bacterial food poisoning are diarrhea and vomiting. Globally, diarrheal illnesses are among the leading causes of death.

Food usually becomes contaminated from poor sanitation or preparation. Food handlers who do not wash their hands after using the bathroom or have infections themselves often cause contamination. Improperly packaged food stored at the wrong temperature can also promote contamination. You will learn more about safe food handling in Chapter 8, "Sanitary Food Handling and Food Safety."

FIGURE 4-11
What if you were visiting another country and saw live chickens roaming though a restaurant? Would you consider that restaurant to be of the same sanitary standards of those in the U.S.? Would you consider eating there?

Case Study

Jennifer woke up this morning feeling awful. Her symptoms included nausea, vomiting, diarrhea, and a stomach ache. The day before, she and some friends had eaten at a new restaurant, and she thinks she may have gotten food poisoning from something she ate there. She calls up her friends who ate there with her, and none of them are sick.

- Should she call the restaurant to let them know that she suspects their food of causing her illness? Why or why not?
- Should she notify the local health department? Why or why not?
- What additional information would be helpful to know when deciding how to answer the preceding questions?

Put It to Use

1. Flatulence, or gas, is a natural by-product of the bacterial action in the large intestine. It's perfectly normal, yet many people are embarrassed by it. What kinds of foods cause the most gas, in your experience? Is there anything you can do to avoid it, other than avoiding those foods? Research your answers online or at your school's library and write up a summary of your findings.

Put It to Use

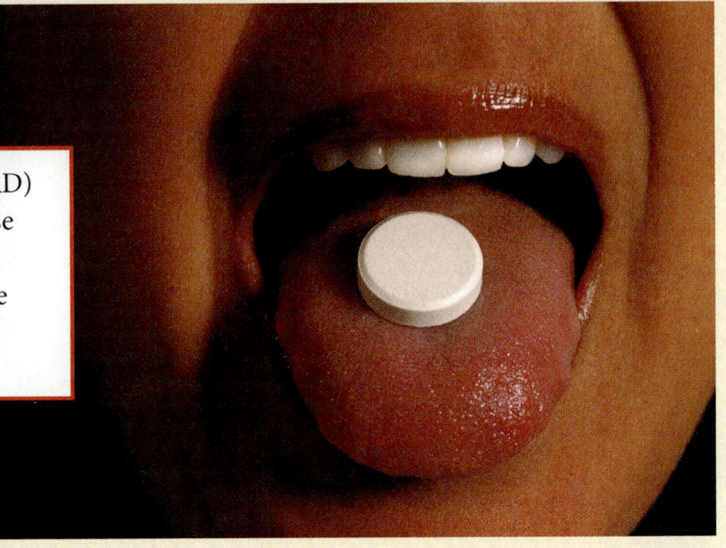

❷ Certain foods are likely to cause acid reflux (GERD) in people who are susceptible to that problem. Use the Internet to research what foods are especially good or bad for someone with GERD. Summarize your findings in a spreadsheet or table.

Write Now

Earlier in the chapter, we mentioned that the large intestine produces vitamin K. Do some research to find out more about this process. How exactly are vitamins produced there? How do those vitamins get where they need to go in the body, rather than being expelled with the waste? Write a paper that summarizes your research. Make sure you cite your sources.

Tech Connect

There are many good tutorials available online for studying the digestive system. Check out some of these resources:

http://www.ahealthyme.com/Imagebank/digestive.swf

http://kitses.com/animation/swfs/digestion.swf

http://kidshealth.org/misc/movie/bodybasics/digestive_system.html

Team Players

Break into teams of between three and six people. As a team, brainstorm a list of foods that one or more people on the team have found to give them heartburn or a stomach ache. Analyze your team's list according to the protein, carbohydrate, and fat content of the items, and the overall level of spiciness. What generalizations can you make about the types of food that are most likely to trigger digestive problems?

Put *It* Together

Match the explanation in column 1 with the term in column 2.

Column 1

- **a.** the chemical process of breaking down or building organic matter
- **b.** the parts of your body that process food
- **c.** the tube that transfers food from the mouth to the stomach
- **d.** rhythmic contractions that move food through your digestive system
- **e.** one of the sections of the small intestine
- **f.** the organ that stores bile from the liver and feeds it into the duodenum
- **g.** one of the sections of the large intestine
- **h.** the common name for gastroesophageal reflux disease (GERD)
- **i.** a sore in the esophagus, stomach, or intestine
- **j.** gluten intolerance resulting in malabsorption of nutrients

Column 2

1. celiac sprue
2. esophagus
3. gallbladder
4. GI system
5. heartburn
6. jejunum
7. metabolism
8. peristalsis
9. sigmoid colon
10. ulcer

5 Special Diets for Weight Loss and Fitness

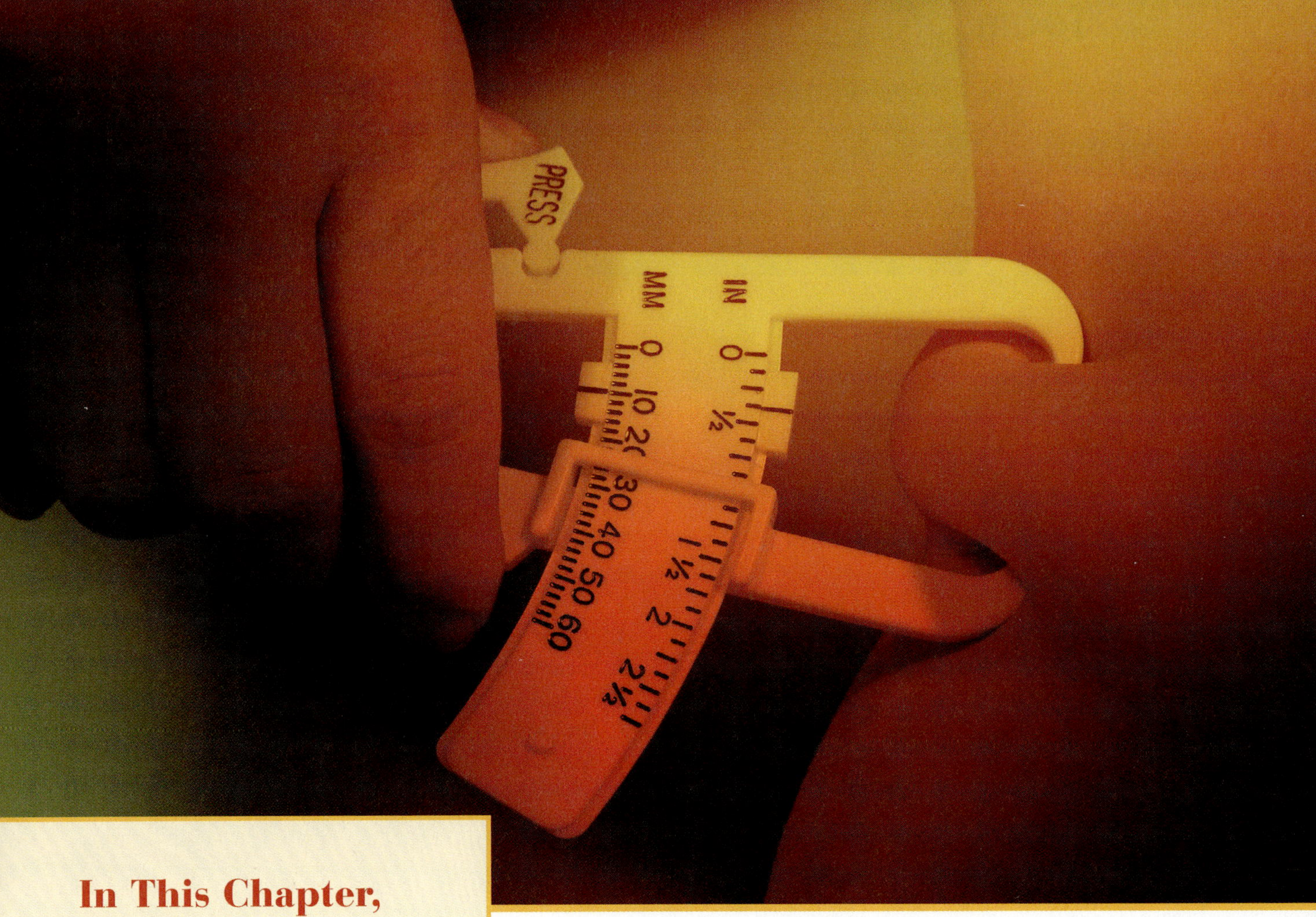

In This Chapter, You Will . . .

- Calculate energy intake and expenditure
- Investigate what factors influence body weight
- Discover how the body stores and accumulates body fat
- Calculate how many calories you should consume per day
- Determine your Body Mass Index (BMI)
- Explore various weight-loss diets
- Examine the special dietary needs of athletes

Why YOU Need to Know This

Fad diets don't work—at least not long term. But that doesn't stop people from trying them, and it doesn't stop the people who create the books and gimmicks surrounding them from getting rich. What are some of the most outlandish fad diets you have heard of? Have you known anyone who lost a substantial amount of weight on one? Did the person gain it all back?

Diet and Body Fat

Everything about the human body serves a useful evolutionary purpose, and body fat is no exception. The purposes of body fat are to insulate the body, cushion the internal organs, and to maintain a supply of back-up fuel, in case the main fuel supply is temporarily cut off. The fat in your body is known as **adipose tissue**. Unlike muscle, fat does not require much energy to maintain it; fat just sits there waiting, like extra fuel in a spare gas tank.

Being overweight can actually provide a health advantage if, for example, a famine were to occur. If two people—one overweight and one very thin—stopped eating and drank only water for two weeks, the overweight person's body would suffer much less from the experience. His body would continue to be nourished from fat reserves. The thin person, in contrast, would experience breakdown of lean muscle tissue, and perhaps even some damage to important organs.

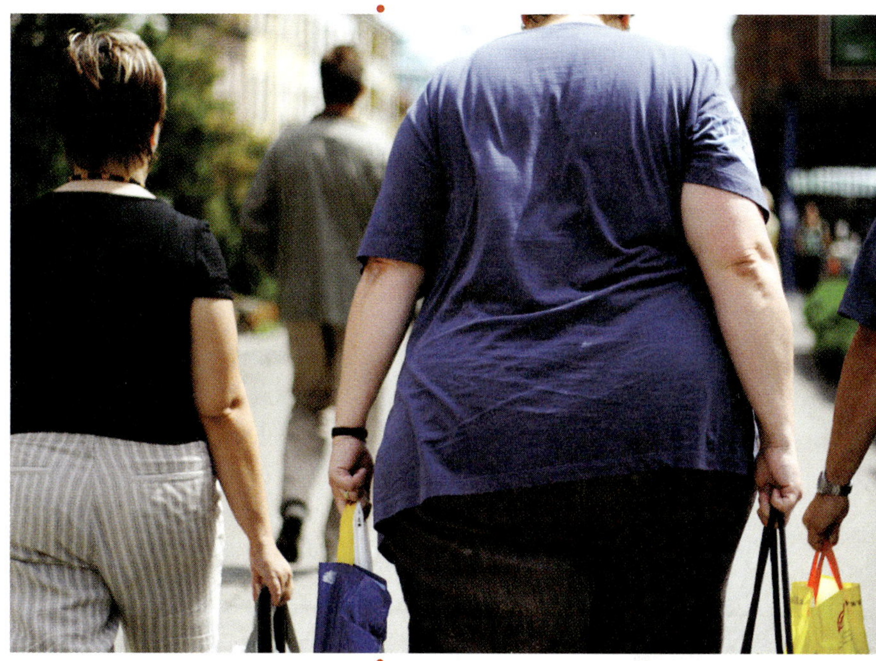

FIGURE 5-1
Obesity is rampant in the United States; some studies estimate that 60% of the American population is overweight or obese. What are some of the drawbacks of being overweight in our culture?

However, the possibility of a famine occurring is fairly remote, so there's no practical advantage to being overweight in our society, and there are significant health drawbacks from it. As a result, many people try to reduce their body fat using various special diets, exercise routines, and even surgeries.

At the basic level, the relationship between diet and body weight is very simple. If you eat more calories than you burn, you will gain weight. If you eat fewer than you burn, you will lose weight.

What determines how many calories your body burns, however, is a complex question that involves many factors, including genetics, exercise, health, and metabolism, as well as, of course, the quantity and type of food you eat. In this chapter, we'll look at the physiological processes of weight loss and gain, and the various types of diet modifications that are commonly used to affect those processes—with varying degrees of success.

When someone is very overweight, they are **obese**. Obesity is considered a chronic, long-term disorder, and can produce health problems and diseases.

Hot Topics

In 2008, Mississippi, West Virginia, and Alabama ranked as the fattest states in the U.S., with obese populations that exceed 30%. Colorado, Connecticut, and Massachusetts were the leanest states, with 20% or under.

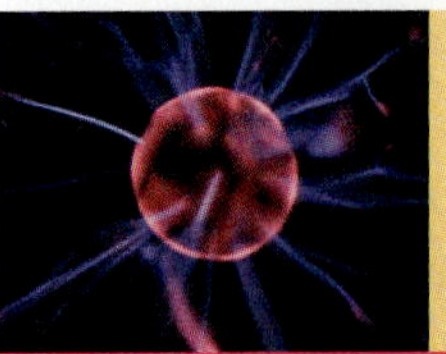

SCIENCE STUDY

When your body needs food, it signals that fact with **hunger**. Some indications of hunger include stomach contractions (growling), lightheadedness, fatigue, and weakness. These physiological signs are triggered mainly by low blood sugar levels. When your blood sugar drops, the hypothalamus in your brain triggers the desire to eat. When you have eaten enough, the hypothalamus turns off your hunger.

The state in which you do not want to eat is called **satiety**. Some factors that create satiety in your body include stomach expansion, nutrient absorption in the small intestine, increased blood sugar, and various hormone releases.

If people only ate when they were hungry, there would be very little obesity. Obviously, there is something else driving our eating habits, and that something is appetite. **Appetite** is psychological, not physical. It is fed by the psychological pleasure of eating food—how the food tastes, how it looks, how it smells, and how it feels in your mouth. Emotions and mood can affect appetite, which is why stress-induced overeating is common.

The next time you eat something, stop for a moment right before you take that first bite and ask yourself: 'Am I hungry? If not, why am I eating?' The answer might surprise you.

FIGURE 5-2
Most foods contain some water—and certain foods contain more water than they do energy-producing nutrients. What are some foods that are mostly water? How can eating these foods help with a weight-loss diet?

Understanding Energy Intake

Energy intake is the amount of energy (kilocalories) in the food you eat. As you learned in Chapter 1, proteins, carbohydrates, and fats are all energy-producing nutrients, which means they have **kilocalories**. A kilocalorie (kcal) is a unit of measurement that describes how much energy the food delivers to the body when it is digested and metabolized. One kilocalorie is the amount of energy needed to raise the temperature of one kilogram of water by one degree Celsius. Technically a kilocalorie is 1,000 calories, but it is common practice to use the word *calorie* to mean *kilocalorie*, and that's what we'll do it in this book from this point on.

You can roughly estimate the number of calories in a food item by multiplying the weight in grams by 4 for proteins and carbohydrates or by 9 for fats. (This is often abbreviated as 4/4/9.) However, this is not as simple a math operation as you might think, because many foods contain a lot of water, which provides no energy, and it is not always obvious how much water a product contains. Therefore the best way to definitively determine the calorie count on a food item is to consult its Nutrition Facts label or use a nutrition reference source such as a book or Web site.

Let's look at a couple of examples. A 1-ounce (28 gram) portion of chicken breast contains the following:

Water: 17.7 grams
Protein: 8.4 grams
Fat: 2.2 grams

A chicken breast is approximately 60% water by weight, so it contains only 60% of the calories that it "should" by weight, had you assumed that it contained pure energy-producing nutrients. Given that information, what would you expect the calorie count to be for a chicken breast, per ounce? 8.4 × 4 = 33.6, and 2.2 × 9 = 19.8. Add those values together, and you come up with 53.4 calories per ounce.

Dry foods such as cereal contain little or no water, so their calorie counts are much closer to the standard 4/4/9. For example, a 1-ounce serving of dry breakfast cereal (28 grams) contains the following:

Water: 4.7 grams
Carbohydrates: 22 grams
Fat: 1.3 grams

How many calories is that? 22 × 4 = 88, and 1.3 × 9 = 11.7, for a total of 99.7 calories per ounce.

Understanding Energy Expenditure

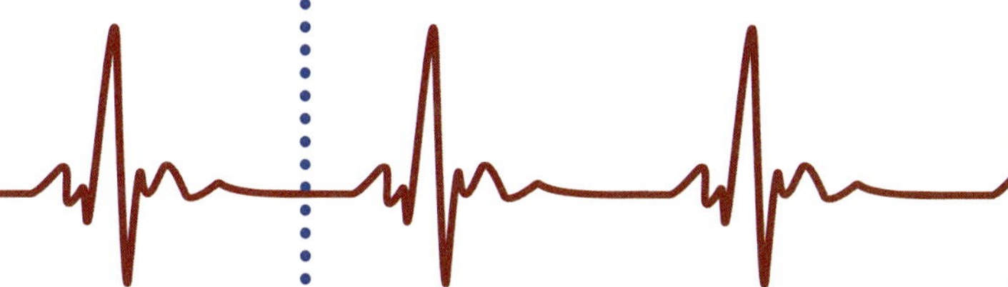

Calories are the only source of energy intake, but there are many sources of energy expenditure—in other words, many ways to burn calories.

Basal Metabolic Rate (BMR)

Your **basal metabolic rate (BMR)** is the number of calories you expend just to maintain your body at rest. Even when you are sitting completely still, your body is still burning calories on things like maintaining body temperature, creating new cells, breathing, circulating blood, and responding to sensory stimulus. The majority of the calories you consume each day (60 to 70%) keep your body up-and-running in this basic way. This is another way of referring to your **metabolism**, which we talked about in Chapter 4.

The BMR varies among people. The main determinant of BMR is the amount of lean muscle you have, because muscle takes more energy to maintain than fat does. Other factors that increase your BMR include greater height, younger age, thyroid function, stress, being male, pregnancy and lactation, and ingesting certain drugs such as caffeine and tobacco. It's not just a coincidence that when people quit smoking, they gain weight. Smoking does increase metabolism. However, its strong negative effects on health, such as cancer and emphysema

FIGURE 5-3
People who have sedentary jobs do not burn many calories each day in addition to their BMR unless they make it a point to exercise. What are some ways that a person working in an office, or sitting in a classroom, can create additional opportunities to exercise throughout their workday?

risk, far outweigh its minor benefits. Conversely, females and people with less muscle, less height, and depressed thyroid function tend to have lower BMR. BMR also decreases with age, by 3 to 5% each decade after age 30. This is partly due to hormonal changes, but most of it is due to the loss of lean muscle mass because of physical inactivity. In other words, it's preventable; if you stay physically fit and active as you age, you can lessen the rate of metabolic slowdown.

Physical Activity

Depending on how active you are, physical activity uses about 20 to 35% of your total daily energy output. This includes not only activities that we would normally consider "exercise," but also incidental movements such as sitting, standing, walking, and talking. Activities that use more muscles, and larger muscles, burn more calories. So, for example, walking is more demanding than talking.

Whereas BMR is not directly affected by weight, the number of calories burned by physical activity is, because a heavier person is transporting a heavier object when they move. Just as it takes more gasoline to drive a semi-trailer up a hill than a sports car, it takes a person weighing 200 pounds more calories to walk up a hill than a person weighing 120 pounds.

To calculate the amount of energy burned by a certain activity, you multiply the person's weight by a fixed value for that activity. Table 5-1 lists some common activities and the energy cost for each one. Each is listed in terms of the calories expended per kilogram of body weight per minute.

Fiction	Fact
Overweight people tend to have lower BMR, or lower metabolism.	Lower BMR may be a factor in some cases, such as a person who has difficulty losing weight due to thyroid problems. However, the amount of lean body mass (muscle) makes the biggest difference. Someone who exercises frequently will likely have a higher BMR than someone who is sedentary, regardless of their weight.

TABLE 5-1: ENERGY EXPENDED FOR PHYSICAL ACTIVITIES

Activity	Energy Cost (kcal per kilogram of body weight per minute)	Calories Burned (per minute for a 120-lb [54 kg] person)
Sitting quietly watching TV or using the computer	0.026	1.4
Cooking or food preparation	0.035	1.9
Walking slowly, shopping	0.04	2.2
Cleaning (dusting, vacuuming, carrying out trash)	0.044	2.4
Stretching, Hatha Yoga	0.044	2.4
Weight lifting, light to moderate	0.052	2.8
Bicycling <10 mph	0.07	3.8
Walking briskly (4 mph)	0.088	4.75
Low-impact aerobics	0.088	4.75
Weight lifting, vigorous	0.105	5.7
Bicycling, 12 to 13.9 mph	0.14	7.6
Running, 5 mph (12 minutes per mile)	0.14	7.6
Running, 6 mph (10 minutes per mile)	0.175	9.45
Running, 8.6 mph (7 minutes per mile)	0.245	13.2

Source: Compendium of physical activities: an update of activity codes and MET intensities. *Med. Sci. Sports Exerc.* 32 (2000); S498-S516.

Let's use Table 5-1 to calculate how many calories will be burned if Janice, a 130-pound woman, does low-impact aerobics for 30 minutes.

Janice's body weight in kilograms: 130 × 0.453 = 59

Energy cost of low-impact aerobics for Janice, per minute: 59 × 0.088 = 5.2

Total energy used for 30 minutes of activity: 5.2 × 30 = 156

Now suppose Janice has an obese friend, Nathan, who takes the same aerobics class. Nathan weighs 260 pounds. He will burn approximately twice the calories that Janice will:

Nathan's body weight in kilograms: 260 × .453 = 118

Energy cost of low-impact aerobics for Nathan, per minute: 118 × 0.088 = 10.4

Total energy used for 30 minutes of activity: 10.4 × 30 = 312

Given that information, it is easy to see how weight loss from exercise is often more dramatic for obese people than for people who are only slightly overweight. The less you weigh, the less energy is required to move your body, and the fewer calories you expend doing so.

Cool Tips

To convert pounds to kilograms, multiply by 0.453. For example, if you weigh 120 pounds, that's 54 kilograms. An alternate way of performing the conversion is to divide by 2.2. The result is nearly the same—within a tenth of a kilogram or so.

Thermic Effect of Food

One final way that the body burns calories is in the processing of the food itself: the **thermic effect of food (TEF).** It is usually 5 to 10% of the number of calories consumed (assuming an average diet with a mix of protein, carbohydrate, and fat sources). So, for example, if you eat something that contains 100 calories, 10 of those calories go toward processing the food, and only 90 of them are available to the body as energy. Different nutrient types have different TEF values. Fat takes very little energy; protein and carbohydrates take more (as much as 25%). Therefore changing your diet so that a smaller percentage of your calories come from fat can result in some weight loss even if the overall calorie count stays the same.

Cool Tips

Drinking ice-cold beverages can burn a few calories! However, the effect is very small—too small to be of much benefit to dieters. As you learned earlier, a calorie (1/1000th of a kilocalorie) raises 1 gram of water in temperature by 1° Celsius. So if you drink a liter (1,000 grams) of ice-cold water, your body must raise its temperature by 37° Celsius, to body temperature. That burns 37 kilocalories.

Genetics and Body Weight

Is it "in your genes" to be thin or fat? Research indicates that 25 to 40% of our body weight can be accounted for by our genetic heritage. Adults who were adopted as children show weights similar to their biological parents, not their adoptive parents, which points to a strong hereditary component. Why is that so? There are several theories, none of them definitively proven. Here are a few of them:

- **Thrifty Gene Theory:** This theory suggests that some people are genetically programmed to be energetically thrifty, so they expend less energy. In other words, they have a lower metabolism than average. Thousands of years ago, this genetic trait probably helped certain populations from dying off entirely during periods of famine, but in today's society—where food is plentiful—the trait is detrimental because it predisposes those people to obesity.
- **Set Point Theory:** This theory suggests that our bodies try to maintain their weight within a narrow range, or at a "set point." When you dramatically decrease the food intake, your body responds by lowering your BMR to try to tip the scale back to "normal." When you increase the food intake, the body responds by raising the BMR. This could explain why overweight people have such a hard time losing weight and too-thin people have such a hard time gaining it; in both cases they are fighting against their bodies' set points.

FIGURE 5-4
Your genes determine all kinds of things about you—not just how efficiently your body stores fat, but also where your talents lie. For example, some people are just naturally more athletically gifted than others. How could a predisposition for athletic success or failure affect a person's weight?

- **Leptin Theory:** Leptin is a hormone produced by body fat (adipose tissue). It inhibits the desire to eat. There is a gene called the **ob gene** (short for obesity gene) that, when functioning properly, produces leptin. When there is a genetic mutation of that gene, however, leptin is not secreted in sufficient amounts and overeating and slower metabolism results. Additionally, there may be problems with leptin receptors or with other hormones produced by adipose tissue. This was all discovered in studies involving mice; studies involving humans have not proved conclusive.
- **Brown Fat Theory:** Regular fat is white, but bodies also have some brown fat as well. (It is more prevalent in infants.) Brown fat's purpose is to generate body heat, so it is burned by the body more readily than white fat. It is theorized that people who have more brown fat are more predisposed to being leaner because their bodies more readily burn stored fat. Some research has shown that the ratio of brown fat versus white fat can be changed in mice via genetic manipulation, so the possibility may exist in humans as well; further research is needed.

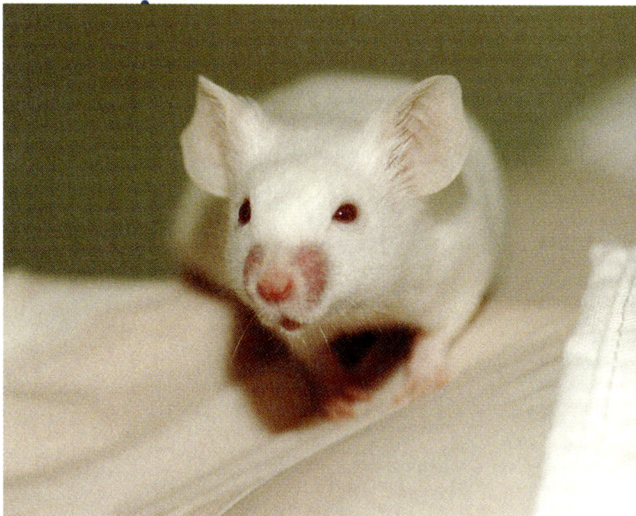

FIGURE 5-5
Scientists are continually at work to find ways of understanding genetics and body chemistry to help people lose weight. One of the frustrations scientists have experienced is that sometimes theories and techniques proven in lab animals do not work with humans. Why do you think this is so?

Environmental and Lifestyle Effects on Body Weight

If heredity accounts for 25 to 40% of our body's weight, then environment—how you take care of your body, including feeding and exercising it—must account for the other 60 to 75%.

Environment begins when we are children. Children who are very physically active and eat healthful diets are less likely to be overweight, just like adults. On the other hand, children who spend most of their time pursuing sedentary activities, and who eat lots of fat and sugar, are more likely to be overweight. Children pick up these habits from the people around them—parents, siblings, friends, and classmates—and those habits often carry over into adulthood.

FIGURE 5-6
Our environment can have a great effect on our body weight. Why is it crucial that children get enough exercise? What is your favorite exercising activity?

The food being advertised and sold in your environment can also influence your choices. Ads for high-fat fast food meals and snacks are everywhere, but there are usually healthful alternatives available if you look for them. For example, your school cafeteria offers low-fat nutritious foods each day, including milk, fruits, and vegetables, and most convenience stores sell not only potato chips and candy bars, but also apples and oranges.

Hot Topics

Though our society makes it easy to eat junk food, a lean body is shown as the ideal in the media. Overweight people are often ridiculed and discriminated against. The double standard is clear: we are encouraged to eat unhealthy, high-fat food, but not look like we ate it. This sets up a near-impossible double bind that drives many people to crash diets, meal skipping, obsessive exercising, and eating disorders such as anorexia nervosa and bulimia. We'll look at these and other disorders, in Chapter 6.

When choosing food from a supermarket, there are so many options available that it can take some work to discern which are the healthful products. Reading the Nutrition Facts labels can help. For example, foods like peanut butter, yogurt, and milk are available in regular, low-fat, and/or low-sugar varieties in most places, and bread is available in high-fiber whole grain as well as white varieties. Even hamburger is usually available in lower-fat versions, made from leaner cuts of meat.

Your friends' diets and activity levels can also be an environmental influence on your own lifestyle. For example, if your friends like to participate in sports, or like to take walks and bike rides together, it will be much easier for you to get enough exercise than if your friends like to play video games and eat pizza together every day.

FIGURE 5-7
Fashion magazines portray the extremely thin woman as the ideal, but very few people actually look like that. What problems does this disconnect cause in our society?

How the Body Stores and Accumulates Fat

Most of the fat in your body is stored close to the surface, between the skin and the muscles and organs. This is called **subcutaneous** body fat. (*Subcutaneous* literally means "under the skin.") A human can have anywhere between 50 and 200 billion fat cells in total. These cells are quite elastic, depending on the number of fat molecules stored in them. When empty, they are very small and flat. When full, they can cause the skin to stretch and even appear lumpy, as the fat cells bulge toward the skin surface. The body also stores some fat around the major organs, such as the liver, heart, and, intestines; this is called **visceral fat**.

"Cellulite" is a media and marketing name for fat that sticks out into the dermis (skin) from the subcutaneous fat layers beneath. It creates a dimpled, "cottage cheese" appearance. Nearly all women have some amount of cellulite on their bodies, and it is more common in white females. It is uncommon in men.

When your body needs food, it tells your brain that you are hungry, and presumably you eat something. The food enters your body and is processed in the small intestine. The nutrients (including amino acids from proteins, fatty acids from fats, and glucose from carbohydrates) then circulate in the blood to nourish the body's cells. Unused carbohydrates or proteins may be stored in the liver or muscles as glycogen. Excess energy may be converted to fat molecules and stored in fat cells. Fatty acids convert easily to fat molecules; glucose and amino acids are about 10 times less efficient in the conversion to fat, but they do convert. This is where the difference in the thermic effect of food (TEF) values come in, discussed earlier in the chapter.

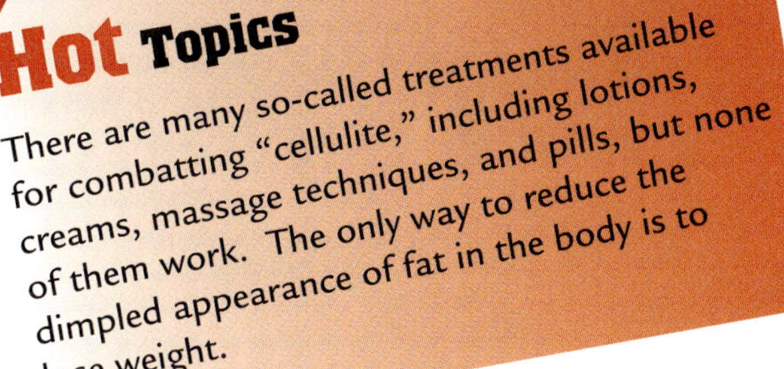

If you *don't* eat, the body turns to its energy reserves for a "meal." The body will first seek energy from the blood sugar and then the glycogen stores in the liver. Then muscles are broken down to meet protein needs. Muscles and fat can both be used to make an energy source called **ketones** to keep your brain going. Finally, fat will be used as energy.

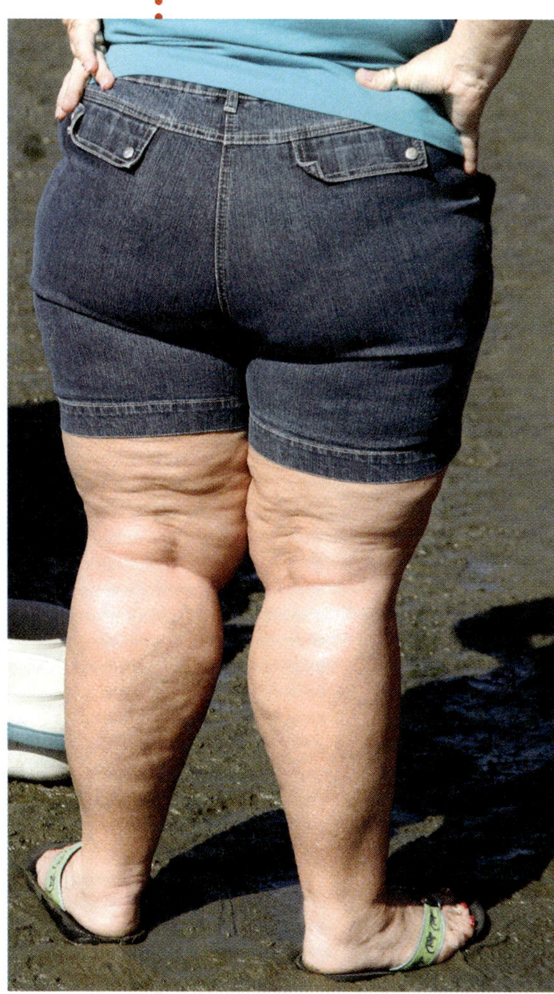

FIGURE 5-8
Even though both men and women can be overweight or obese, it is mostly women who have bulging subcutaneous body fat. Why do you think that is?

How Many Calories Should I Consume?

At this point you may be thinking, "All of this is so complex! Isn't there a simple way I can figure out how many calories to eat per day?" Yes, actually there is. The more accurate you want the results to be, the less simple the calculation process, but let's see if we can strike a happy medium.

1. Convert your body weight to kilograms. To do so, multiply your weight in pounds by 0.453 (or divide it by 2.2). For example, if you weigh 150 lbs, the result is 68 kg.
2. Calculate your Basal Metabolic Rate (BMR), or number of calories burned at rest, per 24-hour day. To do so:
   a. Multiply your body weight in kg by 1.0 kcal if you are male (in other words, no change), or 0.9 kcal if you are female. For example, if you weigh 68 kg and you are male, the result is 68 kcal; if you are female, the result is 61 kcal.
   b. Multiply the result by 24 because there are 24 hours in a day. For the preceding example, the result would be 1,632 for a male or 1,464 for a female.

3. Estimate your activity level of your lifestyle to get an activity multiplier:

	Men (Low/High)	Women (Low/High)
Sedentary/inactive: Sitting, driving, lying down, with little or no walking	0.25/0.4	0.25/0.35
Lightly active: Sitting interspersed with some walking and light lifting	0.5/0.7	0.4/0.6
Moderately active: Lightly active plus intentional exercise such as walking steadily for an hour five days a week, or a job requiring physical labor	0.65/0.8	0.5/0.7
Heavily active: Many hours a day of physical labor, such as carpentry work, roofing, heavy lifting, or digging, or intentional exercise such as jogging	0.9/1.2	0.8/1.0
Exceptionally active: Both a physically active job and intentional exercise; athletes who train for many hours each day	1.3/1.45	1.1/1.3

4. Calculate the additional calories expended over and above your BMR by the activities you participate in. To do this:
   a. Find your activity level in the preceding table, and note the low and high values. For example, for a moderately active man, these values would be 0.65 (low) and 0.8 (high).
   b. Multiply your BMR by the low value. For example, 1,632 × 0.65 = 1,061
   c. Add your BMR to the result. This is the minimum number of calories you need. For example, 1,632 + 1,061 = 2,693
   d. Multiply your BMR by the high value. For example, 1,632 × 0.8 = 1,306
   e. Add the BMR to the result. This is the maximum number of calories you need. For example, 1,632 + 1,306 = 2,938

For our example, a 150-pound man with a moderately active lifestyle needs 2,693 to 2,938 calories a day to maintain his current weight. If he wants to lose or gain weight, he should adjust his caloric intake from there.

Evaluating Body Mass

At present, are you overweight, underweight, or at your ideal weight? There are several ways to define what those words mean. One of the simplest and most widely accepted methods is the Body Mass Index (BMI). BMI is the preferred tool for the National Institutes of Health and the World Health Organization as well as many other health organizations.

BMI is a height-weight ratio, resulting in a single number that evaluates your body weight. The lower the number, the leaner you are; the higher the number, the more overweight. A BMI of 25 or greater is considered overweight; a BMI of 30 or greater is considered obese.

BMI can be calculated in any of these ways:

$$\text{BMI} = \frac{\text{weight in kilograms}}{\text{height in meters}^2}$$

$$\text{BMI} = \frac{\text{weight in pounds} \times 703}{\text{height in inches}^2}$$

$$\text{BMI} = \frac{\text{weight in pounds} \times 4.88}{\text{height in feet}^2}$$

If you want to figure your precise BMI value, use one of the above formulas. If you just want to know approximately where you stand in terms of the BMI recommendations, see Figure 5-9.

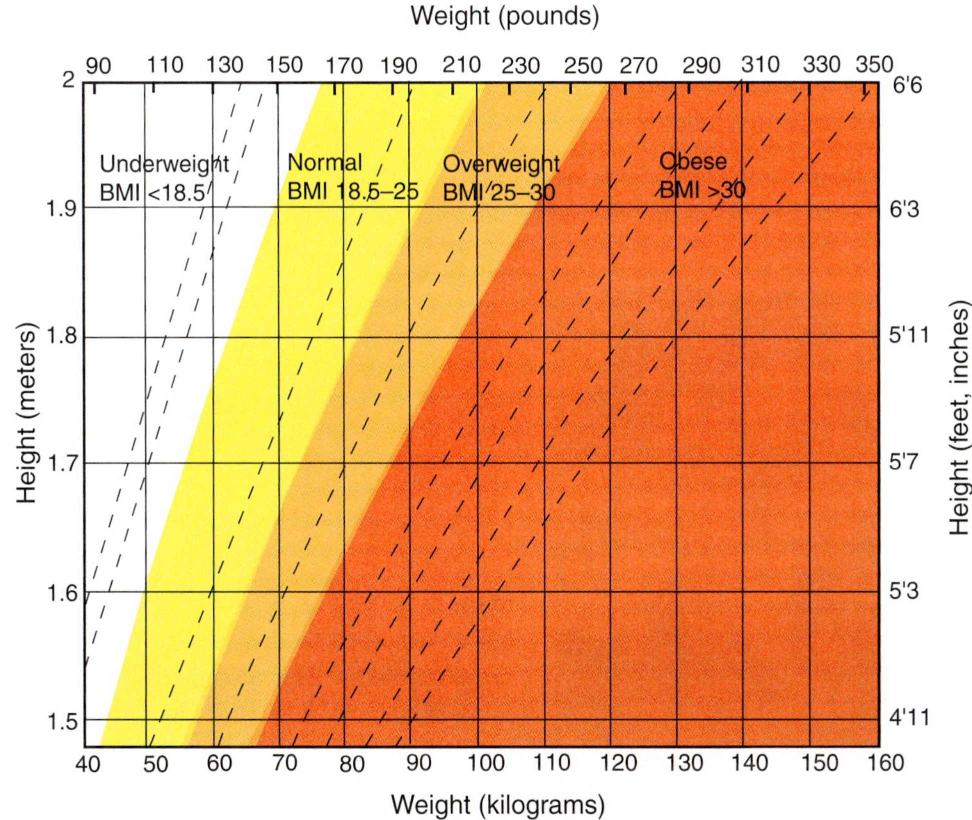

FIGURE 5-9 Evaluation of Overweight/Underweight status based on BMI.

Cool Tips

What about being *underweight*? A BMI of less than 19 is considered underweight. Health problems associated with being underweight include a weakened immune system, slow wound healing, respiratory problems, and malnutrition. Eating disorders can cause a person to be underweight. Some diseases such as hyperthyroidism (overactive thyroid), HIV/AIDS, Crohn's disease, and some types of cancer have underweight as a symptom.

Your BMI provides a benchmark for the likelihood of having weight-related health problems. Having a BMI over 25 can increase your risk of heart disease, high blood pressure, diabetes, strokes, gallbladder disease, sleep apnea, breathing problems, and certain types of cancer. In people with a BMI of 30 or higher, death rates increase 50 to 100% above those of people with BMIs between 20 and 25. As you can see, your BMI is an important number to monitor.

BMI is a handy tool, but it is an imperfect way of evaluating body composition. Since the fat in the body creates the higher health risk—not the body's overall weight—BMI is not a perfect health risk evaluation tool. An athlete who has a lot of muscle mass, for example, will have a higher BMI than average but will have a low percentage of body fat.

In addition, studies have found that race plays a role in how much a high BMI contributes to health risk. Asians tend to have a higher percentage of fat for their BMI than Caucasians, meaning they have a higher risk of health problems; African Americans have less disease risk at a given BMI.

Evaluating Body Composition

Outright obesity is seldom misdiagnosed. However, because BMI deals only with total body mass, it can sometimes deliver misleading results in people with a higher-than-normal amount of muscle. In such cases, another method of evaluating the body may prove useful.

One of the best ways of measuring body composition (that is, the amount of fat versus muscle) is underwater weighing. It works by weighing the body using a normal scale on land, and then comparing that to the underwater weight and the volume of water that the body displaces. Because muscle is heavier than fat, the heavier a person is underwater, the more lean they are. The main drawback of this test is that it requires special equipment that is not widely available.

Another way to evaluate body composition is with calipers that measure the thickness of the fat layer in several locations on the body, such as the abdomen, triceps, and shoulder blades. This method is much more convenient, but using the calipers requires special training, and the results are not as accurate for people who are extremely lean or extremely obese.

FIGURE 5-10
Body composition testing can be an expensive way of determining what is obvious at a glance. For example, a quick look at this body builder will tell you that he is not overweight, even though his BMI would indicate it. When would body composition testing be useful enough to warrant the cost and effort?

Body composition assessment can also be performed with dual-energy X-ray absorptiometry (DEXA). It is essentially a software-assisted series of X-rays that measure body fat, muscle, and bone. Its main drawback is that the machine that performs the test is very expensive.

Bioelectrical impedance analysis (BIA) can also be used. This technique attaches electrodes to a hand and a foot, and then passes a mild, painless electric current through the body. The amount of resistance to the current is measured, and based on that value, the amount of lean tissue versus fat can be calculated. The idea behind this technique is that lean tissue conducts electricity fairly well because it is water-based, whereas fatty tissue conducts electricity poorly because it is lipid (fat) based.

Types of Weight-Loss Diets

Weight loss, like weight gain, is a gradual process. A pound of fat is equivalent to about 3,500 calories, so to lose one pound, you must eat 3,500 fewer calories than your body requires. If you stopped eating entirely (called **fasting**), you would still lose less than a pound a day, on average. In fact, you would probably lose even less than that, because your body would think it was starving and would lower your BMR.

A safe and sustainable weight loss diet has the following qualities:

- **Moderation:** It should reduce your overall caloric intake by a small percentage of the amount you need for **stasis** (that is, to stay at your current weight). For example, your target might be to decrease your calorie consumption by 15%.
- **Balance:** It should provide a balance of nutrients, with fats occupying no more than 20 to 35% and the remainder being split evenly between carbohydrates and proteins.
- **Flexibility:** It should allow for healthy eating choices in a variety of situations, including dining at home with friends, dining out in a restaurant, picking up fast food, taking a lunch to school or work, and eating special-occasion dinners.

Fiction	Fact
A diet that contains no fat at all will help you lose weight faster than a diet that contains a moderate amount of fat.	Everyone needs a little fat in their diet because fatty acids and fat-soluble vitamins are an essential part of body maintenance.

Most diets that eliminate or drastically restrict your consumption of a certain type of nutrient don't work in the long run. Nevertheless, the weight-loss diet industry is a multi-billion dollar enterprise, and new diet plans are constantly being developed. Table 5-2 lists some of the popular weight-loss diet plans. Most **fad diets** (that is, unusual diet plans that come and go in popularity) fall into one of the categories in Table 5-2.

The diet industry is a type of business based upon granting wishes and selling fashion-based diet trend concepts. Like "get rich quick" ideas, "get thin fast" is something for nothing. It does not deliver in the long run. Even if quick weight loss is successful, it usually is fleeting and reversible. Lifetime management is key—and more fun, and less expensive.

Ketosis is one side effect of extremely low carbohydrate diets such as The Atkins Diet. When the body does not receive carbohydrates (glucose) as fuel, the liver ceases to release glucose into the blood as a nutrient, and instead releases fatty acids (from stored fat), which the cells can use as an alternative to glucose. This process also releases ketone bodies, a by-product of converting stored fat into energy. For the safety of the kidneys, the ketones are diluted with water, and the urine volume increases. This creates a water-loss that is perceived as a quick loss of weight. "This diet works!" But, it is really mostly dehydration, and will come back.

Hot Topics

There is disagreement in the medical community about how damaging chronic ketosis is. Some proponents of low-carbohydrate diets insist that the risks are exaggerated and that some people's bodies can safely utilize ketones and fatty acids for energy indefinitely with proper medical monitoring.

Ketosis is hard on the liver and can destroy muscle tissues. People who live for long stretches at a time in a state of ketosis may risk liver damage and muscle deterioration.

TABLE 5-2: POPULAR DIET PLANS

Diet	Description	Results
Fasting or semi-fasting	Eating very little food, or no food at all	Hunger, low energy, lowered BMR, reduction in muscle and water mass as well as fat. Poor growth. Eventual death.
Low-carbohydrate	Eating few or no carbohydrates, without restricting the amount of fat or protein	Ketosis, insufficient fiber, dehydration, higher blood triglycerides and cholesterol.
Low-fat	Eating as little fat as possible, without restricting the amount of carbohydrate or protein	Weight loss is slow. Carbohydrate and protein intake should also be monitored.
Limited food types	Eating only a limited array of foods, such as only soup, or only fruits and vegetables	May not provide adequate nutrition.
Meal replacements	Replacing one or more meals per day with a shake, bar, cookie, or other item	Expensive, leaves you feeling hungry, may be difficult to stick with. Weight loss may be slow unless other meals are also controlled.
Portion-controlled prepared meals	Eating only pre-packaged, portion-controlled entrees, with supplements of fresh fruits and vegetables	Expensive, lacks flexibility, food may not taste as good as home-prepared meals.

Diet Sustainability and the Yo-Yo Effect

Many people lose weight on low-calorie diets, only to gain that weight back again. This is called **weight cycling**, or **yo-yo dieting**. It can increase frustration and decrease feelings of self-esteem. Not only is this frustrating, but it may actually be harmful; according to Medicinenet.com, some studies suggest that weight cycling may increase your risk for high blood pressure, high cholesterol, and gallbladder disease.

The most common cause of weight cycling is using an unsustainable diet to achieve weight loss. If you think of a weight loss diet as a temporary measure, then you are assuming that at some point you will go back to your old eating habits. The trouble with that is your old eating habits are what got you into the overweight state in the first place! A **sustainable diet** is one that you can stick with for the rest of your life. If you never go "off your diet," you will never gain back the weight you have lost. When you reach your goal weight, you can keep the same basic diet plan but add in enough additional calories to stabilize your body weight at that level.

Follow these tips to avoid weight cycling as you lose weight:

- Use the MyPyramid food model to design a diet that you can follow every day.
- Avoid fad diets, diets that promise quick fixes, and diets that omit entire classes of food.
- Adjust your caloric intake so that you are losing no more than 2 pounds per week. A diet on which you lose more than that is not sustainable.
- Exercise regularly using the large muscles in your body (legs, buttocks), in activities such as running, walking, cycling, dancing, and weight lifting. Exercise keeps your metabolism up, burns calories, and builds muscle mass.

Fiction	Fact
Weight cycling will increase the amount of fat tissue in a person, and will make it harder to lose weight in the future.	Researchers have found that after a weight cycle, those who return to their original weights have the same proportion of fat and lean tissue as they did prior to weight cycling.

FIGURE 5-11
One of the problems with yo-yo dieting is the psychological toll it takes on people. If you were starting a weight-loss program, what would you make sure to do, or not do, to prevent the yo-yo effect?

FIGURE 5-12
Studies show that it takes an average of six weeks to establish a new habit. Most people don't stick to their diet plans for that long, so the diets never have a chance to become ingrained. Why do you think people have such a hard time sticking to a diet plan? What can help them be better at it?

Tracking Weight Loss Progress

Staying motivated when following a different diet, especially one in which you eat fewer of your favorite foods, can be a challenge. One way to keep on track is to monitor your progress regularly, and celebrate each small step toward the goal.

To make sure you are eating according to plan, try recording everything you eat in a food journal. Some food journaling Web sites, such as www.fitday.com and www.weightwatchers.com, calculate the calories and nutrients automatically as you enter foods into your journal each day. You can also create your own food journal in a spreadsheet program such as Microsoft Excel, or in a free online spreadsheet such as http://sheet.zoho.com. You can start from scratch, creating your own formulas to calculate values, or you can use a predesigned template. Microsoft offers one for Excel at no charge at http://office.microsoft.com/en-us/templates/TC010684541033.aspx, for example; Figure 5-13 shows it in action.

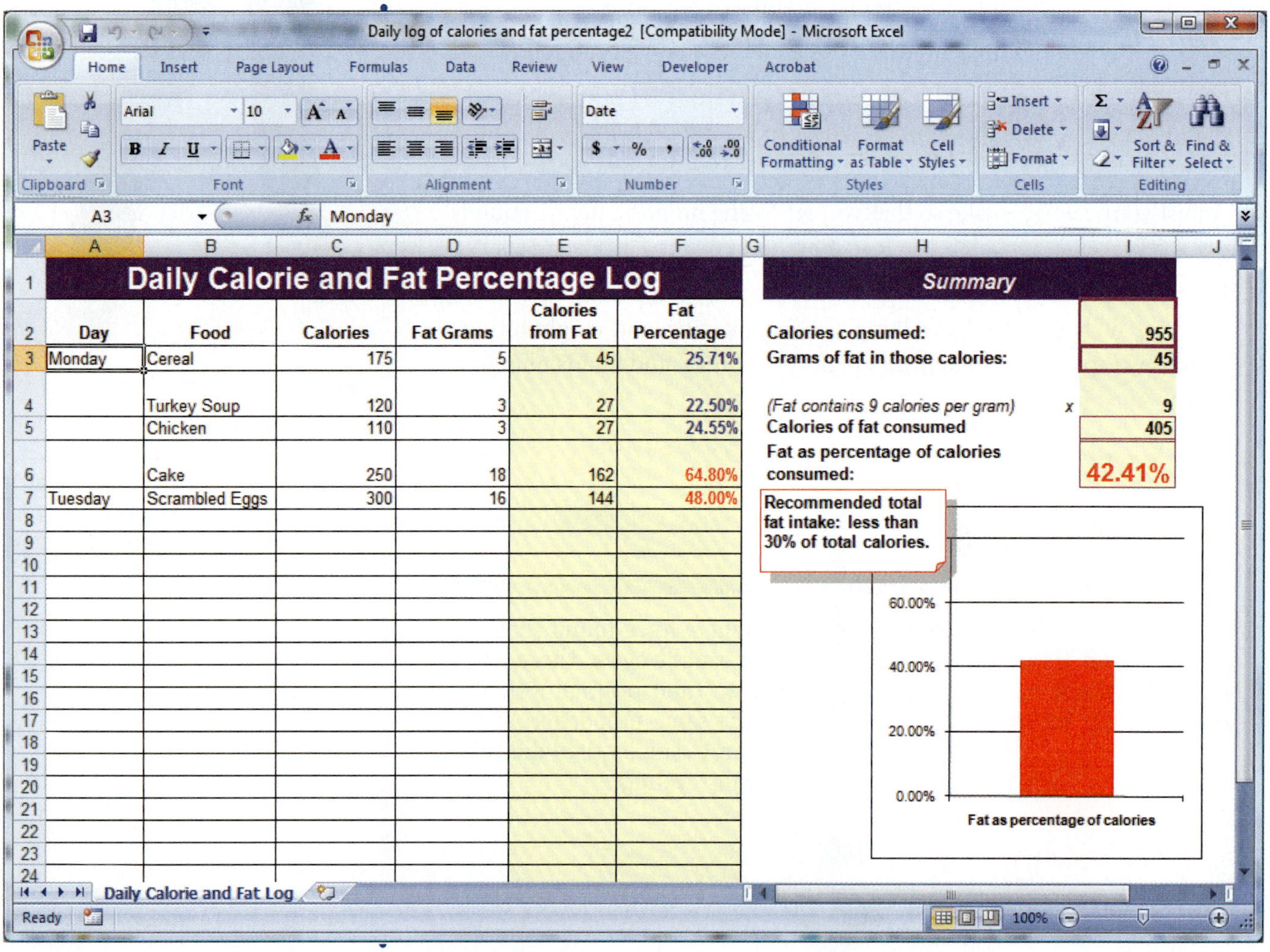

Daily log of calories and fat percentage2 [Compatibility Mode] - Microsoft Excel

Daily Calorie and Fat Percentage Log

Day	Food	Calories	Fat Grams	Calories from Fat	Fat Percentage
Monday	Cereal	175	5	45	25.71%
	Turkey Soup	120	3	27	22.50%
	Chicken	110	3	27	24.55%
	Cake	250	18	162	64.80%
Tuesday	Scrambled Eggs	300	16	144	48.00%

Summary

Calories consumed:	955
Grams of fat in those calories:	45
(Fat contains 9 calories per gram) x	9
Calories of fat consumed	405
Fat as percentage of calories consumed:	42.41%

Recommended total fat intake: less than 30% of total calories.

Fat as percentage of calories (0.00%, 20.00%, 40.00%, 60.00%)

FIGURE 5-13
Daily log of calories and fat percentage in Microsoft Excel. What other software besides Excel could you use to track this information?

Experts recommend that you do *not* weigh yourself every day when trying to lose weight, because your weight will fluctuate daily based on fluid consumption, exercise, time of day, clothing worn, and other factors. Weighing in daily and watching your weight go up and down by a few pounds from day to day can be stressful and does not provide useful information. Weekly weigh-ins should provide enough data to track your weight-loss progress.

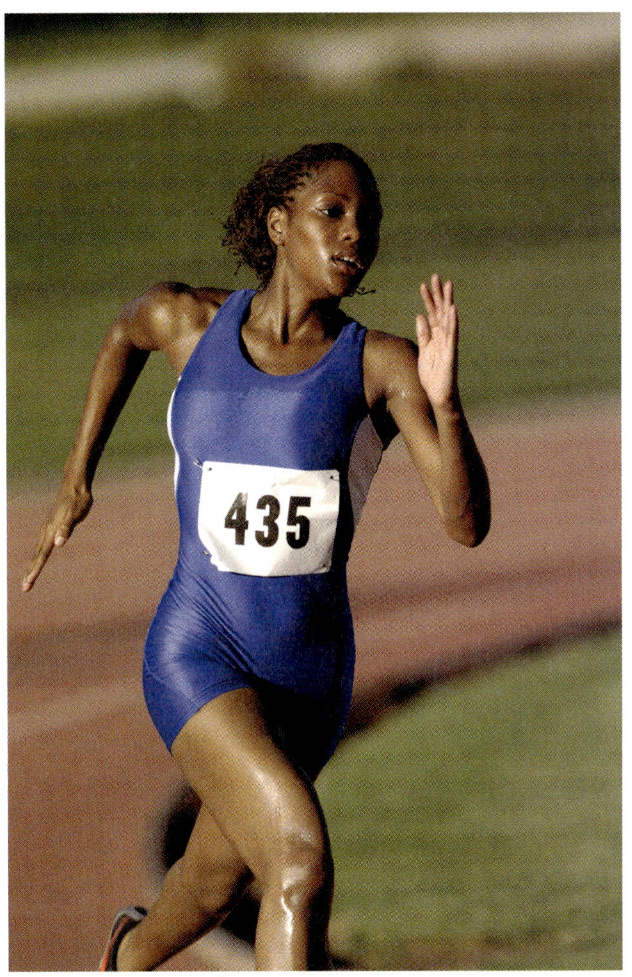

FIGURE 5-14
Most athletes have bodies that are very low in body fat. Why does low body fat contribute to better athletic performance in many sports? Can you think of any sports in which it is *not* critical to have an extremely lean body in order to succeed?

Nutrition and Athletics

Everyone needs exercise, of course, but for some people athletic activity becomes a hobby, a passion, or even a career. Diet can directly affect athletic performance in many ways, giving a burst of energy when it's needed, helping build lean muscle mass, keeping the body lean, and so on. In this section, we'll look at how the body uses fuel during athletic performance, and discuss what types of fuel are best for various athletic pursuits.

How Food Fuels Exercise

As you exercise, three pathways work to provide the fuel to your body:

- **Immediate Energy System:** This is the burst of energy that you use in the first 10 seconds of an exercise. This type of energy is important in sports where you work in very short bursts, such as power lifting, discus throws, and field goal kicking in football. It comes not directly from nutrients, but from a chemical found in almost every cell in your body, ATP, that provides quick energy.
- **Anaerobic Metabolism:** This is a short-duration type of energy that does not require oxygen. (That's what *anaerobic* means.) This energy comes from glucose (from carbohydrates), and it lasts longer than ATP. Anaerobic metabolism provides a burst for the first two to three minutes of exercise. It's essential for sports such as short-distance running and swimming. Anaerobic metabolism can't deliver performance in the long term because when glucose breaks down without oxygen, lactic acid is produced, making the blood more acidic and creating a burning sensation in your muscles.
- **Aerobic Metabolism:** This is the long-duration energy supply, generated when the body has been exercising for a few minutes and enough oxygen has become available. Oxygen allows the body to use glucose and fat provide a steady, maintainable stream of energy. When people talk about **aerobic exercise**, they mean exercise that lasts long enough to get your body into the aerobic metabolism phase.

Each muscle has a limited store of **glycogen** (carbohydrate) in it. It takes between 60 and 90 minutes of intense exercise for the muscle's glycogen supply to run out, but may last as long as five hours in conditioned athletes. That's when athletes get weak, confused, or nauseated; some describe this as "hitting the wall." You can avoid hitting the wall by consuming carbohydrates three to four hours prior to the exercise as part of conditioning. In addition, simple sugar like glucose may be immediately ingested before or during exercise, such as carbohydrate-based sports drinks. As the body digests them, the blood receives an infusion of glucose, so you can continue exercising. The body can eventually draw some energy from stored fat, but that takes longer than processing incoming carbohydrates on-the-spot.

Calories and Nutrient Balance for Athletes

Whereas in a weight loss diet, the primary concern is to minimize the number of calories taken in, an athlete's diet is often the opposite. Athletes who train for several hours a day will need many more calories than an average person—even double or more the calories needed by a sedentary person. That doesn't mean you should eat high-calorie, low-nutrient food, though. The same guidelines apply to athletic diets as to regular and weight-loss diets. There is no magic "special diet" that will dramatically increase athletic prowess, although there are many companies that would like to take your money by convincing you that they have such a product.

Carbohydrates are the most critical nutrient for athletic performance and training. Although athletes use both carbohydrates and fats for energy during their workout sessions, the body has much less storage of carbohydrates on the body than fat. For example, an average 150 pound male has about 2,000 calories of carbohydrate stored on the body, but close to 100,000 calories of fat. So athletes are much more likely to run out of stored carbohydrate. An athlete's diet should be at least 50 to 60% carbohydrates, primarily from complex sources such as pasta, rice, and whole-grain bread.

A **sports nutritionist** or **sports dietitian** is a nutrition professional who focuses on the needs of athletes.

The same career preparation applies for nutritionists and dietitians that you learned about in Chapter 1. In addition, a sports nutritionist or dietitian has special training and knowledge about how athletes' bodies use nutrition to perform in their sports and to build speed, endurance, and muscle mass.

As a sports nutritionist, you might work with a professional sports team a few days a week to assist players in planning their diets, for example. There are very few full-time positions available in sports nutrition counseling, so many people who specialize in this area also work as regular dietitians and nutritionists, or perform dual functions of nutrition counselor and personal trainer or workout coach, for which they receive separate training.

Many athletes increase their intake of complex carbohydrates before an athletic event; this is known as **carb loading**. Carb loading can be effective for an athletic event lasting longer than 90 minutes or for events requiring high-intensity performance where you are in motion the entire time, such as soccer. Some research has shown that carb loading can increase the time it takes for you to become exhausted by up to 20%. That's why some sports teams have pasta dinners the night before a big event. All carbohydrates are not alike, though; avoid carbohydrates from simple sugars such as candy or sugary soft drinks, as these can actually *decrease* performance if you ingest them right before an athletic event, as they may cause a quick spike then drop in blood sugar.

FIGURE 5-15
This plate of spaghetti would be an ideal meal for a marathon runner the night before a big race. What other foods would be good for an athlete to eat before a competition?

BY THE NUMBERS

One method of carb loading is to gradually decrease the duration of your workout as you gradually increase the percentage of your diet coming from carbohydrates during the week before the event. For example, six days before the event, you might be working out 90 minutes and consuming 50% of calories from carbohydrates; you would gradually modify these until one day before the event, you train only 20 minutes and eat 70% of calories from carbohydrates.

Water is the most important nutrient for an athlete to consume, but it is often overlooked. When you exercise vigorously, you can lose several liters of water from your body via sweat. You must replace that loss with an equal amount of water or other fluids before, during, and after exercise. If you are supervising children who are exercising, make sure you offer them water frequently, because children often do not recognize thirst as quickly as adults do.

If you are just working out for an hour or two, plain water is sufficient for hydration. Athletes who work out for many hours at a time, such as marathon runners, may benefit from sports drinks that contain electrolytes. As you learned in Chapter 1, these electrolytes are provided in the form of mineral salts dissolved in the beverages; they help your body maintain its mineral balance when taking in and excreting a high volume of water.

Cool Tips

If you are drinking enough water, your urine should be a pale straw yellow color and transparent. The more dark the yellow color of your urine, the more dehydrated you are. That's because the yellow comes from the pigments and waste products being eliminated from your body. If you are not processing enough water, then the urine is more concentrated. On the other hand, if your urine is completely colorless, you may be drinking too much water or have a medical condition.

What About Protein?

Protein is important in muscle recovery, but it is not a very good fuel for athletic performance. There is a long-standing myth that a high-protein diet will promote muscle growth, but researchers have repeatedly proved this to be false. Only exercise and strength training will create muscle growth.

Protein does support growth and repair of body tissues in general, so an athlete focusing on strength or endurance must have a diet that is adequate in protein to rebuild muscle after exercise. A serious athlete may need as much as two times the amount of protein as a sedentary person (1.2 to 1.7 g of protein per kilogram of body weight per day, compared to the RDA of 0.8). However, most Americans already eat twice the protein they need for healthy body function, so eating adequate protein for maximum muscle growth is simply not a concern for the vast majority of athletes. Excess protein is simply burned by the body as energy or stored as body fat.

Dietary Supplements

As with the population in general, athletes do not need vitamin and mineral supplements if they eat enough food, in the proper balance. It has not been proven that extra vitamins and minerals enhance athletic performance in any way. However, in special situations, such as athletes who are severely cutting back on calories, supplements may be useful. The use of dietary supplements for athletic performance should be monitored by a doctor.

If you've ever been to a vitamin shop, you've probably seen expensive dietary supplements that promise muscle growth (via high-protein powders), quick energy (via high-carbohydrate snack bars and drinks), and other benefits. Collectively these are known as ergogenic aids. *Ergo* means movement (as in the word ergonomics), and genic means positive or improved.

Are these for real? Well, the high-carbohydrate snack foods and meal replacements are indeed high-carb, and they do a good job supplying quick energy, but so do much cheaper snacks. High-protein powders do deliver on their nutrient promises, but high-protein nutrition does little to build muscle mass unless you are deficient in protein already (which is unlikely).

What to Eat After a Workout

After you have exercised for an hour or more, you may feel tired, nauseated, or shaky. To help your body recover, you'll want to have something to eat and drink.

Immediately after your workout, drink water, a sports drink, or fruit juice. Plan on eating a regular, nutritious meal 30 minutes to 2 hours after your workout, including protein, complex carbohydrates, fat, and water.

Recent research indicates that consuming whey protein immediately after a workout may increase the muscles' uptake of amino acids.

Case Study

Heather has been trying to follow a low-fat, low-calorie diet for about 30 days now, but has lost only 2 pounds. Her friend Annie asks what is going wrong, and Heather doesn't know. "I think I'm eating the right things," she says, "and I'm not eating junk food." Annie recommends that Heather keep a food diary so she can analyze where hidden calories might be coming from. The only problem is, Heather is not very proficient on a computer, and has no idea how to set up a spreadsheet to track this.

How could Heather track her progress on paper? Design a paper worksheet she could use to enter the food she eats each day, and to record her weight loss each week. Then choose a software program that she could use to track the same information on the computer.

What else might help Heather stay focused on making the needed changes to her diet? For example, what kind of peer support might be helpful?

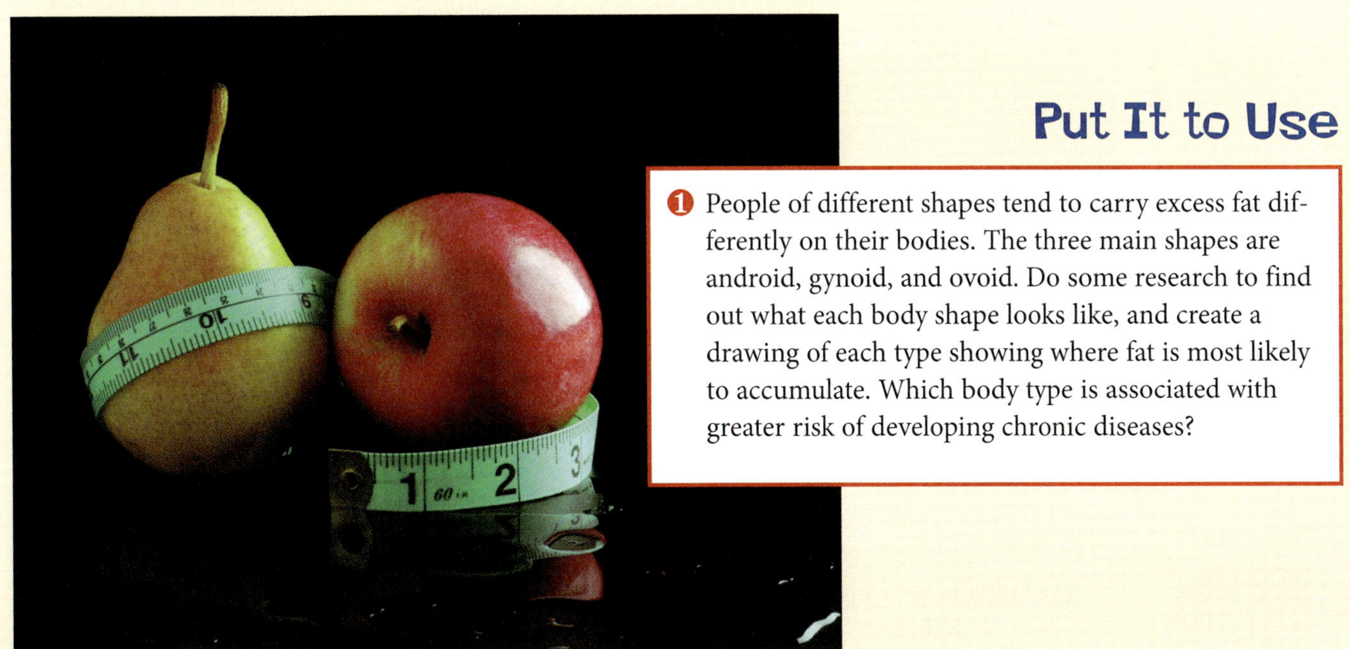

Put It to Use

1. People of different shapes tend to carry excess fat differently on their bodies. The three main shapes are android, gynoid, and ovoid. Do some research to find out what each body shape looks like, and create a drawing of each type showing where fat is most likely to accumulate. Which body type is associated with greater risk of developing chronic diseases?

Put It to Use

❷ Suppose one of your parents would like to lose some weight. Put together a sample menu for one day that includes the proper number of calories for him/her to lose two pounds a week, has the right balance of nutrients, and contains foods that he/she enjoys.

Write Now

As people age, they tend to gain weight, and to have a difficult time losing weight. Write a paper explaining three reasons why this is so, and recommending ways to overcome those difficulties.

Tech Connect

Find several Web sites that contain unbiased dietary information for athletes. Look for non-profit medical associations, sports medicine organizations, and universities.

Then find several Web sites that sell dietary supplements for athletic performance. Based on what you know about athletic nutrition from this chapter and from the unbiased Web sites, evaluate some of the claims and promises on the sites selling supplements. Pick one especially unbelievable claim, and write a report that explains why it is false or misleading.

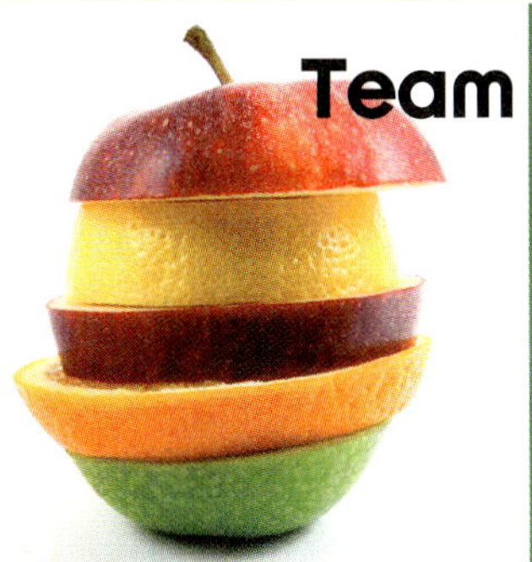

Team Players

Break into teams of between three and six people. Each team will develop a diet plan for their scenario that includes the number of calories to consume per day and nutritious sample menus that contain the right balance of protein, carbohydrates, and fats and as much of the RDA of vitamins and minerals as possible. (Not all of these people are overweight; for some of them, a diet that achieves stasis is the best choice.)

- A 16-year old girl who currently weighs 145 lbs at 5'4" tall. She is lightly active.
- A 27-year-old man who currently weighs 160 lbs at 6' tall and is getting ready to run a marathon 4 days from now. He is extremely active. (Hint: Could he benefit from carb loading?)
- A 45-year-old woman who currently weighs 200 lbs at 5'9" tall. She is moderately active.
- A 60-year-old man who currently weighs 240 lbs at 6' tall and is very sedentary.

Put It Together

Match the explanation in column 1 with the term in column 2.

Column 1

a. the fat beneath the skin on your body
b. physiological desire for food
c. the number of calories you expend just by existing
d. the calories burned by your body as it processes the food you eat
e. the fat surrounding your vital organs, such as the liver
f. a height-to-weight ratio that predicts how lean or fat you are
g. increasing the ratio of carbohydrates in an athlete's diet prior to an event
h. a diet in which you eat nothing at all for a short period of time
i. the repeated losing and regaining of weight
j. a quick burst of energy from burning glycogen during the first 2–3 minutes of exercise

Column 2

1. anaerobic metabolism
2. BMI
3. BMR
4. carb loading
5. fasting
6. hunger
7. subcutaneous fat
8. thermic effect of food
9. visceral fat
10. weight cycling

6 Eating and Nutrition-Related Disorders

In This Chapter, You Will . . .

- Examine the effects of food choices on health and wellness
- Understand the relationship between nutrition and stress management
- Comprehend diabetes and its relationship to insulin
- Learn how the glycemic index of foods affects insulin production and release
- Look at some common and uncommon nutrition-related diseases and disorders
- Learn about eating disorders and food addictions
- Find out how to prevent, treat, and manage diet-related diseases
- Understand how to manage food allergies

Why YOU Need to Know This

What you eat—or don't eat—can dramatically affect how you feel in the moment, and it can also have an impact on your overall health in the long run. For example, being overweight can dramatically increase your chances of developing Type 2 diabetes, which in turn can cause poor circulation (leading to limb amputation, in some cases) and blindness. What other long-term health problems can you think of that are associated with a person's eating habits?

How Your Diet Affects Your Health

Just as a car develops problems if you don't give it the right gas and oil, your body can develop problems if you don't fuel it properly. Diet can affect health in many ways:

- **Short-term energy:** The carbohydrates, fats, and proteins you consume each day deliver quick fuel to move your body and maintain its systems. If you don't eat enough calories, you might not have the energy to get through your day.
- **Body weight:** As you learned in Chapter 5, eating too many calories causes your body to become overweight, bogged down with excess fat deposits. Not only are these heavy to carry around, but they stress almost every system of your body, from the arm and leg joints to the heart and lungs.
- **Vitamin deficiencies:** You need water-soluble and fat-soluble vitamins to maintain the chemical balance in your body. As you learned in Chapter 1, many specific illnesses can be caused by vitamin deficiencies, such as scurvy and rickets.
- **Mineral deficiencies:** Your body relies on minerals to build and maintain teeth, bones, and red blood cells. Without enough calcium, for example, you risk osteoporosis; without enough iron, you may develop anemia (too few red blood cells).
- **Long-term strain on organs:** Over time, unhealthy eating and drinking habits can put stress on some of your internal organs, causing them to fail as you age. For example, chronic heavy alcohol consumption strains the liver, which can eventually become diseased (a condition called cirrhosis).
- **Disease prevention:** Certain food choices provide antioxidant benefits, which some studies have shown can minimize the risk of certain cancers, as well as disorders such as heart disease, diabetes, arthritis, cataracts, and Alzheimer's disease.
- **Age and appearance:** People who eat a healthy, balanced diet tend to stay younger-looking as they mature, because diet affects externally visible systems like the hair and skin. A malnourished person is more likely to have dull, brittle hair and prematurely mottled, wrinkled skin.

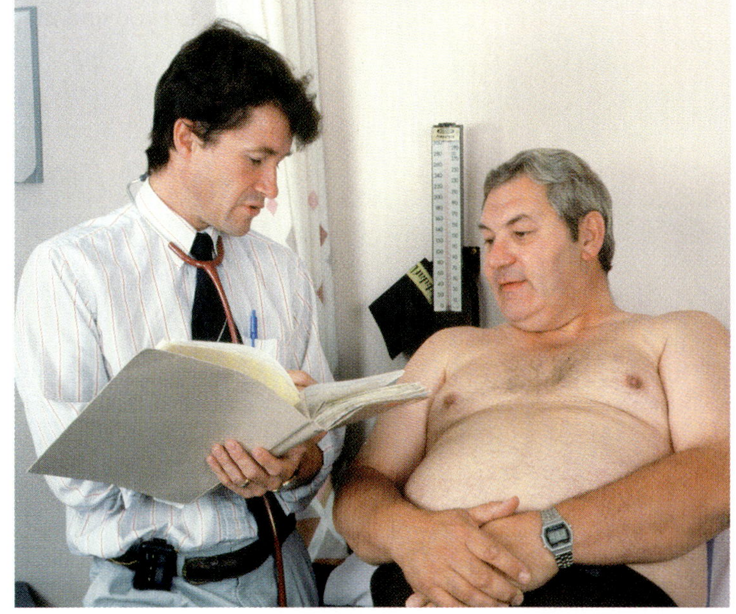

FIGURE 6-1
Some physicians are reluctant to tell overweight patients that they should lose weight, for fear of offending them. Yet weight loss is one of the best ways of lowering the risk for many diseases. What would be some tactful ways that a physician might bring up the subject of weight loss?

Diet and Stress Management

FIGURE 6-2
Diet can affect your stress level either positively or negatively. What foods do you crave when you are stressed? Do these foods end up making your stress better or worse?

Stress disrupts many people's eating habits, making them eat either too little or too much, and often resulting in food choices based on impulse, convenience, or cravings, rather than on good nutrition. In addition, some food choices actually increase stress by causing physiological responses in the body that make you feel anxious, nervous, or nauseated.

Some foods and ingredients that may aggravate stress:

- **Caffeine:** This chemical, found in coffee, tea, some soft drinks, and chocolate, causes your body to release adrenaline, which heightens your senses, makes you more awake, and increases the function of the nervous system and heart. This can result in feelings of nervousness and irritability. Additionally, the energy "boost" does not come from thin air. The body draws upon stored sources of energy to provide the boost. This means there will be an equal and opposite "crash," increased fatigue, tiredness, drowsiness, and even depression. Caffeine can also become an addiction.
- **Alcohol:** Alcohol also increases the release of adrenaline in the body, adding to your body's physiological experience of stress. The body stress produced by alcohol is even worse for you than that produced by caffeine. When your body produces adrenaline, it also produces toxins. The liver normally filters these out, but alcohol limits the ability of the liver to filter properly.
- **Salt:** The body needs a small amount of salt, but many people overdo it. Salt can increase blood pressure in salt-sensitive individuals.
- **Protein:** High-protein foods elevate brain levels of dopamine and norepinephrine, both of which are associated with higher levels of anxiety and stress.

What foods *should* you eat to combat stress? Complex carbohydrates are a good choice. A baked potato or a cup of brown rice or whole wheat spaghetti will nourish you and may make you feel better. In addition, sipping soothing hot drinks like caffeine-free tea can make you feel calmer, not only from the beverage itself but from taking the time to relax a moment while you prepare and drink it.

Taking a daily multivitamin and mineral supplement may also help combat stress. Some studies have shown that the body depletes its stores of certain vitamins and minerals when it is under stress. In particular, make sure you are getting plenty of vitamins A, B, and C, magnesium, calcium, and potassium.

Safe Eats

Caffeine is a drug, and people who consume it develop a physical addiction to it. If you decide to eliminate caffeine from your diet, do so gradually. Abrupt withdrawal can cause severe headaches and fatigue.

Antioxidants and Free Radicals

There's a lot of buzz in the media about the value of antioxidant foods, but what exactly is an antioxidant? For the answer to this question, we have to look at some basic biochemistry.

At a cell level, metabolism is the process of breaking down and building up molecules by losing or gaining electrons. When a molecule loses an electron, it's called **oxidation**. When a molecule gains an electron, it's called **reduction**.

In a stable atom, there are an equal number of electrons and protons. When an electron is lost due to oxidation, the atom may become a **free radical**. Free radicals are considered to be highly chemically reactive (in other words, they react with other molecules very readily) because they are seeking to gain an electron. Free radicals can be harmful to your body because they destabilize other atoms. A free radical atom can steal an electron from a nearby atom, changing its chemical structure, and in turn, generating other free radicals. It reduces (gains an electron), thereby oxidizing the molecule or atom from which it pulled the electron. This can cause a chain reaction of oxidation resulting in various types of cell damage in the body—cell damage that can contribute to the causes of cancer, heart disease, arthritis, Alzheimer's disease, cataracts, and a variety of other illnesses. Free radicals also contribute to the body's physical symptoms of aging, such as deteriorated vision.

Some free radicals are formed naturally as part of the body's normal functioning. They are generated as your body digests food, fights infection, and recovers after vigorous exercise, for example. You can't avoid these free radicals. However, your body also acquires free radicals from environmental sources, and some of these you *can* control—for example, exposure to the sun, radiation, asbestos, or tobacco smoke.

FIGURE 6-3
Free radicals exist in many places in our environment. What areas can you think of that you might want to avoid in order to limit your exposure to free radicals?

To combat free radicals, your body needs **antioxidants,** which are molecules capable of slowing down or preventing oxidation of other molecules. The body generates some antioxidants on its own, and you take in more through the foods you eat. Antioxidants work to minimize the damage from free radicals in several ways:

- Antioxidant vitamins give up their electrons or hydrogen molecules to the free radicals to stabilize them, so they don't produce chain reactions.

- Antioxidant minerals activate enzyme systems convert free radicals to substances that can be excreted.
- Other antioxidants, such as beta-carotene and phytochemicals, help stabilize free radicals.

For example, beta-carotene, also called provitamin A, can be converted into a usable form of vitamin A by the human body. Studies have been mixed as to the benefits of beta-carotene supplements. Some research has shown that consuming a diet rich in beta-carotene foods can reduce your risk for cancer and heart disease. However, there are no RDA recommendations for beta-carotene, and nutritionists do not consider it an essential nutrient. The human body does not appear to show any symptoms due to deficiency of it.

Hot Topics

Too much beta-carotene can cause a problem called "carotenoderma," or orange color to the skin. This is caused by eating too many carrots, tomatoes, or other foods rich in beta carotene. It is rare—you'd have to really eat a lot of carrots to turn orange, but it can happen.

Table 6-1 lists the major types of antioxidants and the foods that contain them.

TABLE 6-1: COMMON ANTIOXIDANT FOODS

Antioxidant	Foods Rich in This Include	
Beta-carotene	Sweet potatoes	Cantaloupe
	Squash	Apricots
	Pumpkin	Mangos
	Kale	Spinach
Vitamin C	Citrus (Orange, Lemon, Lime)	Green peppers
	Broccoli	Green leafy vegetables
	Strawberries	Tomatoes

(continued)

TABLE 6-1: COMMON ANTIOXIDANT FOODS *(CONT)*

Antioxidant	Foods Rich in This Include	
Vitamin E	Nuts (especially almonds)	Vegetable oil and liver oil
	Whole grains	Green leafy vegetables
Flavonoids/polyphenols	Red beans	Cranberries
	Blackberries	Purple grapes
	Pomegranates	Red wine
Lycopene	Tomatoes (cooked)	Pink grapefruit
	Watermelon	
Lutein	Kale	Broccoli
	Kiwi	Brussels sprouts
	Spinach	
Selenium (a component of antioxidant enzymes)	Fish and shellfish	Red meat
	Grains	Eggs
	Chicken	Garlic
	Brazil nuts	

Diabetes, Insulin, and the Glycemic Index

As you learned in Chapter 4, when you eat and digest food, nutrients are released into the bloodstream and carried to the body's cells. This "food" in the blood is in the form of glucose, or sugar. Therefore throughout the day, your blood sugar level rises and falls naturally within a limited range.

Some foods deliver much more glucose to the blood than others, so the body has a built-in regulation system to ensure that the blood sugar level stays within a healthy range. The body regulates the blood glucose level with two hormones, insulin and glucagon, both produced by the pancreas. **Insulin** tells the liver to take glucose out of the blood and store it as glycogen, thereby decreasing blood sugar. It also circulates in the blood to the body's cells, signaling them to open up their membranes to receive the incoming glucose from the blood. **Glucagon** tells the liver to reconvert the glycogen back into glucose and put it back into the bloodstream, raising the blood sugar level.

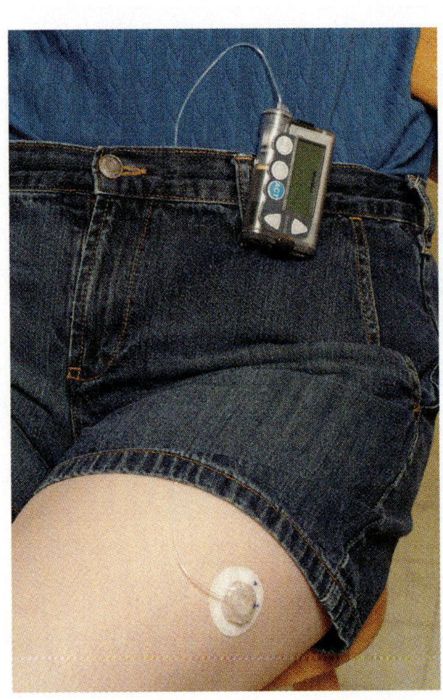

FIGURE 6-4
Insulin pumps, while expensive, allow people with Type 1 diabetes to avoid having to give themselves insulin shots multiple times a day. What types of people would especially benefit from this system of insulin delivery?

What Is Diabetes?

It is important that your blood sugar stay within a certain range, because large fluctuations in blood sugar level damage tissues throughout your body. The condition where the body cannot regulate blood sugar adequately is called **diabetes mellitus**. Diabetes is treatable with diet, physical activity, and medication, but untreated or poorly managed, it can lead to blindness, seizures, kidney failure, heart disease, and even amputation of limbs, coma, and death. Diabetes is rampant in the United States; according to the Centers for Disease Control, approximately 23.6 million people, or 7.8% of the population, have this disease.

There are two types of diabetes: Type 1 and Type 2. **Type 1 diabetes**, which used to be known as juvenile-onset diabetes, affects 10% of the people who have diabetes. It occurs when the body cannot produce enough insulin. When a Type 1 diabetic eats a meal, the pancreas cannot produce the insulin needed to stabilize the blood sugar. The treatment for Type 1 diabetes is daily insulin injections. People with Type 1 diabetes use a **blood glucose meter** to test their blood sugar levels several times a day; then they inject themselves with insulin and eat a balanced diet. The diet and the insulin dosage are developed to meet the needs of the individual patient.

BY THE NUMBERS

Because blood sugar fluctuates throughout the day depending on what you eat, diabetes is most often diagnosed by doing a fasting blood sugar test—in other words, drawing blood after at least 8 hours of no food.

Normal range for a fasting blood sugar test is 70 to 100.

A level of 100 to 125 indicates that the person may have pre-diabetes, or insulin resistance.

A level of 126 or higher is consistent with either Type 1 or Type 2 diabetes.

A glucose tolerance test is also used to diagnose blood sugar diseases. The patient fasts for several hours, then is given a load of glucose (usually in the form of a super-sweet syrupy drink), and blood is drawn. The draws are done at various intervals and the blood glucose is determined. The results are plotted on a graph to see how the patient's blood glucose has responded to the load.

Type 2 diabetes is far more common. In this type, the pancreas still produces insulin normally, but the body cells become less responsive to the insulin. Type 2 diabetes develops slowly over time. In the pre-diabetic phase, the body develops insulin insensitivity, sometimes due to obesity. This makes the pancreas work harder to secrete more insulin. In an insulin-resistant body, there is a high level of insulin circulating in the blood, but the cells are resistant to the message of "open up, here comes the glucose." Eventually, the pancreas may reduce insulin production, either partially or totally, worsening the disease.

Hypoglycemia occurs when blood sugar is too low. People with diabetes can develop hypoglycemia if they inject too much insulin for the amount of food they eat. In addition, people who do *not* have diabetes can develop:

- **Reactive hypoglycemia:** Occurs when the pancreas overreacts, secreting too much insulin after a high-carb meal. Symptoms appear one to three hours after a meal, and may include nervousness, sweating, headache, and rapid or irregular heartbeat. People with this type of hypoglycemia should eat small meals frequently to keep their blood insulin and glucose levels stable.
- **Fasting hypoglycemia:** Occurs when the pancreas keeps secreting insulin even when you are not eating. This is usually caused by another medical condition, such as a pancreatic tumor, cancer, or liver infection or disease.

Cool Tips

Carbohydrates are the main dietary source of glucose. Rice, potatoes, bread, cereal, milk, fruit, and sweets are all carbohydrate-rich foods.

How Lifestyle and Nutrition Impact Diabetes

The most significant risk factor in developing Type 2 diabetes is obesity. Losing weight can decrease your chance of developing diabetes, and can reduce or eliminate its symptoms if you already have it. Exercise also plays an important role; moderate daily exercise can prevent the onset of Type 2 diabetes, especially when combined with dietary changes.

The ideal diet for a diabetic is very similar to the ideal diet for any healthy person, except that it should contain a slightly lower percentage of carbohydrates. In addition, people with diabetes or pre-diabetes (insulin resistance) may benefit from eating a low glycemic index diet, as explained in the next section.

Understanding the Glycemic Index

Have you ever eaten a large amount of a sugary snack and felt a quick energy spike, followed by sluggishness and lethargy? Simple sugars are very similar to glucose, so they don't require much processing to turn into blood sugar. It happens quickly, dumping a lot of glucose into the blood at once. To counteract this, the pancreas reacts by dumping a big load of insulin into the blood. When the sugar hits your blood, you get a burst of energy; when the insulin hits it, you "bottom out." In contrast, a food containing protein, fat, or complex carbohydrates takes longer for the body to convert to glucose, so the release into the bloodstream is more gradual; therefore, neither the blood sugar nor the insulin levels swing wildly.

The **glycemic index** is a measurement of a food's potential to create that quick spike-and-crash cycle. Foods that have a high glycemic index value are not necessarily bad for you, but they are somewhat more stressful on the body's blood sugar regulation system. Foods with a lower glycemic index value cause less stress on the pancreas and less fluctuation in blood sugar. Low-index foods also generally contain more fiber and can increase the level of HDL ("good" cholesterol) in the blood.

Hot Topics

In a normally functioning body, nearly any amount of sugar can be processed efficiently—either by being used as energy or converted to fat stores—with no impact on the pancreas. Diabetes results from a disruption in the pancreas's functionality, not from sugar consumption.

The Exchange System is a system of meal planning created by the American Diabetes Association (ADA) for diabetics, but it can be used by anyone for healthful eating. It groups foods into lists of nutritionally equivalent items, and users create their daily meal plans by selecting a certain number of foods from each list. An **exchange** is a serving, choice, or substitution from a list. Within each list, one exchange is approximately equal to another in its calories, carbohydrates, proteins, and fats.

The Exchange System categorizes foods into six groups: Starch, Fruit, Milk, Vegetables, Meat/Meat Substitutes, and Fat. Each group has its own distinct values for number of calories per serving as well as grams of carbohydrate, protein, and fat. The person's diet might specify, for example, five exchanges a day from the Starch group. They could be any five foods from the list, such as ½ cup oatmeal, 1 slice of bread, or ½ cup of corn or peas. A registered dietitian or nutritionist with special training in diabetic diets calculates the number of exchanges in each category to be assigned per day and provides a list of foods allowed for each exchange. For more information about the Exchange System, including lists of foods in each category from the ADA, see www.diabetes.org.

A food's glycemic index value is based on a comparison to pure glucose. Sugar's value is the highest—100. Foods that the body processes more slowly than that receive a correspondingly lower value. Table 6-2 lists the glycemic index values of an assortment of foods. A value of 70 or greater is considered high.

TABLE 6-2: GLYCEMIC INDEX VALUES

Food	Value
Glucose	100
Potatoes (plain)	85
Cornflakes	83
Jelly beans	78
Doughnut	76
Graham crackers	74
Soda (non-diet)	72
Corn chips	72
White bread	70
Raisins	64
Table sugar	62
Ice cream	61
Potato chips	56
White rice	56
Bananas	52
All-Bran cereal	51
Carrots	47
Orange juice	46
Grapes	46
Baked beans	44
Spaghetti	44
Oranges	42
Apples	38
Skim milk	32
Whole milk	30
Kidney beans	28
Grapefruit	25
Plain yogurt	14

FIGURE 6-5
Many people are surprised that potatoes are one of the highest glycemic index foods because they do not taste sweet. Based on Table 6-2, can you draw any generalizations about the characteristics of high-index versus low-index foods?

Some experts say the glycemic index is not a useful tool for food consumers because simple tables of numbers for individual foods, such as in Table 6-2, are not relevant to the way people actually eat. The glycemic index is based on consuming 50 grams of carbohydrate in the food item, all by itself, on an empty stomach. Hardly anyone really eats that way, and foods are combined in a meal, the values change. It is more relevant to look at the glycemic index of the entire meal, but calculating that value is beyond most people's capability.

Another issue with the glycemic index is that it measures food based on a fixed amount of carbohydrates—and not on typical portion size. Fifty grams of carbohydrates in a heavy, dense food might not even be one serving, but 50 grams of carbs in a lightweight food may be multiple servings.

Fiction	Fact
Foods with a low glycemic index value are "good carbs" and do not increase body fat the way high-value foods do.	The glycemic index is just a tool for estimating how quickly and dramatically a food will affect your blood sugar level; it has limited value in determining whether a food is "good for you" or not. For example, notice in Table 6-2 that a plain potato has a much higher index value than potato chips, but that doesn't mean potato chips are less likely to increase body fat.

Other Nutrition-Related Diseases

In Chapter 1, we briefly looked at how vitamin deficiencies or overdoses can cause certain diseases and disorders in the body. There are many other ways in which nutrition can affect the diseases and disorders you get (or don't get!). Here is a sampling of those.

Obesity

As you learned in Chapter 5, being overweight or obese can cause or contribute to a wide variety of health problems, including heart disease, some types of cancer, joint and circulation problems, and Type 2 diabetes.

Malnutrition

Malnutrition is a broad term that refers to any type of bad nutrition. It can be a deficiency (**undernutrition**) or an excess (**overnutrition**). The symptoms vary depending on what is missing or in excess. Eating too few calories can cause fatigue, confusion, and a destruction of the lean body tissues as the body breaks them down for fuel. Eating too many can cause obesity. In addition, the absence or excess of certain vitamins and minerals can cause specific diseases.

- **Beriberi:** Weakened heart, wasting, and paralysis caused by vitamin B1 (thiamine) deficiency.
- **Ariboflavinosis:** Sore throat, cracked lips, moist scaly skin on genitals, magenta-colored tongue, and decreased red blood cell count, caused by vitamin B2 (riboflavin) deficiency.
- **Pellagra:** Sensitivity to sunlight, aggression, dry and reddened skin with sores, insomnia, weakness, mental confusion, and diarrhea, caused by vitamin B3 (niacin) deficiency.
- **Anemia:** Low red blood cell count, caused by vitamin B6, B9, B12, or iron deficiency, or possibly vitamin E deficiency in infants.
- **Scurvy:** Paleness, depression, spongy gums, bleeding from mucous membranes, caused by vitamin C deficiency.
- **Rickets:** Softening of the bones in children, caused by vitamin D deficiency.
- **Goiter:** An enlarged thyroid gland caused by an iodine deficiency.
- **Osteoporosis:** A loss of bone mass, so that the bones are fragile and break easily. It is caused by a variety of genetic, dietary, hormonal, exercise and lifestyle factors.
- **Decreased taste acuity:** An alteration in the way food tastes, caused by a zinc deficiency.
- **Tooth decay:** A predisposition to cavities caused by excessive sugar intake. Dental fluoride treatments and fluoride in the water or in a supplement can help strengthen teeth. Too much fluoride, however, can harm teeth. Talk to your dentist.

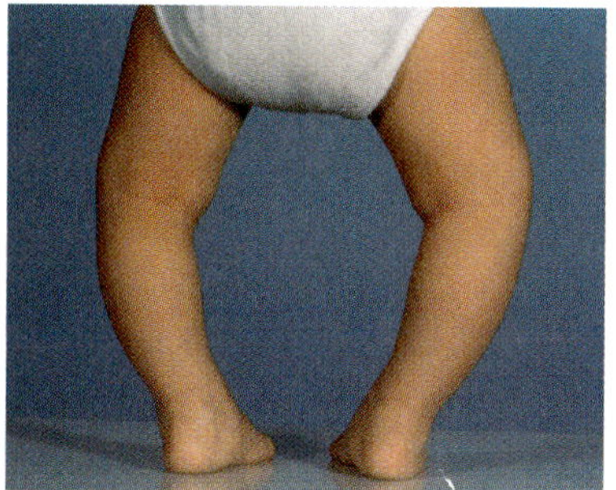

Rickets

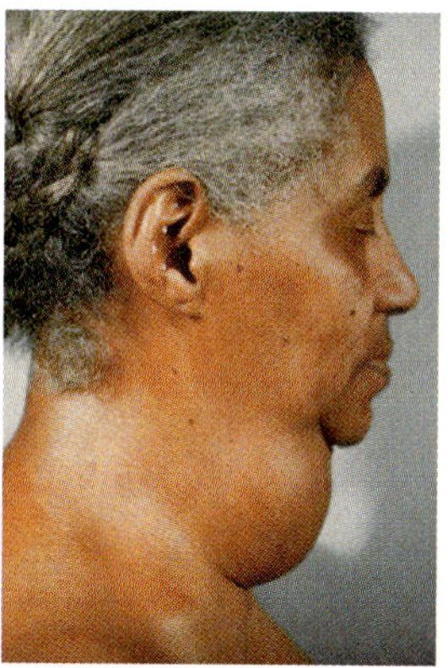

Goiter

Osteoporosis

FIGURE 6-6
Malnutrition-based diseases and disorders can often be cured by simply supplementing the diet to include enough of the missing nutrient.

Cancer

Scientists are still studying the role of nutrition in cancer prevention and treatment, but there appear to be some very likely connections; for example:

- Antioxidant foods, which you learned about earlier in this chapter, may lower your risk of several types of cancers.
- Breast cancer seems to appear more frequently in overweight women.
- People who eat a high-fiber diet rich in fruits, vegetables, and whole grains may have a lower risk of developing colon cancer.
- People who are heavy alcohol drinkers have a higher risk for cancers of the esophagus, mouth, throat, colon, and breast, as well as cirrhosis (scarring) of the liver.
- Some research suggests that a high-fat diet, heavy in red meat, may increase the risk of breast, colon, rectum, and prostate cancers. Conversely, a diet rich in fruits and vegetables is associated lowering the risk of cancer.
- There is some evidence that getting enough selenium may help protect against prostate cancer.

People who counsel individuals with eating disorders typically have a strong background in psychology (up to a Ph.D in the field), as well as nutrition training. Most successful treatment programs are conducted by a team of professionals. Some of the professionals who help people overcome eating disorders include:

- **Psychologists:** Experts in human psychology, usually with a Masters degree or Ph.D., these professionals help patients work through mental problems that may be causing the eating disorder.
- **Psychiatrists:** Medical doctors with extra training in psychology. Because they are M.D.s in addition to being psychologists, they can prescribe psychiatric medications such as antidepressants that may help with treatment.
- **Psychotherapists:** These counselors typically have advanced degrees in social work or psychology, and provide individual as well as group advice and counseling.
- **Dietitians:** Dietitians are experts in nutrition with a college degree in nutrition, a hospital internship, and often a Master's or a Ph.D. They are an integral part of the team that treats eating disorders of all types.
- **Physicians:** Doctors of various specialties, such as endocrinology, cardiology, gastroenterology, gynecology and other M.D.s are part of the team to help treat the effects of eating disorders.

Eating Disorders

Most of the advice about nutrition offered in this book assumes that you are able to make rational choices about your eating. However, for people with food addictions or eating disorders, this may not be possible. People who suffer from these disorders feel compelled toward unhealthy behaviors, including binge eating, purging (vomiting) after eating, or eating too little to maintain a healthy body weight. Eating disorders are both psychologically and physically unhealthy.

What Is an Eating Disorder?

There's a difference between an **eating disorder**—a diagnosable mental health condition—and **disordered eating**—any situation in which a person is eating an unhealthy diet, usually in the short term. For someone to be diagnosed with an eating disorder, their condition and behavior must meet specific diagnostic criteria outlined by the American Psychiatric Association's (APA) *Diagnostic and Statistical Manual of Mental Disorders (DSM-IV).*

Having an eating disorder is not an all-or-nothing condition; there's a continuum (a range) of behaviors, and a person can easily slip from simply having disordered eating to an actual eating disorder. Some of the factors that contribute to the development of eating disorders include:

- Pressure from family and friends to be thin
- Low self-esteem or feelings of inadequacy
- Depression, anger, anxiety, or loneliness
- Exposure to unrealistic media images, depicting overly thin models
- Chemical imbalances that control hunger, appetite, or digestion

Chronic dieting is a common form of disordered eating, but it is not necessarily an eating disorder. With chronic dieting, the person consistently and successfully follows a controlled, restrictive diet to maintain an average or below-average body weight. People who diet chronically may be at risk for poor health and nutrition, and may participate in another type of disordered eating, **yo-yo dieting**. As you learned in Chapter 4, yo-yo dieting, or **weight cycling**, occurs when a person repeatedly loses and gains weight.

Anorexia Nervosa

Anorexia nervosa is a mental condition in which the person develops intense fear of gaining weight, causing them to use unhealthy practices to achieve a body weight of less than 85% of what it should be. Most people with this disease are young females. Anorexia nervosa is a very serious illness, with a high mortality rate. According to the APA, it develops in 0.5 to 1% of U.S. females, and between 5 and 20% of these will die from it.

FIGURE 6-7
Anorexia nervosa most often affects young women. Why do you think young women are more susceptible to this disease than men or older women?

People who have anorexia nervosa are so fearful of being fat that they develop an unrealistic perspective of their weight; they think they are fat when they are—in fact—very thin. They may fast completely, eat only minimal calories per day, or eliminate certain food groups. As a result of the too-low body weight, they may stop having menstrual periods, or, if they have not begun menstruation yet, they may not begin at all.

Anorexia nervosa is so deadly because the body shuts itself down due to starvation. When there is no fat left to burn for energy, the body turns on itself, burning muscle mass and even organ tissue. In addition, people with anorexia nervosa are at great risk for heart problems, including irregular heartbeats, rapid heart rate, low blood pressure, and even heart failure and death. There can also be gastrointestinal problems, such as irritable bowel syndrome, constipation, and stomach pain.

Hot Topics

The organization Overeaters Anonymous deals with food addiction in the same way that related organizations like Alcoholics Anonymous and Narcotics Anonymous do with alcohol and drug addictions. People in this program follow a 12-step model for treatment and recovery. For more information, see www.oa.org.

Bulimia Nervosa

Bulimia nervosa is an eating disorder in which the person eats an abnormal amount (binge eating), and then vomits or uses laxatives to expel it from their system before it is digested, or indulges in excessive exercise to burn off the calories. During the binge portion of the cycle, the person feels unable to control their eating; then during the purge portion, they feel great remorse that leads them to the purge behaviors.

Bulimia nervosa is even more common than anorexia nervosa; it affects between 1 and 4% of women. Many people who have bulimia nervosa also have anorexia nervosa. People who have only bulimia, however, are less likely to die from it; only 1% of those with bulimia die within 10 years of diagnosis.

Bulimia is characterized by recurrent episodes of binge eating, followed by compensating behavior such as vomiting, on a regular and ongoing basis. The health risks include electrolyte imbalance (leading to heart problems), stomach and esophagus inflammation and ulceration, tooth erosion, staining and decay, mouth sores, and swelling of the cheeks or jaw area.

Binge Eating Disorder/Food Addiction

Binge eating disorder occurs when someone has the first part of bulimia nervosa—the binge eating—but not the purging part. Individuals who have it tend to be obese, and may describe their relationship with food as an **addiction**. Between 1 and 2% of the population suffers from this disease. However, it's hard to estimate because not everyone who has it seeks medical or psychological assistance.

How Eating Disorders Are Treated

Treatment for eating disorders is multi-faceted. A variety of behavioral and psychological treatments can be used. Treatment is considered a "success" when harmful behaviors cease, and the patient learns new ways to deal with the emotions or psychological events that trigger the behavior in the first place. Some facets of treatment may include:

- Individual counseling or psychotherapy
- Group counseling sessions
- Behavior modification
- Cognitive/behavioral therapy
- Nutrition study
- Relaxation activities, such as yoga, meditation, or exercise
- Medications (especially in patients who also suffer from depression or anxiety)
- Inpatient care and monitoring (especially in extreme cases of anorexia nervosa, where the patient's weight is dangerously low)

FIGURE 6-8
Group counseling is often one of the treatments for an eating disorder. What do you think the benefits would be of a group counseling situation for an eating disorder, as opposed to one-on-one advice? What are the drawbacks?

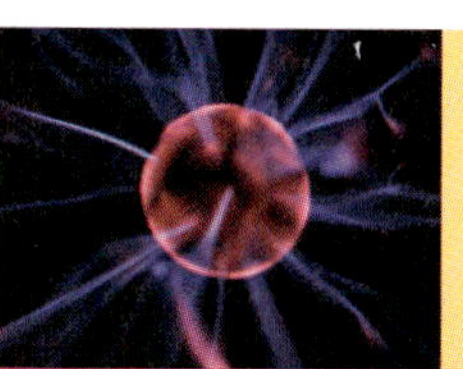

SCIENCE STUDY

When preparing for international travel, especially to Mexico, one piece of advice that many people hear is "Don't drink the water!" Have you ever wondered why the water in one part of a country can be perfectly okay for the natives to drink, but can make a tourist violently ill?

It all has to do with the human body's miraculous ability to adapt to its environment. When a body is repeatedly exposed to a certain pathogen, such as E. coli for example, it develops a certain amount of immunity to it. It is not clear exactly how much exposure is needed to develop immunity or how long it lasts, but it appears that it takes much longer to build up immunity (7 years in one study) than it does to lose it (8 weeks in one study).

Managing Food Allergies and Intolerances

Different people's bodies react differently to certain foods, for a variety of reasons. Some people are born with—or later develop—allergies to certain ingredients; other bodies experience digestive side effects when they eat certain foods, such as milk or wheat. These allergies and sensitivities can be temporary or chronic. As you are planning menus and meals for other people, it is helpful to understand some of the most common allergies and intolerances, so you can think about how you can work around them, if necessary.

Food Allergies

When someone has a food allergy, the body produces antibodies in response to a chemical, (usually a specific protein molecule) in that food which acts as an antigen. When you eat that food, the antigen is recognized, the immune system is activated, histamine is released, and physical symptoms result, such as swelling, difficulty in breathing, skin rashes or hives, intestinal upset (such as diarrhea and cramping), or vomiting. In extreme cases, such as with peanut allergies, the person can even develop anaphylactic shock, a condition in which the blood pressure drops and the breathing becomes shallow. If medical treatment is not obtained immediately, the person can even die from it.

FIGURE 6-9
Because peanut allergies can be so severe, many schools have banned peanuts and peanut butter ingredients in their lunch programs, and most airlines no longer give peanuts as snacks. Do you think schools should offer peanut butter as an option for the students who are not allergic to it, or do you think it is better to not provide it at all? Why?

Foods that many people are allergic to include:

- Wheat
- Eggs
- Chocolate
- Shellfish
- Peanuts
- Tree nuts (nuts that grow on trees)
- Milk

Some people have food allergies as children and then outgrow them; conversely, some people develop food allergies over time, and may not realize they are allergic to the food until well into adulthood. The best way to combat food allergies is to simply avoid that food in your diet. If you accidentally do eat some of the food, a mild reaction can usually be addressed by taking an antihistamine such as Benadryl. More severe reactions will require medical care and it may be necessary for the individual to always carry a syringe of epinephrine, prescribed by a physician, to self-administer if a reaction begins.

Lactose Intolerance

Food intolerance is different from food allergy. If you are intolerant of a certain food, your body does not digest it properly or produces unwanted side effects in its digestion. One of the most common food intolerances is **lactose intolerance**.

Lactose is a sugar found in milk and most milk-based products. A person with lactose intolerance has a deficiency of lactase, an intestinal enzyme. As a result, lactose is not sufficiently digested. Normal bacteria found in the intestine utilize this undigested sugar with gas production as a byproduct. This is what causes that "bloated feeling" you see on TV advertisements. In addition, the undigested lactose draws water into the small intestine causing diarrhea.

Cool Tips

Soy and rice milk are good alternatives for someone who is lactose intolerant. In addition, cow's milk now comes in a lactose-free form at most grocery stores.

Interestingly, it seems that there is a genetic basis for this condition. In some populations, lactase production continues throughout their entire lives. Approximately 15% of Caucasians develop lactose intolerance, usually in middle age, while 80 to 90% of the African American and Asian populations develop this condition to some degree. To avoid this situation, individuals must either avoid milk and other non-cultured or non-fermented dairy products or take an oral form of the enzyme, lactase, before consuming such products. There are also products that can be added to milk to pre-treat it for a lactose intolerant person. These individuals can usually consume yogurt, cheese, acidophilus milk, buttermilk, sour cream and other cultured or fermented dairy foods.

Gluten Intolerance

Gluten is a protein component in wheat, rye, barley, and oats, and is found in almost all pastas and breads. It is not found in rice or corn. Gluten intolerance is less common than lactose intolerance, but still prevalent; an estimated 1 in 7 people are affected. It is difficult to diagnose because the onset of the symptoms is sometimes delayed after eating the food. Some of the symptoms include:

- Stomach bloating, cramping, diarrhea, and flatulence (gas)
- Poor resistance to infection, mouth ulcers, and arthritis
- Skin rashes, eczema, psoriasis, and itching flaky skin

Some people who are sensitive to gluten have a disease known as **celiac sprue**, in which gluten causes damage to the small intestine. However, the majority of gluten-sensitive people are Non-Celiac Gluten Sensitive (NCGS) and simply suffer unpleasant after-effects from eating gluten.

Case Study

Amy is worried that her friend Patrice may have an eating disorder. Patrice eats a lot, but is very thin, and never seems to gain any weight. Patrice always goes to the bathroom right after a meal. Most disturbingly, recently when Patrice spent the night at Amy's house, the next morning an entire pan of brownies they had left in the kitchen had been eaten. Patrice claimed she knew nothing about it, but nobody else except Amy's mom was home at the time, and she doesn't like brownies.

- Do you think Amy has grounds to be concerned?
- What eating disorder may Patrice possibly have?
- What would be a tactful way in which Amy could approach Patrice on this issue?
- Do you think Amy should say something to Patrice's parents?

Put It to Use

1. People often die from the effects of chronic over-consumption of alcohol on their bodies. What specific harmful effects does alcohol have on the body? Research this as needed and write up your findings.

Put It to Use

❷ A friend says that potato chips are a good stress-reducing food. Based on what you have learned about stress and nutrition, do you think he is right? Why or why not?

Write Now

Pick one of the nutrition-related diseases in the section on Malnutrition and write a 2–3 page paper that describes the disease in more detail than presented in the chapter. Discuss the prevalence of the disease today, the countries in which it is generally found, the prognosis for someone who gets it, and the foods would help prevent it.

Tech Connect 

Many good resources are available online to provide people with eating disorders with information and support. Compile a list of at least 10 Web sites that would be helpful for someone with an eating disorder, and write a 1–2 sentence description of each one explaining what it offers.

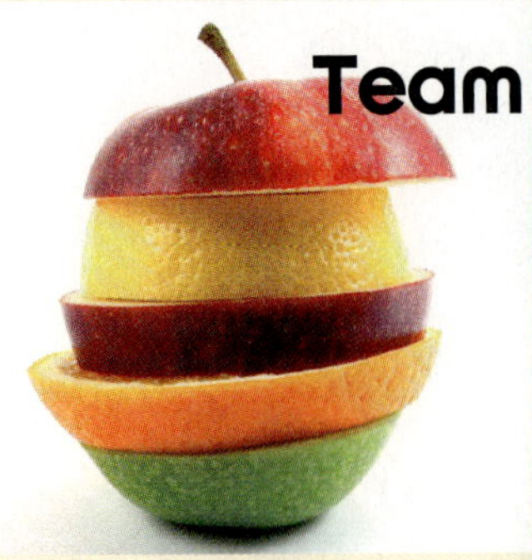

Team Players

In teams of two or more, put together two meal plans:

1. **A menu for a meal that is very high in antioxidants. Use as many ingredients from Table 6-1 as possible, while still keeping a balance among carbohydrates, proteins, and fats. For each item on the menu, indicate which antioxidant(s) it supplies.**
2. **A meal that has very low glycemic index value. Use as many ingredients that have an index value of 60 or less as possible; refer to Table 6-2 as needed, or other sources of index values online or in books. Because the meal is combining foods, the values will not be exact when the meal is consumed, but choose lower-value foods for an overall low effect.**

Put It Together

Match the explanation in column 1 with the term in column 2.

Column 1

- **a.** a drug found in coffee, tea, and soft drinks, causing your body to release adrenaline
- **b.** an atom with an odd number of electrons
- **c.** a chemical that minimizes the damage from free radicals
- **d.** one of several types of antioxidants
- **e.** a blood glucose regulation hormone
- **f.** a disease that results when the pancreas does not produce enough insulin
- **g.** a measurement of a food's potential to affect blood sugar
- **h.** an enlarged thyroid gland caused by an iodine deficiency
- **i.** an eating disorder in which not enough calories are consumed
- **j.** a sensitivity to the sugar found in milk products

Column 2

1. anorexia nervosa
2. antioxidant
3. beta-carotene
4. caffeine
5. diabetes
6. free radical
7. glycemic index
8. goiter
9. insulin
10. lactose intolerance

Part 2

Kitchen Basics

7 Following and Modifying Recipes

In This Chapter, You Will . . .

- Learn how a recipe is structured
- Get tips for finding good recipes
- Understand ingredient quantities and recipe scaling
- Find out what ingredients can be substituted for one another
- Understand cooking methods

Why You Need to Know This

Suppose a friend has a birthday coming up, and has hinted that he would really like to have a made-from-scratch chocolate cake. How would you go about finding a recipe for it? What ingredients and utensils do you think you would need? Brainstorm as a class to construct a list of the possible ingredients and cookware you will need. Then find a recipe in a cookbook or an online database and compare your list to the one in the recipe to see how close you came.

Anatomy of a Recipe

A **recipe** is a written set of directions for making a certain dish. Recipes ensure that a dish is consistently prepared, no matter who is doing the preparation. Recipes can also provide guidance for making dishes with which you are not familiar.

Most recipes contain at least the following information:

- **Title:** A descriptive name for the dish
- **Yield:** The measured output, such as number of servings and serving size
- **Ingredient list:** A list of the items and quantities you will need to make the recipe
- **Method:** Chronological steps detailing how to prepare the recipe

Some recipes also include additional information, such as:

- **Equipment required** (appliances, cookware, hand tools)
- **Nutrition facts** per serving, including calories, protein, fat, and carbohydrates
- **Preparation time** required
- **Tips or hints** on preparation technique
- **Safety tips** that can help prevent spoilage or contamination
- **Service** suggestions, such as garnishing or plating instructions or temperature recommendations

FIGURE 7-1
This girl is following a recipe in a cookbook. Have you ever cooked from a cookbook? What other sources does your family use for recipes?

Figure 7-2, on the next page, shows a sample recipe for blueberry muffins. There will be recipes for you to try in later chapters that follow this same format, so familiarize yourself with the different parts.

Cool Tips

The HACCP designation in Figure 7-2 stands for Hazard Analysis Critical Control Point, a system for maintaining food safety. You'll learn more about it in Chapter 8, "Sanitary Food Handling and Food Safety."

Finding Recipes

Recipes can be found in a wide variety of places, from the Internet to your grandmother's cookbooks and recipe card boxes. You can look to:

- **Magazines:** Many home and family-oriented magazines have recipe sections in each issue, and some even have recipe contests. There are also entire magazines devoted to cooking techniques and recipes.
- **Cookbooks:** Your local bookstore or online bookseller has hundreds, perhaps even thousands, of cookbooks available for nearly every type of cuisine imaginable. Your local library may also have the cookbooks you want.

FIGURE 7-2

A sample recipe. Which types of information are included in this recipe?

TRY IT!

Blueberry Muffins

Yield: 1 Dozen Muffins   Serving Size: 1 Muffin

Ingredients

16 oz (3¾ cups) All-purpose flour (plus 2 Tbsp to coat berries)
1½ tsp Double-acting baking powder
½ tsp Salt
¼ tsp Nutmeg, ground
4 oz (½ cup) Butter at room temperature
8 oz (1 cup) Sugar
1 Egg, large
6 fl oz Milk
½ tsp Vanilla extract
1 cup Blueberries, washed and patted dry
Optional Cooking spray

Equipment

- Appliances: Oven, standing mixer with paddle attachment
- Cookware: Muffin tins, paper muffin tin liners, cooling rack
- Hand Tools: Scale (*optional*), measuring cups and spoons, sifter, mixing bowls, whisk, rubber spatula, 2-oz scoop

Method

1. Preheat the oven to 400° F.
2. Line muffin tins with paper liners or spray them lightly with cooking spray.
3. Sift together 16 oz flour with the baking powder, salt, and nutmeg.
4. Blend the milk, egg, and vanilla extract in a separate bowl.
5. In a standing mixer with a paddle attachment, cream together the butter and sugar until very light and smooth, about 2 minutes.
6. Add the flour mixture in three additions, alternating with the liquid ingredients, mixing on low speed and scraping down the bowl to blend the batter evenly.
7. Increase the speed to medium and mix until the batter is very smooth, another 2 minutes.
8. Mix 2 Tbsp flour with berries to coat them evenly.
9. Fold the blueberries into the batter, distributing them evenly.
10. Fill each muffin cup ⅔ full with batter using the 2-oz scoop.
11. Bake until the top of the muffin springs back when lightly pressed, 18 to 20 minutes.
12. Cool the muffins in the muffin pan on cooling racks for 5 minutes. Then remove them from the muffin pan and finish cooling them on the rack.

Serve warm or at room temperature. If desired, remove paper liner from muffin before serving. Store in an air-tight container with lid.

Recipe Categories

Muffins, Breakfast Foods, Blueberries

Chef's Notes

Coating blueberries with flour keeps them suspended in the batter so they don't all fall to the bottom of the muffin.

Shake baking powder before using. Ingredients can separate and need to be mixed for muffins to rise properly.

Potentially Hazardous Foods

Egg
Milk

HACCP

Keep cold ingredients chilled below 41° F.

Nutrition

Calories	195
Protein	3 g
Fat	9 g
Carbohydrates	26 g
Cholesterol	39 mg

- **Family and friends:** The cooks in your family may have recipes written down that they can share with you, or favorite cookbooks with pages marked. If not, see if you can get someone to teach you how to make some of the family's special dishes, and then write it down yourself to preserve it.
- **Web sites:** Hundreds of Web sites, both large and small, offer free databases of recipes. Many Web sources also have reviews for each recipe, so you can read about other people's experiences with them. Here are a few to get you started:

  www.epicurious.com

  www.allrecipes.com

  www.foodnetwork.com

  www.recipesource.com

  www.cooking.com/Recipes-And-More

FIGURE 7-3
Cookbooks provide expert-tested, easy-to-follow recipes. Identify some unusual specialty cookbooks you have seen or heard of.

Reading a Recipe and Planning Ahead

It's important to read a recipe through completely before you begin preparing it. That way you can make sure that you have all of the appropriate ingredients and equipment on hand. Some of the things to consider when reading over a recipe include:

- **Ingredients:** Do you have all the ingredients you need, in the right quantities? If not, plan for substitutions, or for a trip to a store.
- **Yield:** Will this recipe yield adequate portions? If not, you may need to change the quantities of each ingredient to make a double or half recipe. You'll learn how to do this math later in the chapter.
- **Method:** Do you know how to perform all of the techniques prescribed in the Methods section? For example, in Figure 7-2, Step 5 says you should "cream together" the butter and sugar. Do you know what that means? (If not, see Table 7-8 on page 149.)
- **Timing:** Are you going to be able to complete this recipe in time for the meal or event for which you are preparing? At what time should you start it? At what time does it need to be finished? Do you need to pre-heat the oven or prepare any ingredients in advance?
- **Serving and Holding:** Do you have the appropriate serving equipment, such as plates, bowls, or platters? Do you have the garnishes, sauces, or other accompaniments you need?

Measuring Ingredients

The amount or quantity of each ingredient is clearly spelled out in a recipe; however, the units of measurement may differ between recipes. In this section, we'll look at some common ways that ingredients are measured and described in recipes.

There are two ways of describing the quantity of an ingredient—by weight or by volume (that is, how much space it occupies). The two are related, but not the same, because not all ingredients are equally dense.

A **fluid ounce (fl oz)** is a volume measurement describing how much space one ounce of liquid occupies. When measuring liquids, an ounce and a fluid ounce are the same quantity, but when measuring any other type of ingredient, they may be different due to the differing mass of the substance. Most recipes list measurements by volume, rather than by weight; measuring cups and spoons are much more common in the kitchen than scales. The exception is meat, which is almost always measured by weight in ounces or pounds.

Safe Eats

Never put food directly onto a scale; always use a tray, container, or paper or plastic wrap or barrier.

A **cup** is also a volume measurement; one cup and 8 fluid ounces are two ways of describing the same amount. In the metric system, a **liter** is a standard volume measurement; one liter is slightly more than 4 cups. Rather than fluid ounces, the metric system of measurement uses **milliliters (ml)**. Table 7-1 lists some common measurements by volume for both U.S. and metric systems.

Although less common, some recipes list ingredients by weight. Table 7-2 lists some common weights.

TABLE 7-1: MEASUREMENTS BY VOLUME

Abbreviation	Stands For	U.S.	Metric
tsp	Teaspoon	⅙ fl oz	5 ml
Tbsp	Tablespoon	½ fl oz 3 tsp	15 ml
1 fl oz	Fluid ounce	2 tbsp 6 tsp	30 ml
c	Cup	8 fl oz 16 tbsp	< ¼ liter
pt	Pint	16 fl oz 2 cups	< ¼ liter
qt	Quart	32 fl oz 2 pints	< 1 liter
gal	Gallon	128 fl oz 4 quarts	> 3¾ liters

TABLE 7-2: MEASUREMENTS BY WEIGHT

U.S. SYSTEM

Abbreviation	Stands For	U.S. Equivalent	Metric Equivalent
oz	Ounce		28 grams
lb	Pound	16 ounces	454 grams

METRIC SYSTEM

Abbreviation	Stands For	U.S. Equivalent	Metric Equivalent
mg	Milligram		0.001 grams
g	Gram	1/28 ounce	
kg	Kilogram	2.2 pounds	1,000 grams

FIGURE 7-4
Measuring dry ingredients. Why is it important to level off the top of the measuring cup with a straight edge, rather than allowing the ingredient to be heaped in the cup?

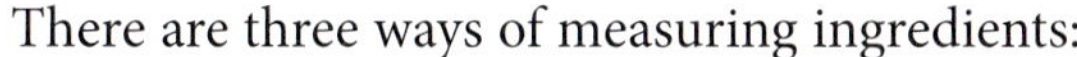

There are three ways of measuring ingredients:

- **Dry volume:** Use a flat-top measuring cup or spoon. Overfill the measuring container, and then tap it lightly to make sure the ingredient has settled and to eliminate any air pockets. Scrape off any excess from the top with a flat object, such as the dull edge of a knife. See Figure 7-4.
- **Liquid volume:** Use a clear measuring cup with measuring lines on it. Set it on a flat surface and then fill to the desired mark, while reading at eye level. See Figure 7-5.
- **Weight:** Set an empty container on the scale and adjust it to account for the weight of the container. (The weight of the container is called the **tare weight**.) Then, fill the container until the scale shows the desired weight.

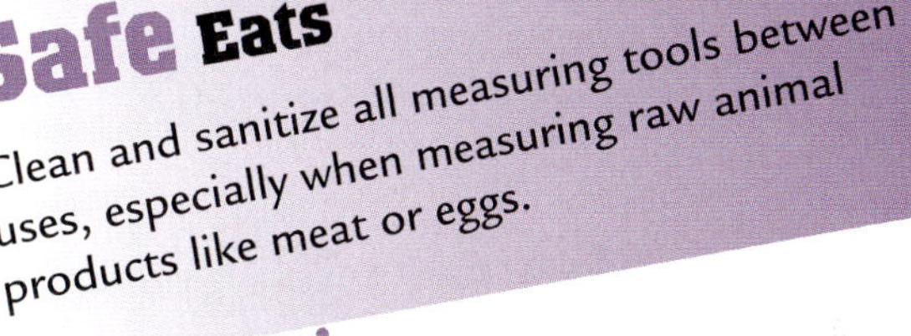

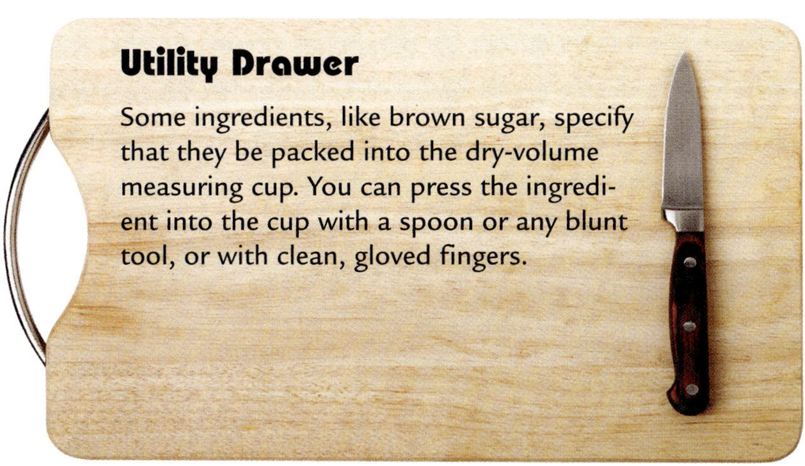

FIGURE 7-5
Measuring liquid ingredients. Why is it important to check the liquid when the cup is at eye level, rather than looking up or down at it?

Adjusting Ingredient Quantities

If you need to make more or less food than the recipe calls for, you must perform some math calculations to **scale** the ingredient quantities. It seems like a simple enough task, to double or halve each quantity, but many cooks have ruined a dish by forgetting to double a few ingredients in an otherwise doubled batch!

To scale a recipe up or down, find the **recipe conversion factor** (RCF) with the following formula:

$$\text{RCF} = \frac{\text{quantity or yield you want}}{\text{quantity or yield of the original recipe}}$$

For example, suppose you want to fix some boiled rice. The original recipe calls for:

5 cups white rice

2½ tsp salt

10 cups water

The yield is 20 ½-cup servings, which is 10 cups of cooked rice. Suppose you only need 5 cups of cooked rice. You could calculate the RCF as follows:

$$\text{RCF} = \frac{5}{10} = \frac{1}{2} \text{ or } 0.5$$

Then you would multiply the quantity of each ingredient by 0.5 (or divide it by 2, whichever is easier for you), to get the new quantities:

5 × 0.5 = 2.5 cups of white rice

2½ × 0.5 = 1¼ tsp salt

10 × 0.5 = 5 cups of water

Suppose you need 40 cups of cooked rice? Calculate the RCF like this:

$$\text{RCF} = \frac{40}{10} = 4$$

Then, you would multiply the quantity of each ingredient by 4 to get the new quantities:

5 × 4 = 20 cups of white rice

2½ × 4 = 10 tsp salt

10 × 4 = 40 cups of water

You can do this math with any type of measurement—cups of food, ounces of food, or even number of servings. For example, let's look back at the recipe for blueberry muffins in Figure 7-2. The original recipe serves 12 people. Suppose that you need to serve 30 people:

$$RCF = \frac{30}{12} = 2.5$$

Therefore you would multiply the quantity of each ingredient by 2.5. For example, in this recipe for blueberry muffins, instead of 3¾ cups of flour, you would use 9.375 cups of flour. (That's about 9⅓ cups.)

Not all ingredients are easily multiplied, so some approximation may be necessary. For example, if the recipe now calls for 2.5 eggs, you could do any of the following:

- Use 3 eggs, keeping in mind that the extra ½ egg may change the texture of the dish slightly.
- Beat 1 egg separately from the others, and then use half of that beaten mixture for ½ an egg.
- Use 3 eggs, but use smaller eggs. (If a recipe does not say what size of eggs to use, you can assume it means large eggs.)

Hot Topics

Some recipes do not turn out as well when they are scaled up or down by a factor of more than 2. When performing a large-scale RCF on a recipe, adjustments may be needed to the equipment and/or the cooking method. For example, when beating egg whites, you may have better results beating a few at a time in a series of small batches, rather than in one big bowl. Some spices and seasonings cannot be scaled up or down by the RCF, or the flavor will be too strong or too faint. This requires some practice, and maybe the advice of an expert.

Here's a look at the full conversion for the recipe at an RCF of 2.5.

Ingredient	Original Quantity	RCF Adjusted Quantity
Flour	3¾ c	9⅓ c
Baking powder	1½ tsp	1 Tbsp plus ¾ tsp
Salt	½ tsp	1¼ tsp
Nutmeg	¼ tsp	½ tsp plus ⅛ tsp
Butter	½ c (1 stick)	1¼ c (2½ sticks)
Sugar	1 c	2½ c
Egg (large)	1	2 eggs plus ½ an egg
Milk	6 fl oz (¾ cup)	15 fl oz (slightly less than 2 c)
Vanilla extract	½ tsp	1¼ tsp
Blueberries	1 c	2½ c

Many restaurants serve portions that are much larger than the "recommended serving." For example, when serving a rice or noodle casserole where the serving size in the recipe is 1 cup, and the MyPyramid serving size is ½ cup, a restaurant might serve 3 or even 4 cups in a single order. If you were the owner or manager of a restaurant, you might need to do an RCF conversion to account for this.

FIGURE 7-6
Not all sugar substitutes are the same, either in potency or in how the substance changes when used in cooking/baking. Based on Table 7-3, which sweeteners could be used when baking cookies?

Ingredient Substitutions and Omissions

You may sometimes need to make ingredient substitutions or omissions for a variety of reasons. For example, in a recipe that calls for peanuts, you might need to substitute a different nut because of a guest's peanut allergy, or you might choose to eliminate nuts altogether. For a low-fat diet, you might decrease the amount of oil in a brownie recipe, or substitute applesauce for part of the oil. (Yes, that really does work!) And sometimes, you might need to make a substitution simply because you are out of a certain ingredient, like an herb or spice.

Any substitution will change the taste of the food. Your diners may or may not notice the difference, depending on how familiar they are with the original dish and how sensitive their tastes are.

Using Sugar Substitutes

For people following a low-sugar diet, you can sometimes use sugar substitutes in recipes. However, be aware that some sugar substitutes break down and lose their sweetness when baked at high temperatures. It is better to add sweetener during the last few minutes of heating or cooking, if possible. Table 7-3 summarizes the sugar substitutes available in the United States.

TABLE 7-3: SUGAR SUBSTITUTES

Sugar Substitute	Calories	Ratio	Notes
Sweet One® (Acesulfame-K)	4 calories per packet	1 packet = 2 teaspoons sugar 12 packets = 1 cup sugar	Can be used in cooking and baking without losing sweetness.
Equal® (Aspartame)	4 calories per packet	1 packet = 2 teaspoons sugar 24 packets = 1 cup sugar	Loses sweetness when baked at high temperatures for a long time.
Equal Spoonful®	2 calories per teaspoon	1 teaspoon = 1 teaspoon sugar 1 cup = 1 cup sugar	Loses sweetness when baked at high temperatures for a long time.
Splenda® (Sucralose)	0 calories	1 teaspoon = 1 teaspoon sugar 1 cup = 1 cup sugar	May not work well in some recipes that rely on sugar for structure. Finished dishes may require refrigeration.
Sweet'N Low® (Saccharin)	4 calories per packet	1 packet = 2 teaspoons sugar 12 packets = 1 cup sugar	Can be used in cooking and baking without losing sweetness.
Brown Sweet'N Low® (Saccharin)	20 calories per teaspoon	1 teaspoon = ¼ cup brown sugar 4 teaspoons = 1 cup brown sugar	Can be used in cooking or baking without losing sweetness.

Fiction	Fact
Saccharin causes cancer in humans.	Back in the 1970s some studies showed a link between saccharin and cancer in rats, and the FDA ruled that saccharin-containing products must carry a warning label and recommended that it be banned entirely from the market. Since then, however, additional research has failed to show a link between saccharin and cancer in humans, and in 2000 the FDA reversed its ruling and recommendation. Saccharin is now considered to be safe for humans.

Trimming the Fat in a Recipe

As you learned in Chapter 5, one of the best ways to lose weight is to trim the amount of fat in your diet. In many recipes, you can simply cut down the amount of fat called for by 25 to 30% without much change in the food's flavor and with only minimal change to the texture. You can also substitute a lower-fat ingredient for part of the fat. See Table 7-4 for some ideas.

TABLE 7-4: FAT SUBSTITUTES

Instead of	Use
1 cup of butter, shortening, or oil (in baked goods)	1 cup applesauce
8 ounces of cream cheese	8 ounces of yogurt cheese (made by straining the excess liquid out of plain yogurt), or light or fat-free cream cheese
1 cup heavy cream (for cooking, not for whipping)	2 teaspoons cornstarch or 1 tablespoon flour whisked into 1 cup nonfat milk
1 cup sour cream	1 cup fat-free sour cream, or 1 cup low-fat cottage cheese plus 2 tablespoons skim milk plus 1 tablespoon lemon juice
1 cup fat for sautéing	4 cups low-fat stock, fruit juice, or wine, cooked down to 1 cup
1 whole egg	2 egg whites
1 cup oil or fat for basting	1 cup fruit juice or low-fat stock
1 ounce baking chocolate	3 tablespoons cocoa powder
1 cup whole-milk ricotta cheese	1 cup low-fat (1%) cottage cheese, or ½ cup whole milk ricotta plus ½ cup either part-skim ricotta or low-fat (1%) cottage cheese
1 cup whipped cream	3 stiffly beaten egg whites, or 1 cup evaporated skim milk, whipped
Sausage	Low-fat ground turkey, or lean ground beef
Mozzarella cheese (whole milk)	Part-skim mozzarella cheese

Avoiding Alcohol in Cooking

Some people prefer not to consume food that contains alcohol, even in trace amounts that have no effect on the body. Recovering alcoholics, for example, may not want food prepared with wine or liqueurs. Table 7-5 lists some substitutions you can make when alcohol is unavailable or off-limits.

TABLE 7-5: ALCOHOL SUBSTITUTES

Ingredient	Use This Instead
Amaretto	Almond extract
Beer	Non-alcoholic beer For light beer: Chicken broth, ginger ale, or white grape juice For dark beer: beef, chicken, or mushroom stock
Brandy	The corresponding fruit juice (for example, for cherry brandy, use cherry juice), brandy extract
Chambord®	Raspberry juice or syrup
Champagne	Sparkling white grape juice or ginger ale
Cognac	Peace, apricot, or pear juice
Cointreau®	Orange juice concentrate
Crème de menthe	Spearmint extract or oil mixed with water or grapefruit juice

Herb and Spice Substitutions

Herbs and spices give many recipes their distinctive flavors, but it may be difficult to keep a full set of them on hand. They also lose their flavor and potency over time, so you can't just buy a full spice rack and expect the spices to still be good after many years.

Therefore, you may sometimes need to substitute one herb or spice for another when preparing a recipe. The overall taste of the dish will most likely be good if you use a complementary substitution. Table 7-6 provides substitution suggestions for many of the most common herbs and spices. If no quantity is given, assume a one-to-one ratio of the original versus the substitute.

FIGURE 7-7
At $40 or more an ounce, saffron is the most costly spice. It is most often used in rice dishes to give the rice a bright yellow hue. If you didn't have saffron, how might you make a rice dish appear yellow?

TABLE 7-6: HERB AND SPICE SUBSTITUTES

Ingredient	Use This Instead
Allspice, 1 tsp	½ tsp cinnamon and ½ tsp ground cloves, OR ½ tsp cinnamon, ¼ tsp ground cloves, and ¼ tsp nutmeg
Aniseed, 1 tsp	1 tsp fennel seed or a few drops of anise extract
Basil	Oregano or thyme
Cardamom	Ginger or cinnamon
Chinese Five Spice, 1 tsp	¼ each of crushed anise seeds, cinnamon, ground cloves, ginger
Chives	Green onions
Cilantro	Parsley
Cinnamon, 1 tsp	¼ tsp nutmeg or allspice
Cloves	Allspice, cinnamon, or nutmeg
Cumin	Chili powder
Curry powder, 1 Tbsp	½ tsp cardamom, ½ tsp cayenne, ½ tsp coriander seed, ½ tsp cumin, ½ tsp ginger, and ½ tsp turmeric
Fennel seeds	Caraway seeds
Ginger, 1 Tbsp	1 tsp allspice, cinnamon, mace, or nutmeg OR ⅛ tsp powdered ginger
Italian seasoning, 2 Tbsp	1 tsp each of basil, oregano, marjoram, rosemary, sage, and thyme
Mace	Allspice, cinnamon, ginger, or nutmeg
Marjoram	Basil, thyme, or savory
Mint	Basil, marjoram, or rosemary
Nutmeg	Cinnamon, ginger, or mace
Oregano	Thyme, basil, or marjoram
Parsley	Cilantro
Poultry seasoning, 1 tsp	¾ tsp sage, ¼ tsp thyme OR ¾ tsp sage, ¼ tsp thyme, ⅛ tsp ground cloves, and ¼ tsp pepper
Pumpkin Pie Spice, 1 tsp	½ tsp ground cinnamon, ¼ tsp ground ginger, ⅛ tsp ground allspice, and ⅛ tsp ground nutmeg OR ½ tsp ground cinnamon, ⅛ tsp ground nutmeg, ⅛ tsp ground mace, ⅛ tsp ground ginger, and ⅛ tsp ground cloves
Red pepper	Dash of bottled hot sauce
Rosemary	Thyme, tarragon, or savory
Saffron, ¼ tsp	1 tsp turmeric (substitutes for color, not for flavor)
Sage	Poultry seasoning, savory, marjoram, or rosemary
Savory	Thyme, marjoram, or sage
Seasoned salt, 4 tsp	2 tsp salt, ½ tsp sage, ½ tsp parsley flakes, ½ tsp onion powder, ¼ tsp marjoram, and ¼ tsp paprika
Tarragon	Fennel seed or aniseed
Thyme	Basil, marjoram, oregano, or savory
Turmeric	Dry mustard

FIGURE 7-8
Sometimes a recipe will specify using an electric mixer; other times you can choose between a mixer and a hand tool, such as a whisk or spatula. What are the advantages and disadvantages of using an electric mixer?

Cutting Up and Combining Ingredients

In each step of a recipe's Method section, you are directed to perform a particular action upon one or more of the ingredients. These actions can involve cutting, combining, or applying heat or cold to the ingredients.

Let's start by looking at actions that don't involve heat or cold. Almost every recipe involves one or both of the following actions:

- **Cutting up an ingredient:** chopping, slicing, or puréeing
- **Combining ingredients:** blending, sifting, or folding

Most methods are easy to accomplish, but you must know what each of the words means in standard recipe terminology. For example, what's the difference between blending and folding? How are chopping and dicing different? In the following sections, we'll examine the various terms and what they mean.

Cutting Up Ingredients

Whole ingredients often need to be cut or divided into smaller pieces. Different words are often used to denote both the style of the separation and the size and shape of the pieces. For example, a *sliced* tomato is different from a *diced* tomato. Table 7-7 summarizes the various actions for creating pieces from a whole item.

TABLE 7-7: TECHNIQUES FOR CUTTING OR SEPARATING FOOD

Action	Meaning
Chop	To cut into small pieces; recipe may specify the size (e.g., finely chopped or coarsely chopped)
Core	To remove the center or core, usually from a fruit
Dice	To cut into square cubes of uniform size, usually about ¼-inch on each side
Grate	To shred or flake, using a grater
Hack	To cut up bone-in meat with a cleaver, so that the bone is cut as well as the meat
Julienne	To cut into very thin strips
Mince	To chop into very small pieces
Peel or pare	To remove the skin, usually from a fruit or vegetable
Puree	To blend or sieve so finely that it becomes a fine-textured paste or liquid
Score	To cut narrow grooves or gashes in the fat of a piece of meat
Shave	To cut wide, paper-thin slices, usually of meat, cheese, or chocolate
Shred	To cut into slivers or slender pieces
Slice	To cut into slices
Zest	To finely grate the colored peel of a fruit, usually citrus

Combining Ingredients

Almost all recipes involve combining ingredients. This combining can take place in various ways—different speeds, different degrees of blending, different amounts of force applied to the mixing process. It's important to follow the recipe's recommendation for best results. For example, in Figure 7-2, the recipe for blueberry muffins instructs you to "fold" ingredients together. This means to very gently and carefully mix them. In this recipe, you need to be careful not to be too rough with the blueberries or they will disintegrate and leak their juice into the batter, resulting in a purple muffin batter. Table 7-8 lists some of the actions you may be instructed to take to combine ingredients.

TABLE 7-8: TECHNIQUES FOR COMBINING INGREDIENTS

Action	Meaning
Beat	To stir a mixture until smooth with a regular, hard, rhythmic movement
Blend	To mix ingredients until smooth and uniform
Coat	To roll foods in a coating (flour, nuts, sugar, etc.) until all sides are evenly covered
Combine	To mix ingredients together (generic)
Cream	To rub, whip, or beat with a spoon or mixture until soft and fluffy; usually refers to combining butter and sugar
Cut in	To mix shortening with dry ingredients using a pastry blender, knife, or fork; usually applies to pastry crust
Dissolve	To mix a dry substance into a liquid
Dot	To scatter small amounts of an ingredient on top of food; for example, you might dot a pie filling with butter before putting on the top crust
Dredge	To sprinkle, coat, or cover with flour, crumbs, or cornmeal
Drizzle	To slowly pour a liquid or glaze in a fine stream over the top of a food
Dust	To sprinkle a food or coat lightly with flour, sugar, cornmeal, or cocoa
Fold	To gently combine ingredients by cutting vertically through the mixture and sliding a spatula or whisk across the bottom of the mixing bowl with each turn
Glaze	To cover with a thin sugar syrup or melted fruit jelly
Knead	To work and press dough with the heels of your hands, so the dough becomes stretchy and elastic
Marinate	To let food stand in a liquid mixture, usually an acid-oil mix; often used with meat or vegetables
Mix	To stir until ingredients are thoroughly combined
Punch down	To deflate yeast dough after it has risen; punch your fist in the center of the dough, then pull the edges toward the center
Scallop	To arrange ingredients in layers in a casserole dish with a sauce or liquid
Scramble	To stir or mix food gently while cooking, such as eggs
Sift	To put dry ingredients through a fine sieve
Stir	To mix, usually with a spoon or fork, until ingredients are worked together
Toss	To tumble ingredients lightly with a lifting motion, as with a salad
Whip	To rapidly beat eggs, butter, or cream in order to incorporate air and expand the volume
Whisk	To beat ingredients with a fork or a wire whisk to mix, blend, or whip

When you prepare food for dry cooking, you can take steps to combat the drying effect of the heat. For example, you can dust food with flour to help it retain its inner moisture as it cooks. Food such as meat or vegetables that you plan to grill or broil can be soaked in oil, flavorful liquids, herbs, and spices before cooking to add moisture. You can also coat food in a batter or bread it before frying it. One of the best ways to maintain moisture in food is to avoid overcooking it.

Cooking Techniques

Applying heat to food is generically known as **cooking**. The term "cooking" is sometimes applied to recipes that do not involve applying heat, but this is not strictly accurate. Cooking changes the texture, taste, and appearance of food, and in some cases, also changes its nutritional value. (For example, raw vegetables may have different amounts of some nutrients than cooked ones.)

There are many ways of cooking food. These can be broadly divided into two groups:

- **Dry heat methods:** The food is cooked by hot air, hot oil, or a hot pan. Examples include baking, broiling, and frying.
- **Moist heat methods:** The food is cooked via water or other liquid (other than oil), either directly by immersion in the liquid or indirectly by steam from it. Examples include boiling, steaming, and poaching.

Dry Heat Methods

In the dry heat methods, heat is transferred, or conducted, into the food in one of the following ways:

- By rays that radiate from a heat source, such as burning coals, flames, or a hot electric element. This is called **radiant heat**.
- By metal that conducts heat from a burner to the food, such as a griddle or frying pan.
- By contact with oil that has been heated by a burner or other heat source.

The heat source in these methods causes the outside of the food to dry out as it cooks. When the surface is dry, it changes color. Often foods prepared using dry heat methods have a golden or deep brown color. As foods brown, the flavor on the outside becomes more intense. The color on the inside of the food also changes as you cook, although not as dramatically as the outside.

Raw foods may not be as digestible as cooked foods. Also, some raw foods contain anti-nutrients like phytic acid or oxalic acid that are deactivated by cooking. The bioavailability of nutrients may depend upon the food being cooked, or on it being raw. Bioavailability is actually more important than the actual nutrient content of the food. If you can't absorb a nutrient and use it, it doesn't matter if it's in the food.

Hot Topics

When food cooks, it loses some of its nutritive value. The longer it cooks, the more value it loses. (One notable exception: cooking improves the bioavailability of lycopene in tomatoes.) Food cooked very quickly with one of the dry heat methods loses relatively few vitamins and minerals. For this reason, some people choose to follow a diet that consists only of raw (uncooked) foods. Others say that the vitamin and mineral loss due to cooking is not significant enough to compensate for the extra time and energy needed to plan and prepare balanced meals that do not involve cooking.

FIGURE 7-9
Cooking onions until their sugar caramelizes makes them golden brown. If the onions are caramelized on the outside, does that mean the insides are done cooking?

Foods that contain sugar change color when sugars on the surface start to brown, or **caramelize**. Protein-rich foods, such as meat, also become brown as they cook; this is known as the **Maillard reaction**.

When the heat comes in contact with the surface of the food, the outer layer of the food stiffens. Sometimes you can see and feel a distinct crust. The crispy skin on a roasted chicken, the crunchy breading on a piece of deep-fried fish, and the crisp outer layer of a French fry are all examples of how dry heat methods change the texture of a food. Eggs, meats, fish, and poultry all become firm as they cook. Other food may become softer; onions, for example, change from a crisp texture to a very soft, almost melting texture when cooked.

There are eight basic dry heat methods of cooking:

Grilling is a dry heat method in which food is placed on a rack for cooking. Grilled foods have a robust, smoky taste. The heat source is located below the rack holding the food; it can be charcoal, gas, wood, or an electric or infrared heading element. The radiant heat from the source heats up the metal in the rack to cook the food (creating the dark grill marks that signify that the food was cooked on a grill rack). The radiant heat also cooks the parts of the food that are not in direct contact with the rack.

Broiling is similar to grilling except the heat source is located above the food. When you put food into a broiler, it cooks from the top. The heat in a broiler is typically a gas flame or an electric or infrared heating element.

Baking is a dry heat technique in which food is cooked by hot air trapped inside an oven. As the hot air comes in contact with the food, the surface of the food heats up and dries out. Eventually the surface takes on a deeper color, and the food texture changes as it goes from raw to cooked. Meats, fish, and poultry tend to become firmer as they cook, while vegetables and fruits become softer. Baking is also used for most cakes, pies, cookies, and other flour-based sweets.

Baking in high altitudes (above 5,000 feet above sea level) can require some minor adjustments to the recipes. The three basic rules are:

- Reduce baking powder. For each teaspoon, decrease by ⅛ to ¼ tsp at 6,000 feet or by ¼ tsp for 7,000 feet or higher
- Reduce sugar. For each cup, decrease by ½ to 2 Tbsp at 6,000 feet, or by 1 to 3 Tbsp for 7,000 feet or higher
- Increase liquid. For each cup, add 2 to 4 Tbsp at 6,000 feet, or 3 to 4 Tbsp for 7,000 feet or higher

Keep in mind that every recipe is different and any or all of these adjustments may be required. Keep notes of how you adjust recipes until you know what works best for your particular location.

Cool Tips

When you sauté food, you should let the pan heat up first, before adding any oil. Chefs refer to this step as **conditioning the pan.** Once the pan is hot, you can add oil. The oil will heat up very quickly, so you can start cooking right away. If you add food to a cold pan with cold oil, it will stick to the pan and your food will absorb more oil, altering its taste.

The heat in an oven is not as intense as the heat generated by a grill, but it still may be too hot for delicate food. One way to control oven heat is to put the food in a pan or baking dish and then set that pan in a larger pan. Then, add enough water to the larger pan to come up around the sides of the smaller pan. This is known as baking foods in a **water bath**. Because water can heat only to 212° F, it insulates and protects the food. A water bath is used to create a creamy, smooth consistency in the food, such as in crème brulee.

Roasting is basically the same thing as baking except that it generally refers to a whole item or a large piece of food. For example, you would roast a whole chicken, but you would bake cut-up chicken parts. When meats are roasted or baked, they are sometimes seared before being placed in the oven. Searing is a type of sautéing, and is covered in the next section. As the food roasts, you can **baste** it to keep it moist. To baste, you spoon or ladle the juices from the bottom of the pan onto the top and sides of the food. After the food has roasted or baked, let it **rest** a few minutes before serving, to allow time for the juices to redistribute, moving back to the outer parts of the food so it is more tender.

Sautéing is a technique that cooks food quickly, often uncovered, in a very small amount of fat in a pan over high heat. Food that is suitable for sautéing is typically quite tender and thin enough to cook in a short time. Food is often coated with seasoned flour before sautéing. Sautéed foods are cooked primarily through contact with the hot pan. The fat helps to keep the food from sticking, and can add flavor if you choose a tasty fat, such as butter or olive oil. Turn sautéed food halfway through cooking. Resist the temptation to move food around unless it is cooking too quickly or getting too dark.

You can vary the steps in sautéing to produce different effects. There are four important variations of sautéing:

- **Stir frying** is very similar to sautéing, but you use a wok (a pan with a round bottom and high sides). Foods for a stir fry are usually cut into small strips, so they can cook quickly. When stir frying, you constantly stir and toss the food as it is cooking.

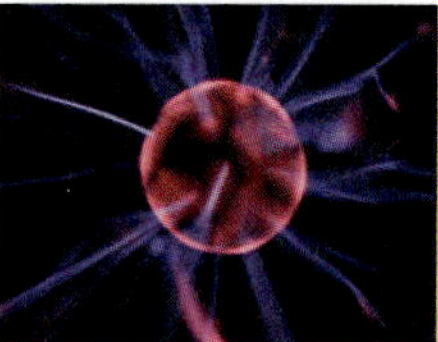

SCIENCE STUDY

Does food really taste different if it is roasted versus broiled versus baked?

For this experiment, you will need two whole chickens. Cut one of them up into pieces; leave the other one whole. Roast the whole chicken at 375° for about one hour, or until done to 165° internally. (Use a meat thermometer.)

At the same time, bake half of the pieces of the other chicken in a glass dish in the same oven, also at 375°, for 50 minutes, or until done to 165° internally.

Then turn the oven to broil and broil the remaining chicken pieces until done (about 20 minutes), turning the pieces frequently so they do not burn.

Sample the results. Do they look different? Taste different? Is the texture of the meat different? Is one more juicy or crispy than another?

- **Searing** means you cook the food in a small amount of hot fat just long enough to color the outside of the food. This can be done to give meat a rich brown color before roasting it. Some foods that can be eaten very rare (almost raw), such as some types of tuna, are also seared.
- **Pan broiling** is very much like sautéing, except that you use no fat. The food is uncovered and cooked over a high heat. Use this method for foods that have a high fat content, such as bacon.
- **Sweating** calls for lower heat than sautéing, searing, or pan broiling. The food, typically vegetables, is cooked uncovered over a low heat in a small amount of fat. The food softens, releases moisture, and cooks in its own juices, but is not allowed to brown. Smothering is a variation of sweating in which the pan is covered.

FIGURE 7-10
This seared tuna is essentially raw in the middle. What meats would not be a safe to eat this way?

Pan frying cooks the food (usually coated or breaded) in hot oil in a pan. The amount of oil you use in the pan is more than that used for sautéing. The oil should be deep enough to come halfway up the sides of the food you are cooking. So, for example, if the food is one inch thick, you need half an inch of oil in the pan. As with sautéing, you turn foods only once as they pan fry. When pan frying, you must heat the oil to the temperature specified in the recipe. Use a thermometer to check the temperature. Most foods are pan fried at about 350° F. If the oil is not hot enough, the food will be pale and will absorb the oil, making it greasy. If the oil is too hot, it will pop and spatter and will cook the outside of the food too quickly. Foods that are most often pan fried are naturally tender and moist. Vegetables, fish, chicken, veal, and pork are common choices.

Pan-fried foods are usually coated before cooking. There are three basic coating options:

- **Seasoned flour:** Coat the food with flour seasoned with salt and pepper.
- **Standard breading:** Dip the food in beaten egg, and then cover it in breadcrumbs or some other coating mixture. It can sit several hours in the refrigerator between breading and frying.
- **Batter:** Coat the food with flour, and then dip it in batter. Fry immediately. There are many recipes for different types of batter, such as beer batter or tempura.

Moist Heat Methods

Moist heat techniques have a built-in temperature control. Most liquids will not rise in temperature much about 212° F, the boiling point of water. This means that food cooked by using a moist heat method will have a different appearance, flavor, and texture than food prepared by a dry heat method.

All foods continue to cook after you take them out of the pan, off the grill, or out of the oven. This process is known as **carryover cooking**. The amount of carryover cooking depends on the size of the food. Big cuts of meat can hold more heat, so they continue to cook longer. The temperature of a food can continue to rise after cooking, anywhere from 2 to 15 degrees. You can't stop carryover cooking, so you need to plan for it. Remove the food from the heat source before it is completely cooked, and let the carryover cooking finish it.

When you cook food with one of the moist heat methods, the food is cooked either through direct contact with a hot liquid or with the steam that rises from it. The heat is conducted from the heat source to the liquid, and then from the liquid to the food.

Because moist heat is fairly low in temperature, the changes to a food's color on the surface are not as dramatic as with dry heat. The color on the outside of the food is often the same as the color on the inside. This plays a role in the way the food tastes. Instead of developing a roasted or caramelized flavor, the foods are often said to have a clean taste.

All the moist heat cooking methods are similar. The major distinctions between them have to do with the foods you choose and the temperatures and amounts of the liquid in which the food is steamed or cooked. There are four basic moist heat cooking methods: steaming, poaching, simmering, and boiling. Additionally there are two important combination methods that require the use of both a dry heat method and a moist heat method. These are braising and stewing.

Steaming cooks the food in a closed pot or steamer. The steam circulates around the food. The food does not come in direct contact with the liquid. Steaming is a gentle technique, and is a good way to retain as many of the food's nutrients as possible. It is a popular technique for preparing many vegetables, but it is also used to prepare tender, delicate meats and fish, such as chicken breasts, whole fish, and shellfish. To add more flavor to foods as they steam, you can add seasonings, flavorings, and aromatics to the steaming liquid. As the liquid heats up, those flavors are released into the liquid and the steam it produces.

Poaching, **simmering**, and **boiling** are almost exactly the same: They all refer to cooking food directly in a hot liquid. The difference is in the degree of heat applied. Poaching is the lowest heat (160° to 180° F). Simmering is medium heat (180° to 200° F). Boiling is high heat (212° F).

FIGURE 7-11
Blanching loosens the outer layer of tomatoes, so that they are easier to peel after being dropped into ice water. What other foods might you blanch to make them easier to peel?

So, how do you determine doneness, when using one of the moist heating methods? That depends on how the food will be used. For example, if you are cooking the food and serving it right away, you will almost always cook the food all the way through. One exception is eggs, which can be cooked to a range of doneness by request. However, if you are using a moist heat method to prepare an ingredient for use in another dish, you won't cook it all the way through. You need to be able to judge when foods are partially cooked.

There are three named degrees of doneness in moist and combination cooking methods:

- **Blanching:** Blanching foods involves cooking them in liquid or steam just long enough to cook the outer portion. You may see a color change; for instance, blanched broccoli becomes bright green. When you use these blanched vegetables, they keep their brilliant colors. Blanching also draws out strong flavors or aromas that might overpower the finished dish. For example, you might blanch a piece of country-cured ham to make it less salty. Blanching loosens the skin of foods such as tomatoes, chestnuts, peaches, and almonds, so they are easier to peel.

To blanch food, bring water to a full boil. Add food directly to the liquid. Let the food cook just long enough for the change you want; then lift the food out and immediately put it in a container of ice water to stop the carryover cooking. Drain the food before you store it or use it in another dish.

- **Parcooking:** This is like blanching, but food is cooked to a greater degree of doneness. Parcooking helps you be more efficient because you only need to cook the parcooked food for the remainder of the time required for the fully prepared food. For example, if it takes 15 minutes to fully cook a food, and you parcook it for 10 minutes, you only need to cook it another 5 minutes at a later time. When you parcook by boiling, it is called **parboiling**.
- **Full cooking:** This food has been cooked all the way through. Be sure to observe the correct temperature for doneness, and remember to allow for carryover cooking.

Appearance is one of the doneness tests you can use, but it is almost always used in combination with another test. When you are only partially cooking food, the tool you use to test doneness is one of the following: a paring knife, a table fork, or a kitchen fork. A parcooked food may be easy to pierce on the outside, but as you continue to push to the center of the food, there is more resistance. When foods are fully cooked, the knife or fork should slide all the way into the food easily. These foods are said to be **fork tender**.

FIGURE 7-12
A crockpot is an excellent piece of equipment for stewing and braising. If you are cooking meat to include in the crockpot meal, be sure to sear it before placing it in the crockpot. In what way does searing improve the dish?

Combination Methods

Braising and stewing are combination cooking methods because they combine a dry cooking method with a moist one. Food is first seared in hot oil (the dry cooking method) to help the food keep its shape as it cooks. This also provides the dish with the rich flavor that develops when food is seared in hot oil. After searing, the food is gently cooked in a flavorful liquid or sauce. When the food is left whole or in large pieces, with enough liquid to partially cover it, it's called **braising**. When the food is cut into smaller pieces and then cooked in enough liquid to cover it, it's called **stewing**. Perfectly braised or stewed foods have a rich, complex flavor and a tender texture.

Braises and stews are usually made from tougher cuts of meat, whole poultry, and firm-fleshed fish or seafood. You can also braise vegetables and beans. Food that is braised must be able to stand up to the long, gentle cooking process without completely falling apart. A good braise or stew has a soft texture. The sauce for a braise or a stew is actually nothing more than its cooking liquid. By the time the food is cooked, it has released a significant amount of flavor and body into the liquid, along with nutrients that may have been drawn out of the food and into the liquid. The result is an intensely flavored, complex sauce.

Case Study

Brenda wants to make a special dinner for her family. At the grocery store, she finds center-cut pork chops at a great price, and buys eight of them to serve as the centerpiece of the meal she will prepare. They are very meaty, well-marbled chops, of average thickness.

- Would a dry or a moist cooking method be more appropriate? Why?
- Which specific cooking technique would you use? Why did you choose that method?
- What side dishes could she serve with the pork chops to make a balanced meal?

Put It to Use

❶ Suppose you are getting ready to prepare a large quantity of shrimp scampi for a restaurant buffet. The recipe calls for 20 pounds of shrimp. However, when you check the refrigerator, you find that there are only 13 pounds of shrimp.

Here is the original ingredient list:

Shrimp, 20 lbs
Minced garlic, 3 cups
Fish stock, 3½ cups
Chopped parsley, 1½ cups
Melted butter, 8 sticks (4 cups)
Chopped green onions, 80
Lemon juice, 1½ cups

Figure out the RCF, and then multiply each ingredient's quantity by that factor to adjust the recipe, so that you can use all the shrimp. You can round amounts to the nearest ¼ cup. Jot down your calculations in the following table.

Ingredient	Original Quantity	RCF	New Quantity
Shrimp	20 lbs		
Melted butter	8 sticks (4 cups)		
Minced garlic	3 cups		
Chopped green onions	80		
Fish stock	3½ cups		
Lemon juice	1½ cups		
Chopped parsley	1½ cups		

Put It to Use

❷ You are going to make lasagna for some friends, but several of them are trying to follow a low-fat diet. Following is the ingredient list from the recipe.

1 pound Italian sausage
¾ pound ground beef
12 lasagna noodles
16 ounces (2 cups) ricotta cheese
1 egg
¾ pound mozzarella cheese, sliced
¾ cup grated Parmesan cheese

Sauce:

½ cup minced onion
2 cloves garlic, crushed
1 (28 ounce) can crushed tomatoes
2 (6 ounce) cans tomato paste
2 (6.5 ounce) cans canned tomato sauce
½ cup water
4 tablespoons chopped fresh parsley
2 tablespoons white sugar
1½ teaspoons dried basil leaves
½ teaspoon fennel seeds
1 teaspoon Italian seasoning
1 tablespoon salt
¼ teaspoon ground black pepper

What substitutions can you make for some of the high-fat ingredients to make it lower in fat, while still preserving the basic taste of the dish?

Write Now

There are many ways of preparing chicken breasts, using dry or moist heat. Pick three different ways, and write a short report comparing and contrasting the cooking methods and the resulting taste and texture of the chicken prepared in each way.

Tech Connect

Many good recipe sources are available online. Compile a list of at least 10 Web sites that contain a good selection of recipes, and write a 1–2 sentence description of each one explaining what it offers.

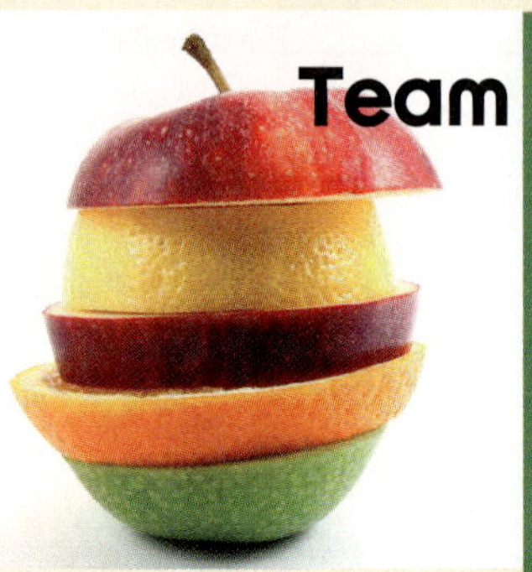

Team Players

Divide the class into two teams.

Team #1 will create a set of bingo cards that list various cooking techniques (from Tables 7-7 and 7-8 and the sections on moist and dry cooking). There should be enough cards for each student to have one.

Team #2 will visit recipe Web sites, pull out recipes at random, and compile a list of action words from the recipes' Methods sections.

Then as a class, with the teacher as the caller, play a bingo game by reading the words at random and filling in the bingo cards.

Put It Together

Match the explanation in column 1 with the term in column 2.

Column 1

a. the volume of one ounce of a liquid
b. a metric measurement of liquid volume
c. the adjustment made to a scale to account for the food's container
d. the amount to multiply a quantity by to adjust a recipe's quantity/scale
e. to cut up an ingredient into very thin strips
f. to finely grate the colored peel of a lemon
g. to sprinkle, coat or cover food with flour, crumbs, or cornmeal
h. a dry heat method that cooks with burning flame or hot coals
i. to cook food in hot liquid at a temperature between 160 and 170° F.
j. to partially cook food in a liquid or steam, just enough to cook the outside

Column 2

1. blanch
2. dredge
3. fl oz
4. julienne
5. liter
6. poach
7. radiant
8. RCF
9. tare
10. zest

8 Sanitary Food Handling and Food Safety

In This Chapter, You Will . . .

- Recognize the types of food-borne contaminants and related illnesses
- Learn how grooming, hygiene, and attire affect food preparation safety
- Clean and sanitize surfaces and equipment correctly
- Identify ways to minimize and control pest infestations in a kitchen
- Understand the procedures for purchasing, receiving, and storing food
- Learn how to safely thaw frozen food, and how to cool hot foods for storage
- Identify procedures for holding and reheating food
- Become familiar with the commercial food inspection process and HACCP

Why You Need to Know This

When dining at a restaurant, you probably take for granted the safety of the food preparation, handling, and storage. Most of the time, this trust is well-founded, because the local health department's mandatory inspection process makes sure all restaurants follow established guidelines. What are some of the basic things that restaurant employees do to ensure the food safety? For example, what safety attire do they wear? How do they keep equipment clean? How do they regulate the temperature of food as it waits to be prepared or served?

Foodborne Contaminants and Related Illnesses

Safe foods are foods that won't make you sick or hurt you when you eat them. Unsafe foods—that is, foods that have been contaminated by various hazardous materials—can make you sick. An illness that results from eating unsafe food is referred to as a **foodborne illness**.

When you are serving food to family and friends, of course you want them to have a safe experience. When restaurants serve food, though, the stakes are higher. Customers who get sick could sue the restaurant, which could generate bad publicity that might even result in its closing. The health department could take away the restaurant's food-handling license as well.

There are three potential hazards that can contaminate food and produce foodborne illnesses: biological, physical, and chemical.

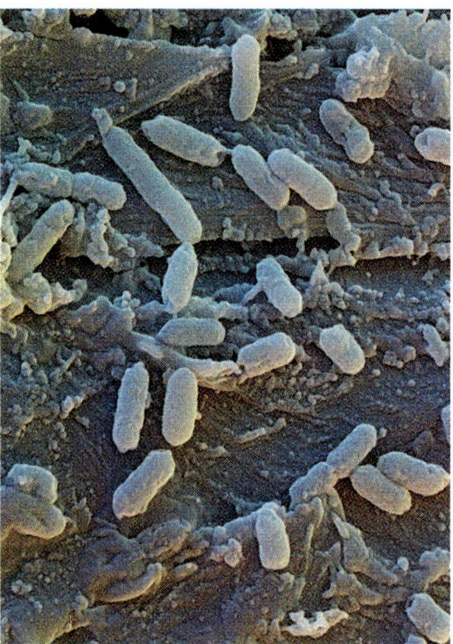

FIGURE 8-1
Bacteria are microscopic organisms, shaped as spheres, spirals, or rods. They are found in most uncooked food and on any surfaces that the food touches. What are some common kitchen activities that can spread bacteria?

Biological Hazards

Biological hazards (also called **pathogens**) are living organisms that exist in or on foods. There are four basic types of pathogens:

- **Bacteria** are single-celled organisms that can live in or on food, water, skin, or clothing. Not all bacteria will make you ill. If there is only a very small amount of bacteria in or on food, you may not get sick. However, a contaminated food contains a great many bacteria. It is the volume of bacteria in a contaminated food that makes you sick.

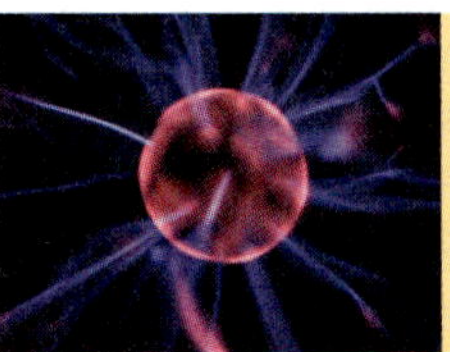

SCIENCE STUDY

You can see the bacteria in your kitchen with a common microscope. If possible, borrow one from your school's biology department, or take some samples from your kitchen classroom to the biology lab. Here are some suggestions for things to look at:

- **A sponge that has been used to wipe off countertops**
- **The plastic wrapper from a package of chicken or other poultry product**
- **A well-worn wooden cutting board with knife marks in its surface**
- **A scraping from underneath your fingernails**

Ask your school's biology teacher to help you identify what you saw.

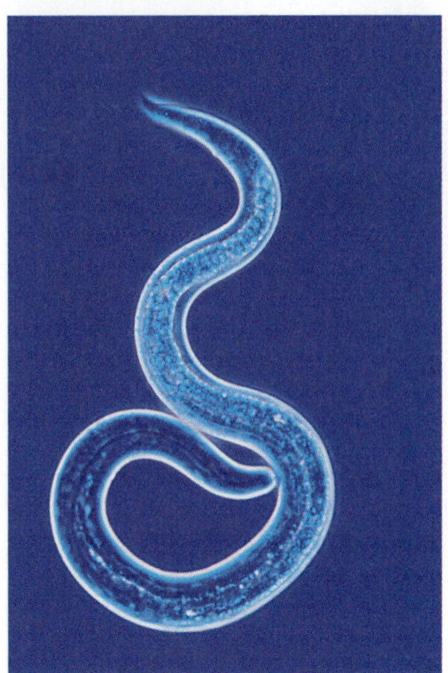

FIGURE 8-2
The most common parasite in the kitchen is the roundworm trichinella spiralis, found in some pork and game meats, which causes trichinosis. Why is it important to cook pork to complete doneness so that there is no pink color in the center?

- **Viruses** invade living cells, including those in foods. Once a virus invades a cell, it tricks the host into making another virus and the process continues. The living cell is known as the host for the virus. A virus needs a host in order to reproduce.
- **Parasites** are organisms with two or more cells that are larger than either bacteria or viruses. Some are actually large enough to see without a microscope. Similar to bacteria, they reproduce on their own. But similar to a virus, they need a host to provide a home and nourishment. Parasites include roundworms, tapeworms, and various insects. When we eat food that contains parasites, the eggs or larvae take up residence in our bodies.
- **Fungi** (the plural of fungus) can be single-celled or multi-celled organisms. Mold and yeast are examples of fungi that you can find in foods. We actually rely on some molds and yeasts to produce foods, such as cheese or bread. However, harmful molds can contaminate foods. As a fungus grows and reproduces, it creates by-products, including various toxins, alcohols, and gases that can cause foodborne illness or foodborne intoxication.

Physical Hazards

If you find a hair, a piece of food packaging, a bandage, or a piece of metal or glass in your food, you've found a physical hazard. Physical hazards are foreign objects, usually large enough to see or feel while you are eating. They are often responsible for injuries, such as cuts or chipped teeth.

Chemical Hazards

Cleaning compounds, bug sprays, food additives, and fertilizer are all examples of man-made chemical hazards. Any of these products, if not used properly, can contaminate food. Symptoms from eating chemically contaminated food can often be felt immediately, and might include hives; swelling of the lips, tongue, and mouth; difficulty breathing or wheezing; and vomiting, diarrhea, and cramps.

Another chemical hazard involves toxic metals. Mercury and cadmium are toxic metals that have found their way into our food and water, often as a result of industrial pollution. The effects of these toxic metals can range from subtle symptoms to serious diseases.

Safe Eats

Don't overlook water as a potential source of pathogens. Food preparation uses water for cleaning the food, washing hands, cleaning equipment and dishes, and blending ingredients. If your water supply is not safe, the food you prepare will not be either.

Common Illnesses Caused by Foodborne Hazards

When you ingest a harmful substance, your body reacts to it—but not necessarily immediately. Some pathogens, especially bacteria, must **incubate** (develop and grow) in your body for anywhere from a few hours to as much as 50 days. Therefore, the relationship is not always clear between eating a meal and getting sick. When someone gets sick from a foodborne pathogen, it is often generically referred to as **food poisoning**, but the exact cause of the illness might be associated with one of many different pathogens, and might come from either food or water. According to the Centers for Disease Control, these are the most common foodborne illnesses:

- **Campylobacter** is a bacterial pathogen that causes fever, diarrhea, and abdominal cramps. It is the most commonly identified bacterial cause of diarrhea illness in the world. These bacteria live in the intestines of healthy birds, and most raw poultry has campylobacter on it. Eating undercooked chicken, or other food that has been contaminated with the juices from raw chicken, is the most frequent source of this infection.
- **Salmonella** is also a bacterium that is widespread in the intestines of birds, reptiles, and mammals, and is often present in raw eggs. It can spread to humans via a variety of different foods of animal origin. The illness it causes, salmonellosis, typically causes fever, diarrhea, and abdominal cramps. In persons with poor health or weakened immune systems, it can invade the bloodstream and cause life-threatening infections.
- **E. coli** is a bacterial pathogen found in cattle and other similar animals. Human illness typically follows consumption of food or water that has been contaminated with microscopic amounts of cow feces. The illness often causes severe and bloody diarrhea and painful abdominal cramps, without much fever. In 3 to 5% of cases, a complication called hemolytic uremic syndrome (HUS) can occur several weeks after the initial symptoms; this may result in temporary anemia, profuse bleeding, and/or kidney failure.
- **Norwalk virus** (calcivirus) is a very common cause of foodborne illness, although rarely diagnosed because the laboratory test is not widely available. It causes acute gastrointestinal illness, usually with more vomiting than diarrhea, that resolves within two days. Unlike many foodborne pathogens that have animal reservoirs, it is believed that Norwalk-like viruses spread primarily from one infected person to another. Infected kitchen workers, who might have the virus on their hands, can contaminate a salad or sandwich as they prepare it.

Hot Topics

Recently, there has been concern about the safety of using animal manures as fertilizers for vegetables and fruits. Cow manure is rich in nutrients and makes an excellent fertilizer, however, it is also a well-known source of food-borne pathogenic bacteria.

There are other common diseases, usually transmitted by alternate routes, which are occasionally foodborne. These include infections caused by *Shigella*, hepatitis A, and the parasites, *Giardia lamblia* and *Cryptosporidia*. Even strep throat has been occasionally transmitted through food.

In addition to diseases caused by direct infection, some foodborne illnesses are caused by the presence of a toxin produced by a microbe living in the food. For example, the bacterium *Staphylococcus aureus*, can grow in some foods and produce a toxin that causes intense vomiting.

The rare but deadly disease **botulism** occurs when the bacterium *Clostridium botulinum* grows and produces a paralytic toxin in foods. These toxins can cause illness (blurred vision, cramps, diarrhea, breathing difficulty, nerve damage, and even death, 12 to 36 hours after exposure) even if the microbes that produced them are no longer there. Botulism most often comes from improperly canned low-acid foods. Botulism poisoning is no longer common in the United States,

Cool Tips

The chemical Botox that some people have injected into their faces to temporarily minimize wrinkles is actually a form of botulism. In fact, the word Botox is a combination of the words "botulism toxin." The toxin paralyzes the muscles in that part of the face so that the areas relax and the wrinkles smooth out. The person may look a bit younger, but their faces take on a lack of expression.

Fiction	Fact
It is not safe to eat raw fish, such as in sushi and sashimi, because of the risk of ingesting parasites.	Many people enjoy sushi and sashimi (raw fish slices) all over the United States every day, with no ill effects. Sushi-grade fish may even be less risky to eat than other fish because of the stringent regulations and standards applied to it. The FDA's guidelines for retailers who sell fish intended to be eaten raw include freezing fish at -31° F for 15 hours or -4° F for 7 days to kill any parasites in it. If you plan on making sushi at home, make sure you buy "sushi-grade" or "sashimi-grade" fish. Such fish has been properly frozen to kill parasites. Regular fish out of your grocer's case has not been properly frozen.

because fewer people can their own food these days, and there are strict government regulations for commercial canning facilities.

Follow USDA approved instructions for canning procedures. High acid foods can be canned safely using a variety of techniques. Low-acid foods require canning in a pressure cooker according to USDA instructions.

The foods most at risk for causing foodborne illnesses are those which spoil easily when not refrigerated, such as milk, egg (especially raw eggs), and meat products. For example, anything made with mayonnaise (raw eggs) is particularly susceptible to spoilage that can cause illness. Many raw meats also carry bacteria that are killed by cooking, which is why eating raw meat is generally not recommended.

FIGURE 8-3
Steak tartare is a gourmet meat dish made from finely chopped or ground raw beef. It is often served with onions, capers, seasonings, and sometimes with a raw egg yoke. It's usually served on rye bread. Many places in Europe serve a version of this, but it is not as common in the U.S., due to obvious health concerns and stricter food guidelines.

How Pathogens Infect Foods

Food can become biologically contaminated in a variety of ways. One way of remembering the various conditions that promote pathogen growth is the mnemonic FAT TOM: Food, Acidity, Temperature, Time, Oxygen, and Moisture.

F Food
A Acidity
T Temperature

T Time
O Oxygen
M Moisture

FIGURE 8-4
The six factors that influence pathogen development in food.

- **Food:** A pathogen needs a food source in order to grow and reproduce. Meat, dairy products, and eggs are all rich in protein, a food source that pathogens love. Cooked beans, grains, pasta, and starchy vegetables can also readily serve as a food source for pathogens, as can sweet foods such as fruits.
- **Acidity:** The acidity of a food is measured on a pH scale. 0 to 7 is acidic and 7 to 14 is alkaline. A neutral pH is 7 (distilled water). Pathogens prefer a pH between 4.6 and 7.5 for growth. This is the same range for most of our foods, making them ideal growth media if they are improperly handled.

- **Temperature:** Pathogens thrive and reproduce at temperatures between 41° F and 135° F, with human body temperature (98.6° F) being the ideal. That is why refrigeration helps stave off spoilage and pathogen development in food—because the refrigerator temperature is lower than 41° F. Cooking food to a temperature above 165° F destroys most pathogens; that's why meats are usually cooked to that temperature or higher.
- **Time:** Pathogens grow rapidly when conditions are right. The length of time they are permitted to grow is a major factor in determining whether there are enough pathogens to make you sick. This is why food that is fresher—and that is served quickly after cooking—is less likely to be harmful.
- **Oxygen:** Some pathogens need oxygen to survive; others do not. When you vacuum-seal a package or container, or simply store food in an airtight covered container, you keep the oxygen out; this can help prevent certain pathogens from reproducing.
- **Moisture:** The more moisture in a food, the friendlier it is to pathogen growth. Water activity (abbreviated a_w) is a measurement of the amount of moisture in a food. The scale runs from 0 to 1.0, with water itself having a measurement of 1.0. Most potentially hazardous foods have a measurement of 0.85 a_w or higher.

Food can become contaminated either via direct contamination or cross-contamination. In **direct contamination**, the food is already contaminated when you receive it. In **cross-contamination**, it becomes unsafe while it is being prepared, cooked, or served via contact with biological, physical, or chemical contaminants.

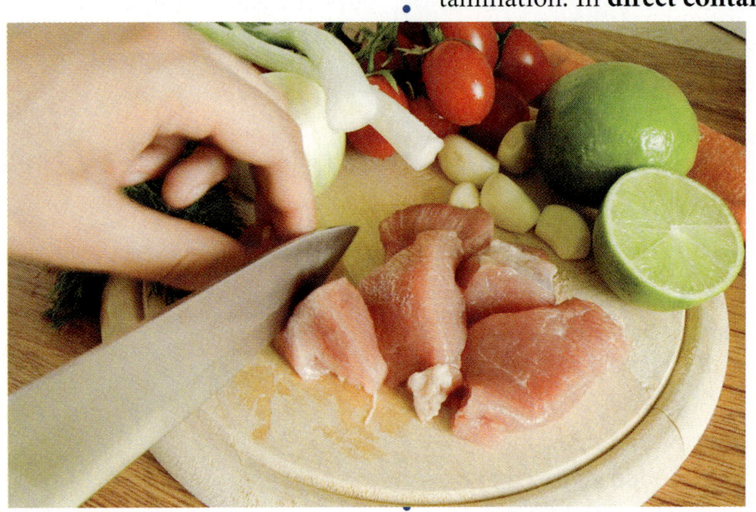

FIGURE 8-5
Raw foods can cross-contaminate other foods. Why is it important to thoroughly wash all cutting boards, counters, and knives that come in contact with raw foods?

Cross-contamination most frequently occurs when pathogens from raw foods, such as raw meats, are transferred to cooked or ready-to-eat foods. This can occur through a cook's contaminated hands, equipment, or utensils. For example, if you use the same knife and cutting board for salad vegetables as you used to cut up raw meat for the entrée, the vegetables become contaminated with the bacteria from the meat. When you cook the meat, the contaminants may be killed in the meat itself, but they persist in the salad, which is not cooked.

Grooming and Hygiene for Safe Food Preparation

To avoid cross-contamination as you are preparing food, it is important to keep yourself clean, well-groomed, and healthy. This includes frequent hand-washing, wearing clean clothing, and not working directly with food when you have an infectious illness, such as a cold or the flu.

Hand-washing is one of the most important ways to ensure that you do not cross-contaminate food. Health regulations specify that a kitchen must have a separate hand-washing sink from the sinks used in food preparation, including both hot and cold water, soap, a nailbrush, and single-use paper towels. (Paper towels are superior to cloth ones for hand-washing stations because they are disposable.)

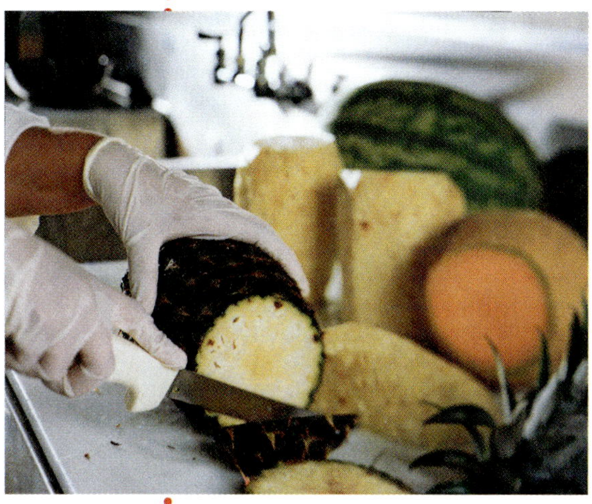

FIGURE 8-6
Gloves act as a barrier between your hands and ready-to-eat foods. Why is it important to wash your hands before putting on gloves?

When working in a kitchen environment, you should wash your hands at the following times:

- When you arrive at work or return after a break
- After using the restroom
- After you sneeze or cough, while covering your mouth with your hand
- After touching your hair, face, or clothing
- After you eat, drink, or smoke
- After taking off gloves
- Before putting on a new pair of gloves
- Before handling food that will not be cooked again prior to serving
- After handling garbage
- After handling dirty equipment, dishes, or utensils
- After touching raw meat, poultry, or fish
- After caring for or touching animals
- Anytime you change from one task to another

Note that wearing gloves does not exempt you from washing your hands! Your hands must be clean *before* you put on the gloves. You'll learn more about gloves in the next section.

Hot Topics

Many food service employees are required to keep their fingernails trimmed very short and not wear nail polish or fake fingernails. Some employees don't like this, because it limits their self-expression, but long or fake nails can be difficult to keep clean, and fingernail polish can chip and flake off into food.

BASIC CULINARY SKILLS

Hand Washing

1. Wet hands using hot running water.
2. Apply soap and work it into a lather.

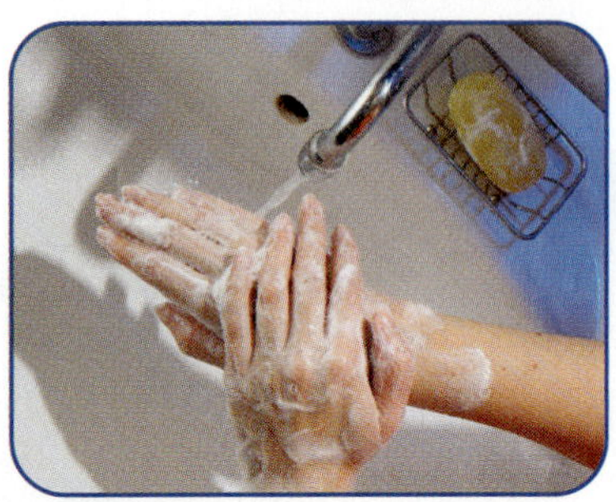

3. Scrub hands, between fingers, and forearms for at least 30 seconds.
4. Scrub under fingernails with a brush.

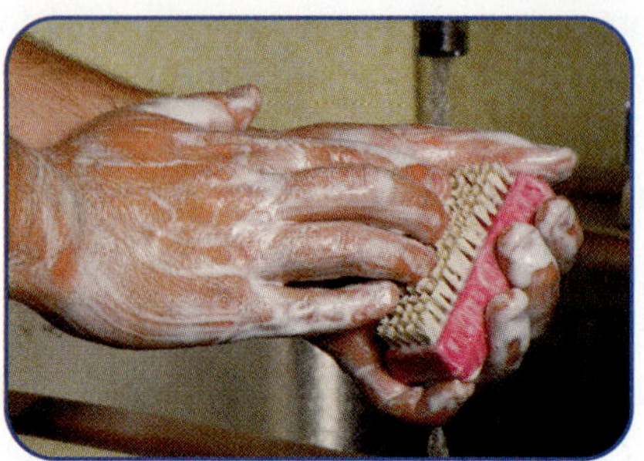

5. Rinse hands and forearms under warm running water.
6. Dry hands with clean single-use paper towels.
7. Turn off water using towel.
8. Open door, using towel if necessary.
9. Discard towel in waste container.

If you are sick with a contagious cold or disease, you should not prepare food for others, because you might pass on your illness to them. If you have any cuts or burns on your skin, cover them completely with bandages, and change the bandages frequently. If they are on your hands, wear gloves at all times in the kitchen to prevent the bandages from falling into the food.

Attire That Promotes Food Safety

Because the human body contains viruses, bacteria, and other pathogens, it's important to maintain barriers between yourself and the food you are preparing. This may not be necessary when you are cooking for family and friends in a home kitchen, but it is a requirement in commercial food service.

Hairnets are required in most food establishments, especially if hair is longer than a certain length. Hairnets prevent loose hairs from falling into the food, and they reduce the need for you to touch your hair while on the job. If you have long hair, you may want to pull it back into a ponytail before putting on a hairnet.

Hats are sometimes worn instead of hairnets, especially for people who serve food but do not come into direct contact with it, like the person who takes your money and assembles your order at a fast food restaurant. They serve the same purpose as hairnets—they keep loose hair out of the food.

Disposable gloves prevent your bare hands from coming into contact with ready-to-serve foods. You do not have to wear gloves all the time, but you do need to wear them when touching food that will not subsequently be cooked.

For example, you do not have to wear gloves when chopping up ingredients that will go into a soup or stew, because they will be cooked. However, if you are chopping onions for a salad bar, you do need to wear gloves.

Beware, however, that the gloves themselves can become contaminated if they touch other foods or a dirty surface. And if your hands aren't clean when you put the gloves on, contamination from your hands can get on the gloves. That's why it's important to wash your hands before putting on gloves and to change them whenever they become ripped or dirty.

If you are handling raw meats, fish, poultry, or eggs, you should change your gloves after you are finished handling them and before you start working with other cooked or ready-to-eat foods. Whenever hands should be washed (see the previous section), you should put on a new pair of gloves. Never reuse or wash disposable gloves.

Safe Eats

Never touch money with gloved hands unless you immediately remove and discard the gloves. Money is highly contaminated from handling.

The clothing you wear can potentially spread pathogens. You should start each shift or cooking project in clean clothing (or a clean uniform, if you are provided one) and wearing a clean apron. Try not to touch your clothing as you work in the kitchen, and if your apron gets dirty or contaminated, change it right away. Do not wipe your hands, tools, or equipment on your apron or clothing.

Most commercial food services do not allow employees to wear jewelry or watches because they can be a physical hazard, falling into food. Jewelry can also be a source of cross-contamination; pathogens can contaminate jewelry, and then transfer from jewelry to food.

Cleaning and Sanitizing Surfaces and Equipment

Everything that comes into contact with food must be clean and sanitary. In a commercial kitchen, there are specific requirements for how this needs to take place; in a home kitchen, it's up to you to decide the best way to provide a balance of sanitation and practicality.

Cleaning a surface is a three-step process:

1. **Clean** the surface by washing it. This involves removing soil or food particles from surfaces such as cutting boards, knives, pots, pans, and other utensils; sweeping the floor; and removing grease and dirt from the stove's ventilation hood, walls, and refrigerator doors. The most common type of cleaner is **detergent**, such as a dishwashing liquid. Other types include **degreasers** (solvents), which are specifically designed to dissolve grease; **acid cleaners**, which remove mineral buildup, and **abrasive cleaners**, which are used to scour dirt or grease that has been burned or baked onto pots and pans.

2. **Rinse** the surface thoroughly with fresh water, removing all traces of the cleaning product you used.
3. **Sanitize** the surface using either heat or chemicals to cut the number of pathogens on the surface to a safe level. You can use hot water (180° F or hotter) or a chemical sanitizer to wipe surfaces. Small tools and dishes can be submerged in hot water or a mixture of water and sanitizer. Larger surfaces and appliances can be sanitized by wiping or spraying them with a sanitizing solution. Sanitizing chemicals include **chlorine**, **iodine**, and **quaternary ammonium compounds**.

FIGURE 8-7
A three-compartment sink is designed to allow for washing, rinsing, and sanitizing. How would you determine how much of the sanitizer solution to use?

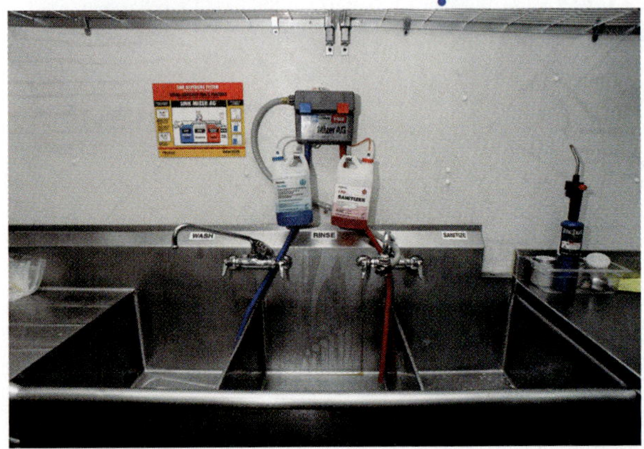

When washing dishes and other equipment by hand, a commercial kitchen typically uses the three-compartment sink, as shown in Figure 8-7. In the first compartment is soapy hot water for cleaning. In the second compartment is clean water for rinsing. In the third compartment is very hot water or a mixture of water and a chemical sanitizer.

Dishwashing machines can also be used to clean and sanitize tools and containers. Some dishwashers use very hot water to both rinse and sanitize; others use chemical sanitizers. Residential dishwashing machines usually rely on hot water for sanitizing.

Once tools and equipment are properly cleaned and sanitized, you should let them air dry before putting them away. In a residential kitchen, you might dry the dishes with a clean, dry towel, but towels should not be used in a commercial kitchen because they might be a source of cross-contamination.

Controlling Pests

Kitchens naturally attract all types of pests, including rodents, flies, and roaches, because of the food available there. These pests can carry pathogens that can cause foodborne illnesses, and their droppings can spread viruses and bacteria.

There are three strategies for dealing with pests in the kitchen:

- Keep pests out of the kitchen if possible.
- Minimize the amount of food left out for them to eat.
- Use pesticides if needed.

Your first line of defense is to make it hard for pests to get into the kitchen in the first place. They can come in through any holes or gaps around doors or windows, and through almost any opening, including roofs and drains. Seal up windows and doors tightly, and use screens to cover any openings that need to remain unsealed for ventilation.

Pests are there for the food, so make it as difficult as possible for them to find it.

- Clean all areas and surfaces promptly and thoroughly after use.
- Wipe up spills immediately and sweep up crumbs.

- Store foods in airtight containers whenever possible.
- Do not store food on the floor or touching the wall.
- Store garbage cans lifted off the ground, with tight-fitting lids.
- Don't pile bags of garbage over the rim of the garbage can or dumpster; the lids must close.
- Cover recycling containers.
- Make sure cans and bottles intended for recycling are rinsed.
- Check all boxes and packaging for pests before using their contents in cooking.
- Get rid of the boxes as soon as you unpack the food.

Sometimes, though, no matter how careful you are, pests can invade the kitchen. At that point, you may need to rely on pesticides to control them, or the services of a professional pest control company. Pesticide chemicals can be dangerous and must be stored well away from the food. There may be local regulations regarding the use of pesticides in commercial food establishments, so check to see what applies to your area.

FIGURE 8-8
Pests are attracted to the crumbs of food left behind after food preparation. Take a look around your kitchen at home—if you were a mouse or a cockroach, where would you hide? Where would you be likely to find crumbs to eat?

Purchasing, Receiving, and Storing Foods

Whether you are buying and serving food in a restaurant or in your own home, there is an orderly progression from the supermarket (or supplier) to your refrigerator, to the oven, and finally to the table for service. Food can become contaminated at any point along this path—or it can be safeguarded against contamination—depending on the precautions you take.

Keeping food at the proper temperature can control many of the biological hazards that can cause illness. In the following sections, we'll look at some best practices for maintaining food safety (including temperature) at each step of the food's progression through the kitchen, from arrival to consumption.

Purchasing and Receiving

The first step in the process is **purchasing**. As an individual, you probably buy your food from the local grocery store. A restaurant would typically order wholesale food in bulk from a reliable supplier.

Next comes **receiving**. If you buy from a grocery store, purchasing and receiving take place in a single step—when you fill up your shopping cart and head for the check-out line. Then when you get home with your purchases and put them away, the receiving is complete. In commercial food service, however, receiving is a separate step. Purchasing involves placing the order, which may occur several days before the truck pulls up with your supplies.

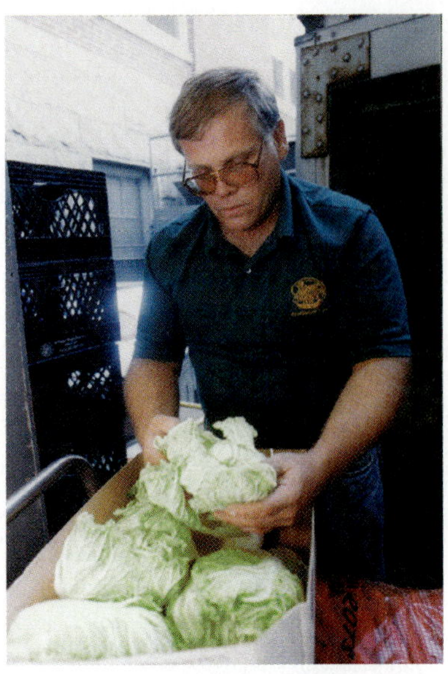

FIGURE 8-9
In a restaurant or other commercial food service, food comes right to your door; you don't have to shop for it at a grocery store. What are the advantages and disadvantages to having food delivered versus picking it out yourself?

If you are responsible for receiving in a restaurant, you must check the food as it arrives to make sure it is in acceptable condition. This can include:

- **Temperature control:** Has the arriving food been kept at the proper temperature to ensure its safety? Is it at the proper temperature now? Frozen foods should be completely frozen (hard solid). Refrigerated foods should be at a temperature no higher than 41° F.
- **Date control:** If the arriving food has a freshness date or expiration date on it, has that date passed?
- **Freshness:** If the arriving food is perishable, such as fruits, vegetables, meat, or dairy products, does it look and smell fresh?
- **Physical damage inspection:** If the goods are canned, do the cans appear in good condition (not bulged or dented)? If the goods are in bags, such as rice or sugar, are the bags sealed? Is the packaging clean and intact for all items? If the food is delicate, such as easily-bruised fruit, does it appear unblemished?

Storing Food

After taking possession of the food, you must immediately place it in proper **storage**. This prevents cross-contamination and spoilage. Perishable goods requiring refrigeration must be transferred immediately to the refrigerator.

As you are storing incoming foods, you should **rotate** the existing stock, so that the oldest food is used first. This may involve pulling any existing food of that type off the shelf and placing the incoming food behind it. This is called "first-in-first-out," or FIFO for short. In other words, the food that was the first into your kitchen should be the food that goes out as prepared servings first. Many food services write the date that the food was received on the packaging with a marker; this is a good practice to follow in your home kitchen as well.

If possible, you should store raw ingredients and prepared foods separately—if not in a separate refrigerator, then in well-defined separate areas of the same refrigerator. If raw ingredients and prepared food must be stored together, always store the raw food in a container, below any cooked or ready-to-eat foods, so if the raw food drips juice, it will not drip on the ready-to-eat food. For example, meats should be stored on the bottom shelf of the refrigerator, below any fruits or vegetables.

The temperature of your refrigerator should be between **35° and 40° F**; check it frequently with an appliance thermometer. However, certain ingredients may require lower or higher temperatures than the norm. In a commercial kitchen, there may be separate refrigerators for different food types, such as a separate produce refrigerator.

FIGURE 8-10
Different areas inside a home refrigerator have different temperatures. Take a look inside your refrigerator at home. Are foods being stored in the appropriate places?

Just because something is refrigerated does not necessarily mean it is being stored safely at below 41° F. In home refrigerators, different spots in the refrigerator have very different temperatures. In the door, for example, the temperature can be as high as 59°.

Here are some other tips for storing items in the correct locations within a home fridge:

- In the door compartments, store only items that do not spoil when warmer than 41°, such as mustard, ketchup, and soft drinks.
- If there is a compartment in the door that has a flap that closes, store butter in it.
- Because heat rises, the coldest area is near the bottom. Store all meats on the bottom shelf, right above the drawers (but not in the drawers).
- Fresh herbs do better at a warmer temperature, so put them on the top shelf—or even outside the fridge, if your kitchen temperature is lower than 70° F.
- Use the bins at the bottom to store most produce.
- Do not store tomatoes or tropical fruit (pineapples, citrus, bananas) in the refrigerator. It makes bananas turn black, and it makes tomatoes and other tropical fruits less tasty.

Thawing Frozen Food

Food should never be thawed by leaving it out at room temperature, because room temperature is in the range for optimal bacteria growth. Here are several ways of safely thawing frozen food:

- **Refrigerator:** Moving food from the freezer to the refrigerator is the best way of thawing, but it is the slowest, taking up to several days depending on the size of the frozen item. Do not unwrap the food; place it in a shallow pan on the bottom shelf of the refrigerator to prevent contamination from dripping.
- **Microwave:** A microwave oven works well for thawing very small portions, such as individual servings. Most microwaves are not large enough to thaw a large item, such as a whole turkey or a case of meat, so their use for commercial thawing is limited.
- **Running Water:** Place the covered or wrapped food in a container under running water. The water temperature should be 70° F or lower, and the water stream should be strong enough to wash loose particles off the food, but not so strong as to splash other foods or surfaces. Clean and sanitize the sink before and after thawing foods under running water.

After you thaw the food, it should be used as quickly as possible. Refreezing is generally safe, but it may adversely affect the quality or flavor.

Cooling Hot Foods for Storage

You can't just put a large quantity of a hot food directly into the refrigerator, for two reasons. One is that the heat may raise the temperature inside the refrigerator to the point where other foods may become unsafe. Another is that it may not cool down quickly enough in the refrigerator. For example, it can take 72 hours or more for the center of a five-gallon stockpot of steamed rice to cool down to below 41° F when taken directly from the stove and placed in a refrigerator.

Cooked food that you plan to store for later use must be cooled down to below 41° F as quickly as possible. There are two guidelines for how quickly this should happen:

- **One-Stage Cooling:** The food should be cooled to below 41° F within four hours.
- **Two-Stage Cooling:** The food should be cooled down to 70° F within two hours, and cooled to below 41° F within an additional four hours.

A small quantity of food, such as leftovers from a dinner party, can cool to room temperature (around 70° F) within an hour of cooking, and can then be refrigerated. Therefore, safe cooling is usually not a big issue with home cooking.

However, a larger quantity of food, such as prepared in a restaurant, may not make it to room temperature before the four hour mark, so you might need to employ special cooling techniques. One method is to use a blast chiller (a refrigerator-like appliance that quickly chills food), if the restaurant has one. A cheaper way is to use a chill wand, which is essentially a large ice pack. You freeze the chill wand, and then place it in a pot of hot soup or other liquid food to quickly chill it.

FIGURE 8-11
A chill wand is a big ice pack (usually plastic) that quickly cools large pots of liquid. How would you clean a chill wand after using it, before returning it to the freezer?

You can also use an ice water bath to chill food. This method requires no special equipment—just a couple of pans (or a sink), some ice cubes, and water.

There are different types of food storage containers, for both home and commercial use.

- **Plastic:** Inexpensive, dishwasher-safe, some types of lids make an airtight seal; but unless specifically designated "microwave safe," generally cannot be used for microwave reheating. Chemicals in some plastics are toxic when heated.
- **Glass:** Durable, easy to clean, microwave-safe; but most lids are not airtight.
- **Metal:** Durable, easy-to-clean; but not usable in a microwave, and most lids are not airtight.

Safe Eats

Family holiday dinners can be prime breeding ground for harmful bacteria because people tend to leave the food sitting out at room temperature for hours at a time while they socialize. Make sure that food is put away shortly after the meal is over to prevent spoilage; you can always get it out again for leftovers later.

FIGURE 8-12
Check the temperature of food being held to make sure that it is within a safe range. Suppose this potato salad's temperature was 37° F. Would that be an acceptable temperature at which to serve this food on a buffet line?

BASIC CULINARY SKILLS

Chilling with an Ice Water Bath

1. Place the food in a stainless steel bowl or container. If solid or semi-solid food, cut it into small portions and spread it in a single layer. Shallow pans work best.
2. Fill a larger pan or tub with ice water. (You can also use a sink.) Place bricks or a rack in the bottom of it, so water will circulate under the bowl.
3. Set the food bowl in the ice water bath. The bath should reach the same height as the food inside the metal container. Adjust the bricks/rack or the amount of ice water in the tub, as needed.

4. Stir the food frequently if possible.
5. Use a thermometer to test the temperature; when it reaches room temperature, transfer the food to a refrigerator for storage.

Safe Eats

If you store food in a lidless container, such as a pan or bin, make sure you cover the food to keep out potential physical hazards and to ensure the food does not dry out quickly. Plastic wrap and aluminum foil are both common inexpensive covers.

Holding Food

Some food is served immediately after cooking; other food may be **held**—that is, prepared earlier and then kept hot or cold until the diners are ready to eat it. Hot food should be held at temperatures above 135° F. Cold food should be held below 41° F. You can use thermometers to check the temperature of food being held. If food has been held at an improper temperature for longer than two hours, you should discard it.

Reheating Food

When reheating food that has been prepared ahead of time, you should move it through the temperature "danger zone" as quickly as you can. Hot food should be reheated to at least 165° F for at least 15 seconds within two hours of serving it.

A steam table, such as at a buffet, maintains reheated foods at a temperature above 135° F. Bring the food to the proper temperature using an oven, burner, microwave, or other heat source, and then put it on the steam table.

Commercial Food Service Inspection

To ensure that all companies that provide food to the public (restaurants, restaurant supply companies, cafeterias, and so on) conform to safety standards, local and regional health departments regularly inspect the storage and preparation areas of these entities and issue health inspection certificates (or not, depending on their findings).

The standards that inspectors follow have been developed by the Food and Drug Administration (FDA), as part of the FDA Food Code. This code is not a law or regulation; it is simply a set of recommendations. It is up to state and local governments to establish their own laws and regulations, and most of them do so by adopting some or all of the recommendations in the FDA Food Code. They can also establish standards that are more stringent.

The inspection that local health department personnel perform is known as a **food safety audit**, or **health inspection**. The number of times a year a business is inspected depends on several factors, including how many meals are served, the types of food, and the number of past violations.

After the inspection, the business receives a report of the result, including notification of any violations. Violations must be corrected within a certain time period or the business risks fines or the revocation of their food service license.

The HACCP System

HAACP is a system of food safety assurance—a step-by-step plan you can follow to ensure food is handled safely. It stands for Hazard Analysis Critical Control Point and is pronounced *HAS-sup*. The HACCP system has been widely adopted by the FDA and the U.S. Department of Agriculture (USDA), as well as most food processing companies and restaurants.

HAACP prescribes methods of auditing a facility for safety and sanitation. The employees at a facility can use HACCP to monitor themselves and their business practices continually, so that when it is time for a health inspection, everything is already in order.

There are seven steps in the HACCP process:

1. **Conduct a hazard analysis.** This step consists of examining the flow of food as an overall "big picture" and identifying potential areas of cross-contamination.
2. **Determine Critical Control Points (CCP).** A CCP is a point in the process where you can prevent, eliminate, or reduce a hazard. For example, if a restaurant typically cooks chicken breasts in advance of a guest ordering them and then reheats them in a microwave as needed, critical control points occur in the temperature at which the cooked chicken is held and the temperature to which they are reheated in the microwave.

Hot Topics

While the FDA Food Code is not a law or regulation, there are many laws and regulations concerning food, nutrition, fitness, and wellness issues. These laws vary from state to state. Why do you think there needs to be regulations and laws? And, why might they be different in different areas? Try and find out some laws and regulations specific to the state where you live. Also, pick another state and see if any of the laws or regulations are different.

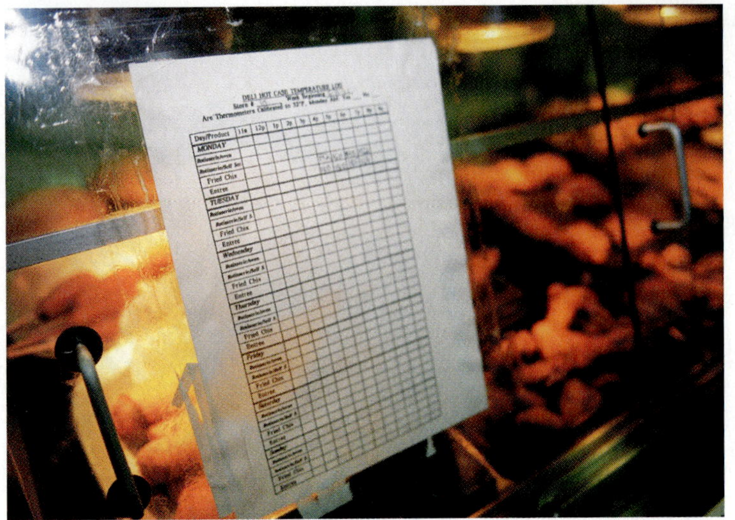

FIGURE 8-13
This simple form provides space for employees to record temperatures at various times of day. If you were the employee recording a temperature, and you noticed it was out of range, what would you do?

3. **Establish critical limits.** The critical limits specify the range of acceptable values for a particular measurement, such as how long food can be held at an unsafe temperature or how long produce may be safely stored before cooking. These limits are typically provided by the local health department based on the FDA Food Code.
4. **Establish monitoring procedures.** In this step, you set up the processes to ensure that the critical limits are maintained at the critical control points. This may consist of a schedule that employees follow to measure the temperatures of foods and record those measurements in a log book or a computer program.
5. **Identify corrective actions.** In this step, you specify what should happen if a measurement taken in the previous step indicates that there is a problem. For example, what should happen to a vat of potato salad if an employee detects that it is being stored at a higher temperature than required?
6. **Establish procedures for record-keeping and documentation.** It is important to keep good written records of all the monitoring and corrections taken in steps 4 and 5. In this step, you create the forms and reports that will be used to document the processes on a daily basis.
7. **Develop a verification system.** Responsibility for the safety standards to which you are conforming should not fall solely into the hands of a single employee. All measurement and correction processes should be periodically verified by a supervisor or inspector to make sure that employees are performing the inspections properly and filling out the paperwork correctly, and that the measurement equipment is functioning as intended.

Waste Disposal and Recycling

The rules and regulations for waste disposal and recycling in a commercial food establishment may vary depending on the state and local laws, the policies of the parent company (if there is one), and the guidelines developed by the facility manager.

The most common container for collecting waste in a commercial area is a dumpster, which is a large covered bin made of metal or plastic that sits outside the establishment. Employees put all the facility's trash in the dumpster, and a trash company comes by to collect it on a specified schedule. There may be a separate container for disposing of used cooking grease/oil.

A **food safety inspector** may work for a local or state Health Department; a large food manufacturing, transportation, or processing company; a restaurant chain; or some other business that serves food to the public or ensures its safety.

Food inspectors perform a variety of tests on food to make sure it has not been contaminated. For example, a food inspector in a private commercial slaughtering plant checks the meat to make sure it did not come from diseased animals and that it is stored and prepared according to the plant's written plans for HACCP and sanitation. A food inspector may also be in charge of regulatory oversight activities, such as working on committees to develop new guidelines.

An entry-level food inspector must have either a Bachelor's degree or job-related experience in the food industry, and must demonstrate knowledge of sanitation practices and control measures used in commercial handling and preparation of food products.

In many areas, recycling is not required, but many facilities do participate in recycling programs by saving used glass, plastic, and metal items and paper products. If your facility has such a program, you may be called upon to separate certain types of waste into recycling containers. Some of the benefits to restaurant recycling include:

- **Cheaper trash fees.** The less you throw away, the fewer times the dumpster needs to be emptied, and the less you will pay for trash pickup.
- **Cash back.** Some governments offer tax credits or other payments for recycling, and trash handlers may have service plans that include recycling that can cost $20 to $50 less per ton than conventional garbage pickup.
- **Reduced purchasing costs.** Restaurants that have effective waste management systems may incorporate reusable items, like cloth cleaning rags and reusable flatware, which reduces the amount of new supplies that need to be purchased.
- **Avoidance of additional fees.** In some cities, recyclable materials such as cardboard and newspaper are banned from landfills, so you have to pay additional disposal fees if you do not recycle these items.

Items that are commonly recycled include aluminum, cardboard, glass, paper, food waste, plastic, steel, and used fryer oil.

Some products should not be placed in the regular dumpster, but should instead be tagged as hazardous waste and taken to (or picked up by) a special facility equipped to handle them. For example, some cleaning supplies, pesticides, electronics and batteries (because of the lead and other heavy metals in them), paints/varnishes, and solvents may require special handling. See the item's MSDS to find out whether it needs special handling for disposal.

Case Study

Ellen made some beef stew to serve to her friends at a party. Most people will arrive between 6 and 7 p.m. and will eat right away, but a few people will not arrive until after 10 p.m., and may not eat until 11 p.m.

After the majority of the people are finished eating, what should Ellen do with the rest of the stew? Should she try to keep it hot until the other guests arrive hours later? Or should she refrigerate the leftovers and warm them up in the microwave later, when the other guests are ready to eat? Or is there some other action you would choose to take?

Put It to Use

1. Ask your school cafeteria manager if your class can tour its kitchen. Ask him or her to explain some of the sanitation procedures followed in the cafeteria to ensure that safe food is served. When you return from the tour, write down your observations, including what matched up with—or was different from—the material presented in this chapter.

Put It to Use

❷ Go home and examine the contents of your refrigerator. Is there any food that is stored in the wrong place? Anything that is growing mold? Any perishable products that have expired? Write a brief report explaining what you found that is not safe to eat, and why. And make sure you throw that unsafe food away!

Write Now

Suppose you were buying supplies for a new restaurant, and you needed to purchase sanitizing chemicals to use in a three-compartment sink. Where would you buy such chemicals? What are your choices in terms of the active ingredients, the various brands and the sizes of containers? How much, on average, does sanitizer cost per sinkful of dishes? Research this topic using any sources that are helpful—the Internet, a local supplier, a restaurant supply catalog, etc. Write up a report explaining what options are available, which one you would choose, and why.

Tech Connect 

Some bacterial pathogens, other than those presented in this chapter, include shigella, staphylococcus aureus, and listeria monocytogenes. Research one of these online to find out what types of foods it commonly comes from, what the symptoms of the illness are, what the incubation period is, and how to minimize the risk of infection by proper food handling.

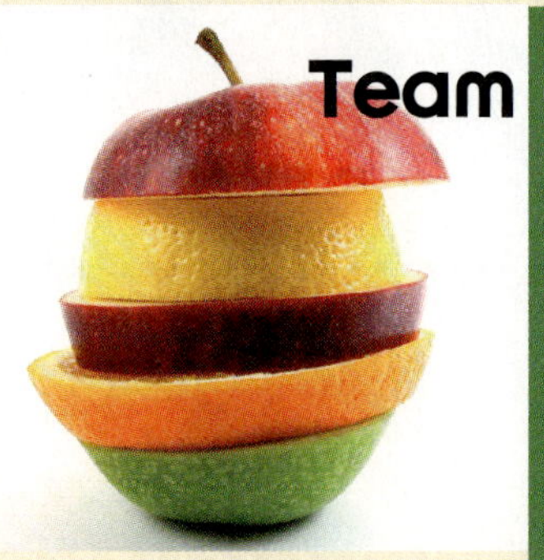

Team Players

Divide the class into three teams. Each team should contact a local restaurant and ask the manager if one or more students (or the entire team) can come in and take a tour of the kitchen. While there, ask the manager about the sanitation and safe food handling practices that are part of that restaurant's HACCP plan. Each team should then give an oral report to the rest of the class describing what they saw and learned. As a class, discuss the differences encountered between teams.

Put *It* Together

Match the explanation in column 1 with the term in column 2.

Column 1

- **a.** a biological hazard that can cause foodborne illness
- **b.** a piece of hair or plastic that ends up in food served to a diner
- **c.** a bacterium that is widespread in the intestines of birds, reptiles, and mammals
- **d.** food that becomes contaminated after you receive it, but before you serve it to diners
- **e.** a chemical that cuts the number of pathogens on a surface to a safe level
- **f.** a method of thawing frozen food
- **g.** a seven-step process for ensuring and auditing a facility for safety and sanitation
- **h.** mold or yeast that can grow on food, such as cheese or bread
- **i.** an infection that invades a cell, tricking the host into reproducing it
- **j.** a pesticide or cleaning compound

Column 2

1. chemical hazard
2. cross-contamination
3. fungus
4. HAACP
5. pathogen
6. physical hazard
7. running water
8. salmonella
9. sanitizer
10. virus

9 Using Basic Kitchen Equipment

In This Chapter, You Will . . .

- Identify various types of food preparation equipment
- Learn how to use and maintain knives
- Find out how to select and use smallware
- Understand the types of cookware available
- Identify and select equipment for holding, serving, and storing food
- Be prepared to handle kitchen fires and injuries
- Learn how to safely handle and dispose of hazardous materials

Why YOU Need to Know This

When a recipe calls for performing an action using a certain type of pan, utensil, or appliance, you are expected to know what that means. As a rule, recipes don't explain the different types of cookware or utensils. Therefore, it's as important to have a basic knowledge of kitchen equipment as it is to understand how to follow a recipe.

For example, there are many types of knives used in a kitchen. How many different types can you name? Can you describe the unique features of each type? Keep a record of your answers, and then try the same exercise again after you have completed this chapter to see what you learned.

Types of Food Preparation Equipment

Commercial kitchens—and many home kitchens as well—are stocked with the equipment that you need to prepare most recipes. In fact, there may be so much equipment that you aren't sure what to use! Too many choices can be overwhelming unless you understand the differences between various pieces of equipment. In the following sections, you'll learn about the basic types of appliances and equipment that you may find in a kitchen.

Hot Topics

Molecular Gastronomy is a trend in cooking where chefs use chemistry, physics, and technology to alter traditional foods and food experiences—think fake caviar made from sodium alginate and calcium; instant ice cream, fast-frozen using liquid nitrogen; and hot soups served as a foam. One chef, Homaro Cantu of Moto, even created edible menus using an ink-jet printer adapted for inks made from fruits and vegetables and paper made of soybean and potato starch. These chefs are using chemicals, lasers, flash-freezing techniques, and even printers to get their desired results. Think about what type of equipment might be necessary in a kitchen preparing these types of food. How would it be different from what you would find in a traditional kitchen? Would you like to try some of these foods?

Chopping, Slicing, and Grinding Appliances

In Chapter 7, you learned about the various cooking techniques, including chopping and slicing. You can chop and slice manually with knives, but for large quantities, it is more efficient to use machines.

One of the most common types of small appliances is a **food processor**. This is a multi-purpose unit that has a bowl, blades, and a lid. Use different attachments to perform a variety of activities including grinding, slicing, mixing, blending, and crushing. There are also various other machines designed specifically for chopping.

An **electric slicer** is most often used for meat, but can also be utilized for cheeses and other solid foods. It has a rotating blade, and a moving metal plate that slides the food past the blade at a precise, adjustable height.

Food processor

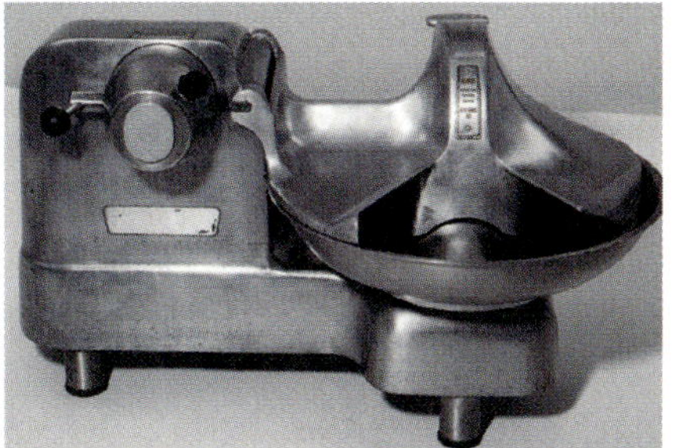

Food chopper

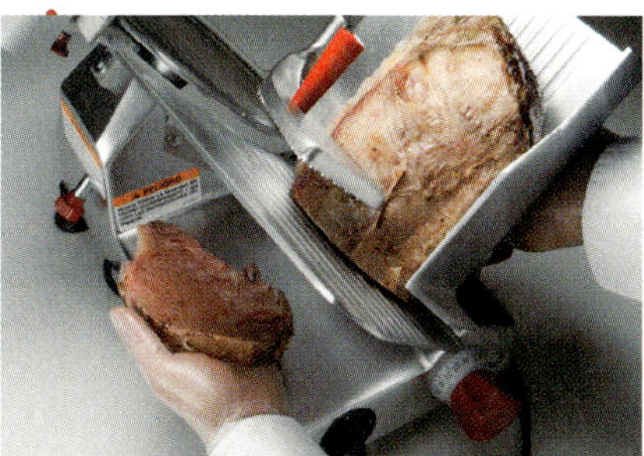

Electric slicer

FIGURE 9-1
Here are some types of chopping and slicing equipment. Which of these would be most essential to have in a home kitchen, and why do you think so?

FIGURE 9-2
A mandoline is an inexpensive alternative to an electric slicer. What kinds of foods might you slice with a mandoline?

A **mandoline** is a manually operated slicer, typically used on vegetables. It is cheaper than an electric slicer, but easier to operate, quicker, and less dangerous than a knife. You can attach different types of blades to make different cut textures. In Figure 9-2, for example, a rippled blade is being used to create a waffle cut.

A **meat grinder** grinds various types of meat to make hamburger, sausage, ground turkey, etc. A dedicated meat grinder (see Figure 9-3), of which there are many types, is a machine that only grinds meat. If you don't have the budget or space for a dedicated machine, you might be able to meet your grinding needs by using a meat grinder attachment that fits on a mixer.

To clean grinders and slicers, unplug and disassemble them to the extent possible, separating the electric parts (which cannot be submerged in water) from the non-electric parts (which can be washed manually using the three-step method you learned in Chapter 8, or washed in a dishwasher).

Mixing and Blending Appliances

Mixers and blenders are very common appliances in both residential and commercial kitchens. A **mixer** simply circulates the ingredients, while a **blender** also shreds or chops the ingredients as it mixes. For example, you might make a fruit smoothie in a blender, because of its ability to chop up the fruit as it mixes. A mixer might use blunt implements like beaters or dough hooks to mix up cake batter or to knead bread dough.

FIGURE 9-3
A dedicated meat grinder (shown here) is needed for a commercial establishment that does a lot of grinding, but a grinder attachment for a mixer can work well for smaller quantities. What would influence a restaurant's choice between the two?

Many home kitchens have a **countertop mixer** that mixes about 5 quarts at once; commercial kitchens typically have **freestanding mixers** that sit on the floor and can mix up to 5,000 pounds of dough in one batch.

A mixer may be categorized by where it sits or by its movement.

- With a **planetary mixer**, the bowl stands still and the tool moves within the bowl. Most mixers are of this type. You can get attachments that can add food processor capabilities to the machine, such as the meat grinder attachment mentioned previously.
- With a **spiral mixer**, the mixing tool holds still and the bowl moves. Spiral mixers are sometimes used for bread dough because they have a gentler action than that of a planetary mixer.

There are also two types of blenders.

- A **countertop blender** has a glass, plastic, or metal container with a blade in the bottom that fits onto a stand. This type of blender is commonly used to make drinks, such as fruit smoothies, and to crush ice for frozen cocktails.
- An **immersion blender** is a stick-shaped machine that you hold in your hand and dip into food to be blended.

Figure 9-4 shows examples of each type of blender.

To clean mixers and blenders, unplug and disassemble them and wash the parts in hot soapy water; then rinse and sanitize, as you learned in Chapter 8. The parts may or may not be dishwasher-safe; follow the manufacturer's guidelines for dishwasher versus hand-washing. You must not submerge the motors; instead, wipe them down with a soapy cloth, then wipe them down with a damp cloth, dry with a clean cloth, and sanitize. Make sure you immediately replace any blade guards or other safety shields after cleaning.

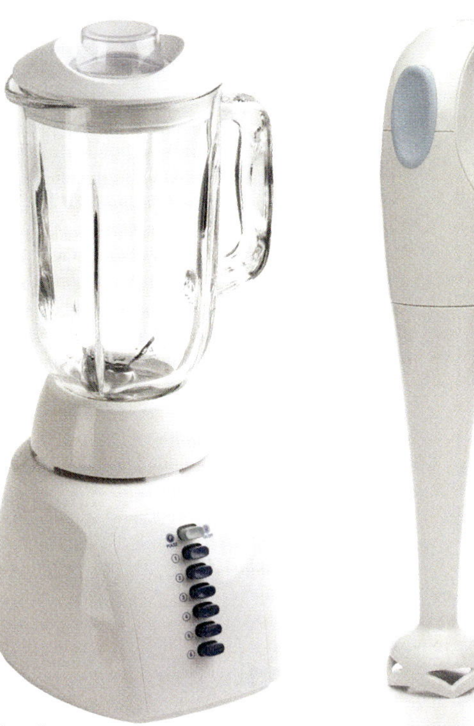

FIGURE 9-4
Countertop blenders (left) include both the blade and the container in one unit. Immersion blenders (right) contain just the blade and the motor that drives it. Which type of blender do you think would be most useful for the type of cooking your family does at home?

Electric Kettles and Steamers

A kettle is a large pot in which you can simmer or boil food. A regular kettle can be set on a burner to heat it up, but there are also electric kettles that supply their own heat. One popular type of electric kettle used in commercial cooking is a **steam-jacketed kettle**. This type of kettle is double-walled; steam circulates within the walls to provide even heat all over the kettle, rather than concentrated heat at the bottom only, as on a burner.

A steamer is an appliance that is designed specifically to steam-cook food. There are two ways in which steamers work:

- A **pressure steamer** heats water under pressure in a sealed compartment; the steam builds up in the pot to cook the food.
- A **convection steamer** pipes steam into the cooking chamber, across the food, and then out a vent; the food cooks by the steam continuously passing over it.

FIGURE 9-5
This steam-jacketed kettle has a tap at the bottom for easy dispensing of the cooked food. What kinds of food might you cook in this type of kettle?

Safe Eats

Always follow safety precautions for the mixing and blending equipment you are using. That includes keeping your fingers away from sharp blades, avoiding electrical shock by turning off and unplugging equipment when cleaning it, and making sure countertop units are not so close to the edge of the counter that they can get knocked off.

To clean a steamer or electric kettle, separate the portion that directly contacts the food from the portion that creates the heat (that is, the electric part). Wash the parts that touch food in a dishwasher or use the three-step manual washing procedure you learned in Chapter 8. Wipe down the non-immersable parts with hot soapy water, rinse by wiping with a damp cloth, and then dry and sanitize.

Ranges, Griddles, and Grills

A **range** is a stovetop with burners on it. It can be covered with a **range hood** that includes a fan to pull steam or smoke away from the range and out through a vent in the ceiling of the kitchen.

Range burners can be either gas or electric. A **gas burner** usually has an open flame, and uses metal supports to hold the pot slightly above the flame. An **electric burner** usually consists of a metal coil that heats up. An electric burner usually does not have metal supports; the pan sits directly on the coil or on a flat stovetop with the burners sealed underneath it.

Hot Topics

For residential flat-top ranges, you can buy one-step cleaners that you wipe on and then wipe off. These products claim to clean and polish in a single step, with no rinsing or soap required. Such products would not be suitable for commercial use, however, because they do not sanitize. You would have to follow up the cleaning with a sanitizer solution. In addition, these products tend to be expensive, and their primary purpose is to keep the cooktop looking shiny and scratch-free; that is not as important in a commercial kitchen.

To clean an open-burner range, turn it off and allow it to cool. Soak the metal supports in hot soapy water, and wash the rest of the range with hot soapy water. After washing, rinse and dry the range and metal supports; reinstall the supports on the range. A flat-top range is much easier to clean. Turn it off and allow it to cool; then clean the surface with hot soapy water, rinse, and dry.

A **griddle** is a flat metal plate, similar to a flat stovetop, that cooks foods such as hot sandwiches, pancakes, and breakfast meats. A commercial griddle is typically large, as in Figure 9-6. A fast

FIGURE 9-6
A large griddle like this one is commonly used in restaurants. If you didn't have access to a large griddle like this, what might you use instead?

food restaurant might use a griddle to fry hamburgers, for example. A griddle for home use may be a small electric-heated platform or a square shallow pan set on top of a range burner.

To clean a griddle, if feasible (such as with a very small griddle), separate the food-cooking surface from the heat source, and wash the food-cooking surface in the dishwasher or use the three-stage manual washing method. Then, clean the heat source portion as thoroughly as you can using a griddle stone (an abrasive stone that helps remove cooked-on food). If the griddle does not come apart from the heat source, clean it in place by allowing it to cool and then using the three-stage cleaning process with cloths—soapy hot water, clean rinse water, and sanitizer.

FIGURE 9-7
A wire brush works well for loosening stuck-on food from the metal rack on a grill. How important do you think it is to clean the wire brush afterward? How would you clean it?

A **grill** consists of an open metal rack over a heat source such as charcoal, wood, or a propane flame. When you grill steaks on your patio, for example, you use this type of product. Some people call a griddle a grill, and call foods cooked on a griddle *grilled,* such as a grilled cheese sandwich, but this is not an accurate description.

To clean a grill, remove and wash/rinse/sanitize the metal rack on which the food cooks. It is important to clean the metal rack after every use. There are commercial products designed specifically for cleaning grills, but a stiff wire brush works very well to loosen any cooked-on pieces. See Figure 9-7. You can then scrub it with hot soapy water, rinse, and sanitize, just as you would any cooking appliance. You do not have to wash and sanitize the parts of the grill where the fuel burns after every use (the charcoal area on a charcoal grill, or the gas/brick area of a propane grill), but—for a casually used home patio grill—you should thoroughly disassemble and clean it at least once a year. For a commercial grill, the amount of use and the manufacturer's recommendations will determine how often a complete cleaning is required.

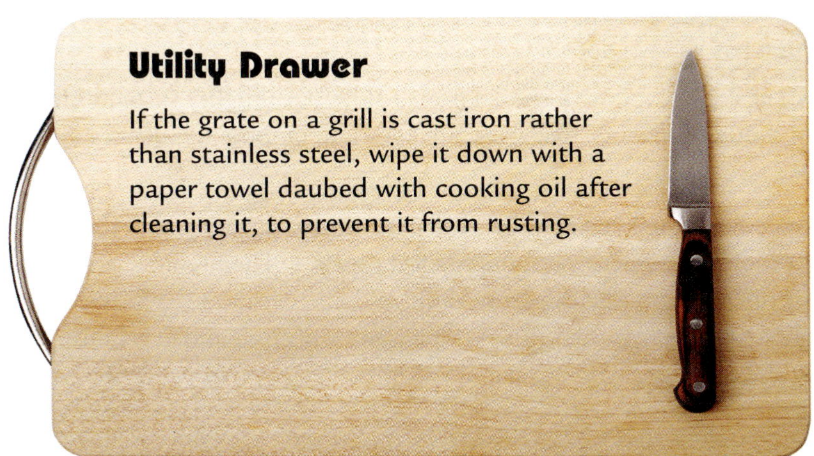

Fiction	Fact
If a gas grill has a "Clean" setting, you can use it to automatically clean the grill.	On most gas grills, the "Clean" setting is designed to preheat the grill, not clean it. This setting will burn up any baked-on food residue, but it is not a substitute for washing and sanitizing.

Ovens and Broilers

An **oven** is a heated box in which you place food to cook. The food cooks as heated air circulates around it. A **conventional oven** heats the interior via electric or gas heat located in the bottom of the oven; the heat rises up to the cooking chamber. Some conventional ovens also have **broilers** in them; these are heating elements that apply heat from the *top* of the oven to the food. Many restaurants have stand-alone broiler appliances.

Some other common types of ovens include:

- A **convection oven** is similar to a regular oven, but includes one or more fans that circulate the hot air in the oven, so that the food cooks more evenly. Some convection ovens also do double-duty as a microwave.
- A **microwave oven** cooks food with microwave radiation. Microwaves cook, reheat, and thaw foods much more quickly than regular ovens, but they do not do a good job of browning foods.
- A **deck oven** is a stack of short ovens, one on top of the other. Pizza restaurants may use deck ovens to cook several pizzas at once, each on a separate shelf that has its own door.
- A **smoker** cooks food with smoke from a slow-burning fire (usually charcoal or wood). Smokers, like grills, are common outdoor appliances in homes.

To clean a conventional, microwave, or convection oven, turn the oven off and allow it to cool. For a microwave, unplug it if possible. Then remove any racks, shelves, or turntables, and clean/rinse/sanitize them. Clean the oven inside and out with hot soapy water; then rinse with a clean wet cloth and dry. Polish the outside of the appliance with a clean dry cloth.

FIGURE 9-8
Why do you think it is important to clean spilled food from the inside of an oven, rather than to simply let it burn away over time as the oven is used?

Conventional ovens sometimes have baked-on food stuck to the bottom and sides that does not come off with plain soap and water. Spray-on oven cleaners are available to help loosen the baked-on food; these typically contain harsh chemicals that you should not breathe or get on your skin. If you use an oven cleaner product, make sure you wear gloves and ventilate the area well. (Look for "fume-free" products to minimize any dangers of breathing fumes.)

Some conventional ovens have a Self-Cleaning feature. Such ovens have textured inner walls rather than smooth shiny ones; these textured walls heat up when in Self-Cleaning mode, burning off food residue. Some oven cleaner products are not for use in self-cleaning ovens, so be sure to carefully read the label.

Fryers

Fryers are used to cook food by immersion in hot oil. Both freestanding and countertop **deep-fat fryers** hold and heat oil in a deep stainless steel reservoir. A heating element keeps the oil at a specified temperature. Food is lowered into the hot oil in a stainless-steel basket.

Deep-fat fryers in a commercial establishment should be cleaned both daily and weekly. On a daily basis, all removable parts (including baskets) should be taken out and washed/rinsed/sanitized. In addition, all exterior surfaces of the fryer should be cleaned with hot soapy water. Once a week, the fryer should be completely drained, and the oil-holding area cleaned. Most commercial fryers have a filter that is used to filter the cooking oil; the exact interval at which this happens varies depending on the amount of usage and the manufacturer's recommendations.

Safe Eats

It is safe to reuse cooking oil if done properly. The greatest hazard is allowing the fat to become spoiled. Rancid oils can contain free radicals that are potentially carcinogenic.

To reuse oil safely, strain it through cheesecloth and store in a cool, dark place in a nonmetallic container.

A small residential-use frying appliance should be cleaned in a similar way. Wash/rinse/sanitize all removable parts after each use. Cooking oil can be stored in the device for reuse for a limited period of time, according to the manufacturer's recommendation.

FIGURE 9-9
A commercial deep-fat fryer.

Using Knives

Knives are an essential part of any cook's kitchen tool set. There are so many different types of knives that it can be difficult for a beginner cook to choose the right tool for a job. For example, some knives have a very wide blade; some have pointed ends; and so on. In this section, we'll take a look at what makes one knife different from another and suitable for a particular task.

Types of Knives

There are eight basic types of knives, as shown in Figure 9-10.

- **Chef's knife:** Also called a French knife. This is an all-purpose knife, with an 8- to 12-inch triangular blade. It can be used for a wide variety of activities, including peeling, slicing, chopping, and dicing.
- **Utility knife:** A smaller, lighter version of a chef's knife, with a 5- to 7-inch blade. It is used for light cutting, slicing, and peeling.
- **Paring knife:** A small knife with a 2- to 4-inch blade, used mostly for peeling and trimming fruits and vegetables.
- **Boning knife:** A thin-blade knife that is used to separate raw meat from bone. The blade is about 6 inches long, and narrow. Some have an upward curve.
- **Filleting knife:** Similar to a boning knife in appearance, but the blade is more flexible. It is designed for deboning (filleting) fish.
- **Slicer:** A long, thin blade with a rounded or pointed tip, used to make smooth slices of large items. The blade can be straight or **serrated** (that is, having a scalloped or toothed edge).
- **Cleaver:** A wide rectangular blade with a blunt end. It can be used for many of the same tasks as a chef's knife, and is particularly good for cutting large pieces of bone-in meat.
- **Scimitar (Butcher's knife):** Similar to a boning knife, a scimitar has a long, narrow, curved blade, and is used for cutting meat into steaks.

Cool Tips

The Chinese cleaver is used mostly for slicing boneless meats, chopping vegetables, and even doubles as a spatula for bringing ingredients to the fry pan. It is similar in shape to a regular cleaver, but is a bit thinner and intended more as a general-purpose knife, rather than a heavy-duty bone-chopping utensil.

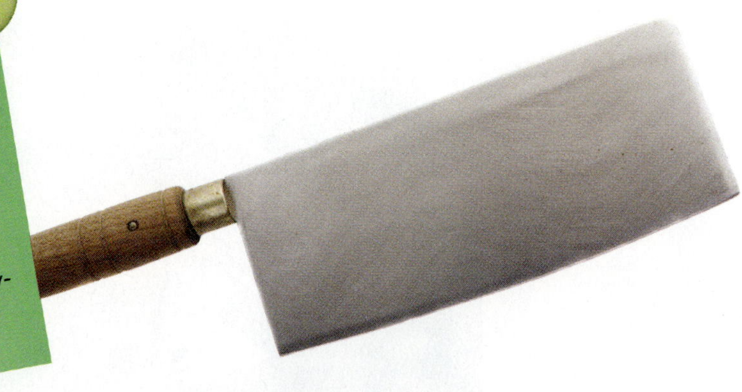

FIGURE 9-10
Some common knife types. Which of these knives would you use for bread? For mincing celery? For peeling potatoes?

Parts of a Knife

A knife has various parts. You can probably guess that the **cutting edge** is the flat metal part of the **blade**, and the **handle** is the part you grip, but some of the other sections may not be as obvious. Figure 9-11 shows the anatomy of a knife.

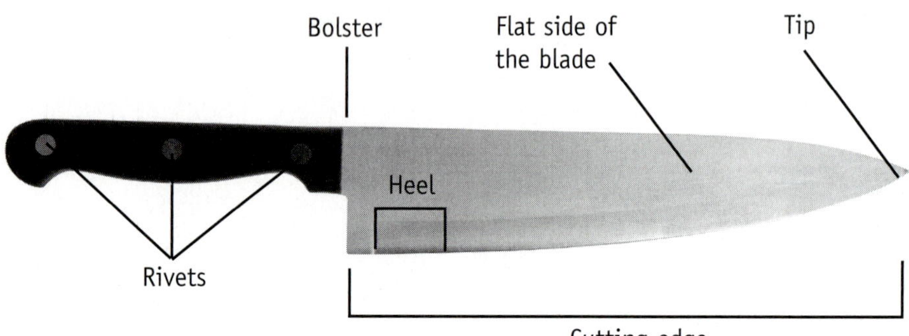

FIGURE 9-11
Parts of a knife.

Different parts of the knife can be used for different tasks in food preparation. It isn't just about the cutting edge! For example, the **tip** can be used for paring and peeling and to cut blemishes out of fruit and potatoes. The **heel** is the widest, thickest part of the blade, and can be used for cutting tasks in which you have to push down forcefully. As you do that, you might push down with your hand on the **bolster** area. The **flat** of the blade can also be used for tasks such as crushing garlic, or as a vehicle to transfer cut ingredients into a bowl or pan.

In addition to the different parts shown in Figure 9-11, each knife has a **tang**. The tang is the part of the metal blade that continues into the knife's handle. If the metal continues all the way to the end of the handle, it's called a **full tang**. If it does not, it's a **partial tang**. A full tang knife is more stable. Figure 9-12 shows the difference.

FIGURE 9-12
A knife with a full tang (right) like this one may be more durable and withstand heavier cutting work than one with a partial tang (left). Take a look at the knives in your kitchen at home. How many of them have a full tang, and how many have a partial tang?

How to Use a Knife

You have probably been cutting food with a knife and fork since you were a pre-schooler, so it might seem odd to talk about how to use a knife. But cutting with a knife as part of food preparation is actually very different from cutting up the food on your plate with a dinner knife.

Because knives are sharp, your first consideration when using a knife must be safety. The hand that is holding the knife is actually very safe; it's nowhere near the blade. Therefore, as you are cutting with a knife, you should be most concerned about where your *other* hand—the **guiding hand**—is.

When cutting a small item, such as a vegetable, tuck the fingers of the guiding hand under the knuckles slightly, and hold the object with the thumb held back from the fingertips. That way, the knife blade rests against the knuckles, making it impossible to cut your fingertips. See Figure 9-13. When cutting something large, like a loaf of bread, you can place your guiding hand on top of the food to keep it from slipping.

FIGURE 9-13
When cutting small objects like herbs, fruits, or vegetables, guide the food with your knuckles rather than your fingertips to avoid cutting a finger. If you need to grip the food you are cutting, you might use a utensil. What type of utensil would work well for that?

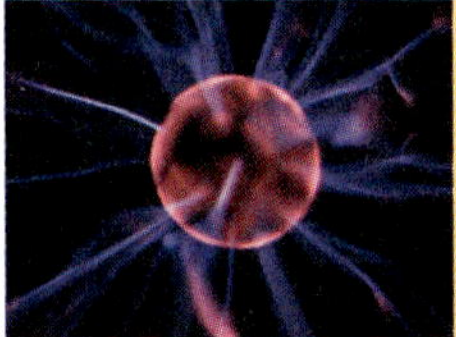

SCIENCE STUDY

The grip you use to hold a knife can affect the power and precision with which you are able to use it. Hmm, sounds like a physics experiment!

Try each of these grips with the same knife as you are cutting up a moderately hard vegetable, like a carrot or radish. Then try each one with a soft, delicate vegetable, such as a tomato.

- **Grip the handle with four fingers and hold your thumb firmly against the top of the knife.**
- **Grip the handle with four fingers and put your thumb against the side of the blade.**
- **Grip the handle with three fingers, rest the index finger against the side of the blade, and hold the thumb against the opposite side.**

Most people find that the first method gives them more power, while the second and third methods provide more control for making precise cuts.

FIGURE 9-14
When peeling, trimming, or slicing food, hold it in one hand and use the knife with the other hand. Do you think the person in the photo above is pulling the knife toward his body or away from his body? Which do you think is safer? How can you minimize the risk of cutting yourself when using a knife to peel?

When peeling or trimming food, you should hold it in one hand, while you work the knife with the other hand, as in Figure 9-14. Some people find it easier to pull the blade toward them as they peel or trim with a paring knife, but this is dangerous. Their hand could slip, and they could cut themselves. Make sure the food, your hands, and the knife handle are all dry to further avoid slippage, or use a peeler utensil rather than a knife.

There are three basic cutting techniques, which you learned about in Chapter 7:

- **Slicing:** Keep the knife straight and even, and stroke it through the food, letting the knife do the work. It's important that the knife be sharp. Choose a long, thin blade for very fine cuts or slices; use a small blade with small food. If there is a large quantity of slicing to be done, consider using a meat slicer or a mandoline.

FIGURE 9-15
Slicing can be done with a knife, as shown here, or with a slicer. If you use a knife, why is it important that the knife be sharp? How will the cuts be affected if you use a dull knife?

- **Chopping/mincing:** To chop or mince, keep the tip of the knife in contact with the work surface, and raise and lower the handle and the back part of the blade rapidly, making repeated small cuts until you get the desired fineness.

FIGURE 9-16
Chopping (left) results in larger pieces than mincing (right). What ingredients can you think of that are usually minced in recipes?

- **Shredding/grating:** To grate hard foods, like vegetables and cheeses, use a grater, or an appliance that has a grating attachment (like a food processor or mandoline). Soft foods can be shredded with a chef's knife.

Sharpening and Maintaining Knives

Every time you make a cut with a knife, it slightly dulls the edge, so over time, even a very good quality knife can lose its effectiveness. A dull knife is actually more dangerous than a sharp one, because you have to use more force to cut; and, using more force often translates into slips and accidents. A sharp knife, in contrast, glides almost effortlessly through most foods.

You can sharpen a knife with a sharpening stone (a **whetstone**). There are various sizes and textures of whetstones, and to sharpen a knife well, you may need multiple stones—first one that is relatively coarse, and then a finer one for finishing. You will also want a **steel**, which is a long steel or ceramic rod with a textured surface, used to straighten the edge of a knife and remove **burrs** (irregularities or bumps) both after sharpening and between sharpenings.

Safe Eats

- Hold a knife by its handle, never by the blade. Even when you hand it to someone, lay it down on a work surface and allow the other person to pick it up.
- If you drop a knife, don't try to catch it.
- Don't set a knife down on a work surface so that the blade hangs over the edge.
- Don't use a knife blade for prying jobs, such as loosening a jar lid or opening a bottle.
- Don't leave a knife loose in a sink filled with water, or anywhere else that it is not clearly visible to others.
- Always cut away from your body.

FIGURE 9-17
This ceramic rod sharpener is an alternative to a whetstone. Each rod is at a different angle and has a different coarseness.

Utility Drawer

For chopping and mincing, choose a knife in which the blade is tall enough that when the cutting edge rests flat on the work surface, your hand can grip the handle without having your knuckles touch down. That way, you can rock the blade back and forth rapidly without smashing your knuckles.

FIGURE 9-18
Sharpen a blade by running it across a whetstone. Why do you think the angle at which the edge connects with the stone is so important?

BASIC CULINARY SKILLS

Sharpening a Knife

1. Choose an appropriate stone; the duller the blade, the coarser the stone you will need to start with.
2. Lubricate the stone with mineral oil or water. You should use the same thing each time with that stone (water or oil); don't mix it up.
3. Point the blade at the stone at a 20-degree angle and slide the entire cutting edge of the blade smoothly across the stone, in one direction only. You can start at either the heel or the tip, as long as you go in the same direction each time. Use your other hand to guide the blade and to keep the pressure consistent.
4. Make 10 passes on each side of the blade.

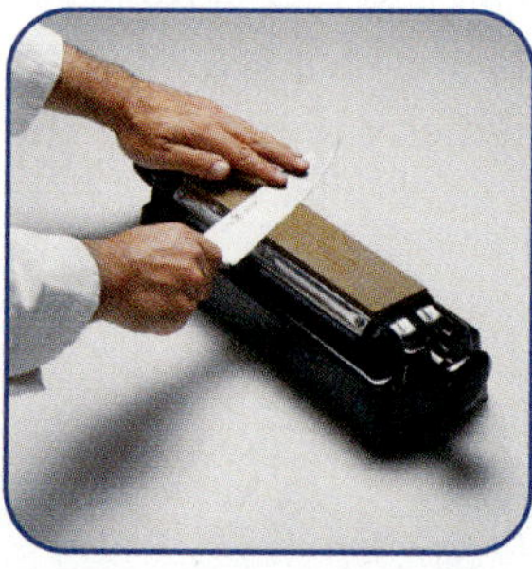

5. Switch to a finer stone, and repeat the process. Repeat again with the finest stone.
6. Hone the knife to remove any burrs. (See the next set of steps).
7. Clean and sanitize the knife, and clean the stone.

Utility Drawer

There are also machines that sharpen knives. Most of these machines are simply automated versions of whetstones, with spinning stone disks and channels that place the blade at the appropriate angle for sharpening.

BASIC CULINARY SKILLS

Honing a Knife

1. Hold the steel vertically with the tip resting on a non-slip work surface.
2. Position the heel of the knife against one side of the steel, near the handle.
3. Using arm action (not wrist action), draw the knife down the shaft of the steel and out from the steel so the entire blade is honed.
4. Repeat a few times for the first side, and then hone the other side an equal number of times.
5. Clean and sanitize the knife.

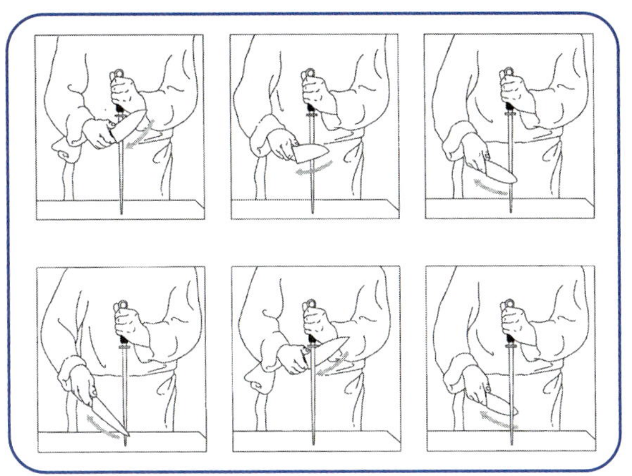

Maintaining Cutting Boards

Do not cut directly on the countertop: Always use a cutting board. There are two reasons for this: one is sanitation (i.e., the countertop may not be sanitized), and the other is that the knife blade could damage the countertop. **Cutting boards** are surfaces designed specifically for cutting with knives. They can be made of wood, plastic, or stone. Wood cutting boards are more difficult to keep clean, because they are softer and tend to get gashes and chips that can hide bacteria.

Safe Eats

Don't test a blade's sharpness by touching it with your finger! Test it by cutting a food.

To avoid cross-contamination, it is important to switch cutting boards when you switch from one type of food to another (from meat to vegetables, for example). Some people have separate cutting boards for different types of food. If you don't have multiple cutting boards, you should stop what you are doing and wash/rinse/sanitize the cutting board before changing the type of food that you are cutting.

Selecting and Using Smallware

Individuals who prepare food use a variety of hand tools, collectively known as **smallware**. Knives are technically considered smallware. Other products might include:

- Tools for taking things apart (cutting/slicing/grating and so on), such as garlic presses, apple corers, and vegetable peelers.
- Tools and equipment for mixing things together, such as scrapers and turners (both of which are sometimes called spatulas), spoons, and whisks.

- Tools for straining, draining, and processing, such as sieves, food mills, and funnels.
- Measuring tools such as scales, measuring cups, and thermometers.
- Bowls and pans for holding the food as you work with it.

Table 9-1 shows some common types of smallware. You will learn the specifics of using some of these tools in later chapters.

TABLE 9-1: COMMON SMALLWARE ITEMS

CUTTING TOOLS

Peeler: Removes the skin, or a thin outer layer, from vegetables and fruits. A safer and more efficient alternative to a paring knife.

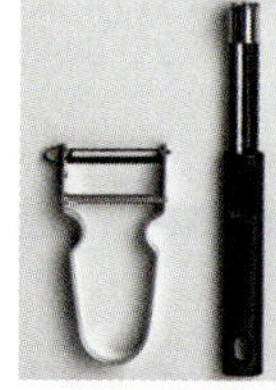

Corer: Removes the core from tree fruit, such as apples and pears.

Scaler: Removes scales from raw fish.

Pitter: Removes the pit from stone fruits, such as cherries and olives. (There are specific pitters for each type of fruit.)

Scissors: Cuts dense or tough items, such as artichoke leaves or grape stems

Melon baller: A scoop that makes melon (or other similarly textured food) into uniform. spheres.

Zester: Finely grates the outer layer of citrus peel.

MIXING TOOLS

Whisk: A set of wire loops used to whip liquids to add more air to them, and also to mix batter into a smooth consistency.

Spoon: Used for mixing, stirring, scooping, and serving; can be wood or metal, and solid, slotted, or perforated.

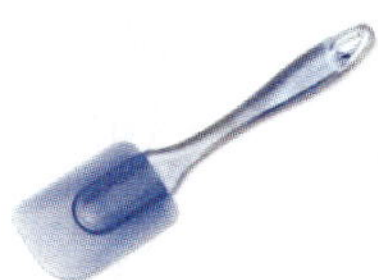

Rubber spatula (scraper): A broad flexible piece of rubber on a handle, used to scrape food from the inside of bowls and pans.

(continued)

TABLE 9-1: COMMON SMALLWARE ITEMS *(CONT)*

COOKING TOOLS

Skimmer: A small hand-held sieve used to skim impurities from liquids or to remove cooked food from hot liquid.

Fork: Used to move meat from a grill or broiler, or to hold large pieces of meat when carving them.

Tongs: Used to pick up hot items or items that you don't want to touch with your hands (like ice cubes when making a drink).

Metal spatula (turner): A flat metal plate on a handle, used to flip or lift hot foods. Sometimes called an offset spatula.

STRAINING, DRAINING, AND PURÉEING TOOLS

Food mill: A hand-crank device that strains and purees food at the same time. It has a flat curved blade that rotates over a perforated plate, pressing the food through the holes.

Sieve: A basket made out of screen, used for draining liquid from food that is too small or fine to be drained with a colander. Can also be used to strain or puree foods.

Colander: A perforated bowl used to strain or drain foods like spaghetti and cooked vegetables.

Funnel: A cone-shaped tool used to pour liquid into a container through a small opening, so that you don't spill any of it.

MEASURING TOOLS

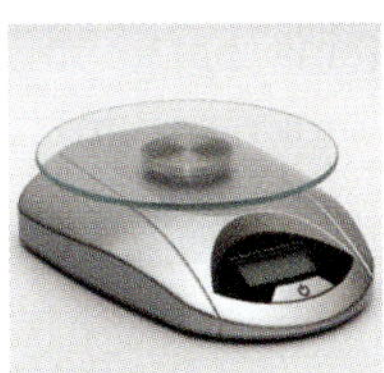

Scale (digital): A meter for measuring the weight of food (or anything else placed on it).

Measuring cup: A container that gauges the volume of an ingredient. As you learned in Chapter 7, measuring cups can be for liquids (typically a clear cup with marks on it) or for dry ingredients (typically a set of flat-top cups).

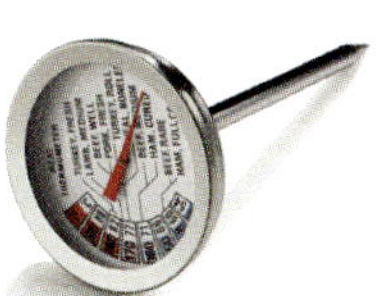

Thermometer: A meter for measuring the temperature of food.

Measuring spoons: A set of spoons with flat tops that gauge small volumes of ingredients.

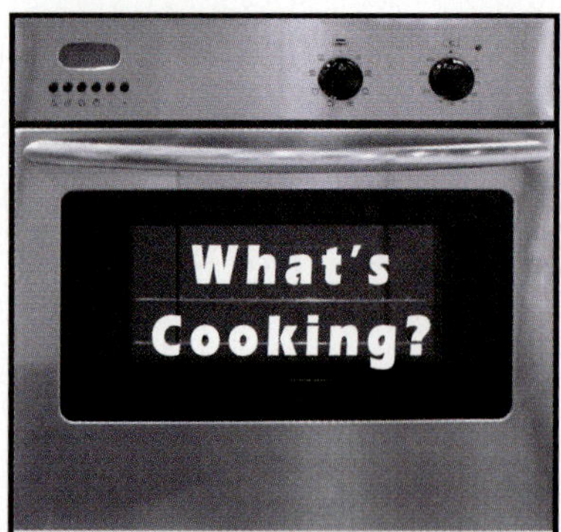

To **strain** something is to remove the solid matter from it, and then use the resulting liquid. For example, you might strain fresh orange juice to make it pulp-free. When you strain, you throw out what's in the strainer.

To **drain** something is to separate the solid from the liquid, like draining the cooking water off of spaghetti. You keep the solid part, discarding the liquid.

To **purée** means to force the food through a fine-holed mesh, breaking up any chunks that are larger than the holes in the mesh. You then keep the food that falls through the mesh, and discard any chunks left in the sieve that can't be broken up.

There is a hierarchy of workers in a commercial kitchen. At the top of the chart is the **executive chef,** who is the manager of the entire kitchen. An executive chef plans the menus, supervises the preparation of the dishes, and manages the budget. At least two years of culinary arts and management training is expected for someone in this position, although some very experienced chefs may rise to this position without formal classroom training.

A **sous chef** is the assistant to the executive chef (the head chef). The sous chef is the second in command in the kitchen, and runs the kitchen when the executive chef is not there. The sous chef is expected to be able to perform all the duties of all the employees in the kitchen, to fill in where required. Being a sous chef is excellent training for taking on an executive chef position in the future.

A **line cook** (a.k.a. a **station chef**) is a person in charge of a particular area of production. For example, there might be a line cook in charge of baked goods, another one in charge of the grill, and so on.

A **commis chef** is an apprentice who is learning the responsibilities of one or more stations.

Types of Cookware

Cookware is another name for pots and pans, the vessels that are typically used for range-top cooking and oven baking. Table 9-2 shows various types of common cookware.

Have you ever wondered what the difference is between a pot and a pan? A **pot** is typically deep with straight sides and two handles. A **pan** is more shallow and has curved sides and a single handle. Most cookware items that go in the oven are considered pans (such as a roaster, a cookie sheet, or a ramekin).

When selecting an appropriate piece of cookware for a task, here are some factors to consider:

- **Size:** The food should fit comfortably without crowding. If you are going to be cooking with a liquid, such as boiling, there should be plenty of room for both the liquid and the food, with extra space at the top, so it doesn't overflow.
- **Material:** Cookware can be made of a variety of metals, including stainless steel, cast iron, non-stainless steel, copper, aluminum, or glass.
- **Sides:** Some cooking techniques work best in a pan with sloped sides, such as a wok. Others work better in straight-sided pots or pans.

TABLE 9-2: POTS AND PANS FOR COOKING AND BAKING

Stockpot

Saucepan

Sautée pan

Skillet

Wok

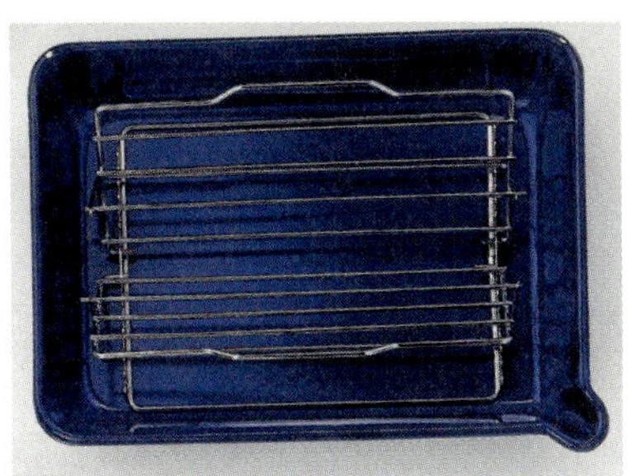
Roasting pan

Sheet pan

Gratin dish

Casserole dish

Safe Eats

Teflon-coated pans are generally safe, but beware of pre-heating a Teflon pan (that is, heating it without any food in it). The coating on the pan begins breaking down when heated above 500° F, and releases a gas that can cause flu-like symptoms in humans. This doesn't happen if there is food in the pan because the heat transfers to the food.

- **Coating:** Some pots and pans have non-stick coating. This coating can prevent the food from sticking to the pan, making cleanup much easier. However, if you use coated cookware, you must choose utensils that won't damage it, such as those made of wood, plastic, or silicone.
- **Range or oven suitability:** Some cookware is designed for baking only; you should not use it on the range-top because the heat is too direct, and can damage the pan. Other cookware may be designed for range-top use only; for example, it might have plastic handles that will not hold up in the oven. Still others are dual-purpose and will work in either place.

Equipment for Holding and Serving Food

In your own kitchen, you will probably hold and serve food in the same pan in which it was cooked. Or you might put the food into a serving bowl, or onto a platter.

In commercial food service, however, there are various types of special equipment, appliances, and pans specifically designed for holding and serving food. For example, in a buffet or cafeteria line, there are pans under heat lamps to keep the food hot while waiting for customers. Some common types of food holding equipment include:

- **Hot plate:** An electric burner that keeps food at a certain temperature, such as a pot of coffee.
- **Holding cabinet:** A cabinet with shelves where you can store multiple trays of hot or cold foods awaiting service.
- **Heat lamp:** A lamp with a special heat-generating light bulb, which shines directly on the food to keep it hot.
- **Steam table:** A large table with a deep, hollow top into which hot water is poured and heated. A grid above the water holds metal pans of food, and the steam generated by the heated water keeps the pans warm.
- **Chafing dish:** A stand-alone covered dish with a heating element underneath it, such as a sterno can or gas burner. As with a steam table, you put water in the bottom of the dish, and then set a serving pan over the top of it. The hot water in the bottom creates steam which keeps the food warm.
- **Ice pans:** Food that should be kept cold can be stored in a larger pan that contains ice, or a mixture of ice and water. On a salad bar, it is common to place containers of perishable ingredients on a bed of ice.

- **Coolers:** For small quantities of food, use of standard coolers is quite common. Lots of people have them, and they are inexpensive and readily available. A cooler can keep food either hot or cold for several hours.

The pans used in a steam table, and in holding and serving many foods, are called **hotel pans,** probably because their use originated in hotel kitchens. There are several sizes available. A full-size pan is 20¾" × 12¾" and can be of various depths, ranging from 2½" to 6". Half-size pans are half of that (10 ⅜" × 12 ¾"). Pans are available using these size proportions all the way down to one-ninth size (6⅞" × 4¼").

FIGURE 9-19
Hotel pans. Why do you think the food service industry tends to use standard sizes for pans?

In places where customers fix their own plates, such as at salad bars and buffets, there are usually **sneeze guards**. These are plastic shields (usually clear) that prevent a sneezing customer from contaminating the food. Customers must reach under the sneeze guard to get the food.

Containers for Storing Food

Most kitchens, both home and commercial, have a variety of storage containers for leftover food, raw food that is being prepped for later processing, and food that is being refrigerated or frozen for later re-heating.

A container can be either covered or uncovered. In most cases, a covered container is superior because it does a better job of preventing contamination. The cover can either be loose (such as a piece of aluminum foil over the top of a pan) or tight (such as a tight-fitting plastic lid that creates a vacuum seal when pressed on). Tight lids seal in the air, which keeps some foods fresher. However, other foods, like certain fruits and vegetables, keep longer when air is allowed to circulate around them.

Storage containers can be stainless steel, glass, or plastic. Glass containers have the advantage of being see-through, so you can tell at a glance what is in them; however, they are prone to breakage, and broken glass can contaminate food. Stainless steel containers are easy to clean, but can't be microwaved, and some acidic foods—like tomatoes and citrus—take on a metallic taste if stored in stainless steel for a long time. Plastic containers are inexpensive and usually have tight-sealing lids, but many are not recommended for microwave reheating.

FIGURE 9-20
Storage containers should be labeled and dated. It's a must in commercial food service, and a good idea at home as well. Why do you think labeling is required in restaurant kitchens?

Whichever type of storage container you choose, clean and sanitize the container before using it. Cover containers with their lids or with heavy plastic wrap or foil. Then—and this step is very important if you are working in commercial food service—attach a label to the container that indicates what the food is and the date of storage.

Kitchen Safety Issues

When you're working in a kitchen, there are two main hazards to look out for—fires and injuries. Fires are a risk because of the heat and open flames used for cooking; injuries are possible because of the sharp blades and implements used to chop and process food.

Fire Safety Basics

There are four important aspects to fire safety:

- **Prevention:** Prevent fires by avoiding the conditions in which a fire can start. Be careful of heating elements, electrical appliances, and other fire hazards. There are many ways to accidentally start a fire in a kitchen. Ranges will set anything aflame that is accidentally left on a hot burner, from a kitchen towel to a cardboard box of supplies. Overheated cooking oil or grease in a pan can spontaneously burst into flame as you cook with it. Electrical wiring can spark when improperly grounded, overloaded, or exposed to water.
- **Detection:** Make sure fire detection equipment is working properly, so that if a fire does occur, it is quickly detected. Smoke detectors are a critical apparatus in all kitchens. Some facilities also have heat detectors, which are activated by a sudden rise in temperature even when there is no smoke. Regularly check the condition of fire detection equipment.
- **Extinguishing:** Make sure fire extinguishing equipment is serviced and ready to go. Some commercial kitchens have sprinkler systems in the ceiling; all kitchens should have fire extinguishers.
- **Training:** Train everyone who works in the kitchen in fire extinguishing procedures, including how to operate the fire extinguishers, which fire extinguishers work on which types of fires, and how to put out a fire when you don't have a fire extinguisher handy. For example, if flames erupt in a pan, put a lid on the pan to smother the flames, and pull the pan off the heat.

Small fires can be put out with fire extinguishers. A "small" fire is one that is no more than three feet wide or three feet tall. If the fire is larger than that, you should call 911.

Fire extinguishers are not all alike; there are different models for different types of fires. Each fire extinguisher has letters on it to indicate the types of fires for which it can be used:

Fire Class	Source of Fire
A	Paper, wood, cloth, or plastic
B	Grease, gas, oil, or liquid stored under pressure
C	Electrical equipment, outlets, cords, circuits, motors, switches, or wiring

A fire extinguisher is a canister with pressurized contents that can extinguish fires. That content can be water, foam, or a dry chemical.

- **Water:** Water-based fire extinguishers douse the fire with water. They are used on Class A fires only and do not work at below-freezing temperatures.
- **Foam:** Foam-based fire extinguishers blanket the fire with a foam that seals it off from the air, so that the fire smothers. They are used on Class A or B fires and do not work at below-freezing temperatures.
- **Dry Chemicals:** Chemical-based extinguishers interrupt the chemical reactions that keep a fire burning. They can be used on Class A, B, or C fires.

Cool Tips

One of the drawbacks of a dry chemical extinguisher is that it leaves a powder residue behind. Halon extinguishers use a halon gas to suck up the air around the fire, suffocating the flame without leaving residue. Sounds great, doesn't it?

Halon is another name for halogenated hydrocarbons, also called chlorofluorocarbons (CFCs), among other things. Unfortunately, halon gas released in the atmosphere is an environmental hazard, so it is illegal to produce new halon fire extinguishers. If you already have one, you can continue to keep it in service until it is discharged. Although halon extinguishers may not have lettered ratings on them, you can safely use them for fires of any class.

FIGURE 9-21
On the left, a water-based extinguisher; on the right, a dry-chemical model. Which do you think would work better for a flaming toaster?

BASIC CULINARY SKILLS

Putting Out a Kitchen Fire with a Fire Extinguisher

1. Confirm that the fire extinguisher is the right type for the fire.
2. Pull the pin.

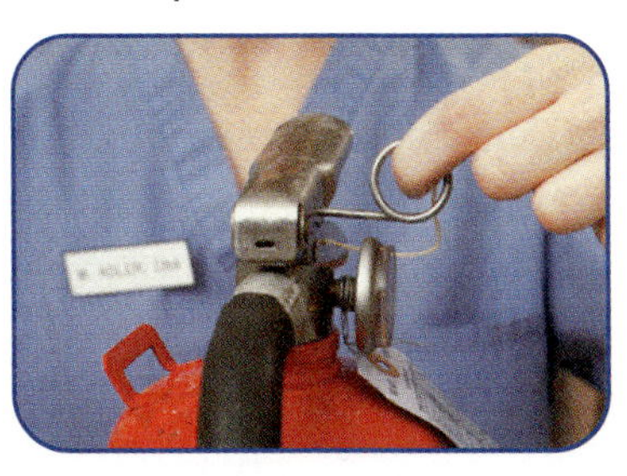

5. Aim at the base of the fire, while standing 6 to 8 feet away from it.
6. Squeeze the trigger.
7. Sweep from side to side until the fire it out.

First Aid for Kitchen Injuries

Whether you are working a small home kitchen or a large commercial one, accidents will happen. If you think about it, a kitchen presents a prime opportunity for accidents, because it has hazards—sharp pointy knives, electrical appliances that heat up, open flames, and flammable grease (that can also cause slippery floors). With all that hazard potential, it's important to be extremely careful.

Burns are a major hazard because of all the hot surfaces used for cooking. The treatment for a burn depends on how severe it is:

- **First Degree:** With this type of burn, the skin is red, sore, and swollen. Treat it by running cool water over it, or covering it with towels soaked in cold water. Do not apply ice directly to a burn.
- **Second Degree:** This type is deeper, more painful, and forms blisters. Cool the skin in the same manner as with a first degree burn, and then seek medical attention. Do not apply ointments or bandages, as this can seal in the heat and allow it to continue to damage the body.
- **Third Degree:** This type occurs when the skin is actually destroyed. It may turn white or black and does not have any feeling because the nerves have been burned. Cover it with a cool, moist, sterile gauze or a clean cloth. Do not apply ointments or water. Seek immediate medical attention.

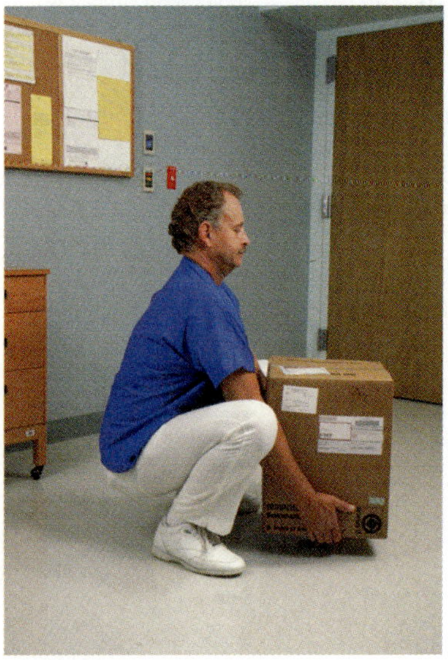

Cuts are the most common kitchen injury, because of all the equipment used for cutting food. For minor cuts, clean the cut thoroughly under cold running water, and apply pressure to stop the bleeding. Holding it higher than your heart may help slow the bleeding. Then, apply an antibiotic ointment and a sterile bandage. Depending on the severity of the cut, you may need to seek medical attention.

Another common kitchen injury is a body **strain** or **sprain** due to tripping or falling. The best way to avoid these injuries is to keep the floor clean and dry; oil spills on the floor can make it very slippery. If you do slip and fall, take a moment to catch your breath and assess the damage to your body. Depending on the severity, sprains (such as a wrist or ankle) may require use of a wrist or ankle brace or require medical attention.

FIGURE 9-22
Bending from the knees, not the lower back, is the correct way to pick up a heavy object.

Cool Tips

"Lift with your knees!" is a common piece of advice, and it really does help to avoid back strain when lifting heavy objects. As you are bending down to pick something up, bend your knees. Then raise yourself up by straightening your knees, keeping your back straight. Do not lift more than about 40 lb without help, machinery, or dividing the load.

Equipment Safety

Occupational Safety and Health Administration (OSHA) publishes standards for safely using and maintaining kitchen equipment. Your employer is required to provide you information and training that includes this information. Employees are also responsible for taking advantage of that information and training so they can keep themselves safe.

Before operating any equipment you should:

- Get trained in its use
- Wear any protective equipment provided by your employer
- Use any machine guarding provided.
- Ask for help if you are not sure how to do something
- Follow the manufacturer's instructions for use and cleaning.
- Be aware of age restrictions that prevent works under 18 from using or cleaning certain equipment, such as electric mixers, grinders, and slicers.

Table 9-3 summarizes some of the OSHA standards for equipment safety.

TABLE 9-3: OSHA SAFETY STANDARDS FOR EQUIPMENT

Equipment	Safety Guidelines
Mincers, Choppers, Slicers, and Dicers	■ Always use push sticks or tamps to feed or remove food from these machines. ■ Do not use your hands to feed smaller pieces of meat through slicers. ■ Make sure you are using any machine guarding that is provided to prevent access to cutter blades. Do not bypass safety guards. ■ Do not open up or put your hands into a running machine to stir contents or guide food. ■ Turn off and unplug the machine before disassembling and cleaning.
Food Processors, Mixers	■ If a foreign object falls into the dough, turn off the machine and unplug it before attempting to remove it. ■ Do not open up the lid to stir contents while food is processing. ■ Make sure the processor is off before opening the lid or adding items. ■ Turn off and unplug the machinery before cleaning or removing a blockage. ■ Use any machine guards provided. ■ Do not wear loose clothing or jewelry that could become caught in the machinery.
Microwave	■ Make sure the microwave is located at approximately waist level and within easy reach, to provide for ease in the lifting of hot foods. ■ Follow manufacturer's instructions for operating microwave ovens. ■ Cover foods cooked in microwaves to avoid splattering. ■ Open containers away from your face because they may be under pressure and be extremely hot. ■ Use appropriate protective equipment such as hot pads when removing foods from the microwave. ■ Make sure door seals are in good condition and free from food or grease buildup. ■ Do not use a microwave if it has a door that is damaged or doesn't lock properly. Damaged ovens may emit harmful radiation. ■ Do not microwave metals, foil, or whole eggs. ■ Keep the interior of the microwave clean to avoid splattering and popping. ■ If you notice any sparking inside the microwave, immediately turn off the microwave, unplug it, report it to the supervisor, and do not use it. ■ Be advised that microwaves may interfere with the workings of a pacemaker. ■ Be aware that food cooked in the microwave can remain hot long after the microwave turns off.

(continued)

TABLE 9-3: OSHA SAFETY STANDARDS FOR EQUIPMENT *(CONT)*

Equipment	Safety Guidelines
Steamers and Pressure Cookers	■ Do not open the door while the steamer is on. Shut off the steam, and then wait a couple of minutes before releasing the pressure and opening. ■ Clear the area around the steamer before opening. ■ Open the steamer door by standing to the side, keeping the door between you and the open steamer. ■ Use oven mitts to remove hot trays from the steamer. ■ Place hot, dripping steamer trays on a cart to transport. If the trays are carried by hand, they will drip on floors and create a slip hazard. ■ If a steamer is stacked, remove the tray from the top steamer first, then the lower one, to prevent burns from rising steam. ■ In a pressure cooker, wait for the pressure to equalize after shutting off the steam supply before opening the lid. ■ Open a pressure cooker by standing to the side and opening away from yourself, keeping the lid between you and the cooker.
Coffee Makers	■ Do not place hot coffee makers close to the edge of counters where people passing by may come in contact with them. ■ Check to make sure the coffee filter is in place before making any coffee. ■ Do not remove the filter before the coffee has stopped dripping. ■ Never stick your fingers into the chamber of a coffee grinder to get beans to drop into the grinder; tapping on the outside of the container will encourage beans to drop into the grinder.

Handling Hazardous Materials

Most of the materials you handle in food services are not hazardous—on the contrary, they are edible! However, in a commercial kitchen you may occasionally need to handle cleaning or pest control supplies that could be hazardous to your own health or the health of others if they are not used and stored correctly.

Each product that you use should have a Material Safety Data Sheet (MSDS) that lists its potential hazards, including whether it is safe to breathe its fumes, allow it to touch your skin, ingest it, or get it in your eyes. MSDSs are available from the manufacturer of the chemical, and an employer must keep the MSDS on file for employees to refer to. Not all MSDSs will look the same, but at a minimum they must contain these facts:

- Product name
- Manufacturer's name and address
- Chemical and common names of each hazardous ingredient
- Name, address, and phone number for emergency information
- Preparation or revision date
- The hazardous chemical's physical and chemical characteristics
- Physical hazards, including the potential for fire, explosion, or reactivity
- Known health hazards
- Exposure limits

- Emergency and first-aid procedures
- Whether any of the ingredients are carcinogens
- Precautions for safe handling and use
- Recommended control measures such as certain work practices or personal protective equipment
- Primary routes of entry
- Procedures for spills, leaks, and cleanup

A chemical may have a multi-colored diamond on its label or on its MSDS, or both, assigning a numeric value to the threat the chemical poses in each of four areas, as shown in Figure 9-23. This labeling system was developed by the National Fire Protection Association (NFPA) and is commonly called an NFPA label. Such labels are required for chemicals in a laboratory, and optional for chemicals in other settings. The four colors are:

- **Blue:** Health
- **Red:** Flammability
- **Yellow:** Reactivity
- **White:** Special

The numbering system goes from 0 to 4 in each of the three colored areas, with 0 representing minimal threat and 4 representing extreme threat. The white area is for any special information. The four symbols you might see there are a W with a line through it, indicating water reactivity, the letters Ox, which signify an oxidizing agent, a radioactivity symbol (indicating radioactivity), or a skull and crossbones (indicating poison), as in Figure 9-23.

Safe Eats

If you are not sure about the safety of a particular chemical, and a MSDS is not available, follow these general tips:

- Read all labels and instructions carefully.
- Don't mix chemicals. In particular, don't mix products containing ammonia with products containing bleach. This releases chlorine gas, which can be deadly to inhale.
- Thoroughly rinse buckets and containers after using chemicals in them.
- Drain cleaners, oven cleaners, and grill cleaners can be caustic and can burn your eyes and skin.
- Try not to inhale the fumes from chemicals. Use them in well-ventilated areas when recommended on the label
- Flush drains before and after dumping the chemical waste. If you aren't sure if the product is safe to dump down a drain, don't do it.

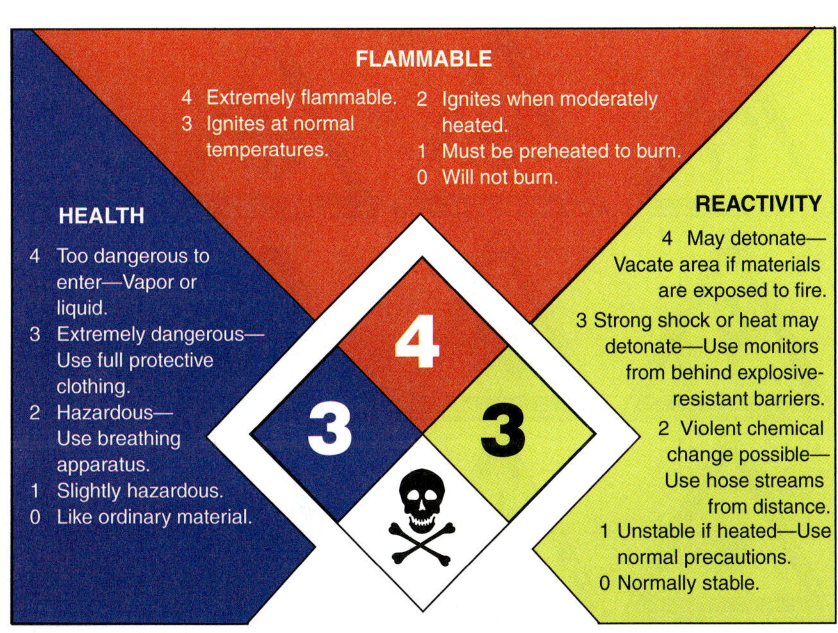

FIGURE 9-23
Some chemicals have an NFPA label that categorizes its threats in several areas.

The OSHA Hazard Communication Standard

According to the Occupational Safety and Health Administration (OSHA), more than 30 million workers are potentially exposed to one or more chemical hazards, and there are an estimated 650,000 existing hazardous chemical products, with hundreds of new ones being introduced annually.

To make sure workers have the knowledge they need to protect themselves, OSHA has developed a Hazard Communication Standard (HCS), also called a "Right to Know" standard, that makes chemical manufacturers responsible for providing accurate and detailed hazard information to consumers, and for employers who require their employees to use these chemicals to train them properly to do so.

The full version of these standards is available at http://www.osha.gov/SLTC/hazardcommunications/index.html. Here are the highlights:

- Chemical manufacturers and importers are required to evaluate the hazards of the chemicals they produce or import.
- Chemical manufacturers and importers are required to prepare labels and material safety data sheets (MSDSs) to convey the hazard information to their downstream customers.
- All employees with hazardous chemicals in their workplaces must have labels and MSDSs for their exposed workers, and train them to handle the chemicals appropriately.

FIGURE 9-24
There are many different warning signs to indicated various hazards. Research what some of these stand for. Which do you think you might be likely to see in a kitchen or restaurant workplace?

Case Study

Robin is preparing a meal in which several recipes call for ingredients to be prepared in specific ways. Robin does not have the right tools for these tasks.

- She needs 1 tsp of lemon zest, but she does not have a zester.
- She needs peeled peaches, but she does not have a peeler.
- She needs melon balls, but she does not have a melon baller.

What ideas can you offer her for preparing the foods without those utensils?

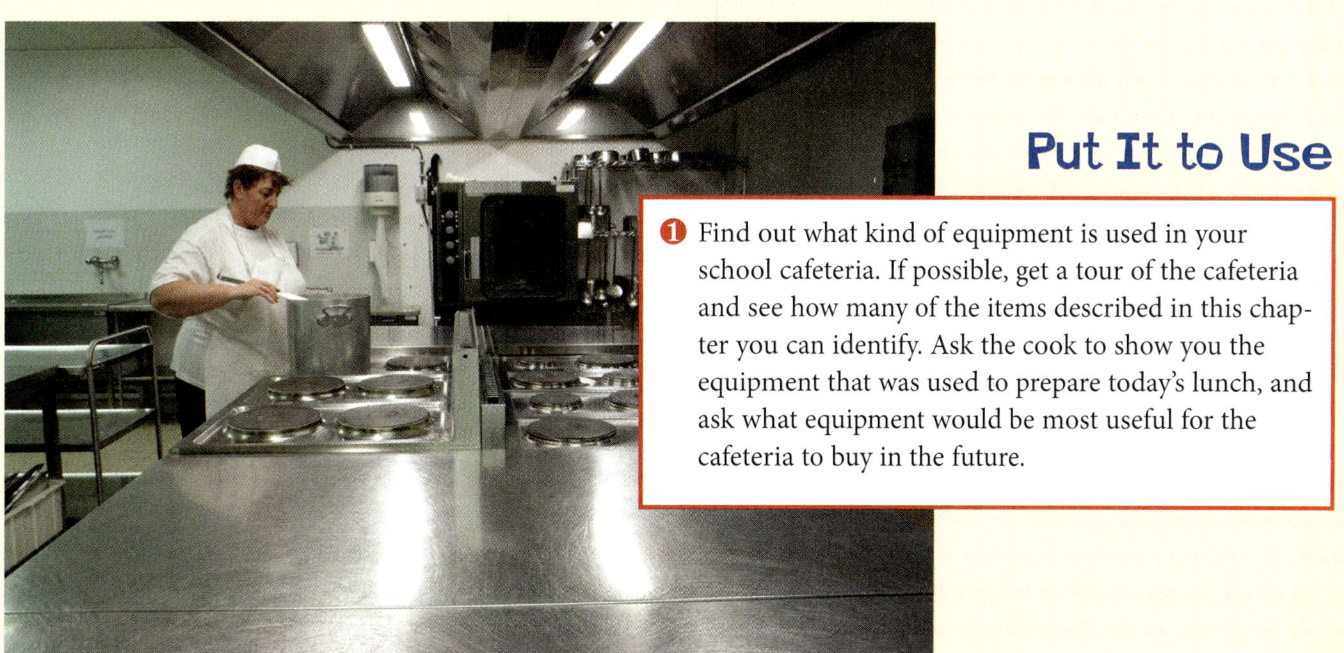

Put It to Use

1. Find out what kind of equipment is used in your school cafeteria. If possible, get a tour of the cafeteria and see how many of the items described in this chapter you can identify. Ask the cook to show you the equipment that was used to prepare today's lunch, and ask what equipment would be most useful for the cafeteria to buy in the future.

Put It to Use

❷ If you have a food processor available for your use (in your classroom, at home, or at a friend's house), try it out by cutting up some raw carrots. Try out each of the blades and attachments on a different carrot to compare the results. Can you identify the types of cuts made by each attachment? For example, can you tell the difference between mincing, shredding, and pureeing?

Write Now

Some cooks have very strong opinions about which type of burner is better to cook on—gas or electric. What is your opinion, and why? Interview some other people and get their opinions; conduct online research to gather additional opinions. Then, create a report that compares and contrasts gas versus electric cooking, including quotes from some of the people or sources.

Tech Connect 

There are dozens of smallware items available that were not covered in this chapter. Look online at kitchen tools and gadgets, and identify six that that did not appear in the chapter. Create a report, complete with pictures, that shows each one, describes what it is used for, and lists its cost and its features.

Team Players

Divide the class into two teams. Each team is responsible for creating a slide show (PowerPoint works well for this) that showcases pictures of different kitchen knives that students have at home. Use a digital camera to take pictures of knives, or find pictures on the Internet of knives that look similar to the ones you have. If possible, include a ruler or tape measure in each picture, so the audience can see is the size of the knife. Each team's slide show should consist of 20 pictures.

Each team will then show their presentation to the other team; that group will try to identify which of the eight basic knife types is represented by each picture. The team that correctly identifies more knives wins.

Put It Together

Match the explanation in column 1 with the term in column 2.

Column 1

- **a.** an appliance that can grind, slice, mix, blend, or crush food
- **b.** a manually operated slicer, typically used on vegetables
- **c.** an appliance that shreds or chops ingredients as it mixes them
- **d.** an appliance where food is mixed in a moving bowl, with the mixing tool held stationary
- **e.** an appliance where food is mixed in a stationary bowl, with the mixing tool moving
- **f.** a perforated bowl used to strain or drain foods
- **g.** a stovetop with burners on it
- **h.** the part of a metal blade that continues into the knife's handle
- **i.** a tool for sharpening knives
- **j.** any of a variety of kitchen hand tools

Column 2

1. blender
2. colander
3. food processor
4. mandoline
5. planetary mixer
6. range
7. smallware
8. spiral mixer
9. tang
10. whetstone

10 Kitchen and Dining Plans and Etiquette

In This Chapter, You Will . . .

- Learn about common floor plan layouts for kitchens and dining rooms
- Discover how to organize a kitchen's storage areas efficiently
- Understand common types of meal service
- Learn how to set a table for various meal types
- Identify the characteristics of appealing menus and dishes
- Learn some etiquette guidelines for serving food
- Practice basic table manners and etiquette for formal dining

Why You Need to Know This

Food service is not just about the food, but about the entire experience. The experience of a cook or a server can be dramatically improved by a well-designed kitchen and dining area. And the experience of a diner can be enhanced by attractive table settings, garnishes, presentation, and service etiquette. Can you think of a time when your eating experience was changed—for the better or for the worse—by something other than the food itself? Discuss some of those experiences as a class, and think about what you can do to create a pleasant dining experience for others.

Understanding Kitchen Layouts

Have you ever tried to work in a home kitchen while other people are also working there? Were you continually bumping into each other? Did you need a utensil or ingredient that was inaccessible because someone was in your way? That's because most home kitchens are not designed for multiple cooks, working at the same time. What may be an optimal arrangement of utensils, pans, and ingredients for a single cook can become a tangled traffic jam when you introduce another person into the process.

Commercial kitchens, on the other hand, are designed for multiple, simultaneous workers. A commercial kitchen is organized into **workstations**, where each work area contains all the tools necessary for accomplishing a certain set of tasks. For example, there might be a frying station, where all fried food for the entire kitchen is prepared.

In a well-arranged kitchen, stations that pass food off to one another are adjacent, and the entire kitchen becomes one big assembly line. For example, a prep station where food is chopped might be next to a grill station where the chopped food is cooked. On the other side of the grill station might be a plating station, where the food is arranged and garnished on a plate.

FIGURE 10-1
Here are two chefs at work in a commercial kitchen. You can see how important it is to have a good workflow layout while many people are working at the same time.

Many studies have been conducted to determine the optimal layouts for commercial kitchens. From these studies, several common work line arrangements have been developed:

- **Straight line:** A single line, perhaps along a single wall, that food passes down, starting with preparation and ending with garnishing. It works well for a narrow space, and for kitchens that produce a limited array of dishes that all pass through a similar preparation process.
- **L-shaped:** A corner arrangement, where two walls are used for workstations. This arrangement works well in smaller spaces because the workers keep out of the center of the room. It also allows workers to step away from their stations to perform other tasks without blocking other workstations.
- **U-shaped:** Like a two-legged L-shape, a U provides a large amount of work surface. However, workers can get in each other's way if the space is not large enough and/or if there are more than a couple of workers sharing the space at once.
- **Back-to-back:** In this arrangement, two straight-line workstations are placed back to back, so that the workers are facing each other, rather than facing a wall. This arrangement is efficient because workers can pass items quickly from station to station without getting in each other's way, but it requires a large amount of space.
- **Parallel:** This is the opposite of back-to-back. There are two parallel straight lines, and the workers for both lines stand in a common aisle between them. The workers stand back-to-back.

FIGURE 10-2
Here are some possible layouts for a commercial kitchen. Which one of these would be most appropriate for a very large kitchen, such as in a banquet hall?

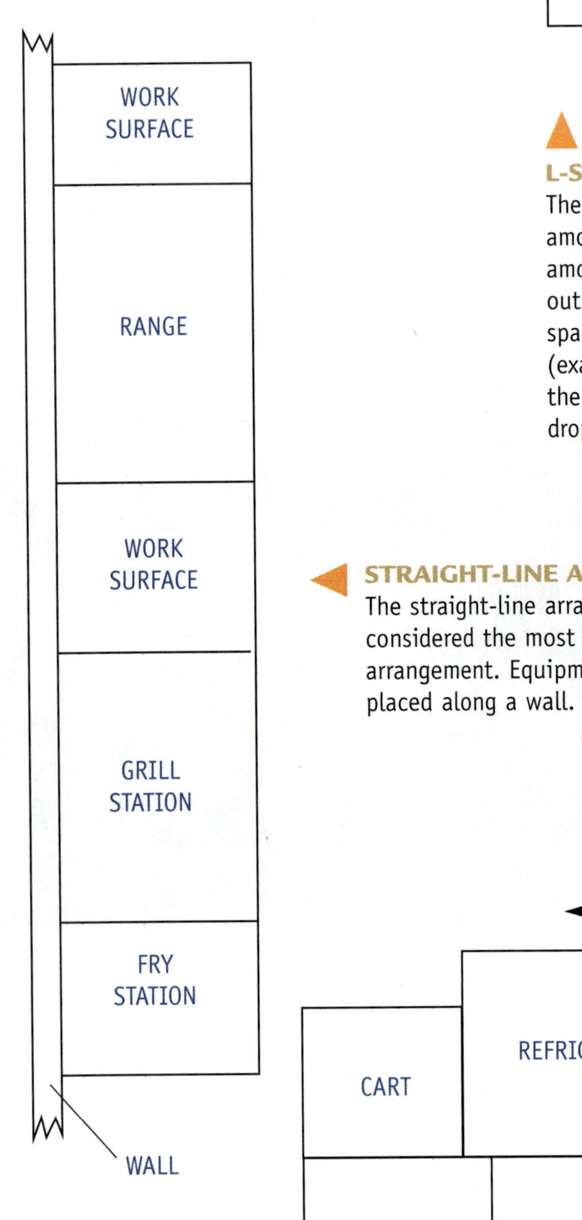

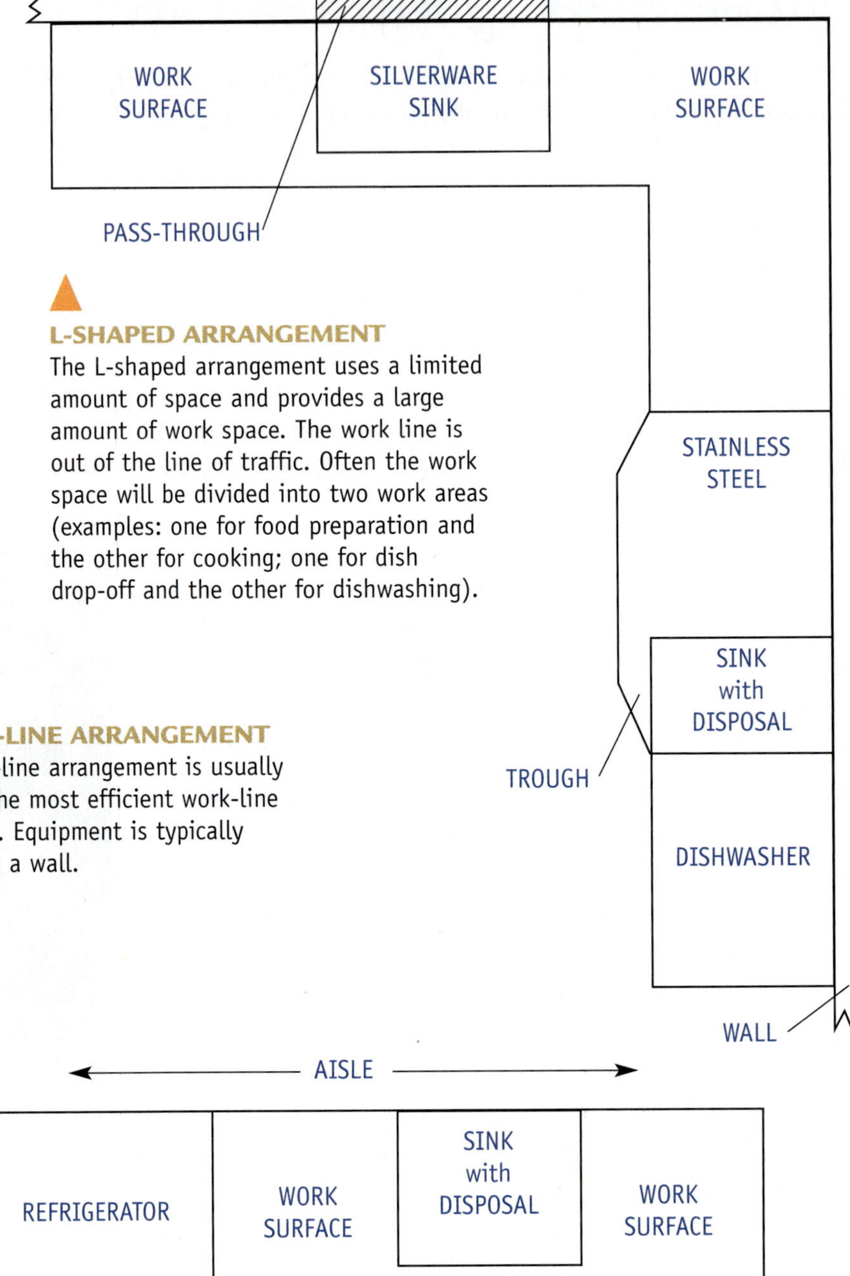

L-SHAPED ARRANGEMENT

The L-shaped arrangement uses a limited amount of space and provides a large amount of work space. The work line is out of the line of traffic. Often the work space will be divided into two work areas (examples: one for food preparation and the other for cooking; one for dish drop-off and the other for dishwashing).

STRAIGHT-LINE ARRANGEMENT

The straight-line arrangement is usually considered the most efficient work-line arrangement. Equipment is typically placed along a wall.

AISLE

CART
REFRIGERATOR
WORK SURFACE
SINK with DISPOSAL
WORK SURFACE

OVEN
FRY STATION
WORK SURFACE
GRILL STATION
RANGE

AISLE

BACK-TO-BACK ARRANGEMENT

The back-to-back arrangement is very efficient, but requires a large amount of space. A shelf or wall may separate the two tables. The back-to-back arrangement is often used in larger restaurants.

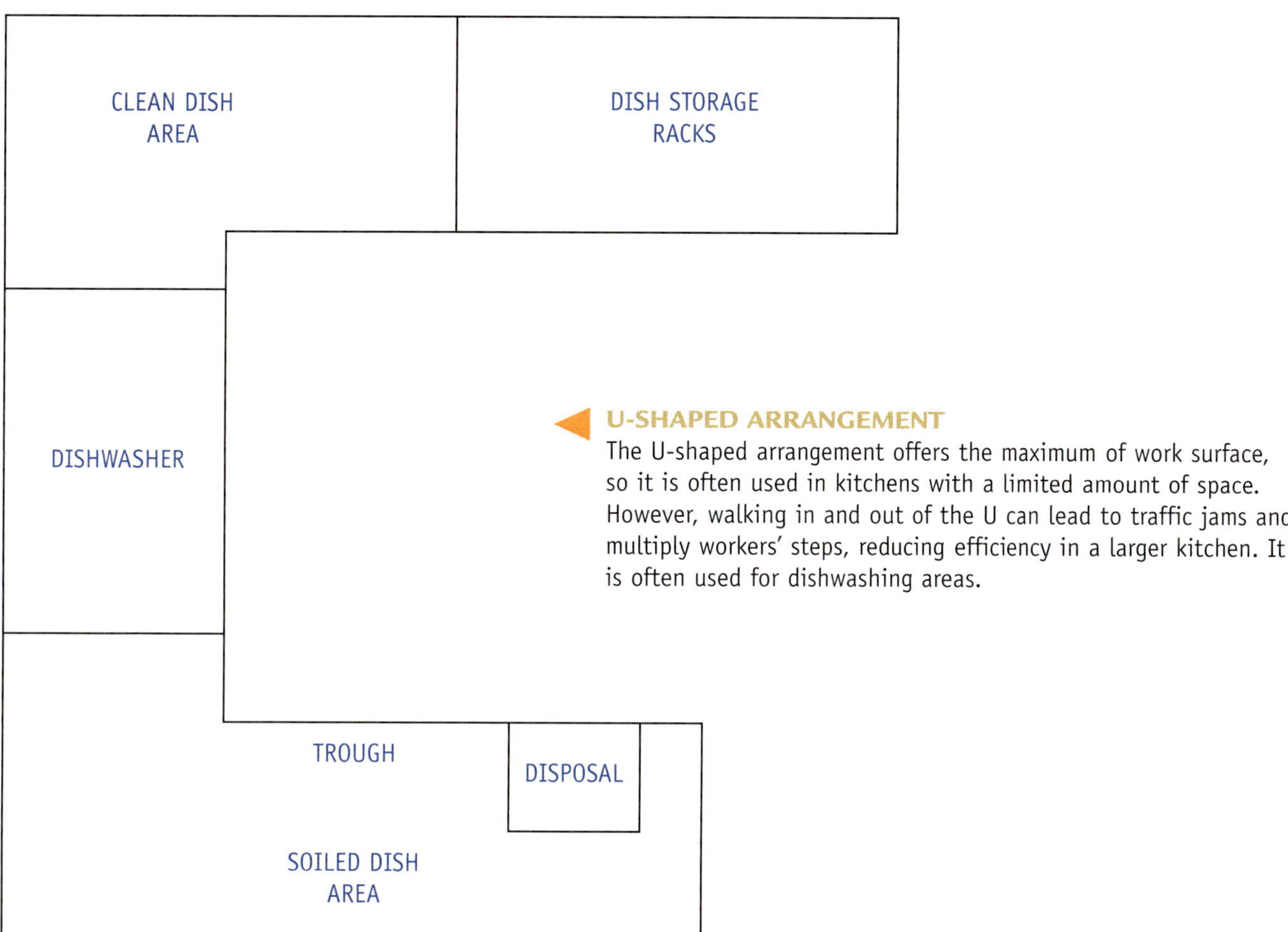

U-SHAPED ARRANGEMENT
The U-shaped arrangement offers the maximum of work surface, so it is often used in kitchens with a limited amount of space. However, walking in and out of the U can lead to traffic jams and multiply workers' steps, reducing efficiency in a larger kitchen. It is often used for dishwashing areas.

PARALLEL (OR FACE-TO-FACE) ARRANGEMENT
In this arrangement, two work lines are arranged face-to-face, with one common work aisle. The parallel arrangement also requires a large amount of space but is quite efficient (particularly when communication between workstations is required).

OVEN	STEAMER	BRAISING	WORK SURFACE	KETTLES

← AISLE →

OVENS	GRILL STATION	RANGE	FRY STATION

Home kitchen layouts are usually similar to commercial layouts, sometimes introducing an island in the middle. An **island** is a freestanding work area, usually in the center of an L or U shaped arrangement.

The classic way to organize a home kitchen is with a work triangle, based on the three main appliances—sink, range, and refrigerator. However, this method, which originated in the 1950s, doesn't take into account the microwave. Some kitchen designers have corrected this omission by placing the microwave above the range.

To create a classic work triangle:

- Arrange the sink, range, and refrigerator in a convenient way that naturally results in a triangle. Don't put any of these adjacent to one another.
- Try to have work areas on both sides of both the sink and the range, so you will have more space to work at each of these workstations.
- The length of the three sides of the triangle, added together, should be no more than 26 feet, with each side of the triangle being between 4 and 9 feet. This provides a work area that is large enough for one person to move around in, but not so large as to make you take multiple steps moving from place to place.
- The main traffic route through your kitchen should not pass through your triangle.
- An island or peninsula should not interrupt the triangle. In particular, there should be no barrier between any of the three major appliances.

Figure 10-3 shows a kitchen layout that conforms to all these rules.

FIGURE 10-3
A home kitchen is usually a U or L shape, often with a center island. What shape is your kitchen layout at home?

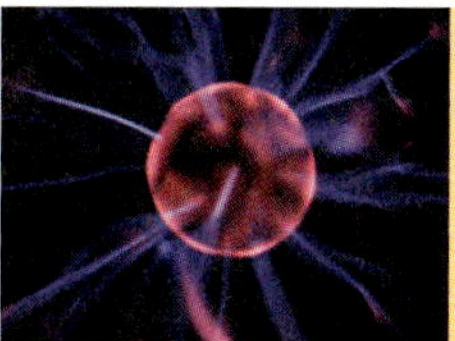

SCIENCE STUDY

The work triangle concept provides a simple way to evaluate a kitchen's usability.

As a class, or in a group of at least four people, create sketches of the kitchen layouts in your homes. Then evaluate each of the sketches to identify the work triangle. Ask the following questions about each drawing:

- **Are the sides of the triangle between 4 and 9 feet?**
- **Do the three sides, added together, equal 26 feet or less?**
- **Are there any barriers between any of the major appliances?**
- **Is someone walking through your kitchen able to do so without stepping into the triangle?**

If the answer to any of the questions is No, figure out a remodeling plan for the kitchen that would solve the problem.

A more contemporary way of looking at home kitchen design is to think about the types of tasks you perform and then design "centers" or mini-workstations for each one. For example, you might have a:

- **Food prep center:** This can be a counter space between the refrigerator and the sink, with a drawer containing knives and other tools for taking food apart, such as food processors and graters.
- **Cooking center:** This can be the range or cooktop and a countertop nearby where food can wait until it is added to the cooking process. This area usually includes cooking-related utensils such as pots and pans, spatulas, cooking condiments, such as spices and cooking oils, oven mitts, and a fire extinguisher.
- **Snack center:** This area may be near the microwave and the refrigerator, and may include other small appliances, such as a coffeemaker and/or water cooler. Plates and bowls used for snack items might be stored here, along with foods that you usually prepare for snacks, such as microwave popcorn.
- **Cleanup center:** This area can include your sink and dishwasher, plus the cleaning supplies you use in and around them. You may also want to store your eating utensils and plates nearby (making it convenient to unload the dishwasher), as well as dish towels and your garbage can.

Cool Tips

Turn off burners, griddles, ovens, and other heat-producing equipment when not in use. It takes a great deal of electricity or gas to maintain a heat source. Exception: you might not want to turn off something that takes a long time to heat up again if you are going to be using it within the next 30 minutes or so.

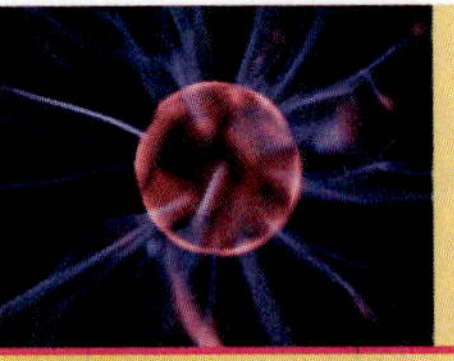

SCIENCE STUDY

Work Simplification is a scientific approach to the study of work processes, with a goal of making people's work as easy and simple as possible, reducing waste, saving time, and raising productivity.

To analyze the effectiveness of a flow, you can use process charts that mark each step of an procedure, such as the preparation of a dish, as one of the following:

- **Operation: Occurs when something is assembled, disassembled, or changes.**
- **Transportation: Occurs when something is moved to a different location.**
- **Inspection: Occurs when an object's state or quality is verified.**
- **Delay: Occurs when an object waits for the next step.**
- **Storage: Occurs when an object is kept and protected.**

Suppose you are making a sandwich in your kitchen, consisting of bread, cheese, and your favorite condiment. Create a flow chart using a combination of the five types of steps above. Then compare your flow chart with one from your classmates who has a different kitchen layout than you do. Whose kitchen is most efficient for the task? What changes could be implemented to simplify the sandwich-making process?

In addition to the work areas of a kitchen, there will also typically be storage areas. These may include walk-in refrigerators or freezers, **pantries** (rooms or closets for storage of non-perishable items), cabinets, and equipment racks.

Food, equipment, and supplies may be stored anywhere that is convenient, provided safety regulations are followed (see Chapter 8). For example, you would not store cleaning supplies or pest control chemicals anywhere near food or appliances that come directly into contact with food. Equipment with blades or other sharp parts should be stored where they will not hurt anyone, and where nobody can accidentally run into them.

Cool Tips

To conserve energy in the kitchen:

- In the wintertime in your home kitchen, leave the oven door slightly ajar after you turn it off, to allow the stored heat in the oven to help heat the kitchen. (Don't leave the door wide open, because it's a safety hazard.)
- Don't run the dishwasher until you have a full load. This conserves water.
- When possible, cook multiple dishes in the same oven simultaneously.
- Keep the refrigerator full, because it holds the cold more easily and requires less energy to operate.

Dining Room Layouts

A home dining room is limited in its layout choices. There is usually a single table, in the center of the room, with chairs on all sides. However, in a commercial dining area, such as a restaurant, there are many ways in which to arrange the dining room(s), depending on various factors, such as the type of meal service.

When planning the layout of a restaurant dining room, designers look at these factors:

- **Function:** How well does the dining room work for the style of dining featured? For example, a good layout for a smorgasbord (meal served buffet-style with multiple dishes of various foods on a table) is different from a good layout for fast food.
- **Capacity:** How many diners can be seated at once? Do you want to pack in as many people as possible, or create a sense of privacy? An average rule-of-thumb is 12 square feet per diner.
- **Flexibility:** How readily can the layout accommodate parties of different sizes? For example, are there enough tables that can be pushed together to seat a large party?
- **Efficiency:** How far must servers walk from each dining area to the kitchen? Can the servers get to all tables easily, even when the restaurant is full?
- **Aesthetics:** How does the dining room look? Is it attractive, or does it seem as if all the diners are in one big "cattle pen"?

Cool Tips

Have you ever noticed that restaurants that have both booths and tables try to fill up the booths before the tables? There are two reasons for this. One is that, for a small party, a booth is often considered more desirable seating because it is more private. So the more people to whom you can give a desirable seat, the more happy guests you will have. The other, more important, reason is flexibility. Tables, unlike booths, can be pushed together to accommodate large parties at a moment's notice. So by leaving as many tables vacant as possible, the restaurant is open to the possibility of accepting a large, unscheduled group of diners.

The type of dining service is a major factor in determining what constitutes a good layout for the dining room. Some of the dining room types include family style (where people sit at large tables and pass around shared serving platters), buffets, plate service (like a catered meal, where a server brings everyone the same meal), individual ordering from a menu at a sit-down restaurant, fast food (or other counter-ordering types), and carry-out/to go food.

When you have dinner parties at home, the type of service you plan to provide may also make a difference in the way you arrange things. For example, if you plan on guests serving themselves in a buffet-style meal, you will probably not set plates on the table; you'll stack them at the beginning of the buffet line. On the other hand, if you are serving a meal family-style, you will set the plates at each chair ahead of time.

Safe Eats

Remember when planning a service method to keep food safety in mind. Make sure food is protected from physical contamination (such as using a sneeze guard on a buffet service and providing tongs for diners to pick up bread), and make sure that hot foods stay hot and cold foods stay cold.

Here are some ideas:

- If you are serving a lot of people, try buffet-style service. It's efficient and serves more people quickly, so food has less time to get cold. You can seat everyone at one big table, or you can set up small café-style tables if you have room.
- If you are serving young children, provide individual plates rather than letting them help themselves to serving dishes. There will be less mess and less wasted food.
- When serving groups of people who have very different appetites or tastes, serve family-style. That way, people who don't like a certain item can unobtrusively avoid it, and everyone can have the portion size they want.

FIGURE 10-4
Restaurant dining areas are carefully designed to balance function, capacity, flexibility, efficiency, and aesthetics. If you were a server in a restaurant, which of these factors would have the biggest impact on your job success? What about if you were the restaurant owner?

Setting a Table

After you decide where the dining tables will go, and what style of meal you will serve, it's time to think about the table settings. Settings can be anything from formal silver-and-crystal to handing someone a plastic fork wrapped in a napkin. Most of the time, of course, you'll use something in between these two extremes.

Place settings might seem intimidating, but there are some simple rules that govern them. These include:

- The dinner plate goes in the middle of the place setting.
- Forks go to the left of the dinner plate, and knives and spoons go to the right.

- If there are multiples of a particular type of utensil (for example, two or more forks), arrange them so that the ones you use first are on the outside.
- Forks go in this order, from closest to the plate to farthest way: dinner fork, fish fork (if used), salad fork.

FIGURE 10-5
Formal place settings can vary depending on the courses and drinks being served. Here are two examples. What do they have in common, and what is different in each?

- Knives and spoons go in this order, from closest to the plate to farthest away: dinner knife, fish knife (if used), soup spoon, fruit spoon (if used).
- Knife blades should face towards the plate.
- Place a folded napkin to the left of all the forks.
- The bread and butter plate goes at 10 o'clock to the dinner plate. The butter spreader (if used) should be placed across the bread and butter plate at a slight diagonal, with the blade pointing to 11 o'clock.
- Dessert forks and spoons should go across the top of the plate. The dessert spoon is closest to the plate, and the handle points to the right. The dessert fork is above the dessert spoon, and the handle points to the left.
- The water glass should be placed just above the tip of the knives, at 1 o'clock in relation to the dinner plate.
- If there are wine glasses, they go to the right of, and slightly below, the water glass. If there is a champagne glass, it goes to the right of, and slightly above, the wine glass(es).

Family meals are seldom formal, and use fewer items in each place setting. For example, an informal family dinner service might simply include a dinner plate, a dinner fork, a dinner knife, a teaspoon, a napkin, and a water glass. These items go in the same positions as the rules dictate for a formal dinner; the only difference is that the other items are not present. Figure 10-6 shows a simple place setting using only those five items.

Many people appreciate having salt and pepper available on the dining table. You can put salt and pepper shakers on the table, or you can use an older, more formal method of placing **salt cellars** within reach of all diners. A salt cellar is a very small glass or silver bowl that holds salt; diners take a pinch of salt out of the bowl with their fingers and sprinkle it on their food by hand.

In addition to the individual place settings, a dining table may have other items on it as well. For example, you might decorate the table with flowers, a centerpiece, candles, and/or a table runner (a narrow piece of decorative cloth running the length of the table, down the center). You may also want to use either a table cloth or placemats. These make the table more attractive and serve to protect the surface of the table. Tables made of finished wood are especially vulnerable to moisture and heat damage; a placemat can protect the table from the heat of the plate and the condensation that may build up on a beverage glass. Some people place a plastic-coated protective pad under a table cloth for the same reason.

FIGURE 10-6
A simple dinner with family and friends may use very few table items. If you wanted to add a salad fork and a soup spoon to this layout, where would you place them?

Some types of meal service have special features or arrangements for table settings. For example:

- For a buffet, dinner plates are stacked at the beginning of the buffet, and not placed at each place setting. That way diners do not have to carry empty plates from the table to the buffet.
- In a restaurant, typically only a basic set of silverware is laid out at the table initially. Depending on what you order, additional silverware will be brought to you. For example, you would not have an oyster fork unless you ordered oysters. Restaurants do this to cut down on the amount of dishwashing they have to do; there is no point in giving someone four forks who is just going to order one course.
- At a luncheon, the size of the main plate is typically smaller than for a dinner.
- At a reception where only desserts or appetizers are being served, dining tables may not be set at all; you might need to pick up silverware and napkins as you pick up your food at a serving table.
- At luncheons, teas, or other daytime events, a coffee or tea cup may be part of the place setting. (At evening meals a cup and saucer are brought only if the person requests coffee or tea.) Coffee or tea cups and saucers are placed to the far right of all the other glasses, including wine glasses.

Cool Tips

A **charger** is a large decorative plate positioned at each place setting. No food is eaten from a charger. Either the charger is removed when the guests are seated, or smaller plates for some courses (such as soup and salad) are placed on top of the charger, and then it is removed prior to the dinner plate being placed.

Presenting Appealing Food

An attractive table is a great way to set the tone for an enjoyable meal. The main event, though, is the food itself. There aren't any fixed formulas for this; creating and serving appealing food is a subjective art, rather than a science. You may come up with your own ideas for a menu that appeals to your guests. However, there are certain generally accepted techniques and concepts that work well in both homes and commercial food services, and you'll learn about some of these here.

Characteristics of Appealing Menus

Generally speaking, people like a plate of food that contains interesting, attractive, and varied components. When you are planning a menu—whether it's just for friends and family, or for a commercially catered dinner or restaurant meal—keep the following suggestions in mind.

- **Vary the colors:** Try to use more than one color on a plate. So, for example, if the main entrée is brown, you could serve a vibrant orange or green vegetable with it.
- **Vary the textures:** Combine textures and food types on a plate. For example, next to a soft noodle dish, you might serve a crunchy garnish such as a pile of julienne-cut raw carrots.
- **Tidy up the plate:** Don't let a stray blob of food spoil the appearance of the plate. If you spill food on the rim of a plate, make sure you wipe it off before service.
- **Use the right size plate:** Don't use an enormous plate for a tiny food item. Some gourmet restaurants do this, but it is considered somewhat pretentious. Similarly, don't use a too-small plate with the food heaped and overflowing.
- **Heat or cool the plate:** When serving a hot meal, warming the plate before you put the food on it can be a nice touch and will keep the food warm longer—especially if it is not placed in front of the diner immediately. Similarly, you can chill a plate before serving cold food.
- **Use attractive plates and bowls:** A pretty plate can improve the look of your food! White plates are fairly standard, but some restaurants use bright red or yellow plates, because those colors are thought to stimulate appetite. You can also use chargers to dress up the table while still using your regular dinnerware.
- **Add accents:** You can easily and inexpensively dress up a plain food with a sprinkling of some sort of topping. For example, powdered sugar makes French toast or pancakes more special, as do a few slices of fruit or a handful of almonds or pecans.

Cool Tips

Once they are cut, some foods discolor quickly. These include apples, bananas, pears, avocados, and potatoes. If you are going to use any of these for garnishes, prevent discoloring oxidation by dipping them in lemon juice or some other citrus-based liquid (like lemon-lime soda pop), then cover and refrigerate them until you are ready to serve. If you make garnishes out of vegetables, such as carrots or celery, ahead of time, they may turn limp. To crisp them up again, soak them in ice water.

Using Garnishes

A **garnish** is a decorative piece of food used to improve the overall appearance of the plate. Garnishes are usually edible items, such as parsley, a strawberry, or a lemon slice; however, garnishes are not usually eaten, especially at business or formal meals. (Garnishes are covered in more detail in Chapter 15.)

Garnishes are often made by carving or molding ordinary foods into unusual shapes. For example, it is common to carve a radish into a rose. Some of the tools you might find useful for creating garnishes include an apple corer, a vegetable peeler, candy molds, a butter curler, a citrus stripper, a cake-decorating bag with tips, a grapefruit spoon, a hand grater, small cookie cutters, a melon baller, skewers and toothpicks, and, of course, knives.

Here are some ideas:

- Carve a radish into a flower by making multiple small cuts in it.
- Use a peeler to make curled thin strips of carrot or zucchini that look like ribbons.
- Cut carrot slices into flower petals and arrange them on a plate with thin strips of celery, herbs, or green beans for stems.
- Cut very thin slices of cucumber, lemon, or lime and twist them.
- Make lengthwise cuts almost to the bulb end of a green onion, or almost to one end of a celery stick. Trim the ends and fan them out in a spray.
- Partially slice a strawberry into several pieces, leaving the stem intact, and splay out the slices in a fan.
- Dip a small cluster of slightly damp grapes into granulated sugar.
- Cut halfway through a cherry with four radial cuts, making it look like a flower when viewed from the top.
- Cut a thin lemon or lime slice into three equal pie slices. Arrange two of them like butterfly wings. Cut the rind off the third piece and arrange the rind pieces like butterfly antenna.
- Shave dark chocolate very finely and sprinkle it over a dessert.
- Melt chocolate and let it harden to a thin layer on a piece of wax paper. Then cut out shapes with a cookie cutter.
- Dip half of a strawberry, dried apricot, cashew, or other small food into melted chocolate.

FIGURE 10-7
A garnish can dress up a plain-looking plate of food. When do you think garnishes would be appropriate to use, and when would they appear to be "overdoing it"?

- Using a pastry decorating bag, squeeze out melted chocolate into fine ribbon shapes on wax paper. Harden them in the freezer and then peel them off the wax paper.
- Use a doily or stencil as a pattern to sprinkle a design of powdered sugar or cocoa.

You can use your own creativity when creating garnishes, but make sure that the food you use does not interfere with the other food on the plate. For example, you would not want to serve a molded chocolate heart on a hot plate because it would melt, and you would not want a garnish whose flavor clashed with that of the main dish. Figure 10-7 on the previous page shows some interesting ideas for garnishes.

Food Service/Presentation Etiquette

If you end up working in food service, you will receive training that explains what a particular restaurant or service requires from you in terms of service etiquette. There is no fixed set of rules; most rules are, in fact, just customs that vary between regions and between types of service.

The cardinal rule of food service, though, never changes—keep the customer satisfied. That means promptly responding to special requests, dealing with any complaints courteously, and using polite words like "Please," "Thank you," and "Excuse me."

While we stated that there are no universal rules, here are some common service customs:

- Whenever possible, place a plate of food in front of the diner from his or her right side. Take finished plates away from the left side (again, when possible). If serving guests who are sitting in a booth, you may not have a choice.
- Serve and clear beverages from the right side of the guest whenever possible, to avoid reaching across their plates.
- Anticipate the diners' needs for condiments, such as ketchup with French fries.
- Remove any utensils, cups, or plates that are not being used. For example, at breakfast, if a diner is not drinking coffee, remove the coffee cup.
- Don't refer to parties that include females as "guys."
- Don't ask open-ended questions like "How is everything?" Instead, engage the diners with specific questions, such as "Is your steak cooked the way you like it?"
- Near the end of the meal, thank the diners effusively for their business and ask them to come back again.
- Instead of taking glasses of iced tea or soda pop away to refill them, bring a new, clean glass of the beverage. (Note: This varies between restaurants; ask your manager how to handle it.)

- Pay attention to who has ordered what dishes, so you don't have to ask "Who gets what?" when you bring out the food.
- If you are expecting a tip and the diners pay in cash, do not ask "Do you want change?" Instead, proceed as if they definitely want change, saying "I'll be right back with your change." If they want to leave you the change as a tip, let them bring it up.
- Do not comment on how much or how little of a dish someone ate. Exception: If they did not eat any of it, you can ask them if they would like a box to take it home; that is an opening for them to tell you if they didn't like it, and for you to ask if they would like something else instead.
- *Never, ever* call a customer by a pet name, such as honey, hon, sweetie, etc. unless they are a close personal friend or relative of yours. It is condescending, and it makes you appear unprofessional.

Table Manners for Diners

There are entire books written about dining etiquette; you may even have read a few of them. Most of these are based on a few simple rules, though—follow the lead of your host/hostess; use the proper utensils; don't call undue attention to yourself (either with noises or gestures); and don't make a mess.

Follow the Lead of the Host/Hostess

If you are at a private dinner party, look to the host or hostess for social cues. Wait to be invited to be seated at the table. Do not begin eating until the host/hostess has taken a bite, and cease eating shortly after he or she does.

Napkin Handling

At a restaurant, put your napkin in your lap as soon as you are seated. At a private dinner party, wait until your host or hostess has put the napkin in his or her lap and follow suit.

If it is a small napkin, unfold it completely across your lap. If it is a large dinner napkin, unfold it only partly and drape it across your knees. The napkin must stay in your lap until the end of the meal; do not wipe your face or clean your silverware with it. If you need to get up in the middle of the meal, drape the napkin to the left of your plate, neither folded nor wadded up. Do not leave it on your chair.

At the end of the meal, place your napkin in that same spot, to the left of your plate. At a private dinner party, wait for the host/hostess to do so first.

When to Eat and When to Stop

In a restaurant, wait until everyone at your table has been served before you begin eating. At a private dinner party, wait until your host/hostess has taken a bite.

At a restaurant, there is no fixed rule as to when to stop eating. It is polite to eat at approximately the same pace as your tablemates, but not required. At a private dinner party, you should stop eating shortly after the host/hostess signals the end of the meal by putting his/her napkin on the table.

Use the Proper Utensils

Many people are concerned about utensil usage, especially at a formal dinner. True, there may be a lot of different forks, but the rules for choosing the right one are fairly simple.

FIGURE 10-8
Holding a knife and fork properly is an important part of good table manners. When someone grips a knife or fork with their fist, what is your impression of them?

If the table is set properly, it's safe to assume that the outermost utensils are the ones to use for the upcoming course. After a course, the utensils for it are removed, so what's "up next" is obvious. It can also help to know the names of the various utensils, because that will tell you what to eat with each one. For example, a salad fork, which has shorter and sometimes fewer tines than a dinner fork, is for use in the salad course.

Do not grip utensils in your fist; instead, hold them gently between the thumb and the side of the index finger. The thumb and the first two fingers are closed around the handle two-thirds of the way up the shank.

A fork can be used as a flat platform, on which you place food to convey it to your mouth, or as a spear, to pick up and convey chunks of food. Put only as much on your fork as will fit in one sensible mouthful; do not strain to open your mouth wider to accommodate a large bite, and do not take multiple bites off of a single forkful.

When eating ice cream or cereal, you can use a spoon in a dipping motion toward you, but when eating soup, you should always dip away from you. Unlike with a fork, it is permissible to sip a small amount from a spoonful gradually; you need not put the entire spoon in your mouth at once. If you are sipping from a spoon, take care not to make any noise.

There are two ways to use a knife and fork together. The American way is to cut with the knife in the right hand and the fork in the left; then to set the knife down, and put the fork in the right hand for eating the cut morsel. The European way is to cut with the left hand and keep the fork in the right hand. Either is acceptable in the United States. If you are at private dinner party, it is

While resting.

When finished.

FIGURE 10-9
Setting your utensils on the plate signals to the server or host/hostess whether or not you are finished with your meal.

nice, but not required, to mimic the style of the host/hostess. Either way, as you are cutting with a knife and fork, point your index finger down the shaft of each implement. Never hold the fork in your fist as you use it to steady the item being cut. Do not cut more than one or two bites at a time.

When you are resting (that is, not done eating yet), set your knife and fork down pointing toward the center of the plate at 10 o'clock and 2 o'clock, as in Figure 10-9. When you are finished eating, set the knife and fork parallel to one another across the plate, pointing to 10 o'clock.

Sit Correctly

Sit up straight at the table, and do not fidget with your hands. When you are not eating, you may rest your forearms—or even an occasional elbow, if you are leaning forward engaged in conversation—lightly on the table. However, when you are eating, there are absolutely no elbows allowed on the table! When eating using one hand, put the other hand in your lap.

Don't Call Attention to Yourself

Part of being a pleasant dinner companion is not embarrassing your fellow diners by making people at other tables notice you in any way. Avoid laughing or talking loudly, belching, slurping, or making any other unnecessary noises.

Hot Topics

When dining with others, don't use cell phones, pagers, music or video players, games, or other electronic devices at the table. It is very rude to take a cell phone call at the table, and even worse to send a text message, because it implies that your first priority is not your dining companions. If you absolutely must take a phone call or respond to a text message, excuse yourself and leave the table to do so.

Don't Make a Mess

Try your best to keep the food on your plate or in your mouth. Don't spill food on the tablecloth or the floor if you can help it. Unobtrusively wipe your fingers on your napkin frequently throughout the meal, so you do not leave fingerprints on water glasses. Never wipe your fingers on your clothing.

Ordering, Tipping, and Paying in a Restaurant

There are many types of restaurants, of course, so there isn't one fixed set of etiquette rules that covers all restaurants. However, in most sit-down restaurants, there are some general guidelines that you can follow.

Ordering Etiquette

Ordering first is considered an honor, so you should offer that honor to the person at the table with the highest precedence:

- In business, clients and customers before employees of the company.
- Women before men.
- Older people before younger people (if the same sex).
- In business, people of higher rank before those of lower rank.
- If two business people have the same rank, the one with greater seniority orders first.

After the first person has ordered, the order in which the remaining guests order is less significant. A server may request that the remaining guests order according to where they are seated, to make it easier to keep track of who ordered what, especially if it is a large party.

There is also etiquette involved in deciding what to order. It all boils down to these general rules:

- **If you are paying**, put others at ease by acting as if the cost of the meal is not important. You can do this by ordering something that is mid-range price or higher, by offering to order an appetizer for the table to share, and by encouraging guests (by your example) to order extra courses, such as soups or salads.
- **If you are not paying**, be aware of how much the food costs that you are ordering, and try not to overspend or appear greedy. Do not order something that is more expensive than what the person who is paying for the meal orders, and do not order extra courses unless invited to do so.
- **Do not talk about the prices.** When choosing what to order, be aware of the prices, but do not talk about them. If you are paying for the meal, talking about the prices indicates that you are concerned about them. If you are not paying for the meal, talking about the prices implies that you are worried about your host's ability to pay.

Hot Topics

When a man and women are dining out together socially, should the man order for them both? Some say yes, that it is a charming old custom that emphasizes the man's protectiveness and chivalry toward his date, much like holding a door or walking nearer to the curb than the woman. Other people object to it on the grounds that it sends the wrong message—either that the man is in charge of what the woman eats, or that the woman needs to be shielded from having to interact with the wait staff.

- **If the man is ordering for the woman**, he should ask her what she wants before the waiter comes to the table—he doesn't get to arbitrarily decide what she is going to eat unless she specifically asks him to do so. Each individual couple may do as they please in this matter when dining socially. However, in business, a man ordering for a woman is not appropriate because it implies that he is focusing on her as a woman rather than as a business colleague.

Payment and Tipping Etiquette

At very casual restaurants, the server may bring a bill to your table that you then go to a cash register at the entrance to pay on your way out. At more upscale restaurants, you pay your server, who then carries the bill to the cashier for you, or processes the payment and makes change himself/herself. If you are not sure which is the custom at a particular restaurant, ask the server.

When the bill comes, do not whip out a calculator and divide the bill out to the nearest penny by figuring out who ordered what. That is considered tacky because it appears that you are overly concerned about money. Instead, plan on dividing the bill up evenly between all diners unless one person has offered to pay for everyone. However, if you ordered much more expensive food than others did, you should chip in enough extra money to cover the difference.

In the United States, tipping the server is expected at any restaurant where a waiter or waitress performs a service for you, such as taking your order, bringing your food, bringing your drinks, and keeping your drinks full. For average or better service, 15 to 20% of the total bill is the norm. (If you are using a coupon or gift certificate, tip based on the amount that the bill would have been if you had *not* had the discount.) An exception: at a buffet-style restaurant, where the server only fetches your drinks, tipping less is acceptable. (However, do not tip less than $1 per person at the table.) A tip may be left on the table in cash, or you can add a tip to your credit card charge. Most servers prefer cash tips, but either a cash or charge tip is acceptable.

Case Study

Patrice wants to help her grandmother reorganize her kitchen. There are several problems with the current organization. For example, she keeps canned goods in a lower cabinet, where it is difficult to find anything without getting down on her knees. She also stores her pots and pans in a cabinet over the range, and sometimes pans and lids fall down as she is trying to pull out the right one.

Her grandmother has the following cabinets in her kitchen:

- An upper and lower cabinet between the refrigerator and the range.
- A lower cabinet under the workspace between the range and the sink.
- An upper and lower cabinet next to the dishwasher.
- Two upper and two lower cabinets on an adjacent wall, two steps away from the work areas.

Make some suggestions as to what she should be storing in specific locations to optimize efficiency in preparing food, cooking, and cleanup. The items she needs to store include canned goods, plates, silverware, glasses, pots and pans for cooking, cooking utensils, kitchen towels, and knives.

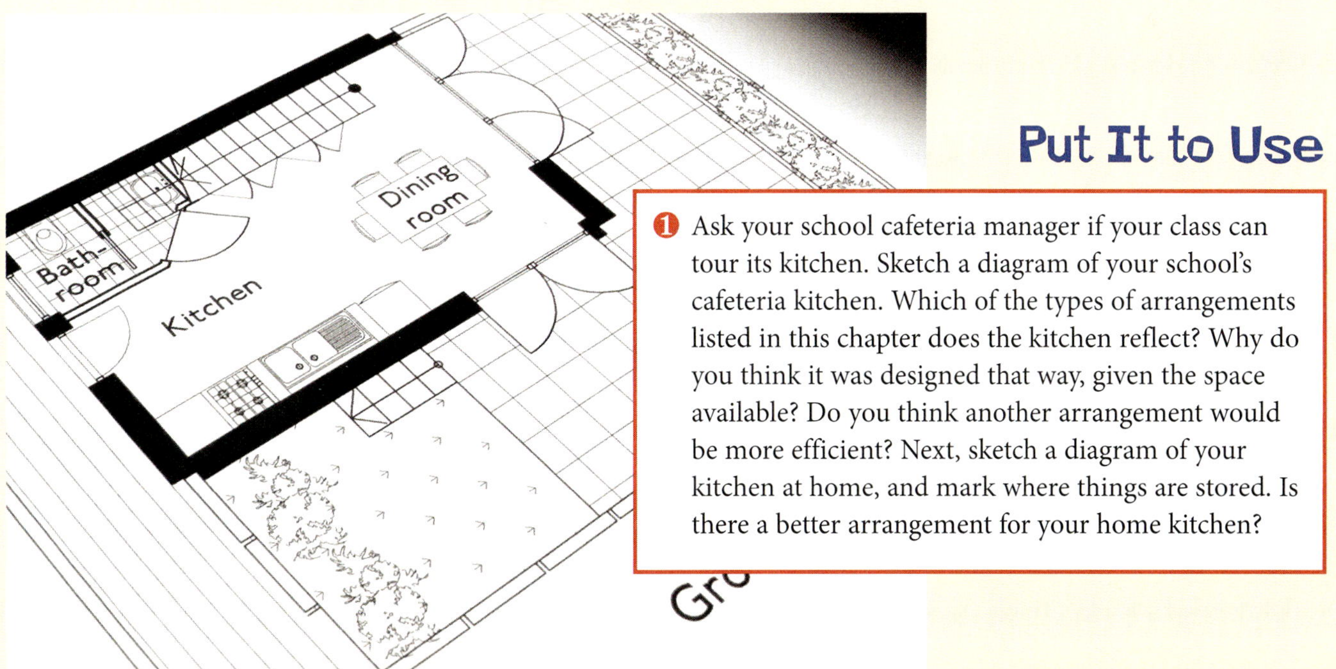

Put It to Use

1. Ask your school cafeteria manager if your class can tour its kitchen. Sketch a diagram of your school's cafeteria kitchen. Which of the types of arrangements listed in this chapter does the kitchen reflect? Why do you think it was designed that way, given the space available? Do you think another arrangement would be more efficient? Next, sketch a diagram of your kitchen at home, and mark where things are stored. Is there a better arrangement for your home kitchen?

Put It to Use

2. Suppose an organization to which you belong has decided to have a formal high tea for a group of retired women. Research to find out what dishes, cutlery, and glassware would be appropriate for such an event, and create a diagram showing a sample place setting.

Write Now

Dining etiquette in other countries may differ from what was presented in this chapter. Choose three other countries, and research the dining etiquette in each. Write a paper explaining how those countries' dining customs are similar to or different from those in the United States.

Tech Connect 

Using the Web, find pictures and instructions for making various types of garnishes. Create a PowerPoint presentation (or other slide show) showing some of the most attractive and unusual garnishes you found.

Team Players

Divide the class into three teams. Teams A and B are responsible for preparing two different desserts for everyone to eat. When they finish, these teams will sit down, and Team C will function as waiters/waitresses, taking the orders for one dessert or the other, serving it, and clearing the dishes afterwards.

Afterwards, discuss what went well in the service of the dessert. What did Team C do well? What could they have done better? If you were a server, were the diners polite to you? Did they use good dining etiquette? If you were a diner, did the servers take your order efficiently and politely? Did you get the item you ordered?

Put It Together

Match the explanation in column 1 with the term in column 2.

Column 1

- **a.** an area of the kitchen arranged for a certain task, such as grilling
- **b.** a freestanding work area, usually in the center of an L or U shaped arrangement
- **c.** the fork that goes closest to the left side of the plate in a formal dinner service
- **d.** the fork that sits horizontally along the top of the plate in a formal dinner service
- **e.** the utensil that sits to the right of the knives in a formal dinner service
- **f.** the item that is placed just above the tip of the knives in a formal dinner service
- **g.** a decorative, edible item on a plate of food
- **h.** the style of eating with a knife and fork, where you transfer the knife between hands with each bite
- **i.** the style of eating with a knife and fork, where the fork stays in the left hand all the time and the knife is used with the right hand
- **j.** the item that should go to the right of, and slightly below, the water glass in a formal dinner service

Column 2

1. American
2. dessert fork
3. dinner fork
4. European
5. garnish
6. island
7. soup spoon
8. water glass
9. wine glass
10. workstation

Part 3

Nutrition and Cooking

11 Grains, Pasta, and Legumes

In This Chapter, You Will . . .

- Identify types of grains
- Learn how to prepare grains
- Identify types of pasta
- Learn how to prepare pasta
- Identify types of legumes
- Learn how to prepare legumes
- Identify the nutritional value of grains, pasta, and legumes

Why YOU Need to Know This

Oatmeal for breakfast, a sandwich on a whole grain roll for lunch, and Spaghetti Bolognese with a side of peas for dinner! Grains, pastas, and legumes are relatively inexpensive, filling, tasty, and highly nutritious. They come in a wide assortment of types, shapes, and flavors, so you can eat them frequently and not get bored. You can prepare them in different ways to create a wide variety of main courses, side dishes, appetizers, and even desserts. They are easy to grow, easy to store, and easy to cook. In fact, on a global scale, grains provide people with more nutrients and calories than any other food source. As a class, discuss the importance of a nutritious food source that is easy to grow, store, and prepare.

Types of Grains

Grains are the seeds of **cereal grasses**—grasses grown by people for use as food. Grains are basic foods that provide the foundation for a healthy diet in most cultures of the world. Common grains include:

- Wheat
- Rice
- Oats
- Corn
- Rye
- Buckwheat
- Barley

After harvesting, grain is processed so that it can be used as food. For example, grains may be cut, crushed, rolled, or ground. The process varies depending on the type of grain and its intended use. We can make grain easier to digest; improve its storage life; or change the flavor, texture, or color of the grain.

Hot Topics

If processing removes the nutrients from grain, why is it done? When the bran is removed, the grain becomes lighter in color and quicker to cook. It may be called a pearl grain. When the germ is removed, so is the oil. Since the oil tends to go bad quickly, removing the germ increases the shelf-life of the grain.

Whole Grains

Whole grains are minimally processed. Just the husk or hull is removed, leaving the bran and germ intact. Whole grains are considered healthier than other grains, because they retain their dietary fiber as well as the vitamins and minerals stored in the germ.

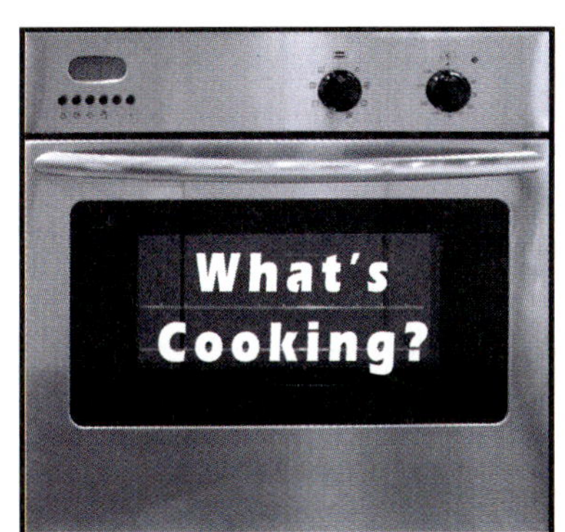

To make whole grains easier and quicker to cook, they may be **parcooked**—partially cooked by boiling or steaming. They can then be dried or toasted before packaging.

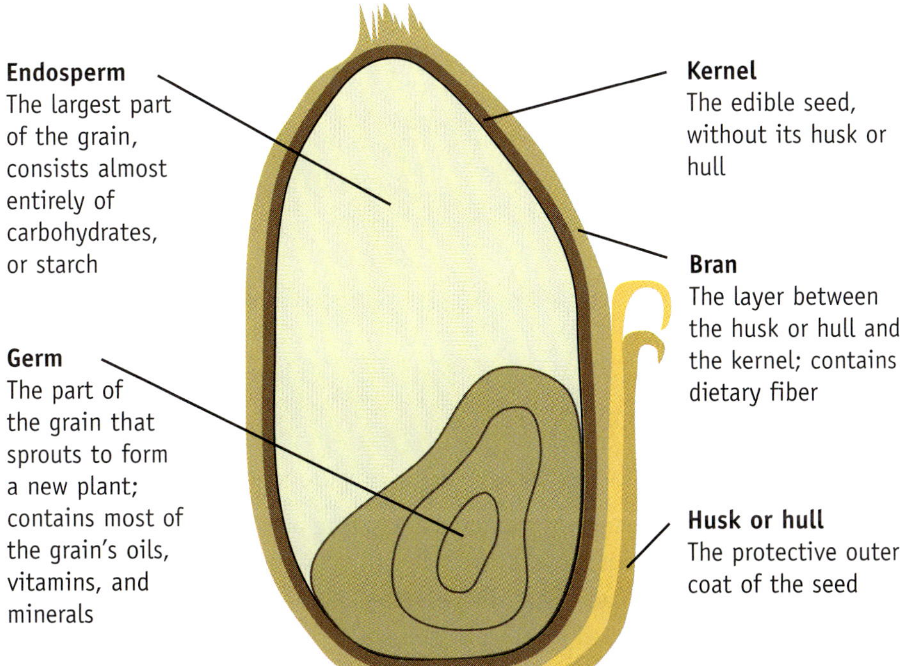

FIGURE 11-1
Parts of a grain. When grain is processed, some of the nutrients may be removed. To make sure you get the nutritional benefits, what type of grain should you eat? Why?

Check the Label

Hominy A whole dried corn kernel that has the hull and germ removed

Oat groats The whole grain of the oat, with the hull removed

Oatmeal Coarsely ground oats

Rolled oats Steamed groats that are rolled into flat flakes (also called old-fashioned oats)

Quick-cooking oatmeal Rolled oats cut into smaller pieces, so they cook faster

Barley A grain that looks like a doubled grain of rice

Rye berries The whole grain of rye (if the berries are rolled, the result is rye flakes)

Quinoa A high protein grain (pronounced KEEN-wah)

Bulgur A quick-cooking form of whole wheat, usually durum, which is parcooked, dried, and then ground

Amaranth A high-protein grain

Popcorn A type of corn that explodes from the kernel and puffs up when it is exposed to dry heat

Refined Grains

Grains that are highly processed are called **refined grains**. The bran and germ are removed, taking the fiber, vitamins, and minerals with them and leaving the grain with less nutritional value than whole grains. White rice is the most common refined grain.

Cracked Grain

When the kernel of grain is cut or crushed into pieces, it is called **cracked grain**. Cracked wheat is made from minimally processed wheat.

Meal

Processed grain that is ground into fine particles is called **meal**. The grinding is done using a process called **milling**, during which the grain is rolled between steel drums, or between stone wheels. If the grain is parcooked before milling, it may be rolled to produce flakes.

Hot Topics

Allergies to grains and legumes are on the rise. Estimates indicate that close to 3 million Americans are allergic to peanuts or tree nuts, and that 1 out of every 133 Americans have celiac disease—an intolerance to the wheat protein gluten. Exposure can result in severe illness, or even death.

The government passed the Food Allergy Labeling Consumer Protection Act (FALCPA), which went into effect in 2006. It requires that easy-to-read labels be placed on food and other products noting the presence of milk, egg, fish, shellfish, tree nuts wheat, peanuts, and soybeans. Look for the labels on everything from cookies to ice cream.

Grain Products

Grain products fall into four main categories: breads, cereals, pasta, and rice.

- Breads cover a wide range of products, including breads that you might have with breakfast (such as muffins or waffles), with lunch (such as whole wheat or rye sliced bread), with dinner (such as rolls), and also with dessert (such as cake). You'll read more about breads in Chapter 21.

Flour is the main ingredient in most breads, and can be milled from many different types of grains (even some starches that are not grains, such as beans or roots). Wheat flour is the most widely used, though it is now becoming easier to find flours made from other grains such as rice, corn, oatmeal, or sorghum.

- Cereals are another type of grain product that are produced mainly from the bran portion of a grain, which is high in fiber. Rice, wheat, corn, oats, barley, and millet are the most widely used cereal grains.

Today, most of us think about cereals in the form of breakfast cereals, such as those shown in Figure 11-3, but cereal grains can be used in other ways. For example, bulgur wheat or rice can be easily prepared with a dinner meal.

FIGURE 11-2
Flour is a fine powder made from cereal grains. How do you think flour is made in non-industrialized countries?

FIGURE 11-3
Breakfast cereals are sold as uncooked or ready-to-eat. Uncooked cereals include oatmeal and cream of wheat. Ready-to-eat cereals include a wide array of products made from corn, oats, wheat, and rice that have been rolled, flaked, puffed, or shredded. What's your favorite breakfast cereal?

Check the Label

White flour Flour from processed grain—usually wheat—from which the bran and germ have been removed

Bleached flour White flour that is bleached to lighten its color

Whole wheat flour Wheat flour that includes the bran and the germ; usually coarser than white flour

All-purpose flour A blend of flour from different wheats

Self-rising flour Flour that contains a leavening agent such as baking powder or baking soda

Cake flour Flour made from soft wheat; usually lower in gluten and higher in carbohydrates than all-purpose flour

Rice is a grain product that is used worldwide. Most of us are familiar with white rice and brown rice. Wild rice, which is closely related but a different plant, is not really a rice at all, but a cereal grain.

We often refer to rice in terms of short, medium, or long grain. Generally, the smaller the grain, the stickier the rice. Long grain is what most of us have as a side dish with lunch or dinner. Medium grain is more popular in meals such as paella or risotto because it creates a denser consistency. Short-grain rice sticks together extremely well, which is why it is used to make sushi.

FIGURE 11-4
Rice is a primary source of nutrients and calories for many populations around the world. It comes in three sizes—short grain, medium grain, and long grain. While usually served as a side dish, rice may also be used as an ingredient in other recipes, such as rice pudding, a filling in stuffed cabbage or grape leaves, or in soups and salads. Why is rice an important food source?

Check the Label

Converted rice Partially cooked rice from which the hull is removed. It is then dried and milled to produce either white or brown rice.

Enriched rice White rice fortified with vitamins and minerals such as thiamin, niacin, and iron. It is more nutritious than white rice that hasn't been enriched, but less nutritious than brown rice.

Wild rice The seed of a marsh grass that grows in areas of North America. It is not really a rice at all, but a cereal grain.

Fiction	Fact
Brown rice and white rice come from different varieties of grain.	White rice is processed to remove all of its bran, which lightens the color of the grain. Brown rice retains some or all of its bran. All types of rice can be processed as either white or brown.

- Pasta is a grain product that is usually made from wheat flour, though corn and rice pastas are becoming more accessible for people who are gluten intolerant and cannot eat wheat.

  Pastas come in many different varieties across the world—from lo mein noodles in China to fettucini in Italy. Depending on how they are made and cooked, they can vary quite a bit in flavor in consistency. Pasta is covered extensively in the next section.

FIGURE 11-5
Starchy foods, like grains, potatoes, and legumes, are sometimes referred to as **farinaceous.** Do you think foods made from these items—like pasta and bread—would also be called farinaceous? Why?

Preparing Grains

You make grains tender enough to eat by boiling or steaming them in liquid. When boiling, you drain the excess liquid once the grain is thoroughly cooked. When steaming is complete, all of the liquid should be absorbed into the grain. The ratio of liquid to water depends on the type of grain, but is usually at least two parts liquid to one part grain.

You cook cereal—such as oatmeal—by stirring the grain into a simmering liquid. Because of the high levels of starch, you must stir the cereal constantly to keep it smooth and free of lumps. **Gelatinization**, which is the thickening or formation of a starch gel, occurs when starches are cooked. The goal is to end up with a soft gel, not a pasty one.

Hot Topics

When you boil or steam food, nutrients such as folic acid and vitamin C seep out of the food into the boiling liquid. USDA studies show that boiling causes more nutrient loss than steaming, probably because boiling liquid is usually thrown out, while steaming liquid goes back into the food.

Pilaf is a grain dish in which the grain is sautéed in a pan before adding a hot liquid. Vegetables, pastas, and nuts may be added to enhance the flavor.

Risotto is an Italian rice dish that becomes creamy when cooked. The rice is sautéed, and then liquid is added gradually while the mixture is stirred constantly.

BASIC CULINARY SKILLS

Boiled or Steamed Grains

1. Measure liquid according to recipe or cooking instructions. Bring liquid to a boil.
2. Add grain all at once. Stir to separate grain.
3. Reduce heat to a simmer. Cover pot if steaming.

4. Simmer until grain is fully cooked and tender.

5. Drain grain if necessary.

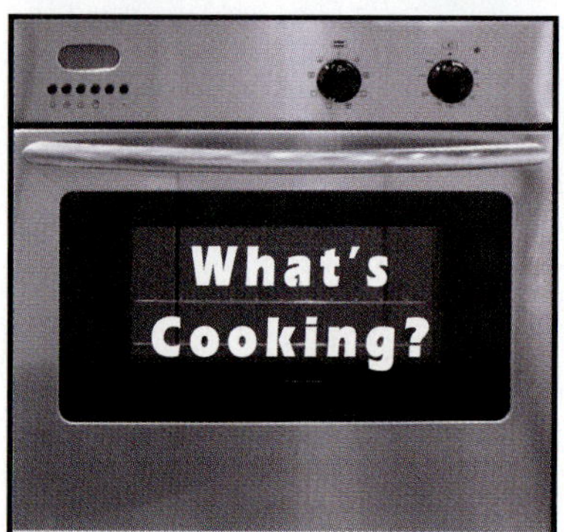

Can you guess the primary component of most grains? It's starch. When starch is heated, it absorbs water, swells, and becomes thick and sticky. It also takes on a milky color. When cooked right, starch loses the milky color and becomes translucent.

Utility Drawer

A rice cooker is a kitchen appliance designed specifically for steaming rice and other grains. Rice cookers simplify the process by automatically controlling the heat and timing. They don't necessarily make better rice, they just make cooking it easier.

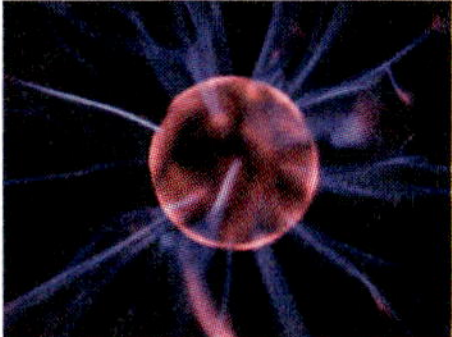

SCIENCE STUDY

If you've ever cooked brown rice, you know it takes a lot longer to cook than white rice. That's because the germ and bran are removed during the processing of white rice, leaving it softer and more absorbent.

You can test this just by reading the cooking instructions on packages of different types of rice. Or, try cooking them.

Boil two cups of water in one pot, and two cups of water in a different pot. Add one cup of brown rice to the first pot and one cup of white rice to the other.

Lower the heat, cover, and simmer. The white rice will be done in about 20–25 minutes. It will take at least an additional 15–20 minutes for the brown rice to cook.

What, if anything, can you do to speed up the cooking time of the brown rice? What about the cooking times of other types of grains? What grains do you think will take the longest time to cook?

Types of Pasta

Pasta means dough in Italian. It is made from flour and water, and then turned into various shapes which are usually boiled. No matter where you are in the world, you are likely to find some form of pasta available. It might be called noodles, udon, wonton, vermicelli, macaroni, or spaghetti; although the name may be different, the basic ingredients—flour and water—are the same.

Fresh Pasta

You can make or purchase fresh pasta, which is pasta made by blending flour with a liquid, such as water or eggs. The dough should be soft enough to knead by hand, but firm enough to hold its shape. Fresh pasta cooks faster than dried pasta, because it is already moist. Pay attention when cooking fresh pasta so that you do not overboil it.

FIGURE 11-6
Vegetables, herbs, and spices can be added to pasta to enhance the flavor and provide color. What might you add if you want to make pasta green? How about red? Is there any color or flavor you think people might not want to eat? Why?

Cool Tips

Pasta cooked **al dente** is not too hard or too soft. It has a bit of resistance when you bite it. The term al dente literally means *"to the tooth"* in Italian. It is sometimes translated as *"to taste."*

Dried Pasta

Dried pasta has the same basic ingredients as fresh pasta. The dough is stiffer, though, so it can be shaped using machines. The shapes are dried until they are hard and brittle. They become soft when cooked in boiling water.

FIGURE 11-7
Pasta comes in many shapes and sizes. Most pasta shapes are made using the same ingredients and techniques. Can you think of reasons you might choose to use one shape over another for particular recipe?

Check the Label

Soba Japanese noodles made from buckwheat flour

Udon Japanese noodles made from wheat flour

Lo mein Chinese noodles made from wheat

Spaetzle Austrian or German dumplings made from dough dropped into boiling water

Gnocchi Italian dumplings

Pierogi Polish dumplings

Fiction	Fact
All pasta is made from wheat flour.	Italian pasta is usually made from wheat flour, but noodles may be made from other types of flour, as well. In Asia, noodles are often made from rice flour, or even bean flour.

BASIC CULINARY SKILLS

Boiling Pasta

1. Bring a large pot of water to a boil. Use at least 1 gallon of water for every pound of pasta.
2. Add pasta and stir until it is softened, submerged, and separated.

3. Cook fresh pasta or filled pasta such as ravioli, at a simmer.
4. Cook dried pasta at a full boil.
5. Cook until done, stirring occasionally.

6. Drain pasta immediately.

Fiction	Fact
Adding a splash of oil to boiling water before adding pasta keeps the pasta from forming clumps.	Pasta clumps because it becomes sticky as the starch it contains breaks down during cooking. Oil just floats on the surface of the water, and has no impact on the stickiness at all. To keep the pasta from clumping, try adding an acid like vinegar or lemon juice to the boiling water. The acid stops the breakdown of the starch, making the pasta less sticky.

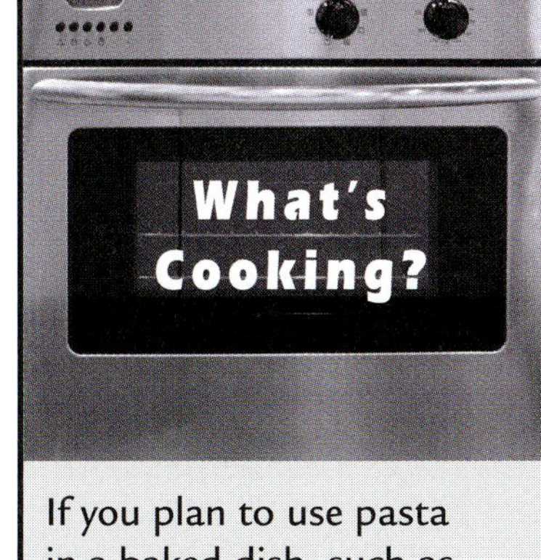

If you plan to use pasta in a baked dish, such as lasagna, undercook the pasta during boiling. It will finish cooking as it bakes.

Restaurant workers, owners, and managers must be prepared to answer questions about the items on the menu. It is important to know how to pronounce the names of dishes and ingredients, and how to explain the differences between them.

Practice pronouncing the names of different types of grains, pastas, and legumes. Review the differences between types of grains, shapes of pasta, and types of legumes. In pairs, act as a customer and a server and discuss the different items.

Types of Legumes

A **legume** is a plant with a double-seamed pod, containing a single row of seeds. If people eat the seed together with the pods, like they do with green beans, then the legume is treated as a vegetable. If people eat only the seeds, like they do with peas, the legume is treated as a grain.

- Dried legumes that are longer than they are round are called **beans**. Examples include navy beans, kidney beans, and fava beans.
- Round legumes are called **peas**. Examples include black-eyed peas, green peas, and chickpeas.
- Disk-shaped legumes are called **lentils**. They may be brown or red.
- Dried legumes are sold in bags, boxes, or packages. Look for packaging that has no rips or tears. Store dried legumes in their original packaging, or in airtight containers in a cool, dry location.

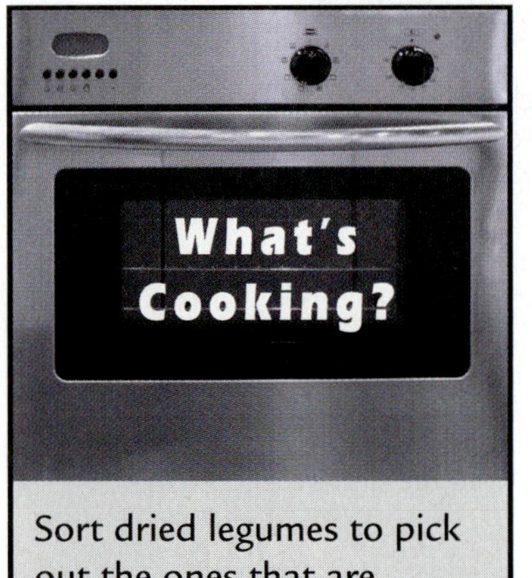

Sort dried legumes to pick out the ones that are shriveled, cracked, and otherwise unappetizing.

Chickpeas are also called garbanzo beans, or ceci—pronounced cheh-chee.

FIGURE 11-8
Rinse dried legumes by placing them in a container and adding enough cold water to cover them. Throw away any that rise to the surface before draining and rinsing the legumes. Why do you think you should throw away legumes that rise to the surface of the water during rinsing?

Preparing Legumes

Dried legumes require a bit of work before they are ready to use in a recipe. They should all be sorted and rinsed, and most types then need to be soaked.

Safe Eats

Avoid buying canned food if the can has dents, leaks, or bulges. Damage to a can might let in air, which can encourage the growth of harmful microorganisms like botulism, or cause rust to form inside the can.

BASIC CULINARY SKILLS

Quick-Soaking Dry Legumes

1. Sort and rinse legumes.
2. Place in a pot and add cold water to cover by about 2 inches. Bring to a boil.
3. Remove the pot from the heat, cover tightly, and let soak for about 1 hour.
4. Drain and rinse legumes before cooking.

BASIC CULINARY SKILLS

Long-Soaking Dry Legumes

1. Sort and rinse legumes.
2. Place in a pot and add cold water to cover by about 3 inches.
3. Refrigerate for the required soaking time:
   - 4 hours for black beans, chickpeas, great northern beans, kidney beans, whole peas, pinto beans
   - 12 hours for fava beans
4. Drain and rinse legumes before cooking.

BASIC CULINARY SKILLS

Cooking Dry Legumes

1. Sort and rinse legumes.
2. Soak legumes if necessary.
3. Place in a pot and add cold water to cover by about 2 inches. Bring to a boil, stirring occasionally.
4. Boil until nearly tender, skimming foam off the top as necessary.
5. Add salt and other ingredients as called for by your recipe.
6. Continue cooking until tender enough to mash easily, but skins are still intact.
7. Drain and use, or cool and hold in the cooking liquid for later use.

BY THE NUMBERS

Dried—or dehydrated—foods increase in volume and weight when properly prepared. Use a measuring cup to determine the volume of a 1-pound batch of beans, such as navy beans or black beans. Prepare the beans correctly using the long-soak method. Calculate the percentage of increase in volume and weight of the cooked beans.

Hint: After cooking, measure the batch to determine the increase in volume and weight, then convert the increases to a percentage.

Safe Eats

If dry legumes or grains become damp, they might develop a coating of mold. If so, throw them out. The mold can develop **aflatoxin**, which is a highly poisonous substance believed to cause certain types of cancer.

Nutritional Value of Grains, Pasta, and Legumes

Grains, pasta, and legumes fall into the Grains category of MyPyramid. They tend to be low in fat and high in carbohydrates.

Whole grains—and grain products, like pasta and flour—are full of thiamin, riboflavin, and niacin, along with other vitamins and minerals such as iron and phosphorous. They contain two types of carbohydrates: starch, which supports energy; and fiber, which promotes healthy digestion.

Most grains provide incomplete protein, because they lack some of the essential amino acids. Legumes, however, are an excellent source of protein. Therefore, dishes like red beans and rice that combine grains and legumes tend to be well-balanced and highly nutritious.

Hot Topics

Soy beans are a versatile legume and a key ingredient in many processed foods, including tofu and soft-serve ice cream. There are two types—fermented and unfermented. Some people recommend avoiding unfermented soy products because they may inhibit your ability to digest protein, and because the high concentration of phytic acid blocks the body from absorbing minerals such as iron. In moderation, however, both fermented and unfermented soy can be part of a well-balanced healthy diet.

FIGURE 11-9
Some sprouted raw legumes, such as mung beans (above), chickpeas, or lentils, are excellent on salads. Look for them in your grocery store, or try sprouting them yourself. After soaking dried chickpeas overnight, rinse twice daily, morning and night, until they sprout. Store in the refrigerator during the whole process, to inhibit mold. You should see a root sprout out of them in about 1 to 3 days.

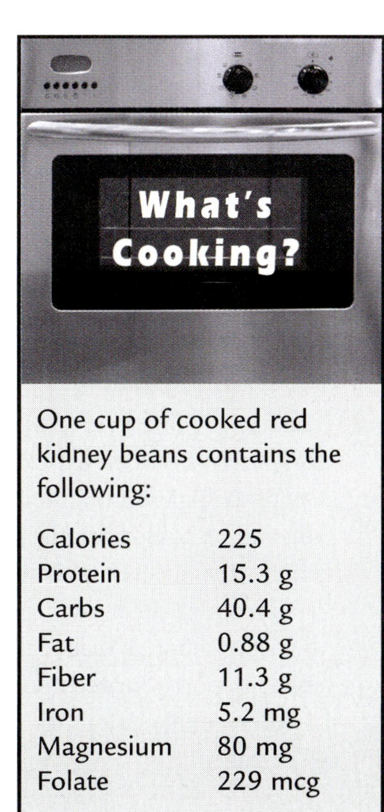

One cup of cooked red kidney beans contains the following:

Calories	225
Protein	15.3 g
Carbs	40.4 g
Fat	0.88 g
Fiber	11.3 g
Iron	5.2 mg
Magnesium	80 mg
Folate	229 mcg

Case Study

Recently, Shawna started feeling tired, and noticed that she has lost some weight, even though her diet and exercise routines have not changed. Whenever she eats cereal for breakfast, she has pain in her abdomen, and her stomach seems bloated. Shawna thought she might be having a problem digesting milk, so she switched from cereal to whole wheat toast. But, the problems persisted.

One day the abdominal pain was accompanied by diarrhea, so Shawna went to the school nurse. Based on the symptoms, the nurse thought Shawna might have developed celiac disease, which is a condition triggered by eating the protein gluten. She called Shawna's parents and advised them to consult a doctor.

- If Shawna does have celiac, how will Shawna have to change her eating habits?
- What problems might Shawna encounter if she does not change her diet?
- Do you think it would be difficult to stop eating foods containing gluten? Why?

Put It to Use

1. Visit the breakfast cereal aisle of a supermarket and compare the nutritional values of common cereals. Pay particular attention to the serving size, so that you will be comparing the same amounts. Decide which type of cereal you would recommend to someone on a high-fiber diet. How about a low-sodium diet? How about a diabetic? How about a young child? Report your recommendations to your class, including explanations for why you selected each type of cereal.

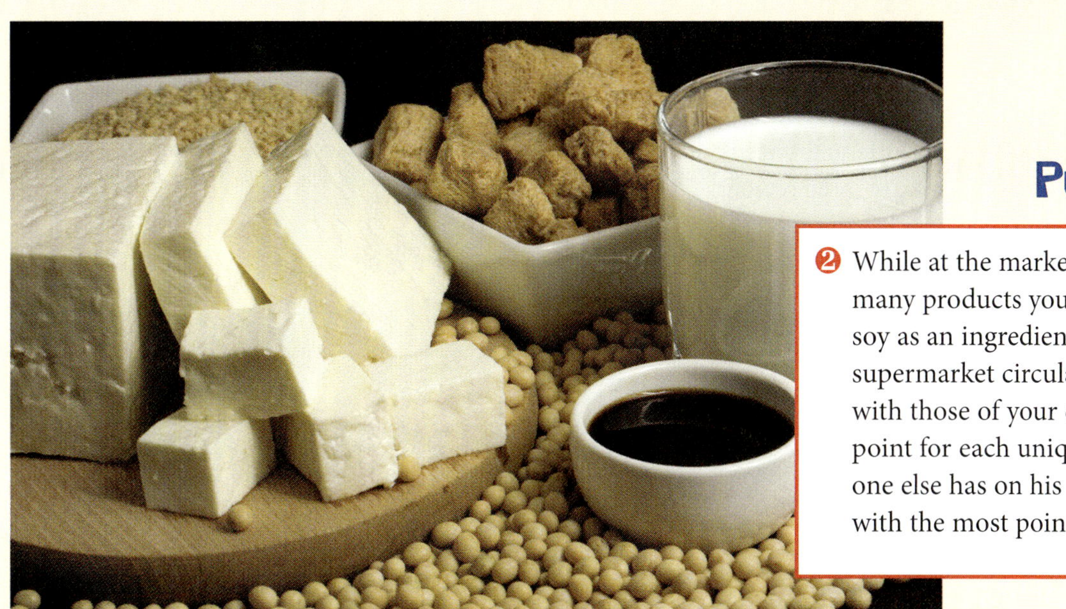

Put It to Use

❷ While at the market, keep a list of how many products you can find that contain soy as an ingredient. (You can also use a supermarket circular.) Compare your list with those of your classmates. Award a point for each unique item—an item no one else has on his or her list. The person with the most points wins.

Put It to Use

❸ Individually, or in groups, create a cooking chart for various grains, pastas, and legumes. Include the food name, the amount to cook, the amount of liquid to use, the cooking time, and the yield. For extra credit, include the nutritional value, as well.

Write Now

Imagine that a disease kills the rice crop in Asia or the corn crop in North America. What would be the effects of such an occurrence? Write a story about an incident in which a major crop of grain is destroyed. Include information about who would be affected, how, and why. At the end, add a paragraph or two discussing ways in which technology could be used to help avoid such a disaster.

Tech Connect 

Pilaf is a dish that can be made using different types of grain. Use the Internet to locate at least three recipes for pilaf, each using a different type of grain. Select one recipe to prepare, and share it with your class.

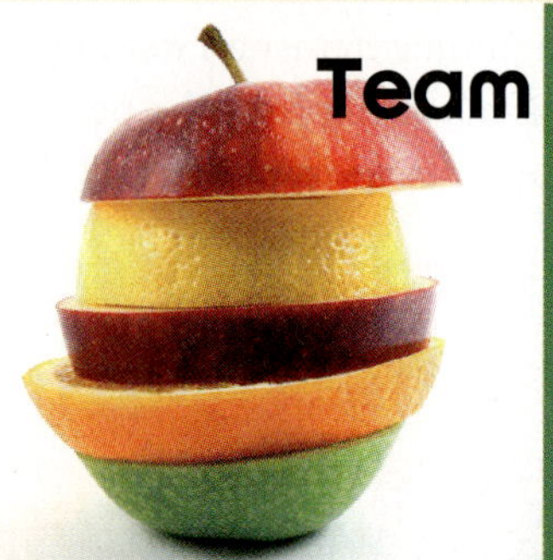

Team Players

Pasta of one type or another is enjoyed in most countries. In small groups, select a country and research how they make and eat pasta. Find out what type of flour they use, what shapes they make, and the name given to the dish. Researach the history of the dish, whether it is eaten regularly, or if it is part of a special occasion. Prepare a presentation about your topic, including nutritional information, and share it with the class. You might even cook the dish, and create a video about the process.

Put It Together

Match the explanation in column 1 with the term in column 2.

Column 1

- **a.** a condition triggered by eating the protein gluten
- **b.** the edible seed of a grain, without its husk or hull
- **c.** the part of a grain that contains dietary fiber
- **d.** the part of a grain that contains most of the oils, vitamins, and minerals
- **e.** the seed of a marsh grass that grows in areas of North America
- **f.** the only grain we eat both fresh and dried
- **g.** the primary component of grains
- **h.** Japanese noodles made from buckwheat flour
- **i.** Italian dumplings
- **j.** a plant with a double-seamed pod, containing a single row of seeds
- **k.** a highly poisonous substance believed to cause certain types of cancer

Column 2

1. germ
2. starch
3. legume
4. celiac disease
5. wild rice
6. soba
7. aflatoxin
8. gnocci
9. kernel
10. corn
11. bran

TRY IT!

Polenta

Yield: 10 Servings Serving Size: 1 cup

Ingredients

80 fl oz Chicken stock (more as needed)
1 tsp Salt
¼ tsp Black pepper, freshly ground
1 lb (3 cups) Yellow cornmeal, coarse

Method

1. Bring the stock or water to a boil in a 3-qt sauce pot.
2. Add the salt and pepper.
3. Pour the cornmeal into the boiling stock in a steady stream, stirring constantly until you have added all the cornmeal.
4. Simmer over low heat, stirring often, until done, about 45 minutes.
5. When done, polenta will pull away from the sides of the pot and will be soft in texture.
6. Remove the pot from the heat and finish as desired.
7. Adjust the consistency with stock or water, if necessary.
8. Season with salt and pepper, if necessary.

Recipe Categories

Grains

Chef's Notes

Polenta can be poured onto a greased half-sheet pan and refrigerated until cool and firm.

Shapes can be cut out and sautéed, grilled, or baked.

Potentially Hazardous Foods

Cooked Cereal

HACCP

Cool to below 41° F within 4 hours (1-stage cooling method) or within 6 hours (2-stage cooling method).

Nutrition

Calories	136
Protein	3 g
Fat	4 g
Carbohydrates	23 g
Sodium	592 mg
Cholesterol	8 mg

TRY IT!

Macaroni and Cheese

Yield: 6 Servings   Serving Size: 8 oz

Ingredients

1 16-oz package Macaroni, uncooked
1 Tbsp Butter
1 lb sharp Cheddar cheese, sliced
12 fl oz evaporated milk
Salt and pepper to taste

Method

1. Preheat oven to 350° F.
2. Bring a large pot of salted water to a boil.
3. Add the macaroni and return to a boil.
4. Cook the macaroni until al dente, about 7 to 9 minutes. Do not overcook.
5. Drain the pasta and shock in cold water to stop cooking.
6. Grease a 2-quart casserole dish.
7. Place ¼ of the macaroni into the bottom of the dish.
8. Layer ¼ of the cheddar cheese evenly on top of the macaroni.
9. Dot with butter, and season with salt and pepper to taste.
10. Repeat to create three layers of macaroni and cheese.
11. Pour the evaporated milk evenly over the top layer.
12. Bake, uncovered, for one hour, or until the top is golden brown.

Recipe Categories

Pasta

Chef's Notes

For variety, you can use different cheeses, such as Gruyere, or add minced peppers, onions, or garlic. For a creamier consistency, double the evaporated milk.

Potentially Hazardous Foods

Dairy, Cooked Pasta

HACCP

Cook to an internal temperature of 145° F or higher.

Maintain at 135° F during service.

Nutrition

Calories	692
Protein	33.5 g
Fat	33 g
Carbohydrates	64.2 g
Sodium	454 mg
Cholesterol	101 mg

TRY IT!

Refried Beans

Yield: 9 Portions Serving Size: ½ cup

Ingredients

1 lb dried Pinto or Red beans
1 large Onion, quartered
3 cloves Garlic
½ tsp Cumin, ground
3 drops Hot Pepper Sauce

Method

1. Cull, wash, and soak beans overnight.
2. Drain.
3. Place beans in large sauce pan and add water to cover by 2 inches.
4. Add onion and garlic.
5. Bring to a boil.
6. Cover and cook over low heat for 2 hours, or until the beans are very soft, adding water to keep the beans covered, if necessary.
7. Remove the onion and garlic, and discard.
8. Mash beans with a potato masher.
9. Season with cumin and hot pepper sauce.
10. Serve immediately, or hold hot for service.

Recipe Categories

Beans and Other Legumes

Chef's Notes

You can leave some of the beans whole or partially mash them to provide additional texture.

As a garnish, top with onions and peppers or with grated Monterey Jack cheese.

Potentially Hazardous Foods

Stocks, Cooked Beans

HACCP

Maintain at 135° F during service.

Nutrition

Calories	180
Protein	11 g
Fat	1 g
Carbohydrates	34 g
Sodium	6 mg
Cholesterol	0 mg

12 Meats

In This Chapter, You Will . . .

- Identify types of meats
- Identify the nutritional value of meat
- Learn how to prepare and serve meat in a variety of ways

Why YOU Need to Know This

The flesh of an animal which is eaten by other animals is called **meat**. Meat is rich in protein and other vital nutrients, and most people enjoy its flavor. For these reasons, if meat is available, humans will eat it. Meat is usually the most expensive part of a meal. As a chef or a consumer, it is in your interest to understand how to make the most of the meat you prepare, serve, and eat. Our ancestors had to risk their lives hunting animals for meat. We can still hunt if we want to, or we can head out to the butcher shop or grocery store to purchase our meat. As a class, discuss the advantages and disadvantages of hunting for your food as compared to purchasing it at a retail outlet.

Types of Meat

Traditionally, meat raised for food and sold commercially includes beef, veal, pork, and lamb. Less traditional meats, such as bison, are popping up on supermarket shelves and restaurant menus. They are high in protein and low in fat, and tend to be more expensive than the traditional meats.

In general, the flavor, color, and texture of meats depend on factors that affect the animal before slaughter. For example, the amount of exercise the animal gets affects the muscle; the age of the animal affects the tenderness of the meat; and the type of feed the animal eats affects the flavor. The breed may also have an impact on the quality of the meat. Beef from Black Angus cattle usually costs more than that of other breeds because it is thought to be more tender and flavorful.

Cool Tips

By law, meat must be inspected by the government before it can be sold. Usually, three inspections take place: live animals are checked for disease, farms and ranches are checked for safety and cleanliness, and the meat is checked for wholesomeness.

Wholesomeness encompasses the qualities that make a food good for you. For example, wholesome meat is free of pathogens and contaminants that could make you sick.

Beef Rib

Rib Roast

Boneless Rib Eye Roast

Strip Loin (top view)

Tenderloin (top view)

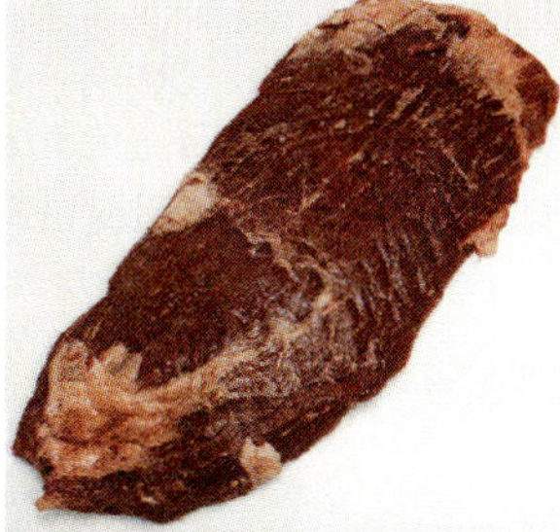

Flank Steak

Top Round

Bottom Round

Beef Brisket

FIGURE 12-1
Beef Primal Cuts. There are eight USDA grades of beef. From highest quality to lowest, they are Prime, Choice, Select, Standard, Commercial, Utility, Cutter, and Canner. The grades most commonly sold in retail outlets are Choice and Select. Why do you think there are more primal cuts from beef than from the other meats?

Check the Label

Beef Meat from cattle over one year old

Veal Meat from cattle three to fourteen weeks old

Pork Meat from hogs or pigs, usually less than one year old

Lamb Meat from sheep less than fourteen months old

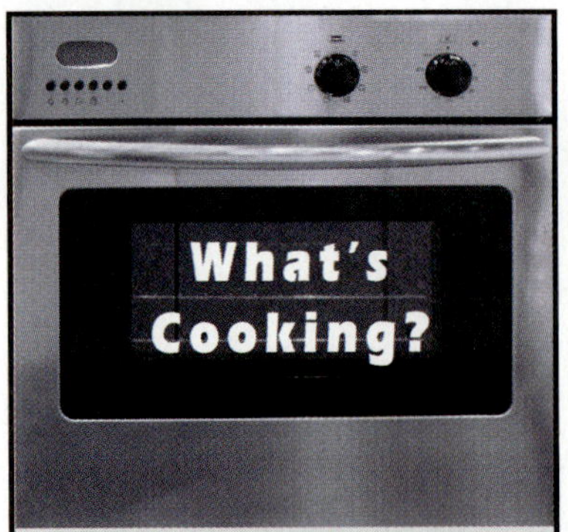

Game meat is the meat of wild animals hunted for food. It includes small mammals like rabbit, as well as deer, elk, caribou, buffalo, and even wild boar.

About Butchering

After slaughtering, meat is **butchered**, or cut up. Large animals are usually cut into halves, called **sides**, and then into quarters. The front quarter is called the **forequarter** and the rear quarter is called the **hindquarter**. Smaller animals, like veal, are usually cut into **saddles**. The **foresaddle** includes the front portion of the animal, and the **hindsaddle** includes the rear portion of the animal.

The large pieces are divided into **primal cuts**, which are portions that meet uniform standards for each type of meat. For example, beef is cut into eight primals, veal into six, and both pork and lamb into five. The primals are divided into **subprimal cuts**, which in turn are divided into the **retail cuts** you purchase at the store.

Cuts of Meat

Before it can be sold to consumers, meat is cut into different shapes and sizes suitable for handling and cooking. All cuts of meat are made up of muscle tissue, connective tissue, and fat. Some cuts contain bone. The industry has standardized the way in which meat is cut and labeled to make it easier for consumers to know what they are buying, no matter what animal is used. Cuts are identified by the primal cut and the retail cut. For example, if you buy a pork, lamb, or beef loin chop, you know you are purchasing a portion of the rib, called a **chop**, cut from the tender area along the lower part of the back, called the **loin**.

Have you ever noticed a purple stamp on the meat you buy at the market? It's a USDA grade stamp, made from nonpoisonous dye. Unlike inspection, grading meat for quality is not required by law. The USDA has grading standards, but meat packers may assign grades based on their own standards, instead. Grades are assigned to a carcass, and then applied to all cuts from that carcass.

Hot Topics

Variety meat, or **offal**, is meat from areas other than the primal cuts, such as organs and muscles. Liver, tripe (the lining of the stomach), kidneys, **sweetbreads** (the thymus glands or the pancreas), brains, tongue, intestines, and even the heart may be cooked using a variety of methods. At some tables, these are considered a delicacy.

FIGURE 12-2
Beef variety meat—oxtail. Oxtail is used worldwide to make deliciously rich soups and stews. Would you eat oxtail stew? Why, or why not?

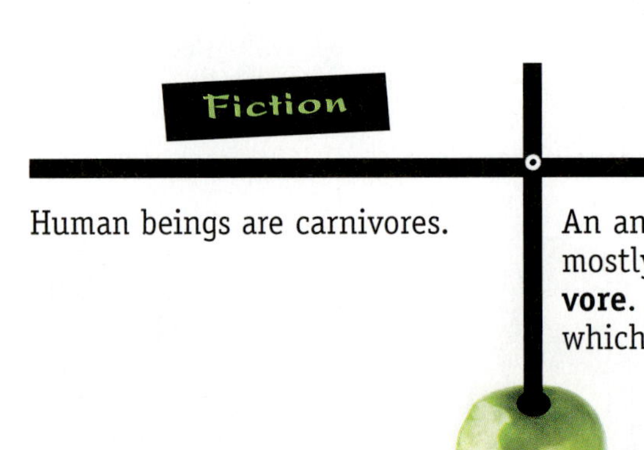

Fiction: Human beings are carnivores.

Fact: An animal or plant that eats mostly meat is called a **carnivore.** People are **omnivores,** which means we'll eat anything.

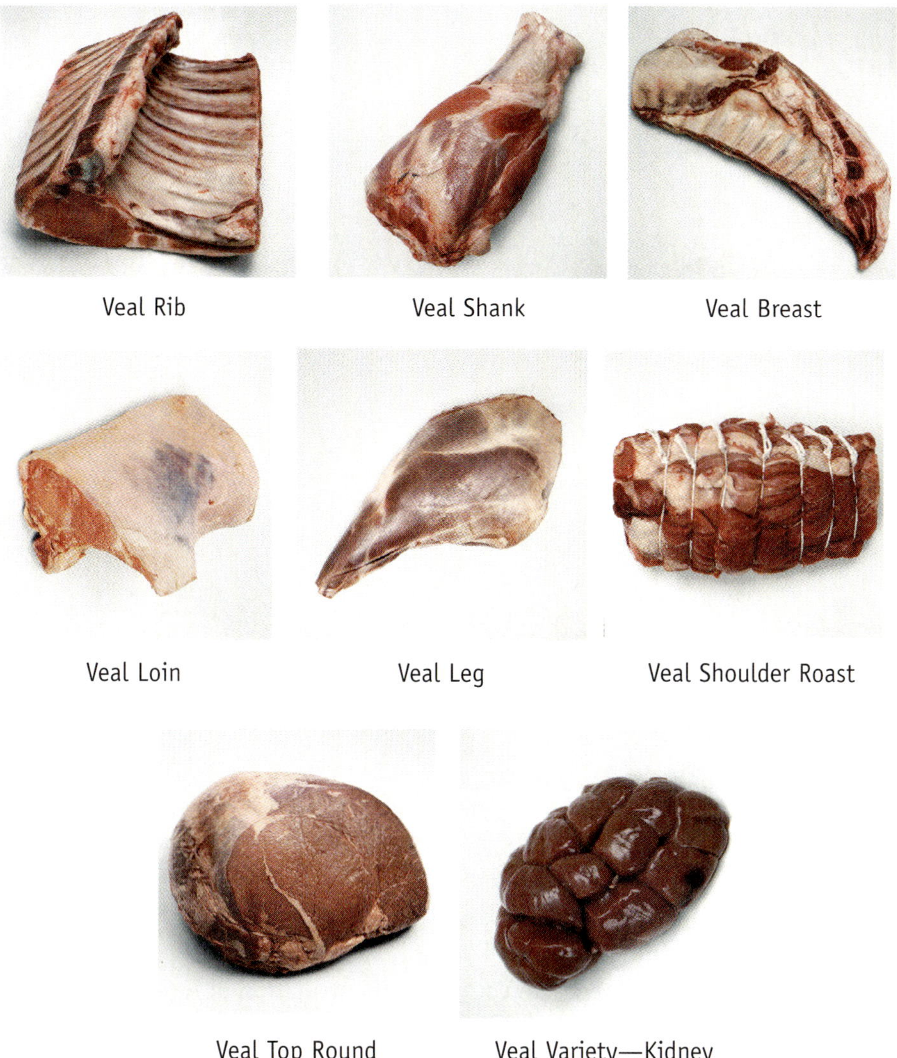

FIGURE 12-3
Veal Primal Cuts. There are six USDA grades of veal: Prime, Choice, Good, Standard, Utility, and Cull. Only Prime and Choice grades are used in restaurants or sold in retail outlets. What differences can you find between primal cuts of veal and primal cuts of beef?

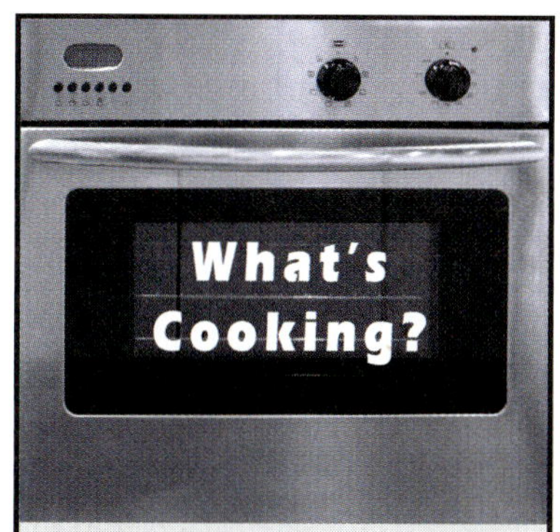

Aged meat is meat that has been stored between 34° F and 38° F for anywhere from a few days to a few weeks.

The meat undergoes a natural tenderization process during aging. Enzymes break down the **collagen**, or structural proteins, that make meat tough.

If you're in a rush to tenderize the meat, you can sprinkle it with powdered tenderizer made from an enzyme extracted from papaya, called **papain,** before cooking.

Buying Meat

You can identify good quality meat by looking at its color, marbling, and texture. **Marbling** is the amount of fat in the meat. High quality meat usually has a lot of marbling throughout. The fat should be white or creamy-white and pretty firm. If the meat is coarse, very dark in the lean area and has yellow fat, it is not good quality. You can use the color of the bone to help judge the age of the animal: porous and pink means the animal is young, and should be tender; gray and brittle means the animal is older, and may be tough.

Store meat in the refrigerator at less than 41° F. Keep it wrapped in its original packaging, or loosely covered to allow airflow around it. If you plan to freeze the meat, separate it into individual servings and use a double layer of moisture proof wrapping. Once you let the meat thaw, you should use it immediately. Do not refreeze it.

Cool Tips

Mutton is meat from a sheep that is over 16 months old. It is tougher than lamb, and has a strong flavor.

FIGURE 12-4
Pork Primal Cuts. The USDA puts pork into two quality levels: Acceptable and Utility. Acceptable quality meat is graded on a scale from 1 to 4, based on the amount of meat compared to the amount of fat or bone. What do you think utility quality pork is used for?

Boston Butt — Pork Loin — Pork Tenderloin

Slab Bacon — Spareribs — Ham (top view)

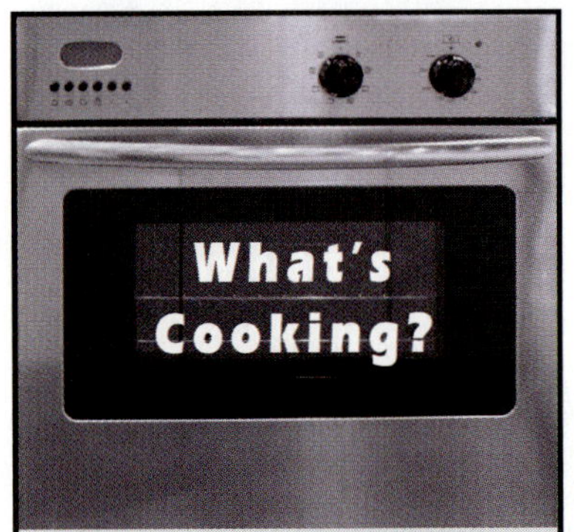

Compare the fat, cholesterol and calorie content of a 3-oz serving of 85% ground beef to a 3-oz pork loin chop:

	Hamburger	Pork Chop
Calories	180	134
Total Fat	12 g	5.7 g
Sat Fat	6 g	1.97 g
Trans Fat	0 g	0 g
Cholesterol	57 g	48.4 g

Source: USDA National Nutrient Database for Standard Reference

Nutritional Value

Meat falls into the Meat and Beans category of MyPyramid. It is a rich source of complete protein and iron, and one of the most reliable sources of the B vitamins, like thiamin, riboflavin, and niacin. In addition to iron and copper, which are vital for maintaining red blood cells, meat contains phosphorous, which helps to build bones and teeth and supports the oxidation of foods in the body.

Don't count on meat as a primary source of calcium, or vitamins A or C. You should include dairy, fruits, and vegetables in a meal with meat in order to maintain a balanced diet.

The fat content of meat varies depending on the type and cut. Fat provides energy as well as flavor. Most retail outlets label ground beef with the percentage of fat and percentage of lean. A rating of 85% means it has 85% of lean and 15% of fat. Usually, the higher the fat content the lower the cost.

Complete protein contains all of the essential amino acids required by your body. Essential amino acids are the ones your body cannot make on its own.

Lamb Shoulder (square cut)

Full Rack of Lamb

French Rack (and single chop)

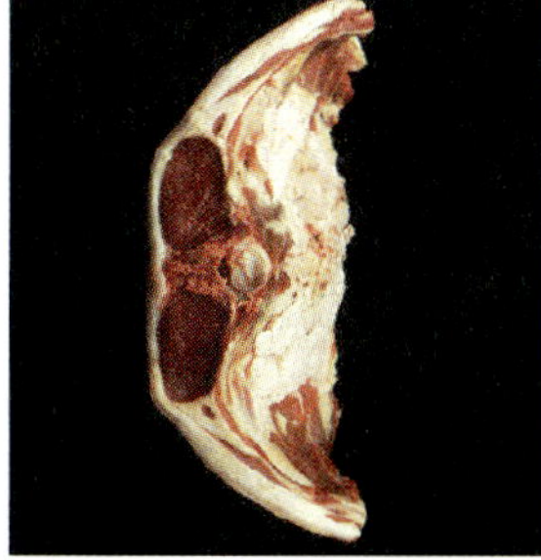

Lamb Loin Saddle

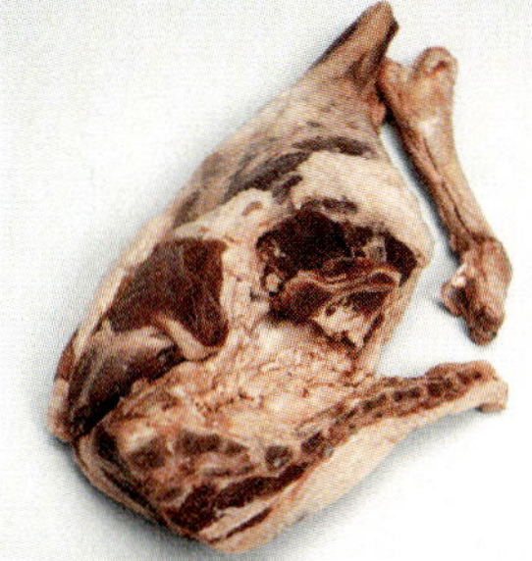

Leg of Lamb (with shank)

FIGURE 12-5
Lamb Primal Cuts. Lamb for the restaurant industry or retail sale is either Prime or Choice. Lower grades—Good, Utility, and Cull—are only used commercially. Why do you think lamb is usually tender?

But—and this is a big but—there is no standard for measuring the amount of fat. Most butchers simply estimate the amount of fat by looking at the meat and assessing the amount of fat they trim off before grinding.

Besides fresh meats, there are also many ready-to-eat or ready-to-cook meats available in most supermarkets. They include cold cuts, like ham, bologna, and salami, as well as canned meats and meat products, such as beef stew and corned beef hash. You can buy frozen meat products, like meatballs or meat-filled raviolis.

These items are usually convenient, and may come in handy when you are in a rush, but they are also likely to be more expensive and less nutritious than meat that you buy and prepare yourself.

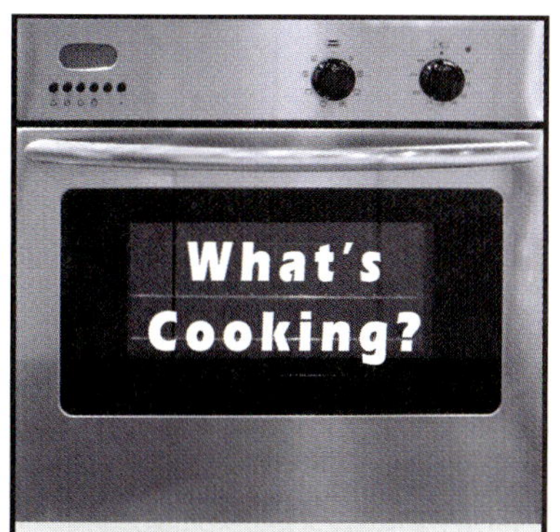

Would you eat raw ground beef mixed with onions, spices, and raw egg yolk? It's actually a dish called **Steak Tartare**, and is usually served with toast as a first course or main meal.

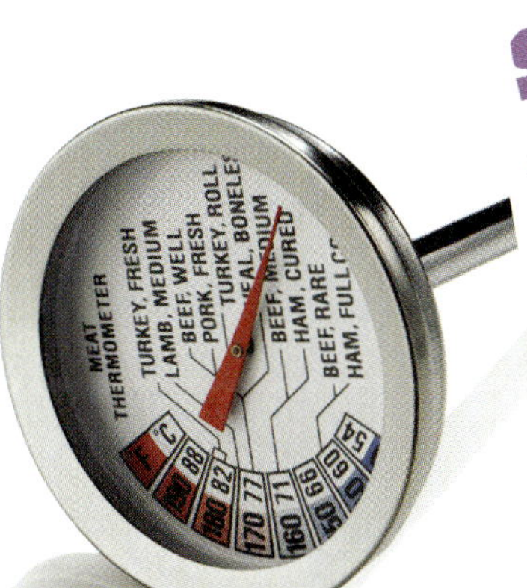

Safe Eats

Although all meats can harbor disease-producing microorganisms like E-coli and salmonella, pork may also contain the parasite Trichinella spiralis, which, if you eat it, causes the potentially fatal trichinosis. Cook pork to an internal temperature of 160° F to be sure it is safe to eat.

Check the Label

How do you want your steak?

Rare Red in the middle, 120°F

Medium rare A small amount of red in the middle, surrounded by pink, 145° F

Medium Pink in the middle, no red, 160° F

Medium well Very little pink in the middle, 165° F

Well done No pink in the middle, 170° F

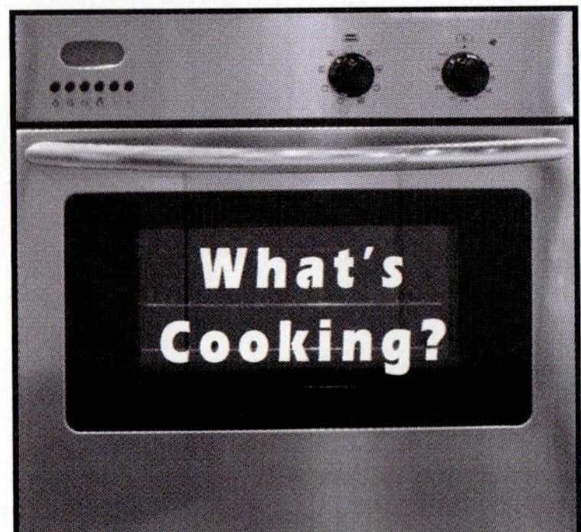

Just because you take meat out of the oven, doesn't mean it's done. Most meats continue to cook after you take them away from the heat source, and the internal temperature may rise by as much as 10 degrees! Called **carry-over cooking**, it's caused by heat transferring from the hotter exterior of the meat to the cooler center.

If you let the meat "rest"—or sit—after cooking, the temperature stabilizes and juices which may have been forced out during cooking are reabsorbed, resulting in a juicier, more tender, more flavorful meal.

FIGURE 12-6
An herbed, boneless leg of lamb being carved, with parsley and roast potatoes on the plate. Judging from the color, do you think this roast is cooked rare, medium, or well-done?

BY THE NUMBERS

Most chefs assume a serving of meat is about ¼ lb. per person. Of course, if the cut has a bone, you will need more. How many people will a 1¾ lb. boneless leg of lamb serve?

Hint: Divide the total weight by 0.25.

Preparation

You cook meat in order to make it more tender, to kill off harmful organisms, and to make it taste good. The proper type of cooking method depends on the cut of meat. Tougher or larger cuts benefit from longer cooking times and added moisture; tender and smaller cuts can be cooked quickly, using dry-heat methods.

Preparation Skills

Most cuts of meat require very little preparation before cooking. You may have to trim the fat off of some cuts before cooking. Roasts may be tied with a strong piece of string to keep them from losing their shape and to make sure they cook evenly. Cutlets may be pounded so they have an even thickness over their entire surface. This makes them ready for quick sautéing or pan-frying. You may want to cut meat into cubes to prepare it for stewing or grinding.

Safe Eats

To keep meat fresh and safe, store it in the refrigerator at less than 41° F. Store meats separately—you don't want your pork coming in contact with your beef. If you must remove the original packaging, rewrap meat in paper that allows airflow; airtight containers may promote the growth of bacteria.

Cooking with Dry Heat

Use roasting, grilling, broiling, and pan-broiling to cook tender cuts of meat. Roasting is particularly good for cooking large, tender cuts, such as a beef rib roast, lamb loin roast, or pork blade loin roast. Broiling or pan-broiling works well for smaller cuts, like beef steaks, lamb chops, and bacon.

Cooking with Moist Heat

When you have a less tender cut, like a brisket or beef short ribs, you should use a moist heat method of cooking such as braising, simmering, or stewing. These methods add liquid, which helps soften the proteins in the connective tissues, making the meat more tender and flavorful.

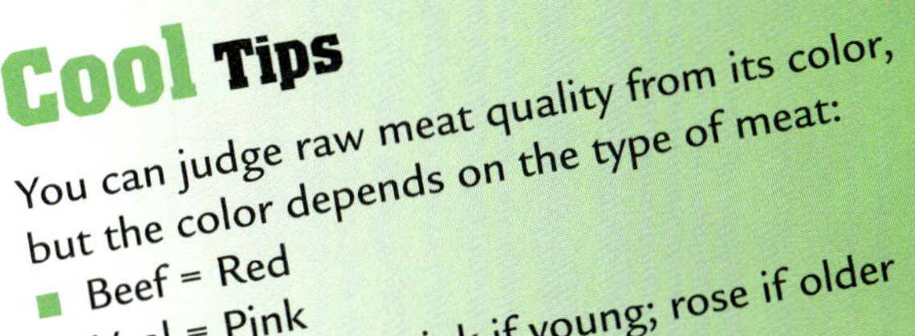

Cool Tips

You can judge raw meat quality from its color, but the color depends on the type of meat:

- Beef = Red
- Veal = Pink
- Pork = Grayish pink if young; rose if older
- Lamb = Dark red

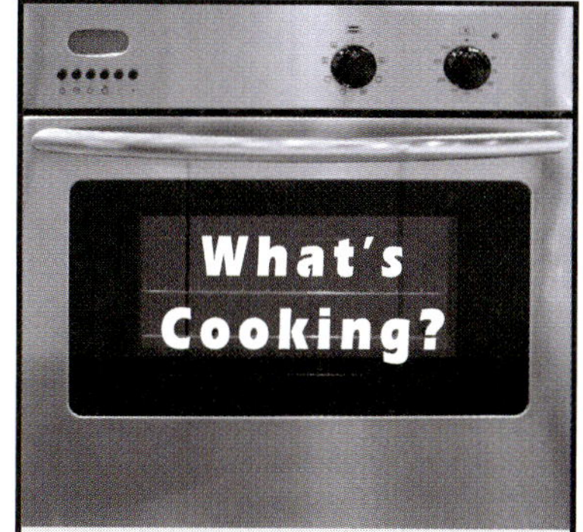

What's Cooking?

Barbeque is not the same as grilling, but both are good methods of cooking meat.

Barbeque uses a low heat and smoke to cook meat for a long time. It's great for cuts like brisket, ribs, and pulled pork. Grilling is done quickly, over a high heat. Use it for steaks, chops, and hamburgers.

BASIC CULINARY SKILLS

Grilling and Broiling

1. Oil the grill or rack.
2. Heat the grill or broiler.

3. Place the food on the grill or rack, with the presentation side facing down. Brush with sauces or glazes if your recipe calls for them.

4. When the first side is done, turn the food.

5. Cook on the second side until properly cooked.
6. Serve very hot, on heated plates.

BASIC CULINARY SKILLS

Pan Frying

1. Coat the food as directed by your recipe.
2. Heat oil or cooking fat in a pan. You want to use an amount equal to half the thickness of the food to be cooked.
3. Carefully add the food to the hot oil. Do not allow the pieces to touch.

4. Pan fry the first side until a golden crust forms.
5. Turn and finish cooking.
6. Drain excess oil and blot with a paper towel before serving.

BASIC CULINARY SKILLS

Braising and Stewing

1. Heat a small amount of fat or oil in a pan.
2. Add food to the pan.

3. Sear until evenly colored.

4. Remove seared food from the pan.
5. Add aromatic ingredients and cook in the hot fat or oil.
6. Add cooking liquid to the pan, stir well, and bring to a simmer.

7. Return the food to the cooking liquid and cover the pan.
8. Cook at a low simmer until very tender. Turn braised food as it cooks. Add more liquid if necessary, to keep food appropriated covered.

9. Skim grease and other impurities from the cooking liquid.

Case Study

Three days a week after school, Luis and Manny meet at the gym to work out. They spend about 40 minutes on cardio activities, and then lift weights. When they are done, they meet their friends at the local fast food joint for an afternoon snack.

According to Manny, they have earned their burgers, sodas, and fries by exercising regularly. Neither of them gains weight, so they figure they have achieved a good balance between how many calories they take in, and how many calories they burn.

- Do you agree that the boys earn their burgers, sodas, and fries with their workouts?
- Does the fact that they don't gain weight mean that they are successfully balancing a nutritious diet with an active lifestyle?
- Would you recommend any changes in their diet and/or exercise routine? Why?
- How does their routine compare to yours?

Put It to Use

1. Meat is an excellent source of complete protein, as well as many vitamins and minerals. It is also high in fat. Create a publication, such as a flyer or brochure, that explains the benefits and risks of eating meat. Include suggestions of how to include meat in a healthy, well-balanced diet.

Put It to Use

❷ You are hosting an elegant dinner party for 10 friends on New Year's Eve. Create a menu that includes meat as the main course, and then develop a shopping list. Be specific about the type and cut of meat, as well as the amount you need to serve all of your guests.

Write Now

Humans have been hunting animals for food since the dawn of time. Imagine you are living in a primitive village in an ancient time. You must hunt and kill the meat you need. Write a story about the hunt. Include information about where you live, the types of animals in your environment, and the weapons or strategies you will use during the hunt. Explain how you will use the meat you catch.

Tech Connect 

Braising is a moist heat cooking method suitable for less tender cuts of meat. Broiling is a dry-heat cooking method suitable for quickly cooking small, tender cuts of meat. Use the Internet to locate a recipe for braising a cut of meat, and a recipe for broiling meat. Compare the cooking times, number of ingredients and overall costs of the two recipes.

Team Players

The government inspects meat for quality and safety at three critical stages. As a team, write, prepare and perform a skit showing how inspections help insure that the meat we buy and eat is safe.

Put It Together

Match the explanation in column 1 with the term in column 2.

Column 1

a. meat from cattle three to fourteen weeks old
b. meat from hogs or pigs, usually less than one year old
c. the highest quality USDA grade of beef
d. raw ground beef mixed with onions, spices and raw egg yolk
e. the amount of fat in meat
f. powdered meat tenderizing made from papaya
g. meat from cattle over one year old
h. a potentially fatal disease caused by eating under-cooked pork
i. meat from sheep less than fourteen months old
j. meat from a sheep more than 16 months old
k. variety meat from areas other than the primal cuts

Column 2

1. marbling
2. trichinosis
3. veal
4. lamb
5. offal
6. beef
7. mutton
8. papain
9. steak tartare
10. prime
11. pork

TRY IT!

Broiled Sirloin Steak

Yield: 10 Servings   Serving Size: 1 Steak

Ingredients

10 Beef sirloin steaks, boneless (about 8 oz each)
1 fl oz Extra virgin olive oil
1 oz (4¼ tsp) Salt
1 Tbsp Black pepper, freshly cracked

Method

1. Trim the steaks of any silverskin and excess fat.
2. Brush each portion with olive oil.
3. Season generously with salt and pepper.
4. Grill over high heat until the first side of steak is lightly charred, about 3 minutes.
5. Turn the steaks and finish grilling on the second side, another 3 to 6 minutes, depending on the desired doneness.

Recipe Categories

Meat, Beef, Main Entrees

Chef's Notes

If desired, top each steak with a compound butter (butter with a flavor, or additional ingredient(s) added to it) and then flash them briefly under a broiler until the butter is lightly melted, about 30 seconds.

Potentially Hazardous Foods

Meat

HACCP

Cook to at least 145° F for at least 15 seconds.

Nutrition

Calories	840
Protein	60 g
Fat	65 g
Carbohydrates	0 g
Sodium	1,150 mg
Cholesterol	240 mg

TRY IT!

Pan-Fried Veal Cutlets

Yield: 8 Servings Serving Size: 5 -oz Cutlet

Ingredients

2½ lb Boneless veal top round
½ cup (2½ oz) All-purpose flour
1 cup (5 oz) Breadcrumbs, dried
2 Eggs, large, beaten (for egg wash)
To taste Salt and black pepper, freshly ground
16 fl oz Vegetable oil or olive oil (for frying)

Method

1. Trim any surface fat or silverskin from the veal.
2. Cut the veal into 8 equal portions of about 5 oz each.
3. Pound the cutlets between two pieces of plastic wrap to an even thickness of about ¼ inch.
4. Place flour on one plate and bread crumbs on another.
5. Place egg wash in a bowl.
6. Blot cutlets dry and season generously with salt and pepper.
7. Press each cutlet into the flour and shake off excess.
8. Dip each cutlet into egg wash and drain off excess.
9. Press each cutlet into the breadcrumbs and hold on a separate plate.
10. Heat oil to a depth of about 1/8 inch over moderate heat until oil is hot, about 350° F.
11. Add the breaded cutlets to the hot oil.
12. Pan fry on the presentation side for 2 to 3 minutes or until golden brown and the cutlets release easily from the pan.
13. Turn the cutlets once and finish cooking on the second side, about 2 to 3 minutes.
14. Cook the remaining cutlets, following the above procedure.
15. Remove the cutlets from the pan and drain briefly on absorbent paper toweling before serving.

Recipe Categories

Meat, Veal, Main Entrees

Chef's Notes

To test whether heated oil is the right temperature, drop a few breadcrumbs into the oil. They should sizzle but not turn black or sink to the bottom of the pan.

If the oil starts turning dark while cooking and is full of burned bits, it should be refreshed.

Potentially Hazardous Foods

Meat, Eggs

HACCP

Cook to at least 145° F for at least 15 seconds

Nutrition

Calories	589
Protein	12 g
Fat	15 g
Carbohydrates	1 g
Sodium	587 mg
Cholesterol	114 mg

TRY IT!

Baked Pork Chops

Yeild: 4 servings   Serving Size: 1 chop

Ingredients

3 tablespoons olive oil
4 thick cut boneless pork chops—about 1 lb total weight
2 tablespoons dark brown sugar
½ teaspoon salt
½ teaspoon ground black pepper
½ teaspoon dry mustard
1 tablespoon fresh lemon juice
10 oz tomato sauce
¼ cup water

Method

1. Preheat oven to 350° F.
2. Heat olive oil in a medium skillet over medium heat.
3. Place pork chops in the skillet, and brown about 5 minutes on each side. Remove from heat.
4. In a small bowl, mix brown sugar, salt, pepper, and dry mustard.
5. Arrange pork chops in a medium baking dish. Sprinkle with lemon juice, season with brown sugar mixture, and cover with tomato sauce. Pour water into the baking dish.
6. Cover, and bake 1 hour in the preheated oven, to an internal temperature of 160° F.

Recipe Categories

Meat

Chef's Notes

The pork can be served with potatoes, pasta, or a whole grain, if desired, with the sauce.

Potentially Hazardous Foods

Meat

HACCP

Cook to an internal temperature or 160° F or higher.

Maintain at 135° F during service.

Refrigerate at 41° F or below.

Nutrition

Calories	482
Protein	38.9 g
Fat	30.4 g
Carbohydrates	12.2 g
Sodium	797 mg
Cholesterol	107 mg

13 Poultry

In This Chapter, You Will . . .

- Identify the types and forms of poultry
- Identify the nutritional value of poultry
- Learn how to prepare and serve poultry in a variety of ways

Why YOU Need to Know This

Poultry, which refers to any domesticated bird raised specifically for use as food, falls into the Meat and Beans group on MyPyramid. It supplies complete protein—good for the muscles and lots of other essential vitamins and minerals. It's also tasty, versatile, and available pretty much everywhere throughout the world. Although chicken may be the most well-known and common type of poultry served in homes and restaurants, there are many other types and varieties. As a class, see how many types of poultry you can list in less than five minutes.

Types of Poultry

Poultry includes any bird raised on a farm as food. You are probably most familiar with chicken, turkey, ducks, and geese. You may also have heard about—or eaten—farm-raised **game** birds, such as pheasant, quail, Guinea hens, Cornish Game hens, and squab. Game is the term used for animals that are hunted for sport.

Hot Topics

Because the class of poultry usually includes a recommended cooking method, you can use the information to help you select the best bird for the recipe you have in mind.

Poultry may be classified by type, age, weight, and even sex. It is often labeled according to the best cooking method. For example, older birds are going to be less tender than younger birds, so they are classified as stewing birds. Younger, more tender, birds may be classified as fryers or roasters.

FIGURE 13-1
As with most bird species, the male chicken, called a *rooster*, is usually more colorful than his female counterpart, called a *hen*. Do you think the chicken in this picture is male or female?

FIGURE 13-2
An ostrich (left), and emu (right). What similarities do you see?

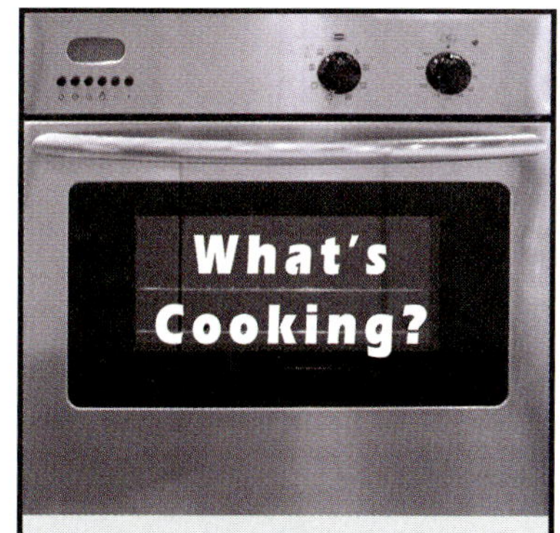

Recently, members of the family of flightless birds referred to by their Latin name, ratites—pronounced RAT-ites—have become popular menu items. Ratites include the ostrich, emu, and rhea. Unlike chicken and turkey, the meat of ratites is a rich red color. It is lean and low in fat, so many people consider it healthier than traditional poultry.

Forms of Poultry

All types of poultry are usually sold ready-to-cook, in a variety of forms. They may be fresh, frozen, or processed.

Cool Tips

Look out for freezer burn if you decide to buy a frozen bird. Freezer burn occurs when an item loses moisture because it is wrapped improperly, or frozen too long. It causes the item to appear dry and discolored. It doesn't make the food unsafe, but it does cause a loss of flavor and a rubbery texture.

- Whole birds are cleaned, and the head and feet are removed. They usually contain a small bag within the cavity, called a giblet bag. The **giblet bag** holds the liver, stomach (called a **gizzard**), heart, and neck.
- Whole birds may be cut into individual pieces. The pieces are usually breasts, drumsticks, thighs, and back.
- Breasts may be whole or halved. They may be sold bone-in or boneless, with or without the skin.
- Whole legs, which are both the thighs and drumsticks, usually include skin and bones.
- Thighs, or the upper part of the leg, may be sold bone-in or boneless, with or without the skin.
- Drumsticks, or the lower part of the leg, are usually sold bone-in, with or without the skin.
- Ground poultry, which is similar to ground beef, may be used as an ingredient in items such as burgers, soups, or stews.
- Processed poultry is used to create processed items such as patties, nuggets, sausages, and bacon.

Check the Label

Fresh Poultry that has never had an internal temperature below 26° F

Frozen Poultry that has a temperature of 0° F or below

Hard chilled An outdated term for poultry that has a temperature between 0° and 26° F

FIGURE 13-3
Drumstick, thigh, wing, breast. Poultry is often sold in pieces. What is one advantage of buying poultry in pieces, rather than whole? What is one disadvantage?

Buying Poultry

When you look for poultry to buy, you want plump breasts and meaty thighs. The skin should be intact with no tears or punctures. In addition, take note of the packaging. There shouldn't be any holes in the wrapper, and the grade assigned by federal inspectors should be on the label. Only buy poultry from a reputable vendor, because if it hasn't been properly chilled during processing and storage, it might harbor harmful bacteria, such as **salmonella**, which can cause food poisoning. Keep uncooked poultry in the refrigerator at a temperature of 40° F. If you don't use it within two days, freeze it.

For safety, poultry should be kept chilled to below 32° F during storage. Raw poultry must be chilled to 26° F during processing. Poultry chilled to less than 26° F is labeled "frozen" or "previously frozen."

Poultry labeled "free-range" is healthier and more nutritious than regular poultry.

According to the USDA, any poultry with access to the outdoors can be labeled "free-range."

Check the Label

Certified Implies that the USDA's Food Safety and Inspection Service and the Agriculture Marketing Service have officially evaluated the product for class, grade, and other quality characteristics

Free-range Poultry has been allowed access to the outside

Halal, or Zabiah Halal Poultry handled according to Islamic law, or under Islamic authority

Kosher Poultry prepared under Rabbinical supervision

FIGURE 13-4
In some parts of the world, people buy their poultry live. What sanitation issues might this present? What preparation steps might be required before you could use the poultry in a recipe?

FIGURE 13-5
Why is it important to make sure raw poultry does not touch your other food? What could happen to you if you contracted salmonella? Do you know someone who has ever had food poisoning?

Safe Eats

When you store poultry, it is a good idea to put it in a drip pan before placing it in the refrigerator. That way, if it drips or leaks, it will not contaminate other food stored nearby.

Check the Label

Natural Poultry that contains no artificial ingredients or added color and is only minimally processed

No antibiotics added Poultry raised without antibiotics

Organic Poultry produced without the use of sewer-sludge fertilizers, most synthetic fertilizers and pesticides, genetic engineering (biotechnology), growth hormones, irradiation, and antibiotics

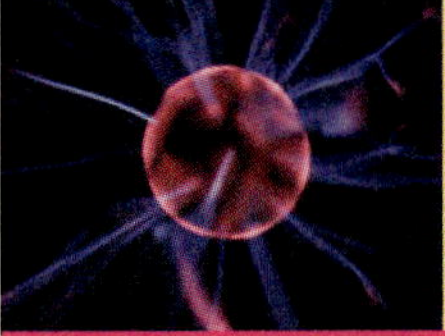

SCIENCE STUDY

Did you know that acids, like vinegar, remove calcium carbonate from compounds such as bones? You can test this using the so-called "Rubber Chicken Experiment." You'll need a cooked drumstick bone, a glass jar with a lid that is large enough to hold the bone, and a quart of white vinegar.

First, roast a chicken drumstick and remove as much meat from the bone as possible. Try to bend the bone. You can't because the calcium keeps the bone strong and firm.

Put the bone in the jar, fill the jar with enough vinegar to cover the bone, and put on the lid. After two days, remove the bone and try to bend it again. It should be more flexible. (If it's not, dump out the vinegar and replace it with fresh vinegar.)

Put the bone back in the jar, and test it again every few days. By the end of the week, most of the calcium will be removed from the bone, and it should flop around like it's made out of rubber!

What does this experiment prove about bones and calcium? What foods can you eat to insure your bones remain strong and firm?

Nutritional Value

Poultry falls into the Meat and Beans category of MyPyramid. Because it contains complete protein, less fat than most meat, and no carbohydrates, it can be used as a healthy substitute for meat in most menus. It is also a good source of vitamins and minerals. For a complete, nutritionally balanced meal, include vegetables, grains, and salad with a main course of poultry.

Although the nutritional value varies depending on the specific type of poultry, without the skin, most poultry is considered low in fat. Breast meat is generally lower in fat than other parts of the bird. In addition, most of the fat content is unsaturated, leaving only about 1 gram of saturated fat in a skinless breast. A 3.5-oz boneless, skinless chicken breast has only about 110 calories before cooking.

The USDA assigns grades to poultry that passes a mandatory quality inspection: Grade A means the bird is full-fleshed and meaty. Grade B means less meaty, but still good. Grade C has less flesh and fat. Restaurants and retail outlets purchase Grade A poultry.

You are planning Thanksgiving dinner for 12 adults and five children. You want to have 1.25 pounds of turkey per adult, and 0.75 pounds per child. How large a turkey do you need?

Hint: Multiple the number of adults by 1.25. Multiply the number of children by 0.75. Add the products to get the answer.

FIGURE 13-6
Breast of duck, with rosti potato and cassis jus. Would you think this is poultry if you were served it in a restaurant? How does the color of duck breast differ from chicken?

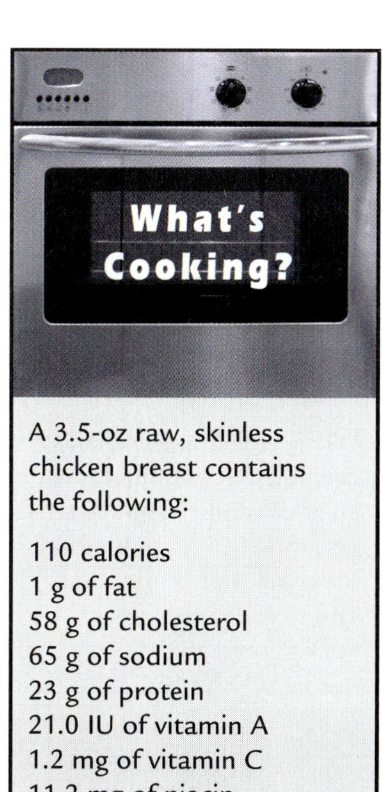

A 3.5-oz raw, skinless chicken breast contains the following:

110 calories
1 g of fat
58 g of cholesterol
65 g of sodium
23 g of protein
21.0 IU of vitamin A
1.2 mg of vitamin C
11.2 mg of niacin
11.0 mg of calcium
0.7 mg of iron

Source: USDA National Nutrient Database for Standard Reference

Preparation

One thing that makes poultry unique is its versatility. It can be roasted, grilled, baked, broiled, fried, or stewed. It can be used as a main course in a family dinner, or at a formal restaurant. It makes a great sandwich or salad for lunch or snack, and is often used as an ingredient in other recipes, such as casseroles and appetizers. Ground poultry can be shaped into burgers, loafs, or meatballs. When processed, it can be turned into nuggets, fingers, and even sausage. You could probably eat poultry every day for year and never have the same recipe twice!

Most cooking techniques that can be used for one type of poultry can be used for all types of poultry. That means if you have a recipe that calls for chicken breasts, it's a good bet that it will work for turkey breasts, or even ostrich breasts. Of course, you will have to adjust the cooking times to account for larger or smaller portions.

Safe Eats

Undercooked poultry is unappetizing at best and unhealthy at worst. Bacteria such as Salmonella and Campylobacter jejuni thrive on raw poultry and can cause food poisoning and diarrhea. To kill bacteria, poultry should be cooked to an internal temperature of at least 165° F.

When selecting a cooking method, keep in mind the class of poultry you have. Use moist cooking methods for older, less tender birds, and dry methods for young, tender birds.

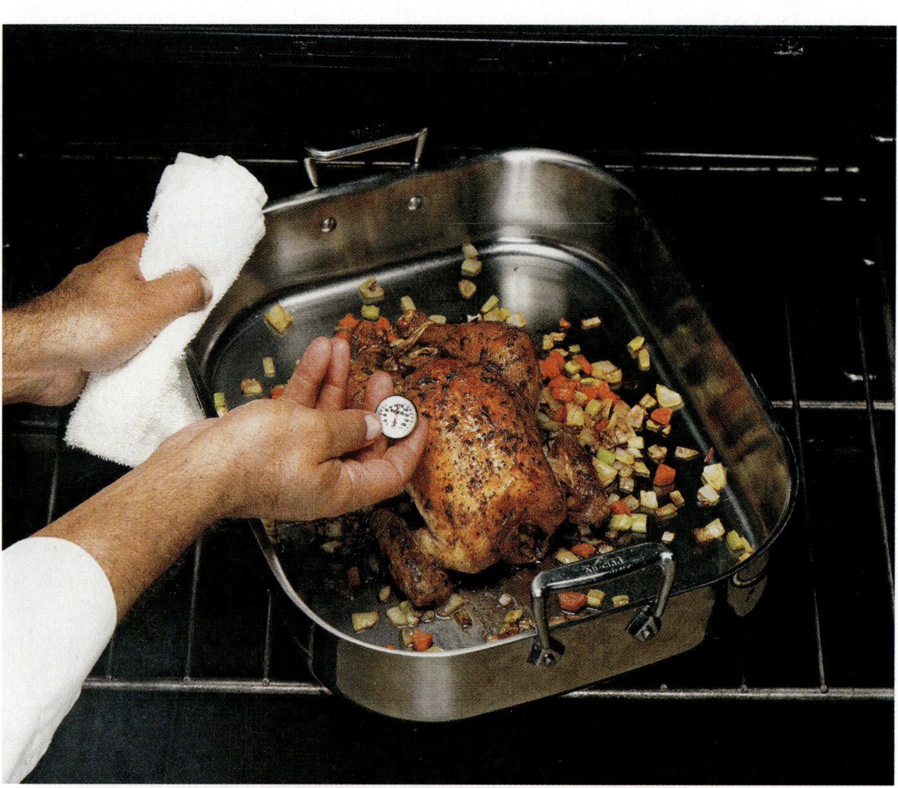

FIGURE 13-7
A chef checks the temperature of a roasted chicken. When poultry is cooked properly, the juices run clear, with no trace of pink. What is the internal temperature of a properly cooked chicken?

Preparation Skills

Before cooking, you should always rinse poultry in cold water, and then pat it dry with paper towels. If you have a whole bird, be sure to remove the giblet bag and rinse the cavity. If it's frozen, let it thaw in the refrigerator for at least 24 hours.

One of the most important skills required for cooking whole birds is trussing. Trussing means tying. The object of trussing is to give the bird a smooth, compact shape so it cooks evenly and retains moisture. Check out the Basic Culinary Skills feature on page 285 to learn the proper steps for trussing.

Another necessary skill is disjointing, or cutting up, poultry. You can disjoint a whole bird before or after cooking. You can cut it into halves, quarters, or eighths, depending on what you plan to do with it. For example, you might want to grill chicken halves, fry quarters, or use eighths in a stew. You can learn how to disjoint a bird in the Basic Culinary Skills feature on page 286.

Poultry is high in protein. Protein becomes tough when cooked at high temperatures. So, be careful not to use too high a heat when cooking your poultry!

Fiction	Fact
You should never cook stuffing inside the cavity of a bird.	It is safe to cook the stuffing in the cavity if you loosely stuff the bird immediately before cooking, and cook it until the internal temperature of the stuffing reaches 165° F. You should never stuff the bird and wait to cook it until later.

Utility Drawer

It is important to use the right knife when disjointing poultry. Smaller birds require more delicate, precise cuts, so you should choose a smaller blade. Larger or older birds are generally tougher. They require a heavier blade and greater pressure to break through the joints. In any case, make sure the blade is sharp!

FIGURE 13-8
Does this grilled chicken look appetizing to you? Some people like the burnt flavor that can come with grilling, but what might this do to the quality of the meat?

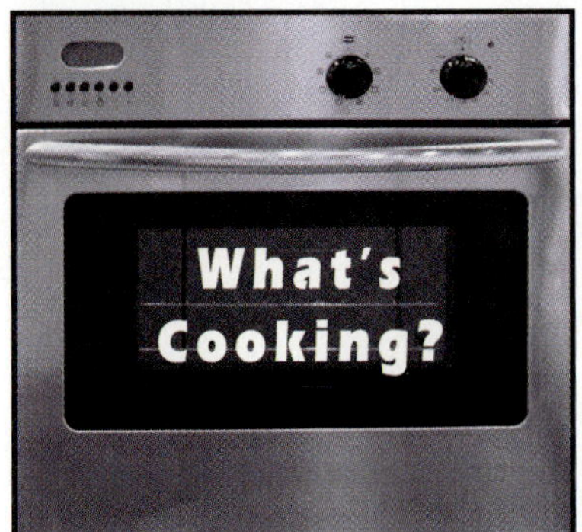

To insure a moist, flavorful bird, you can brine it before cooking. Brining means soaking it in a salt-based solution. A basic brine is made from one quart of water and ½ cup of coarse, or Kosher, salt. Mix up about one quart of solution for each pound of poultry, immerse the bird in the brine, and refrigerate for about one hour per pound.

Cooking with Dry Heat

How do you prepare turkey on Thanksgiving? Roast it, of course! Because most poultry is tender, meaty, and moist, it can be cooked using dry heat methods such as roasting, baking, grilling, and frying. Exceptions include poultry such as stewing chickens, which are less tender than broilers or fryers.

Cooking with Moist Heat

Any moist heat method or combination heat method is suitable for poultry. This includes steaming, poaching, simmering, stewing, and braising. Steaming and poaching are often used for lean, tender portions. Shallow-poaching is popular for breast portions because the cooking liquid can serve as the basis for a sauce that can add moisture and flavor to the breast when it is served.

Restaurant owners and managers must know how to save money while still offering the highest quality products. One way to save money with poultry is to purchase it whole, even if you plan to use it in pieces, or without bones. A well-trained chef will be able to disjoint and remove the bones quickly and easily.

Safe Eats

Unless you want to risk coming down with food poisoning or something worse, take care in how you handle raw poultry. You don't want to give bacteria the opportunity to infect equipment, other food, or your body, so wash everything that comes in contact with the bird in hot soapy water before you use it for anything else! That includes knives, cutting boards, countertops, the sink, and your hands.

BASIC CULINARY SKILLS

Trussing Poultry

1. Remove giblet bag.
2. Cut off the first wing joints, and any pockets of fat from the bird's cavity.
3. Stretch skin to cover the breast meat.
4. Pass the middle of a long piece of string underneath the joints at the end of the drumstick.
5. Cross the ends of the string to make an X.

6. Pull the string toward the tail and begin to pull the string back along the body.

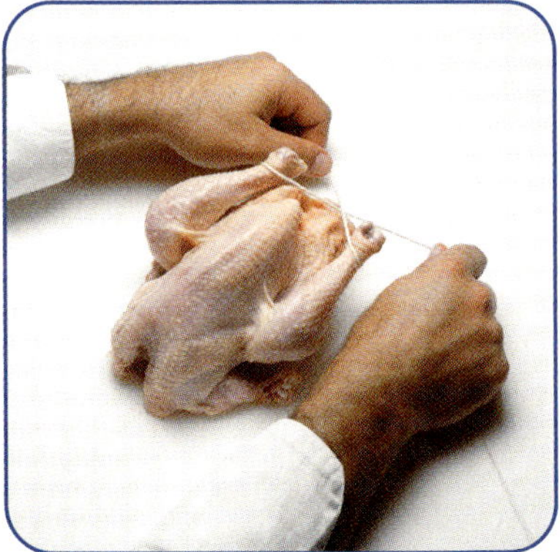

7. Pull the string tightly across the joint connecting the drumstick and thigh.
8. Then, pull it along the body toward the bird's back, catching the wing underneath the string.

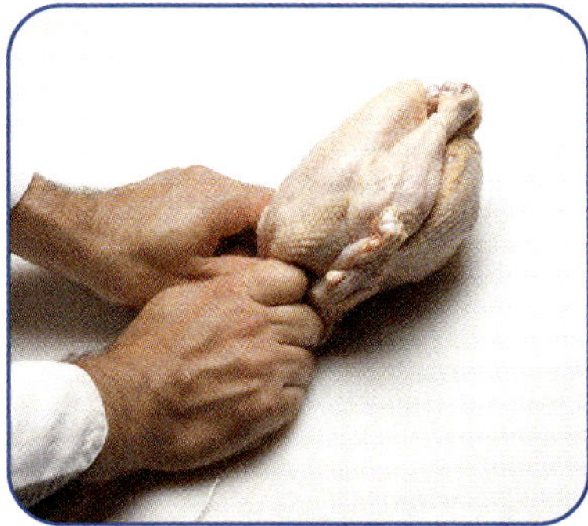

9. Pull one end of the string underneath the backbone at the neck opening.

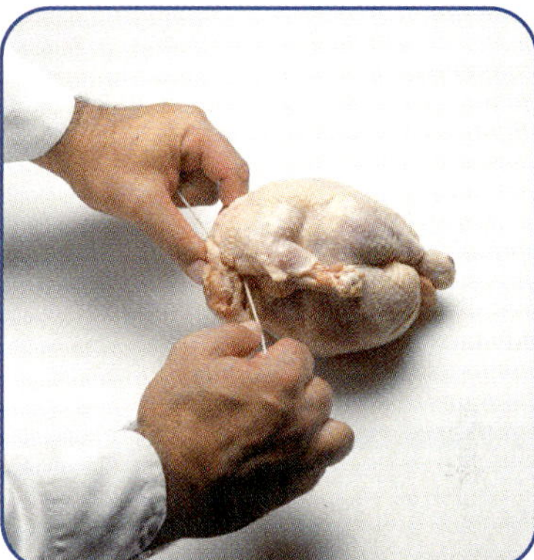

10. Tie the two ends of the string securely.

BASIC CULINARY SKILLS

Disjointing Poultry into Quarters

1. Remove the backbone by cutting along both sides of it.
2. Remove the keel bone by pulling it away from the chicken. The keel bone and the cartilage behind it join the two halves of the breast.
3. Cut the chicken into halves by making a cut down the center of the breast to divide the bird in half.

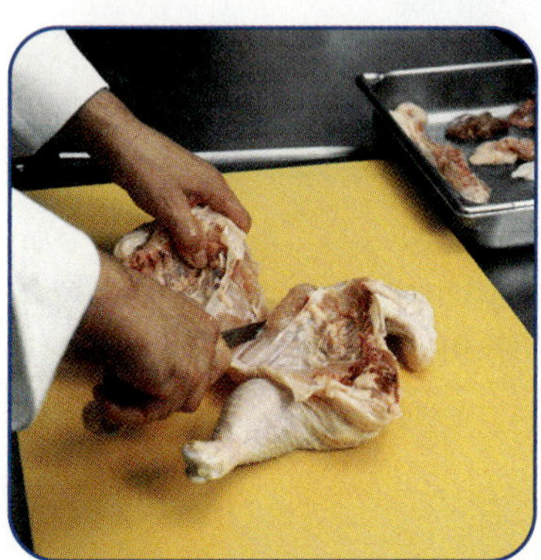

4. Separate the leg and thigh from the breast and wing by cutting through the skin just above where the breast and thigh meet.

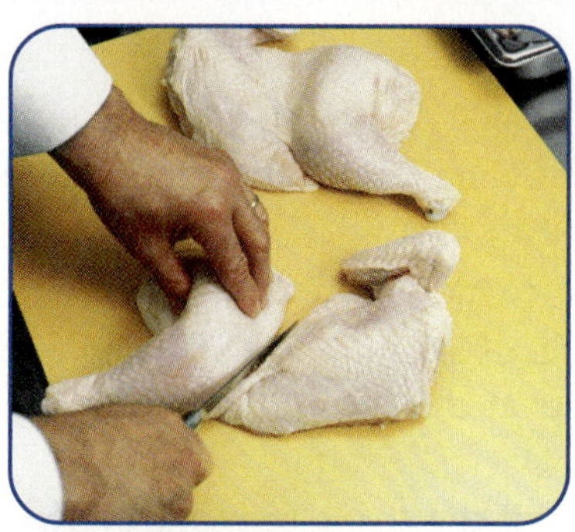

Note: To divide into eighths, cut at joints to separate leg and thigh, wing and breast.

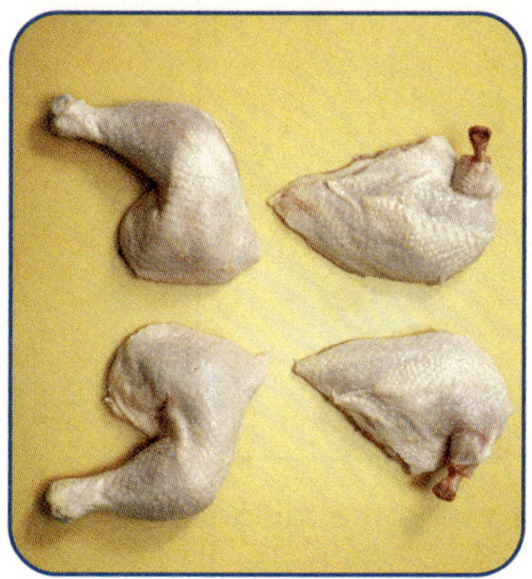

Case Study

Kaitlyn's father is on a low-sodium diet. Most nights, boneless, skinless chicken breasts are the main dish for dinner at their house. Kaitlyn is getting tired of eating the same thing over and over. She complains to her parents, and gets angry when they try to explain it is for her father's health.

Kaitlyn has started skipping family dinner and eating snack food or fast food instead.

- Do you think Kaitlyn is right to complain and get angry at her parents?
- What do you think might be a more effective method of communicating with her parents?
- What would you recommend the family do to introduce more variety into their dinners?

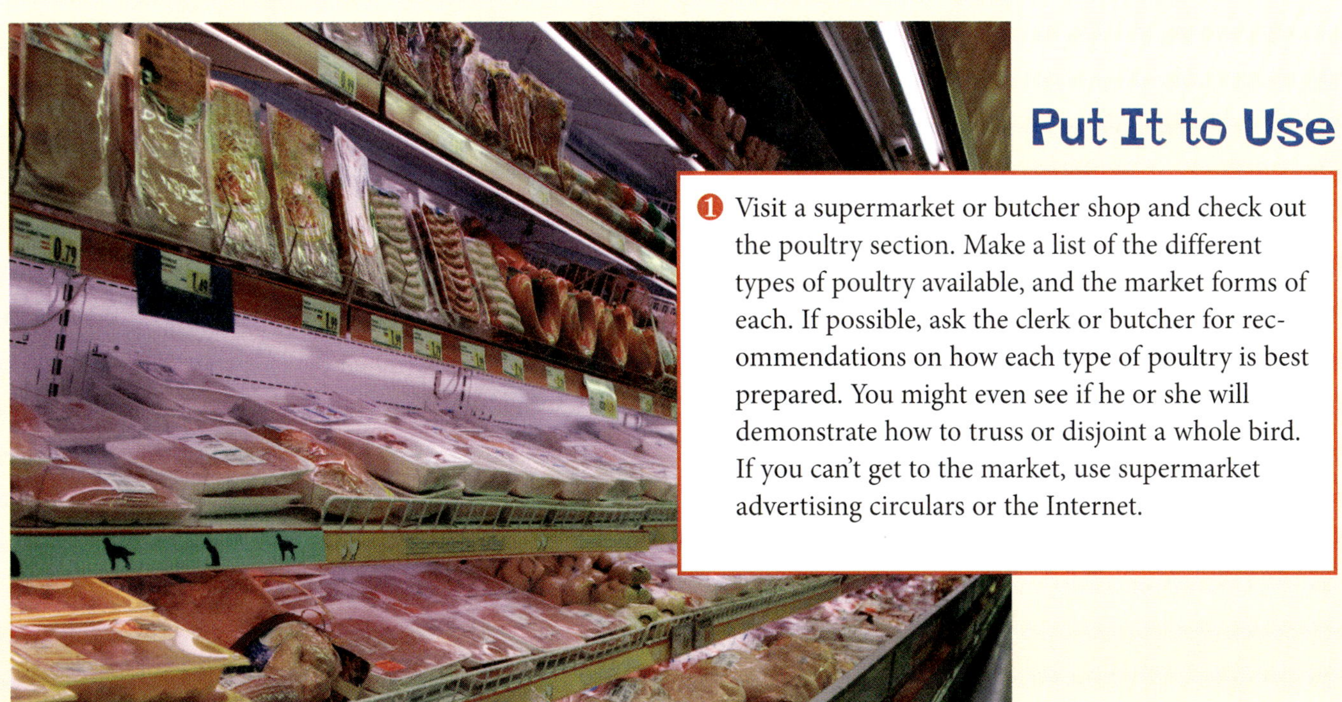

Put It to Use

1. Visit a supermarket or butcher shop and check out the poultry section. Make a list of the different types of poultry available, and the market forms of each. If possible, ask the clerk or butcher for recommendations on how each type of poultry is best prepared. You might even see if he or she will demonstrate how to truss or disjoint a whole bird. If you can't get to the market, use supermarket advertising circulars or the Internet.

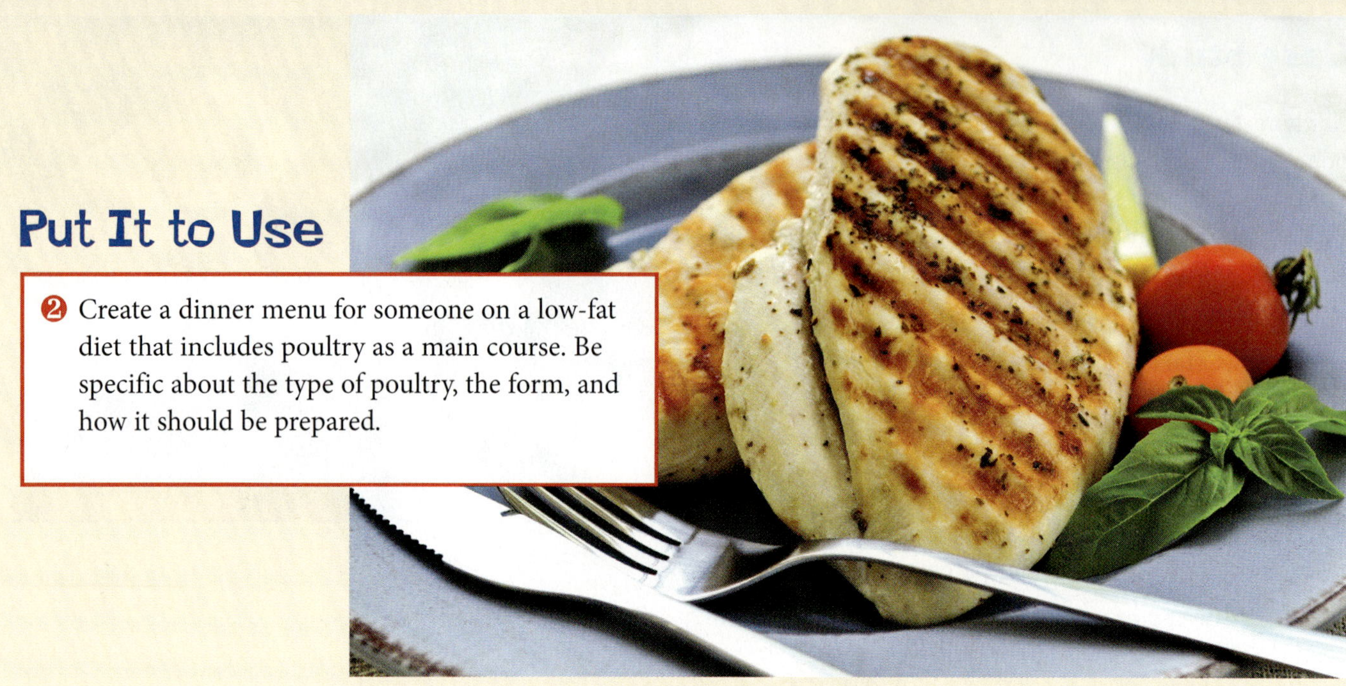

Put It to Use

❷ Create a dinner menu for someone on a low-fat diet that includes poultry as a main course. Be specific about the type of poultry, the form, and how it should be prepared.

Write Now

Think about the terms *free-range* and *organic*. If necessary, look up the definitions in a reference book or on the Internet. When you have the information you need, write an essay about whether or not you think it is important to eat free-range or organic poultry, and why. Use facts to support your thesis.

Tech Connect 

What has more fat—a 3-oz breast of chicken, a 3-oz breast of turkey, or a 3-oz breast of duck? Use the Internet to look up the answer. Expand your search to include ostrich, emu, and pheasant.

Hint: Try the Food and Nutrition page of the USDA.gov Web site.

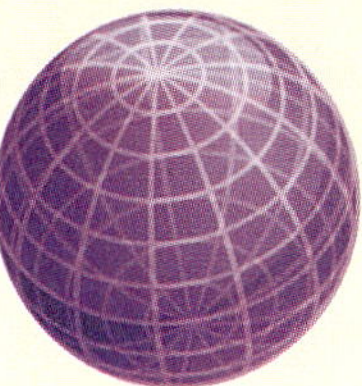

Team Players

Poultry is eaten in almost every country in the world, from small villages in Africa, to sophisticated restaurants in Europe. In small groups, work together to research the history of poultry in a country or area that interests you. Learn about the types of poultry they use, how they prepare it, and whether it is associated with any traditions or holidays. Prepare a presentation about your topic, and share it with the class.

Put It Together

Match the explanation in column 1 with the term in column 2.

Column 1

- **a.** the maximum temperature of poultry that is labeled frozen
- **b.** the stomach of a bird
- **c.** any domesticated bird used for human consumption
- **d.** the upper part of the leg of a bird
- **e.** bacteria that can cause food poisoning
- **f.** a method of tying a whole bird for cooking
- **g.** a small bag containing the liver, stomach, heart, and neck of a bird, stored in the cavity of a whole bird
- **h.** a method of cutting a whole bird into pieces
- **i.** the internal temperature of a poultry breast that is properly cooked
- **j.** a family of flightless birds
- **k.** the lower part of the leg of a bird

Column 2

1. poultry
2. giblet bag
3. gizzard
4. drumstick
5. thigh
6. ratites
7. salmonella
8. trussing
9. disjointing
10. 165° F
11. 0° F

TRY IT!

Roast Chicken with Rosemary

Yield: 6 Servings Serving Size: 1/2 lb

Ingredients

1 Whole roaster chicken (3 lb)
To taste Salt and black pepper, freshly ground
1–2 tsp Olive oil
1 Onion, small quartered
1/4 cup Rosemary, fresh chopped

Method

1. Preheat the oven to 350° F.
2. Rinse the chicken thoroughly and blot dry.
3. Season the inside and outside of the chicken generously with salt and pepper, and rub with olive oil.
4. Place the onion and rosemary inside the cavity of the chicken. Truss the chicken.
5. Place the chicken on a rack in a roasting pan.
6. Roast for 2 to 2 1/2 hours or until an instant-read thermometer inserted at the thickest point of the thigh reads 165° F.

Recipe Categories

Poultry

Chef's Notes

The internal temperature of the chicken will rise due to carryover cooking as it before carving.

Potentially Hazardous Foods

Poultry

HACCP

Cook to an internal temperature of 165° F.

Maintain at 135° F during service.

Cool to below 41° F within 4 hours (1-stage cooling method) or within 6 hours (2-stage cooling method).

Reheat to an internal temperature of 165° F.

Nutrition

Calories	291
Protein	30.8 g
Fat	17.2 g
Carbohydrates	1.3 g
Sodium	94 mg
Cholesterol	97 mg

TRY IT!

Chicken Stir-Fry

Yield: 4 Servings Serving Size: About 4 oz

Ingredients

1 lb Boneless skinless chicken breasts
3 Tbsp Cornstarch
2 Tbsp Reduced-sodium soy sauce
½ tsp Ginger, ground
¼ tsp Garlic powder
3 Tbsp Vegetable oil
2 cups Broccoli florets
1 cup Celery, sliced
1 cup Carrots, thinly sliced
1 Onion, small and cut into wedges
1 cup Reduced-sodium chicken stock or broth

Method

1. Cut chicken into ½-inch strips.
2. Combine soy sauce, ginger, garlic powder, and corn starch and whisk together.
3. Toss sauce with the chicken to coat, and refrigerate, covered, for 30 minutes.
4. In a large skillet or wok, heat 2 Tbsp of oil.
5. Stir-fry chicken until no longer pink, about 3 to 5 minutes. Remove chicken from pan and place in a clean bowl.
6. Add remaining oil to pan and stir-fry broccoli, celery, carrots, and onions for 4 to 5 minutes, or until tender crisp.
7. Add chicken stock or broth.
8. Return chicken to pan and cook until heated through and sauce becomes thick and bubbly.
9. Serve immediately.

Recipe Categories

Poultry

Chef's Notes

Serve with rice or chinese noodles.

Potentially Hazardous Foods

Poultry

HACCP

Cook to an internal temperature of 165° F.

Maintain at 135° F during service.

Cool to below 41° F within 4 hours (1-stage cooling method) or within 6 hours (2-stage cooling method).

Reheat to an internal temperature of 165° F.

Nutrition

Calories	306
Protein	30 g
Fat	14 g
Carbohydrates	18 g
Sodium	239 mg
Cholesterol	73 mg

TRY IT!

Chicken Fajitas

Yield: 4 Servings Serving Size: 2 Fajitas

Ingredients

4 Chicken breasts
1 Tbsp Olive oil
2 Tbsp Lemon juice, freshly squeezed
To taste Salt and black pepper, freshly ground
½ tsp Cumin seed, ground
1 cup Tomato salsa
1 cup Guacamole
1 cup Cheddar or jack cheese, shredded
1 cup Romaine or iceberg lettuce, shredded
1 cup Tomatoes, diced
½ cup Sour cream
½ cup Scallions, sliced
¼ cup Cilantro, chopped
8 Flour tortillas

Method

1. Trim the chicken breasts and pat them dry.
2. Combine the olive oil, lemon juice, salt, pepper, and cumin in a shallow dish.
3. Add the chicken breasts and turn to coat them evenly.
4. Marinate for 30 to 45 minutes.
5. Preheat a grill.
6. Cook the chicken until it is cooked through, turning occasionally.
7. Assemble all the other ingredients.
8. Heat the flour tortillas by steaming them over a little simmering water or wrapping them in foil.
9. Heat the tortillas on the grill as the chicken cooks.
10. Remove the chicken from the grill and slice it into strips on the bias.
11. Serve the chicken on a platter with the garnishes and heated tortillas.

Recipe Categories

Poultry

Chef's Notes

Sautéed chicken can be substituted for grilled chicken.

Other accompaniments might include sliced or diced red onion, sautéed pepper strips, black olives, different salsas or relishes, or diced avocado.

Potentially Hazardous Foods

Poultry, Dairy, Raw Scallions

HACCP

Cook to an internal temperature of 165° F.

Maintain at 135° F during service. Cool to below 41° F within 4 hours (1-stage cooling method) or within 6 hours (2-stage cooling method).

Reheat to an internal temperature of 165° F.

Nutrition

Calories	376
Protein	22 g
Fat	13 g
Carbohydrates	42 g
Sodium	400 mg
Cholesterol	44 mg

14 Fish

In This Chapter, You Will . . .

- Identify types of fish and shellfish
- Identify the nutritional value of fish and shellfish
- Learn how to prepare fish and shellfish

Why YOU Need to Know This

There are literally thousands of different types of fish living in oceans, lakes, rivers, and ponds around the world. Fish is low in calories and saturated fat, and high in protein, vitamins, and minerals. It comes in a wide range of tastes and textures, and can be prepared in many different ways. Although it is usually served as the main course, some varieties make a great appetizer or salad, and most can be used as an ingredient in other dishes. As a class, discuss the types of fish you like to eat. How many different types can you think of? Categorize them based on whether they come from saltwater (the ocean) or from freshwater (like a lake or river).

Types of Fish

It is nearly impossible to list all of the different types of fish that we use as food. In general, they are classified as either **finfish**, which have scales and fins, like tuna, trout, and halibut, or **shellfish**, which are enclosed in a hard shell, like clams, scallops, and lobster.

Both finfish and shellfish are inspected by the National Marine Fisheries Service (NMFS), part of the National Oceanic and Atmospheric Administration (NOAA). Inspectors conduct three evaluations:

- Type 1 evaluates quality and wholesomeness.
- Type 2 evaluates the accuracy of the labeling and weight.
- Type 3 evaluates the sanitation of the processing facility.

Cool Tips

Most fish live either in freshwater or saltwater, but there are a few species that live in both. Called **anadromous** fish, they are born in freshwater, spend most of their lives in saltwater, then travel back to freshwater to **spawn**, or reproduce. Salmon and steelhead trout are examples.

Forms of Fish

Shellfish are available fresh or frozen. Finfish are available in many forms:

- **Whole fish** includes the head and stomach.
- **Drawn fish** is a whole fish without the stomach.
- **Headed and gutted** is a whole fish without the stomach or head.
- **Pan-dressed fish** has the fins removed. The head and tail may also be removed.
- **Cross-cuts** are large pieces of a drawn fish.
- **Steaks** are single portion-sized cross-cuts. They usually include at least part of the spine and other bones.
- **Fillets** are boneless pieces of fish. Some fillets have the skin removed. Fillets may be sold fresh or frozen.

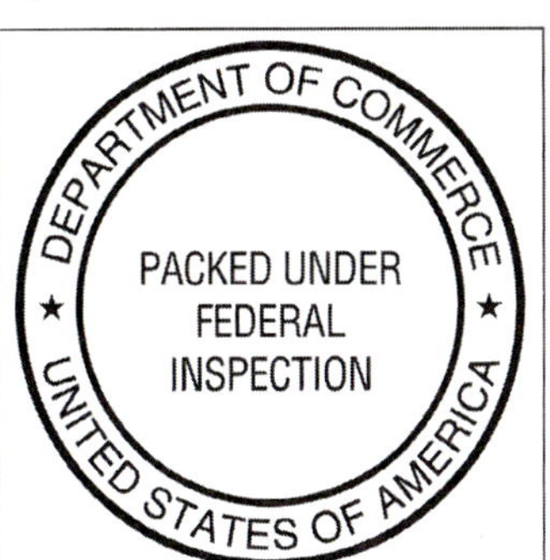

FIGURE 14-1
A PUFI mark is a seal awarded to a facility that passes a Type 1 inspection. PUFI stands for Processed Under Federal Inspection. Do you think it is important to purchase fish that has a PUFI mark? Why?

Check the Label

Round Fish Have eyes on both sides of their heads, swim upright, with the belly down and the back up; examples: haddock, bass, tuna, and salmon

Flat Fish Have both eyes on the same side of their heads, swim near the bottom, and are wider than they are thick; examples: sole, halibut, and fluke

Non-Boney and Other Fish Don't fit into the above categories; may have cartilage instead of bones; examples: eel, catfish, sardine, tilapia, and shark

FIGURE 14-2
Salmon fillet (left) and salmon steak (right). What's an easy way to tell the difference?

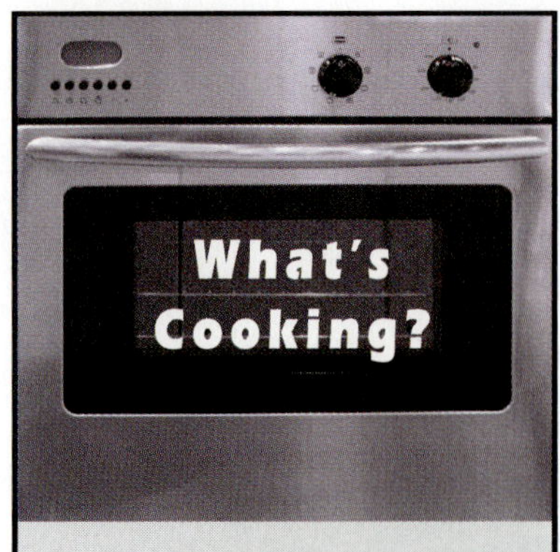

Finfish are often classified as *lean*, *moderately fatty* or *fatty* based on fat content because the fat affects the color and taste:

More fat = darker flesh and stronger flavor

Less fat = light-colored flesh and milder flavor

- **Glazed fish** is a whole fish dipped in water and then frozen several times to build up a layer of ice.
- **Canned fish** is completely cooked and packed in cans.
- **Cured fish** is preserved by salting, smoking, or pickling. Salted fish is usually a fillet soaked in salty brine or coated with salt and then dried. Pickled fish is cured in a brine that contains vinegar and pickling spices. Fish smoked between the temperatures of 120 and 180° F is called **hot smoked**. If the temperature is between 80 and 90° F it is called **cold smoked**. Cold smoked fish is preserved—not cooked.

Buying and Storing Fish

It is very important to purchase fish from a reputable dealer. Fish must be handled properly from the minute it is taken from the water, or it will spoil.

Selecting Finfish

Fresh finfish has a clean, sweet smell, a healthy overall appearance, and is stored at less than 41° F. The flesh is soft, yet firm, and does not hold a mark when pressed. The gills have not turned brown. Fillets and steaks should be packaged in clean containers with only a small amount of liquid at the bottom. Pieces should be cut neatly, and be of an even size. The flesh and skin should look moist and have no cracks, tears, or punctures.

Safe Eats

There is no mandated grading system for fish. On a voluntary basis, fish may be graded as Grade A, Grade B, or Grade C. Grade A is the highest grade and is given to fish from facilities that pass the Type 1 evaluation. Grades B and C are used primarily for processed or canned products. Look for the grade labels on packaging, or on the display case where fresh fish is sold.

FIGURE 14-3
A fresh fish display in a U.S. grocery store (top), and in a fresh market in Beijing, China (bottom). Throughout the world, our fresh markets display what is native to the waters of our homelands. Can you identify any of the fish in the U.S. store? What about in the Chinese market? What types of sea life appears to be indigenous to that area of the world?

BASIC CULINARY SKILLS

Selecting Whole Fish

1. Smell the fish. It should have a clean, sweet smell.
2. Check the temperature. It should be at a temperature of 41° F or less.
3. Check the appearance. There should be no cuts or bruising. The fins should be pliable. The slime should be clear. If there are scales, they should stick tightly to the body. If there is a head, the eyes should be full and not shrunken or dried out.
4. Check the texture by pressing on the body. The flesh should rise quickly after being pressed and should not hold the mark.
5. Check the gills and the belly. The gills should not be brown or slimy. The flesh should stick to the bones, especially along the backbone.

Storing Finfish

Fish will go bad quickly if not stored properly. It should be stored in the refrigerator at 41° F or below, but even under ideal circumstances, fish will only last a few days.

FIGURE 14-4
Pack the belly of a whole fish with ice, then store it belly down in a perforated pan of shaved or flaked ice. Why do you think a perforated pan is better than one with no holes? Why is shaved or flaked ice better than cubes?

Safe Eats

Do not buy frozen fish that has white frost on the edges. It has freezer burn, which indicates that it was not packed well, or that it has thawed and been refrozen.

There are two types of shellfish:

Mollusks have soft bodies and no skeletons. Examples include clams, oysters, mussels, scallops, octopus, squid, and abalones.

Crustaceans have jointed exterior shells. Examples include lobsters, crabs, shrimp, and crayfish.

Professional kitchens are required to keep certificates and invoices for the fish they receive. This helps validate the freshness and quality of the fish, and also provides a trail to the source of the fish if any problems arise.

Selecting and Storing Shellfish

Only purchase mollusks that have been **depurated**—placed in tanks of fresh water to remove impurities.

Fresh shellfish is frequently sold "live." It should have a sweet, sea-like aroma, and tightly closed shells. Live mollusks must be stored between 35 and 40° F. Keep them tightly closed in their bags. Shucked mollusks should appear plump and creamy in color, and the liquid around them should be odorless and clear.

Live crustaceans should be packed in seaweed or damp paper, and they should be moving. The tail of a lobster should snap back quickly after you flatten it out. Store them in the refrigerator until you are ready to cook them.

Hot Topics

The FDA requires shellfish harvesters and processors of oysters, clams, and mussels to put a tag on sacks or containers of live shellfish and a label on containers or packages of shucked shellfish. These labels identify the dealer, note when the shellfish was harvested, and when it was shipped. You have the right to see the tag before making a purchase.

Check the Label

Shrimp is sold by the **count**—the number of shrimp per pound. The larger the shrimp, the fewer there are in the count:

Colossal	10 or fewer
Jumbo	11–15
Extra-large	16–20
Large	21–30
Medium	31–35
Small	36–45
Miniature	about 100

BY THE NUMBERS

You are planning to serve shrimp scampi at a dinner party for six people. You would like to present five extra-large shrimp per person. How many pounds of shrimp should you buy?

Hint: You need 30 shrimp. The count for extra-large shrimp is 16 to 20. Divide 30 by the count to calculate the number of pounds you should buy.

Nutritional Value of Fish

Fish is an excellent source of complete protein. It is usually low in saturated fat and cholesterol. In fact, many types of fish, including salmon, trout, mackerel, herring, and sardines, contain **omega-3 fatty acids**—a family of unsaturated fatty acids thought to promote heart health and reduce the risk of heart attack and high blood pressure. Among shellfish, mussels, clams, and oysters contain some of the highest levels of omega-3 fatty acids.

Most fish are a good source of the B vitamins. Fatty fish are a good source of vitamins A and D. The mineral content varies depending on the type of fish, but most provide iodine, phosphorous, and calcium. Some mollusks—specifically mussels, clams. and oysters—are a good source of iron.

Most varieties of shellfish contain only 50–150 calories, 2–4 grams of total fat, less than a gram of saturated fat, and less than 100 milligrams of cholesterol per serving. Shrimp, however, contain two to five times more cholesterol than other shellfish.

What's Cooking?

All fish is fully cooked when the internal temperature reaches 145° F. The flesh will be firm and opaque. If flaky, the flakes should slide apart and appear moist.

Preparing Finfish

The techniques used to prepare fish vary depending on the type of fish. Leaner fish require a more delicate cooking method, such as sautéing over a low temperature. Pan frying and deep frying are suitable for lean and some moderately fatty fish. All fish can be grilled, broiled, baked, roasted, or steamed.

In general, fatty fish do well when cooked using dry-heat methods such as broiling, while lean fish use moist-heat methods, such as steaming. Recall that fatty fish include whitefish, mackerel, catfish, salmon, and trout; lean fish include swordfish, red snapper, halibut, haddock, and flounder.

Tips

Fish is one of the only natural sources of the mineral iodine, which is necessary for the production of thyroxine in the thyroid gland. Thyroxine helps control the rate of metabolism in your body cells.

FIGURE 14-5
Sautéed fresh cod with green beans and lemon slice. Cod is a mild white fish that thrives in the cold waters of the northeast. What is your favorite fish?

BASIC CULINARY SKILLS

Sautéing Fish

1. Heat pan and cooking fat over moderate heat.
2. Dust fish with flour.
3. Add fish to the pan, careful to avoid splashing.
4. Sauté on the first side until golden.
5. Turn fish once and finish cooking on the second side.
6. Remove fish from the pan and serve with seasoning or sauce, according to recipe.

BASIC CULINARY SKILLS

Broiling Fish

1. Lightly butter or oil a broiling pan.
2. Season fish according to recipe, and brush with butter.

3. Add a breadcrumb topping or sauce, according to recipe.
4. Broil a few inches from the heat source until the top is browned and the fish is cooked through.
5. Serve at once.

Preparation Skills

Like cooking, preparation techniques depend on the type of fish. Common preparation skills include filleting and trimming.

- Fillet a whole fish to create pieces suitable for cooking and serving. Round fish produce two fillets, one on each side of the back bone. Flat fish may be cut into two or four fillets, called quarter fillets.
- Trim a fish fillet to remove the skin as well as the belly and pin bones. **Belly bones** are found along the thinner edge of the fillet. **Pin bones** are found in the middle of the fillet.

Safe Eats

Some fish and shellfish are served raw, but eating raw fish may be risky. Shellfish, in particular, are likely to collect viruses from the water, including Hepatitis A. Bacteria and parasites may also be present in raw seafood. To minimize risk, eat raw fish obtained only from a reputable source, or that has been previously frozen, as freezing kills parasites (but not all microorganisms). Or, cook fish thoroughly to 145° F.

BASIC CULINARY SKILLS

Filleting Round Fish

1. Place fish on cutting board with the backbone parallel to the side of the board, and the head on the same side as your dominant hand.
2. Cut behind head and gill plates, using a fish filleting knife. Angle the knife down and away from the body. Cut to the backbone only. Do not cut off the head.
3. Turn knife, without removing it, so the cutting edge points toward the tail.
4. Run the blade down the length of the fish, cutting against the backbone. Do not saw back and forth (see picture at right).
5. Remove fillet and lay it skin-side down on the cutting board.
6. Repeat on the second side.

Utility Drawer

Use a grilling cage or basket when grilling lean, delicate fish. It encloses the fish so you can lift and turn it without the risk of it falling apart.

BASIC CULINARY SKILLS

Filleting Flat Fish (Quarter Fillets)

1. Place fish on cutting board with the head away from you and tail toward you.
2. Make a cut on one side of the backbone from head to tail.

3. Cut along the bones, working from the center to the edge. Keep the blade angled slightly so the cut is very close to the bones.

4. Remove the first quarter fillet and lay it skin-side down on the cutting board. Trim away any internal organs attached to the fillet.

5. Turn the fish so the tail is toward you.
6. Cut along the bones, working from the center to the edge to remove the second fillet.
7. Turn the fish over.
8. Repeat on the second side.

FIGURE 14-6
Use needle nose pliers or tweezers to pull pin bones out of the middle of a fillet. Pull in the direction of the head so you don't rip the flesh. Why is it important to remove all pin bones before serving the fish?

FIGURE 14-7
Use a sharp fillet knife with a thin blade to remove the skin from a fish fillet. Do you think it is always necessary to remove the skin before cooking fish?

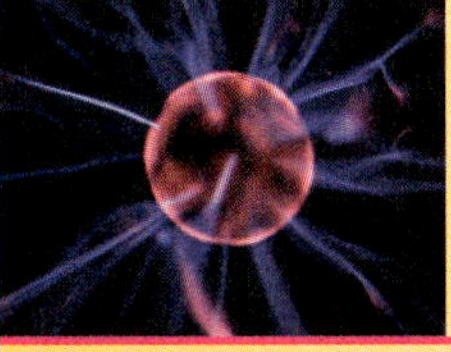

SCIENCE STUDY

High mercury levels in fish and seafood pose a risk to people, particularly children, senior citizens, and pregnant or nursing women.

Mercury occurs naturally in the environment, and nearly all fish contain mercury. The risk comes from elevated levels, which are usually found in areas where there is industrial pollution. The FDA sets the safety limit of mercury for human consumption at 1 part per million (PPM).

Some species have higher levels than others, particularly those that are large, and live a long time, like shark, swordfish, and Ahi tuna.

Use the Internet or your library to look up the affects of eating too much mercury. Find out how much is too much, what problems can be caused by mercury poisoning, what types of fish to avoid, and what types of fish are safe. Present your results to the class.

Fiction	Fact
Fish you catch yourself is safer to eat than fish you buy.	Some wild fish accumulate environmental contaminants, such as PCBs—polychlorinated biphenyls—which can be harmful to eat. The contaminants are usually stored in the fat, so you can lesson your exposure by trimming the fat and skin off the fish before cooking. Check with local agencies before eating self-caught fish.

Paupiette A thin fillet that is rolled before cooking, giving the fish a neat appearance and helping it cook evenly; usually made from lean fish, such as flounder or sole. May be stuffed

Tranche A slice cut from a fillet on an angle; a tranche slice has more surface area than a straight slice

Goujonette A straight slice cut from a fillet; sometimes called a fish finger, it is usually the width of a thumb

FIGURE 14-8
A piece of sushi (top) and sashimi (bottom).

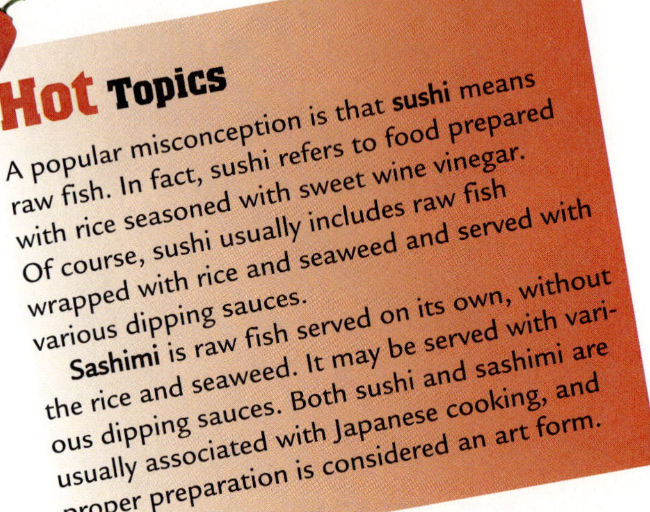

Preparing Shellfish

Although shellfish is sometimes served raw, it may also be boiled, steamed, poached, grilled, fried, sautéed, stir-fried, or baked.

Most shellfish may be cooked in the shell, although many recipes call for shucked shellfish. Once cooked, mollusk shells open, making it easy to remove the flesh from the shell.

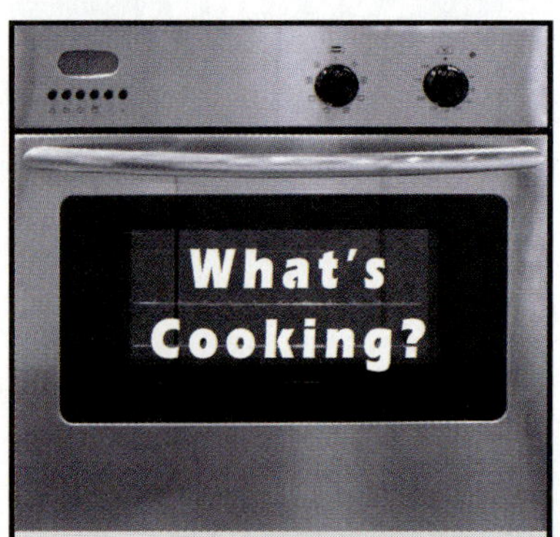

Shrimp turns pink when it is cooked. Lobster turns red. The flesh of both becomes pink and opaque.

Scallops are done when the flesh turns milky white, opaque, and firm.

Mollusks cooked in the shell are done when the shells open. Throw away any that do not open.

Preparation Skills

Mollusks should be thoroughly cleaned before use. Scrub them with a brush under cold running water.

Mollusks are often purchased already shucked. If not, you may have to shuck them yourself for use on a raw bar, or in a recipe.

Shrimp may be purchased cleaned and deveined, or you may have to clean them and devein them yourself.

FIGURE 14-9
Remove the beard from a mussel by pulling it away from the shell. The mussel will die once the beard is removed. Why is it important to remove the beard from a mussel immediately before cooking?

BASIC CULINARY SKILLS

Opening Clams

1. Wear a wire mesh glove to hold the clam.
2. Place the clam in your hand so the hinged side is facing outward.
3. Work the side of a clam knife into the seam between the upper and lower shells.

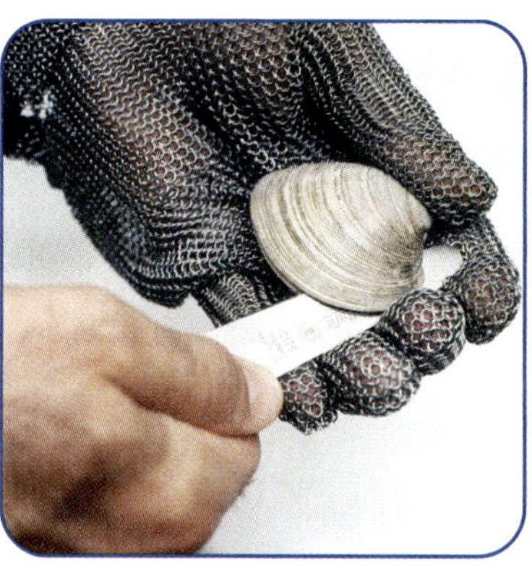

4. Twist the blade slightly like a key in a lock to pry open the shell.

5. Slide the blade over the inside of the top of the shell to release the clam.
6. Slide the blade under the clam to release it from the bottom shell.

BASIC CULINARY SKILLS

Opening Oysters

1. Wear a wire mesh glove to hold the oyster.
2. Place the oyster in your hand so the hinged side is facing outward.
3. Work the tip of an oyster knife into the hinge, holding the upper and lower shells together.
4. Twist the blade like a key in a lock to break open the hinge.

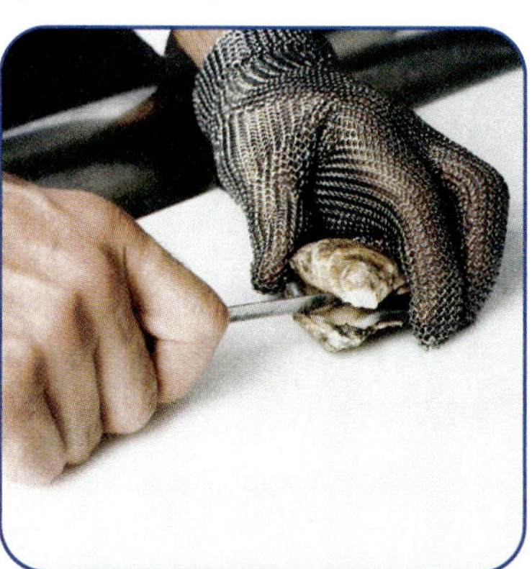

5. Slide the blade over the inside of the top of the shell to release the oyster.
6. Slide the blade under the oyster to release it from the bottom shell.

BASIC CULINARY SKILLS

Peeling and Deveining Shrimp

1. Pull the shell away from the shrimp, starting on the underside where the feathery legs are located.
2. Placed shelled shrimp on a cutting board with the curve outer edge of the shrimp on the same side as your cutting hand.
3. Make a shallow cut on the curved outer edge using a paring or utility knife.

4. Scrape out the vein using the tip of the knife.

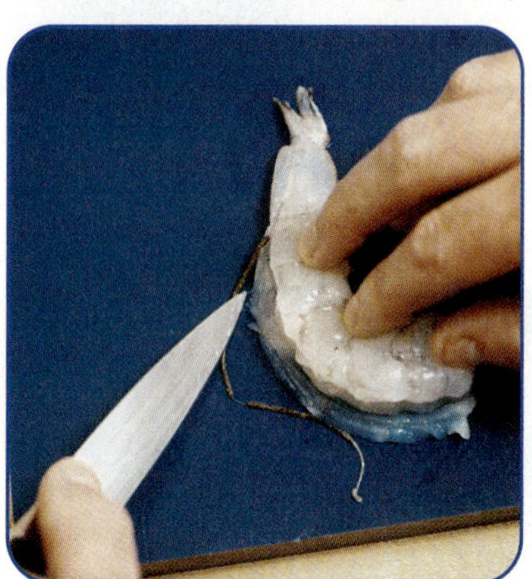

Safe Eats

If a mollusk shell appears open, tap it lightly. If it does not close, the mollusk is dead and should be thrown away. You should also throw out any that have cracked or broken shells.

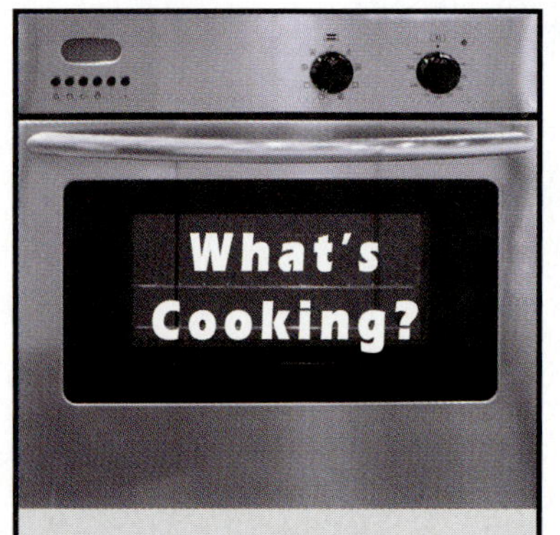

Shucked shellfish have been removed from the shell. Scallops are almost always sold shucked.

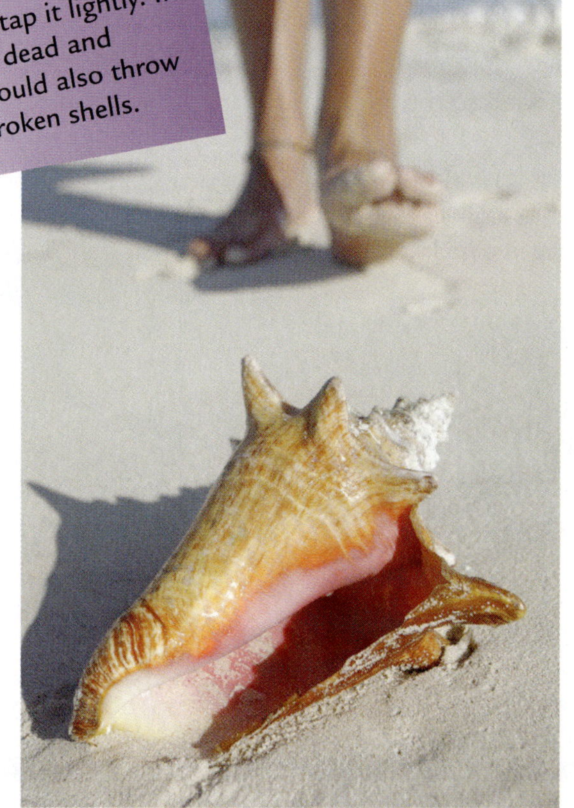

FIGURE 14-10
In many parts of the world, such as the Caribbean, conch is used to make wonderful fritters and chowder. Conchs are large snail-type mollusks. You've probably seen a conch shell before, but would you have guessed that the creature that lives inside it is edible? Have you ever eaten one?

Case Study

Penny Cho and Michelle Stein were classmates assigned to work together on a science project. They agreed to meet Saturday afternoon at Penny's house.

While they were working, Michelle noticed an unfamiliar smell coming from the kitchen. Penny said her mother was cooking Doenjang Chigae, a Korean stew made with bean paste and seafood. She invited Michelle to stay for dinner. Michelle refused. She had never eaten Korean food before, and did not think she would like it.

- How do you think it made Penny feel when Michelle refused her invitation to dinner?
- Do you think Michelle did the right thing? Why or why not? What could she have done differently?
- Have you ever been invited to try new or unusual food? What did you do?
- Are there any traditional foods that your family enjoys that you think people from other cultures might find unusual?

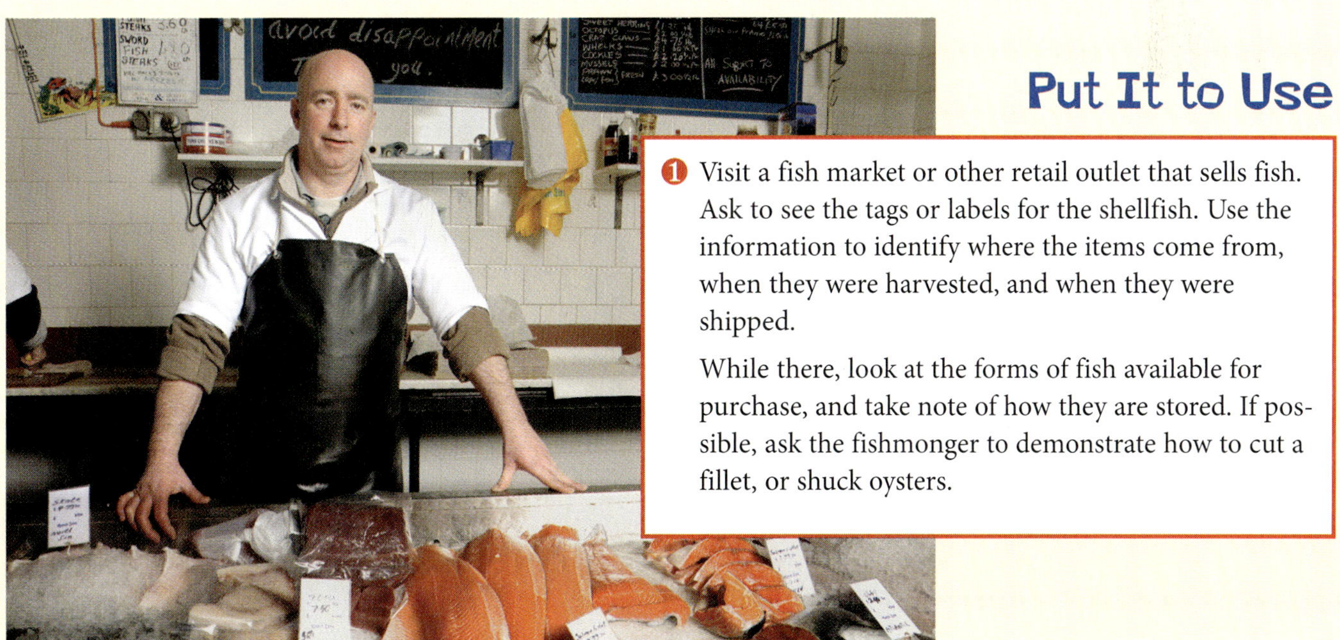

Put It to Use

1. Visit a fish market or other retail outlet that sells fish. Ask to see the tags or labels for the shellfish. Use the information to identify where the items come from, when they were harvested, and when they were shipped.

   While there, look at the forms of fish available for purchase, and take note of how they are stored. If possible, ask the fishmonger to demonstrate how to cut a fillet, or shuck oysters.

Put It to Use

❷ Clam chowder is a popular stew made from clams and vegetables. However, there are regional differences in how clam chowder is prepared. Look up recipes for New England style clam chowder, and for Manhattan style clam chowder. Compare the similarities and differences between the two. If possible, prepare both so you can compare the appearance and taste. As a class, vote on which you prefer.

Write Now

Some people don't like the practice of farming fish. They believe farm fish are less nutritious than wild fish and may contaminate wild fish populations. Other people think that farm fish are a healthy, environmentally friendly alternative to overfishing our wild fish resources. What do you think? Research the topic and form an opinion. Then, write an essay on the topic stating your opinion, and backing it up with facts and figures. Share your essay with the class.

Tech Connect 

The government regulates fishing in many ways, including setting fishing seasons, issuing fishing licenses, and setting limits on the size and number of fish you can catch. Use the Internet to learn about fishing regulations in your area. When is the fishing season? Does it cover all types of fish? How much does a license cost? How long is it good for? Are there restrictions to the size or number of fish you can catch? Report your findings to the class.

Team Players

As you learned in this chapter, fish can be classified in many ways. In small groups, work together to identify different fish and classify them. Use the Internet or your library to locate information and pictures of the fish, and then create charts, posters, or a presentation showing the different classifications. In addition to the classification, include information about whether the fish is lean or fatty, where it comes from, and how it might be prepared. Share your results with the class.

Put It Together

Match the explanation in column 1 with the term in column 2.

Column 1

- **a.** fish that live in saltwater but spawn in freshwater
- **b.** shellfish that have soft bodies and no skeletons
- **c.** shellfish that have jointed exterior shells
- **d.** a whole fish without the stomach
- **e.** a boneless piece of fish
- **f.** shellfish that has been removed from its shell
- **g.** shellfish that has been placed in a tank of fresh water to remove impurities
- **h.** a slice cut from a fillet on an angle
- **i.** a straight slice cut from a fillet
- **j.** Japanese-style raw fish prepared with rice seasoned with sweet wine vinegar and seaweed
- **k.** Japanese-style raw fish served on its own

Column 2

1. crustaceans
2. tranche
3. sashimi
4. goujonette
5. anadromous
6. drawn fish
7. sushi
8. shucked shellfish
9. depurated shellfish
10. fillet
11. mollusks

TRY IT!

Sautéed Trout Meunière

Yield: 10 Servings   Serving Size: 1 Trout

Ingredients

10 Pan-dressed trout (about 10 oz each)
To taste Salt and black pepper, freshly ground
As needed Flour
2 fl oz Butter or oil (for sauteing)
10 oz (1¼ cups) Butter (for the sauce)
2 fl oz Lemon juice
3 Tbsp Parsley, chopped

Method

1. Rinse the trout.
2. Trim the trout as necessary, removing the head and tail if desired.
3. When ready to sauté, blot dry and season with salt and pepper.
4. Dredge the fish in flour, shaking off any excess.
5. Heat a sauté pan to medium-high.
6. Add the butter or oil.
7. Sauté the trout until the flesh is opaque and firm, about 3 minutes per side (145° F).
8. Remove the trout from the pan. Keep the trout warm on heated plates while completing the sauce.
9. To begin preparing the sauce, pour off the excess fat from the pan.
10. Add whole butter (about 1 oz per portion).
11. Cook until the butter begins to brown and has a nutty aroma.
12. Add the lemon juice.
13. Swirl the pan to deglaze it.
14. Add the parsley and immediately pour or spoon the pan sauce over the trout.
15. Serve immediately.

Recipe Categories

Fish

Chef's Notes

When you add the lemon juice to the pan, the sauce will foam up.

Potentially Hazardous Foods

Seafood

HACCP

Refrigerate at 41° F or below.

Cook to an internal temperature of 145° F or higher.

Nutrition

Calories	333
Protein	19 g
Fat	17 g
Carbohydrates	24 g
Sodium	705 mg
Cholesterol	76 mg

TRY IT!

Broiled Lemon Sole on a Bed of Leeks

Yield: 10 Servings Serving Size: 6 oz

Ingredients

3¾ lb Sole fillet
1½ fl oz Lemon juice
½ tsp Salt
¼ tsp Black pepper, freshly ground
1 fl oz butter
6 oz (2 cups) Bread crumbs, white, fresh
2 oz (¼ cup) Butter, unsalted (for the sauce)
1½ lb (6 cups) Leeks, julienned
4 fl oz Heavy cream

Method

1. Preheat the broiler.
2. Cut the fish into ten equal 6-oz portions (or two 3-oz pieces per portion).
3. Season the fish with the lemon juice and half of the salt and pepper.
4. Brush the fish with ½ fl oz of the butter.
5. Work the remaining ½ fl oz of butter into the breadcrumbs to moisten them slightly.
6. Coat the top of the fish with the breadcrumbs.
7. Place the sizzle plate 4 inches under the broiler.
8. Broil undisturbed for about 4 minutes or until the fish is done and the topping is browned.
9. Melt the butter in a large sauté pan.
10. Add the leeks.
11. Cover.
12. Stew gently until the leeks are tender, about 6 to 8 minutes.
13. Season the leeks with the remaining salt and pepper.
14. Add the cream.
15. Reduce slightly, about 2 minutes.
16. Serve the fish on a bed of 4 oz of stewed leeks.

Recipe Categories

Fish

Chef's Notes

Other flaky fish, such as cod or sea bass, could be substituted for the sole.

Potentially Hazardous Foods

Seafood, Dairy

HACCP

Refrigerate at 41° F or below.

Cook to an internal temperature of 145° F or higher.

Nutrition

Calories	211
Protein	34 g
Fat	6 g
Carbohydrates	4 g
Sodium	150 mg
Cholesterol	96 mg

TRY IT!

Mussels Marinière

Yield: 10 Servings Serving Size: 6 oz

Ingredients

4 oz (1/2 cup) Butter, unsalted, cut into small cubes
5 oz (3/4 + 2 tsp) Shallots, minced
4 fl oz fish or chicken stock
To taste Salt and black pepper, freshly ground
2 Thyme sprig
4 lb Mussels, cleaned and debearded
3 Tbsp Parsley, chopped

Method

1. Melt 1 oz of butter in a large heavy saucepan over medium-high heat. Add the shallots.
2. Cook until soft and translucent, 1 to 2 minutes.
3. Add the stock, a sprinkling of the pepper, and the thyme.
4. Allow the mixture to simmer 2 to 3 minutes. Add the mussels.
5. Cover. Cook over high heat, shaking the pan often so all the mussels open at about the same time, 2 to 3 minutes.
6. Take off the cover. Remove the mussels as they open and place them on a warm serving platter.
7. When all the mussels have opened, strain the cooking broth through a fine sieve into a clean saucepan.
8. Bring the liquid to a boil.
9. Cook briefly over high heat until syrupy, about 1 minute.
10. Remove the saucepan from the heat.
11. Gradually add the remaining butter to the broth, whisking to incorporate.
12. Adjust the seasoning, if necessary.
13. Pour the sauce over the mussels. Serve hot.
14. Sprinkle with chopped parsley.

Recipe Categories

Fish

Chef's Notes

Additional herbs and spices (such as dill, mustard seed, coriander, or cayenne pepper) can be substituted for or added to the thyme in Step 3.

Potentially Hazardous Foods

Seafood

HACCP

Refrigerate at 41° F or below.

Cook to an internal temperature of 145° F or higher.

Nutrition

Calories	302
Protein	13 g
Fat	5 g
Carbohydrates	52 g
Sodium	154 mg
Cholesterol	9 mg

15 Spices, Herbs, and Garnishes

In This Chapter, You Will . . .

- Identify types of spices, herbs, and garnishes
- Learn how to buy and store spices and herbs
- Learn how to use spices, herbs, and garnishes when preparing food

Why YOU Need to Know This

For most people, eating is more than just consuming the calories and nutrition we need to survive. It is a social activity that uses all of our senses. Different tastes, smells, textures, and even the way the food looks either enhance or detract from the experience. To create food that is appealing, you enhance its flavor by adding different forms of seasonings, and you add garnishes to adorn the plate. Although seasonings and garnishes have little or no nutritional value, they have the power to turn a plain meal into something special. As a class or in small groups, take turns blindfolding each other and trying to identify different herbs and spices based on smell. Use herbs such as oregano, basil, rosemary, thyme, or dill. Use spices such as cinnamon, pepper, cloves, or ginger.

Types of Spices

Spices are **aromatic**—or sweet-smelling—natural ingredients that you add to food to instill a specific flavor. They are the seeds, bark, roots, stalks, buds, fruit, and even the flowers from a wide variety of plants. Some of the most common spices include those shown in Table 15-1.

TABLE 15-1: MOST COMMON SPICES

Picture	Description
	**Allspice:** The nearly-ripe fruit or berry of an evergreen tree. The berries are dried to a dark reddish-brown and then cracked or ground. They may be available whole. The flavor is a mixture of cloves, cinnamon, nutmeg, and ginger. It is used in both savory and sweet recipes.
	**Caraway Seeds:** The seeds from the caraway plant, which is an herb. They have a nutty taste and are used in baked goods—including rye bread—and dishes.
	**Cardamom:** Long light green or brown pods containing a single seed. It has a strong musty flavor, and is available in whole pods or ground. It is used frequently in Indian cuisine.
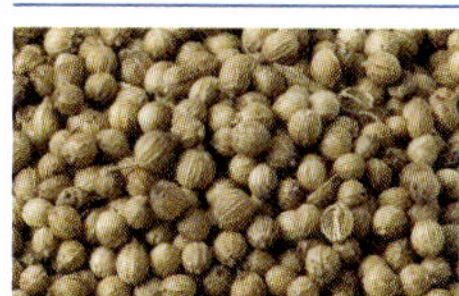	**Coriander:** The dried, ripe fruit of the cilantro plant. Coriander is available whole or ground, and is used frequently in Spanish and Mexican cuisine.
	**Cinnamon:** The bark of a small evergreen tree. It tastes and smells sweet. The color is reddish brown, and it sold in sticks or ground. It is used in a wide variety of sweet and savory recipes.
	**Cloves:** The dried unopened bud of a tropical evergreen tree, available whole or ground. Whole cloves are shaped like nails. They have a strong aroma, and are sweet. They are used in baked goods, and to add flavor to many dishes, such as glazed ham, stews, and gravies.
	**Cumin:** The crescent shaped seed of a plant in the parsley family. It is available whole or ground. It has a strong, earthy flavor and is often associated with Mexican, Indian, and Middle Eastern cuisine.
	**Dill:** The seeds of the dill plant, which is a member of the parsley family. They are flat, oval, and brown, and are commonly used as a flavoring for pickles, as well as for rye and pumpernickel breads.

continued

TABLE 15-1: MOST COMMON SPICES *(CONT)*

Picture	Description
	**Fennel:** Most commonly available as whole fennel seeds, the plant leaves are also used fresh or dried. The flavor is similar to licorice. The seeds are commonly used in Italian and Central European cuisine.
	**Ginger:** The root of a tall tropical plant. It is available fresh and whole or dried and then ground. Whole ginger must be peeled before use. Its sweet, peppery flavor makes it an ideal ingredient in many sweet and savory recipes. It is commonly used in Asian and Indian cuisine, and also in baked goods.
	**Mustard:** The seeds of a plant from the cabbage family which are available whole or ground. (The leaves are eaten as a vegetable.) Mustard comes in yellow, red, and black varieties, ranging in flavor from mild to hot. Whole seeds are used frequently in Indian cuisine.
	**Peppercorns:** The berry of the pepper vine, available in white, black, and green. Black peppercorns are unripe berries. White peppercorns are the ripe berries which are allowed to dry and then have their husks removed. Both are sold ground, whole, or cracked. Green peppercorns are unripe berries that are pickled or freeze-dried.
	**Poppy Seed:** The dried seed of the poppy flower. They have a nutty flavor, and are used as a filling for baked goods, and are sometimes sprinkled on the top of rolls and breads.
	**Saffron:** The dried inner part of the crocus flower. Saffron has a combination sweet and bitter flavor. It is used both to flavor foods and color them deep yellow. It is available in threads, which are usually crushed before use, and as a powder. It is commonly associated with Indian cuisine, and rice.
	**Turmeric:** The dried root of a plant in the ginger family. It is bright yellow and slightly bitter. Turmeric adds color, flavor, and aroma to savory foods. It is an ingredient in prepared mustard and curry powder. It is usually sold ground.
	**Vanilla:** The unripe pod of an orchid plant, cured and dried. Vanilla is available as whole pods, or as an extract, which is a blend of alcohol and the oils pressed from the plant. It is very sweet, and is used frequently in baked goods and candies.

The Significance of Spices, Herbs, and Flavorings

Through history, certain spices have had significant value because they were rare and in high demand. In the Middle Ages, spices such as cinnamon and cloves were grown in secret locations in Asia, and traded for goods.

FIGURE 15-1
Nutmeg is the seed of the nutmeg tree; **mace** is the lacy coating that surrounds the seed. Both are sweet and fragrant. Nutmeg is available whole, and both are available ground. They are used in sweet and savory dishes. Can you think of any other plants that produce more than one spice? How about plants that provide spices and herbs, or are used as vegetables?

Now we can purchase most spices at our local markets, but some are still rare and can be quite expensive. Saffron is the most expensive spice in the world. It takes about 75,000 blossoms or 225,000 hand-picked stigmas from the Saffron crocus flower to make a single pound of the spice!

Along with spices, herbs and herb combinations are used worldwide to enhance the flavor of food. They are frequently associated with specific types of cuisine:

- Basil and oregano: Italian
- Tarragon and chives: French
- Cilantro and parsley: Chinese
- Oregano and mint: Greek

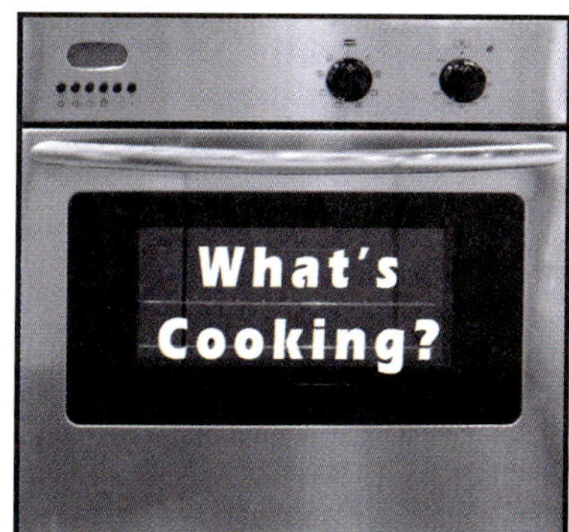

You might see pink peppercorns sold alongside black and white pepper. Pink peppercorns are not peppercorns at all. They are the dried berries of a South American rose, and have a bitter flavor.

Cool Tips

What's the difference between an herb and a spice? They both come from plants—sometimes even the same plant! In general, herbs are the soft part of a plant, like the leaves and flowers, and spices are the hard parts, like bark, berries, and seeds.

FIGURE 15-2
Curry powder is made by blending up to 20 different herbs and spices, including cinnamon, fennel seeds, ginger, cumin, and peppers. What benefits might there be to using a combination of spices?

Check the Label

Spice blends Combine a variety of spices—and sometimes herbs—into one package. You can purchase blends already made, or make your own by mixing ground spices. Curry powder, chili powder, and Asian five-spice powder are examples of spice blends.

Rubs, or dry rubs Made by combining spices and herbs. They are then applied directly to meat, poultry, or fish before cooking.

Extracts A blend of alcohol and the oils pressed from aromatic plants, such as vanilla. They are added in small amounts to dishes before or during cooking.

Flavored oils or vinegars Made by infusing the liquid with a spice, herb, or other aromatic, such as garlic, ginger, or lemon. They are usually used uncooked to enhance the flavor or aroma of a dish.

In addition to herbs and spices, aromatic fruits and vegetables, aromatic liquids, and cured foods are sometimes used to add flavor and fragrance to dishes.

Aromatic vegetables include plants in the onion family, mushrooms, and celery. Fruits include citrus, such as lemons, limes, and oranges, as well as dried apricots or raisins. Aromatic liquids include extracts and infused oils. Cured foods include smoked ham, bacon, and salted anchovies.

Ingredients or mixtures that enhance the flavor of foods are also called **condiments**. The most common condiments in the United States are ketchup and mustard. Other condiments include salsa, relish, wasabi, chutney, soy sauce, Worcestershire sauce, and Tabasco sauce.

Condiments may be overused, in which case they hide the flavor of the food item they are meant to enhance.

Hot Topics

Herbal tea—a drink made from the leaves of herbs such as mint and chamomile—is not really tea at all. Teas must be made from the leaves of the camellia sinensis plant, otherwise known as the tea bush. You make tea by steeping the leaves in hot water. It may be served hot or iced. It is the second most popular beverage in the world—after water.

Coffee is made from the seeds of the coffee plant. Commonly called coffee beans, the seeds are roasted, ground, and brewed with water.

FIGURE 15-3
Spices for sale at a bazaar in Provence, France. What kind of spices might you expect to see commonly sold in France?

BASIC CULINARY SKILLS

Sachet d'Epices

1. Measure peppercorns, thyme, and parsley.
2. Wrap the ingredients in a square of cheesecloth.

3. Tie the cheesecloth with string to make a bag.

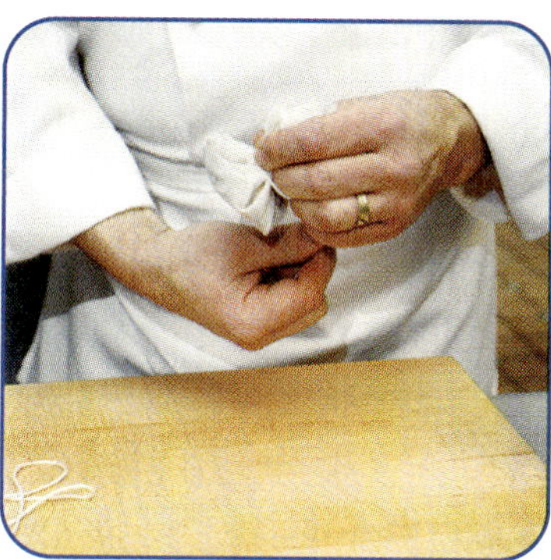

4. Add to the dish according to the recipe.
5. Simmer until the dish is aromatic, or as specified in the recipe.
6. Remove the bag and throw it away.

Types of Herbs

Herbs are the leaves, stems, and flowers of aromatic plants. They are sold fresh or dried. It is important to be careful when substituting dried for fresh herbs in a recipe. Most dried herbs have a more concentrated flavor than fresh herbs, because they contain less water. In general, you can substitute one teaspoon of a dried herb for every tablespoon of a fresh herb called for in a recipe. The most common herbs are listed in Table 15-2, shown on pages 320 and 321.

FIGURE 15-4
A **bouquet garni**—pronounced boo-KAY GAR-nee—is similar to a sachet d'épices, but includes only fresh herbs and an aromatic vegetable. A traditional bouquet garni include sprigs of fresh thyme, fresh parsley stems, rosemary, citrus peels along with leeks, garlic, or scallions. Why might you use a bouquet garni instead of a sachet d'épices?

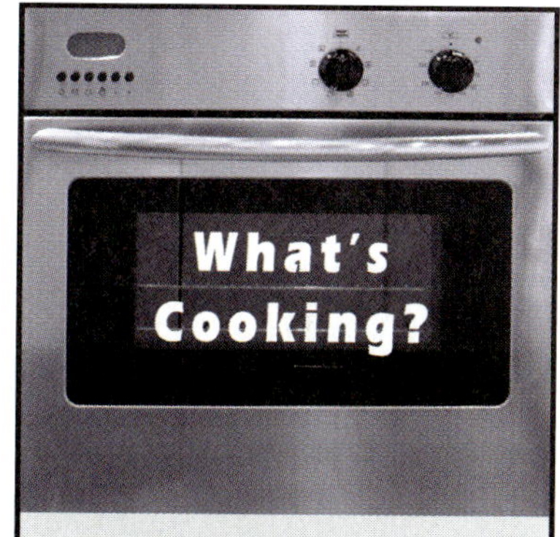

A **sachet d'épices**—pronounced SAH-shay DAY-pees—is a combination of fresh and dried herbs and dried spices, tied up in cheesecloth, and cooked with a dish. It literally means *bag of spices* in French. A traditional sachet d'épices includes peppercorns, dried thyme leaves, and fresh parsley stems.

TABLE 15-2: MOST COMMON HERBS

Picture	Description
	**Basil:** A member of the mint family. It has pointed green leaves available in large- or small-leaf and purple varieties. It combines well with tomatoes. Use it to flavor sauces, stews, soups, and salad dressings, or in chicken, fish, lamb and pasta dishes. A popular flavoring for oils and vinegars.
	**Bay Leaves:** Leaves from the evergreen sweet-bay or laurel tree, they are usually sold dried, and are removed from foods before serving. Use them to flavor long-cooking foods, such as stews and soups.
	**Chervil:** A member of the parsley family. It has dark green, curly leaves, with a sweeter flavor than parsley. Dried chervil is much less flavorful than fresh. Use it in soups, sauces, salads, and stuffing.
	**Chives:** A member of the onion family. Chives grow as long, hollow stem. They are usually used fresh or frozen, because dried chives lose a lot of flavor and color. Use chives minced or snipped as an ingredient or garnish for salads, omelets, cheese dishes, dips, and sauces. They also combine well with fish, chicken, potatoes, and rice.
	**Cilantro:** Similar to flat-leaf parsley, it has a tangy and sharp flavor. Cilantro is used commonly in Asian, South American, and Central American cuisine. It is also called Chinese parsley.
	**Dill:** A member of the parsley family. It has feathery leaves, a strong aroma, and a tart flavor. Use it to flavor sauces and stews, as well as salads, egg dishes, and vegetables. It is a common ingredient in Central and Eastern European cuisine.
	**Lemongrass:** A tropical grass with a long, greenish stalk and serrated leaves. It is commonly used in Asian cuisine.
	**Mint:** A family of plants that includes many varieties. Common types include peppermint and spearmint. The leaves are usually textured and deep green. It is used around the world in both savory and sweet dishes.
	**Garlic:** A member of the lily family, fresh garlic is the small sections of the root, called **cloves**. Garlic is used chopped, minced, crushed and whole in a variety of recipes. It may also be dried and ground and used as a spice.

continued

TABLE 15-2: MOST COMMON HERBS *(CONT)*

Picture	Description
	**Horseradish:** The root of the white horseradish plant, which is grated, minced, or chopped. It is sometimes colored with beet juice to make red horseradish. It has a zesty, powerful flavor and aroma.
	**Marjoram:** A member of the mint family, with short, oval, pale green leaves. It is available dried or ground, as well as fresh. Use it in all but sweet recipes. It is common in Mediterranean cuisine.
	**Oregano:** A type of marjoram with small oval leaves. Use it to flavor soups, stews, and sauces, as well fish, meat, and poultry. It is common in Italian and Greek cuisine.
	**Parsley:** Available in curly and flat-leaf, parsley has a mild smell and pleasant flavor. It is one of the most commonly used herbs because it blends well with many foods and flavorings, and provides an attractive garnish. Use the leaves and stems in simmered and long-cooking dishes. Flat-leaf parsley is also called Italian parsley, and is usually a bit spicier than the curly variety.
	**Rosemary:** The leaves of an evergreen shrub, which is related to the mint family. It has needle shaped leaves and is very fragrant. It goes well with many foods, including vegetables, meats, and poultry. Dried rosemary is nearly as flavorful as fresh.
	**Sage: A** member of the mint family. Sage leaves are oval and covered with soft threads. Use it whole in simmered dishes, or chopped to flavor roast meats or poultry. Dried sage is sometimes called rubbed sage.
	**Savory:** Similar to thyme and rosemary, savory has small, narrow, gray-green leaves and a bitter flavor. Use it fresh or dried.
	**Tarragon:** Narrow, dark green leaves with a strong licorice flavor. Use the stems in simmering dishes and sauces. Chop the leaves as a final flavoring ingredient for poultry, fish, veal, and egg dishes. It is commonly used in French cuisine.
	**Thyme:** A member of the mint family, thyme has very small gray-green leaves. Use whole sprigs or chopped leaves to flavor soups, stews, and sauces, as well as to season meats, poultry, and fish. Dried thyme retains almost as much flavor as fresh thyme.

Salt

Although not a spice, salt is one of the most important seasonings in the world. It is used in all types of cuisines, in all countries. You can add it to foods before cooking, during cooking, and after cooking. Like many ingredients, when used in small amounts salt enhances flavor. In larger amounts, it changes the flavor.

Salt is a natural mineral compound found underground and in sea water. It is composed of sodium chloride, and its chemical symbol is NaCl.When stored properly, it can last forever. Unrefined, natural salt exists in many varieties, each with its own unique flavor. Refined salt is processed for human use.

BY THE NUMBERS

How much is a pinch of salt? By definition, it is the amount of salt you can pinch between your thumb and first two fingers. Of course, that amount might vary from person to person. You can measure your own pinch so that you know how much it really is.

- Take a pinch of salt, and measure it using a scale. How much does your pinch weigh?
- Now, measure a new pinch using a measuring spoon. What is its volume?
- Is your pinch the same as your classmates'? How might you account for the differences when cooking?

FIGURE 15-5
Salt harvesting on the Italian island of Sicily. Salt harvesting is usually done in September when the sun is very hot. The salt is harvested completely by hand, which is an extremely difficult job. The salt is unrefined and contains more magnesium and potassium, and less sodium chloride, than regular salt. Why do you think harvesting is done when it is hot?

Check the Label

Table salt Is refined to remove other minerals or impurities. It is processed to give it a fine, even grain, and a small amount of starch is added to keep it from forming clumps.

Iodized salt Is table salt that has been enriched with iodine as a nutritional supplement.

Sea salt Is extracted from the ocean, using evaporation techniques. It is usually not refined, so it contains additional minerals and other elements found in sea water, which affect the flavor. One variety, **Fleur de Sel**—literally, flower of salt—is harvested by hand off the coast of France, and costs about $20.00 per pound!

Kosher salt Has no additives, so it has a purer flavor than table salt. It is usually coarser than table salt, which means it has larger crystals, although it is also available with finer grains. To substitute kosher salt for table salt, use twice as much kosher salt as called for in the recipe.

Rock salt Is less refined than table salt, and is not meant to be eaten. It is used in ice cream makers and as a bed for certain items, such as oysters or clams served on their shells.

Types of Garnishes

A **garnish** is an edible ingredient used to add flavor, color, and texture to individual items, individual dishes, composed platters and trays, and to buffets. A successful garnish draws attention to the food, but does not overwhelm or detract from the food in any way.

Use a bright piece of vegetable such as a tomato wedge, carrot curl, or apple peel to garnish a salad. Add a colorful accent to a beverage using mint leaves, a lemon twist, or even a scoop of sorbet. Keep in mind that garnishes may be eaten, so they should be fresh, clean, and tasty as well as attractive.

Cool Tips

Use color when creating a garnish to affect the mood or tone of a dish.

- Green: freshness and vitality.
- Brown or gold: warmth, comfort, richness.
- Orange or red: intensity, desire, hunger.

Use the following guidelines to help you select and use garnishes effectively:

- **Function.** Use garnishes to create a visual impression and to add a taste experience.
- **Flavor.** Garnishes should taste fresh and complement the taste of the dish.
- **Visual appeal.** Garnishes should be visually attractive. The placement, shape, and color should enhance the dish.
- **Texture.** Garnishes should add a different texture to the dish.
- **Size.** Garnishes should be in proportion to the dish. If they are too small, they will be overlooked. If too large, they will compete for attention with the main item.
- **Effects.** Garnishes can be cut, shaped, and molded to add special effects to a dish. For example, use a fan cut on strawberries to garnish a breakfast or dessert, or a radish rose to garnish a salad or plate of crudités.

FIGURE 15-6
Did you know some flowers are edible? You can use them as spectacular garnishes in salads like this, or even on cakes and sweets. Would you eat these flowers?

In most professional kitchens, the **garde manger**, or pantry chef, is responsible for planning and selecting garnishes. He or she should take care to make sure the garnishes are suitable for the menu items, are an integrated part of the dish or presentation, and are not boring or overused.

BASIC CULINARY SKILLS

Making a Fan Cut

1. Place item on its side on a work surface.
2. Cut in paper-thin slices from tip to stem, leaving the flesh at the base of the stem still attached.
3. Use your fingers to spread the slices out like a fan.
4. Lift the fan carefully onto the plate or platter using a knife spatula, or palette knife.

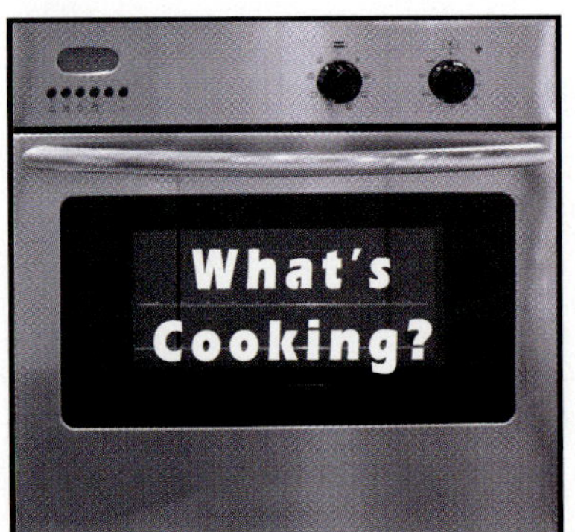

Most spices are available whole or ground. Ground spices tend to lose their aroma more quickly, so it is best if you purchase whole spices and grind them yourself as you need them. You can use an electric spice grinder or an old-fashioned mortar and pestle.

Buying and Storing Spices and Herbs

Spices, seasonings, and dried herbs tend to lose their flavor, aroma, and color over time. Although not a safety risk, old products will not work as well as new ones. Buy them in small amounts, as you need them, so they will be effective in your recipes. Store the containers tightly closed in a cool, dry, dark place.

You can identify fresh herbs and spices by smell, color, and texture. When you rub them between your fingertips, the aroma should be strong and pleasant, not musty. Fresh herbs should be soft and fresh, and the leaves should be intact. If they are dry, brittle, or stale, if they appear to be bruised or wilted, or if the color is off, throw them away. Take note of any roots still attached to the herbs. They should be dry, not soft or wet.

Like spices and dried herbs, buy small amounts of fresh herbs and use them as you need them.

Herbs contain volatile oils (another term for essential oils) which give them their characteristic aroma and flavoring. The oils dissipate when exposed to air. If you place fresh herbs in the refrigerator in a tightly closed container with a little water in the bottom, they should keep for about two weeks. Otherwise, wrap them loosely in a damp paper towel, place them in a loosely closed plastic bag in the fridge, and use them within a few days.

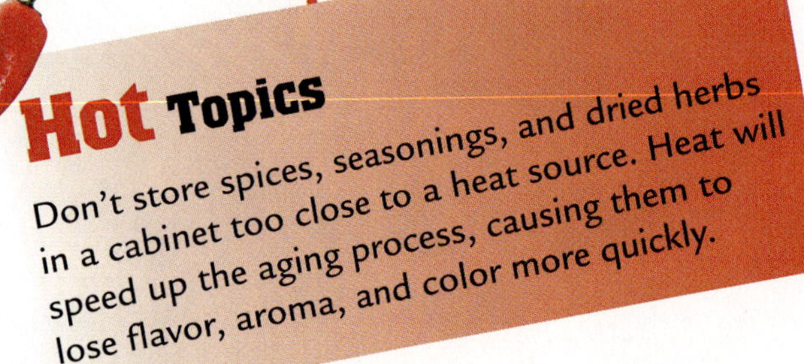

BASIC CULINARY SKILLS

Freeze Fresh Herbs

1. Wash the herbs and pat them dry.
2. Spread the herbs in a single layer on freezer paper.
3. Roll the paper tightly, and seal each end with freezer tape.

Note that the leaves will become dark when the herbs thaw.

Preparing Foods with Spices and Herbs

The method of adding spices and herbs to a dish depends on the recipe. Sometimes they are added whole, and then strained out. Sometimes they are toasted, ground and sprinkled into a dish as it cooks, or cooked in oil first to distribute the flavor evenly through the dish.

They may be applied directly to food before cooking, such as a rub applied to ribs before smoking, or used fresh or uncooked as a finishing flavor. Whole sprigs and stems of fresh herbs are usually added to a dish at the start of cooking, so the herb can gently flavor the entire dish. Chopped or whole fresh leaves may be added at the end of cooking, in order to infuse intense flavor.

When cooking with spices and herbs, keep in mind the basic reasons for seasoning food:

- **To enhance natural taste.** Salt, in particular, is frequently used to enhance the natural flavor of food.
- **To balance taste.** Seasonings can overcome strong flavors such as sour, sweet, or bitter tastes. For example, you can add a sweetener, such as sugar or honey, to lemonade to cut the sourness of the lemons.
- **To cut richness.** Seasonings can help reduce the richness or oiliness of fatty food. For example, use an acid such as lemon juice to improve the flavor of an oily dish, such as one that contains mayonnaise.

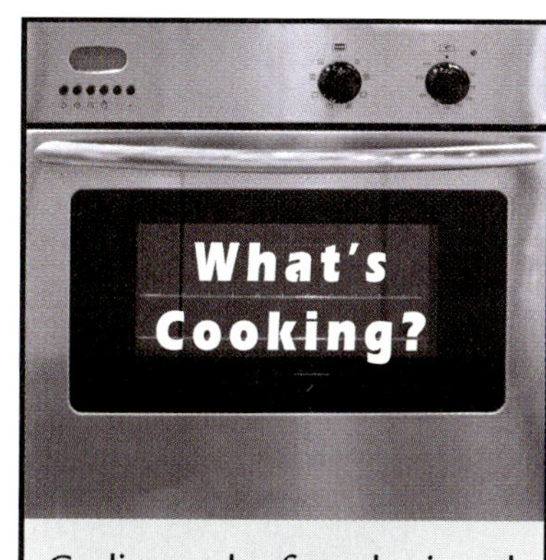

Garlic can be found minced or chopped and sold in jars. Use about ½ teaspoon of jarred garlic for every clove of garlic called for in a recipe.

Safe Eats

On occasion, food-borne illnesses have been traced to fresh herbs, such as cilantro, chives, parsley, and basil. Remember to use standard sanitary practices:

- Wash hands with hot soapy water before and after handling herbs.
- Rinse or wash the herbs in water before cooking or eating.
- Do not cross-contaminate!

FIGURE 15-7
This knife, known as a *mezzaluna*, has one or two curved blades with handles on each end. It is specifically designed to chop herbs more efficiently. Why do you think the cutting board has a curved bowl?

Combining Herbs and Spices with Foods

- You can use spices to enhance the flavors of fruits whether they are cooked or uncooked. Many fruit recipes include such spices as allspice, cinnamon, cloves, ginger, nutmeg, poppy seed, rosemary, and vanilla.
- Herbs and spices that complement grain dishes include basil, bay leaves, cumin, curry powder, dill, garlic, marjoram, oregano, pepper, poppy seed, rosemary, saffron, sage, tarragon, thyme, turmeric, and chives.
- Add flavor to milk drinks using allspice, cinnamon, cloves, nutmeg, and vanilla.
- Cheese and recipes based on cheese may benefit from such spices as basil, caraway, chives, dill, garlic, marjoram, oregano, parsley, pepper, rosemary, sage, tarragon, and thyme.
- Spice up egg dishes using basil, chives, dill, marjoram, oregano, parsley, pepper, sage, tarragon, thyme, and turmeric.
- Meat, poultry, and fish are so versatile that they can be combined with a wide variety of herbs and spices. You'll find recipes that include allspice, basil, bay leaves, caraway, cinnamon, cloves, cumin, curry powder, dill, garlic, ginger, marjoram, nutmeg, oregano, parsley, pepper, rosemary, saffron, sage, tarragon, and thyme.
- You'll find that you can enhance the flavor of vegetables by adding herbs and spices such as basil, caraway, chives, dill, garlic, ginger, pepper, poppy seed, rosemary, saffron, sage, tarragon, and thyme.

Utility Drawer

A **mortar and pestle** is a tool used to crush, grind, and mix substances. Developed in ancient times, it is still used in modern kitchens, pharmacies, and laboratories. The mortar is a bowl in which the substance is placed, and the pestle is a stick with a blunt end used to crush the substance. In the kitchen, it is useful for cracking whole spices, such as peppercorns and cloves; for fine grinding you might prefer an electric spice mill.

FIGURE 15-8
Mortars and pestles come in many shapes and sizes. Some are purely functional, and some are quite beautiful. Many reflect a design of the culture in which they were crafted. Do you think the material used to make the tool makes a difference on how well it works?

Case Study

Jake's grandfather died of a heart attack at the age of 58. Jake's mom takes medicine for high blood pressure and high cholesterol. To help keep herself and Jake healthy, she tries to serve meals that minimize the amount of sodium, trans-fats, and cholesterol consumed.

She packs Jake a bag lunch every day. She puts in a sandwich on whole grain bread, a piece of fruit, and a granola bar. Jake usually gives away the granola bar and buys himself a bag of potato chips and a soft drink to go with the sandwich and fruit.

- What problems do you see with what Jake eats for lunch?
- What would you recommend he do differently?
- Do you think Jake's mom could do more to encourage Jake to eat a healthy diet?
- Is there anything besides diet that Jake and his mom should consider as part of a healthy lifestyle?

Put It to Use

1. Test different methods of grinding your own spices. Start with whole spices, such as peppercorns or cloves. Use a mortar and pestle to grind one batch, and an electric spice mill to grind a separate batch. Which is easier to use? Which produces a finer grind? Compare both to ground spices purchased at the store.

Put It to Use

❷ You can keep fresh herbs, such as parsley, available all the time by freezing individual servings in ice cube trays. Chop the fresh herbs and measure them into the divided ice cube trays, in a standard amount, such as one teaspoon. Fill the trays with water, and then freeze. Use them in a recipe such as a soup or stew that calls for fresh herbs by dropping in as many frozen cubes as you need—one if the recipe calls for one teaspoon, two for two teaspoons, etc.

Write Now

What's your favorite seasoning? Write a descriptive essay about your favorite herb, spice, seasoning, or condiment. Imagine the person reading the essay has never had the opportunity to taste it before. Describe its flavor and texture and how it can be used to enhance food. Share your essay with the class.

Tech Connect 

Many spices and herbs are thought to have medicinal value. Sometimes these beliefs are based on myth or folklore, while other times it has been studied scientifically. Use the Internet to locate information about herbal medicine. Select one herb, and research its use as medicine, including its history, how it is used, and for which ailments.

Team Players

Herbs are often used to flavor spreads, such as butter or olive oil. Divide the class into small groups, and have each group select a different herb, such as chives, oregano, tarragon, thyme, or rosemary. As a team, work together to create an herbed butter. Finely mince the herb, and then mix it with 2 tablespoons of butter at room temperature. Spread the butter on toast and cut it into enough pieces for each member of the class. Before serving, add a garnish to the plate. Have each student taste and rank the samples from most favorite to least favorite. Tally the results and see which team wins. If you have time, repeat the exercise using spices, such as pepper, cinnamon, cloves, ginger, or cumin.

Put It Together

Match the explanation in column 1 with the term in column 2.

Column 1

- **a.** the dried, ripe fruit of the cilantro plant
- **b.** the dried unopened bud of a tropical evergreen tree
- **c.** the most expensive spice, derived from crocus flowers
- **d.** the lacy coating that surrounds the seed of the nutmeg tree
- **e.** a combination of fresh and dried herbs and dried spices tied up in a piece of cheesecloth
- **f.** a combination of fresh herbs and an aromatic vegetable tied up in a piece of cheesecloth
- **g.** a member of the onion family that grows as a long, hollow stem
- **h.** a type of marjoram with small oval leaves, common in Italian and Greek cuisine
- **i.** a type of sea salt harvested by hand off the coast of France
- **j.** an edible ingredient used to add flavor, color, and texture to dishes or plates
- **k.** the chemical symbol for salt

Column 2

1. mace
2. chives
3. saffron
4. Fleur de Sel
5. NaCl
6. coriander
7. garnish
8. bouquet garni
9. cloves
10. oregano
11. sachet d'épices

TRY IT!

Pesto

Yield: 16 fl oz Serving Size: 1 oz

Ingredients

1½ oz (⅓ cup) Pine nuts, toasted
3 Garlic cloves, minced
½ oz (1 Tbsp) Salt
10½ fl oz Olive oil
4 oz (2¾ cups) Basil leaves, washed and dried well
2 oz (⅔ cup) Parmesan cheese, grated

Method

1. Place pine nuts, garlic, half of the salt, and half of the olive oil in a blender or a food processor fitted with the blade attachment.
2. Blend to a paste, about 1 minute.
3. Begin adding basil leaves gradually.
4. Blend on and off to incorporate basil into the emulsion.
5. Add the additional oil gradually until the paste is thoroughly combined.
6. Adjust seasoning with salt as needed.
7. Add the Parmesan cheese and blend just before serving.
8. Pesto should be stored under refrigeration with a layer of oil across the surface.

Recipe Categories

Garde Manger,
Dressings & Dips

Chef's Notes

Walnuts may be substituted for the pine nuts.

Potentially Hazardous Foods

Raw Garlic

HACCP

Store cold sauces containing raw garlic below 41° F.

Nutrition

Calories 1	68
Protein	2 g
Fat	18 g
Carbohydrates	1 g
Sodium	339 mg
Cholesterol	3 mg

TRY IT!

Salsa Fresca

Yield: 1 qt Serving Size: 1 oz

Ingredients

17½ oz (3 cups) Tomatoes, seeded and diced
3¼ oz (¾ cup) Onion, minced
2⅔ oz (½ cup) Green pepper, diced
2 Garlic cloves, minced
1 Tbsp Cilantro, chopped
1 tsp Oregano, chopped
2 Limes, juice only
1 Jalapeño
2 Tbsp Olive oil
¼ tsp White pepper, ground
2 tsp Salt

Method

1. Combine all the ingredients and adjust seasonings.
2. Hold the sauce under refrigeration.

Recipe Categories

Garde Manger,
Dressings & Dips

Chef's Notes

To decrease heat, use less jalapeño.

Potentially Hazardous Foods

Raw Garlic

HACCP

Store cold sauces below 41° F.

Nutrition

Calories	28
Protein	0.59 g
Fat	1 g
Carbohydrates	5 g
Sodium	180 mg
Cholesterol	0 mg

TRY IT!

Boiled Parslied Potatoes

Yield: 10 Servings Serving Size: ½ cup

Ingredients

30 Potatoes, new
2 Tbsp Salt
2 oz (¼ cup) Butter, unsalted
1 oz (¾ cup) Parsley, chopped
½ tsp Black pepper, freshly ground

Method

1. Peel the potatoes and cut into equal-sized pieces (for example, medium dice or tourné).
2. Place the potatoes in a large pot with enough cold water to cover them by about 2 inches.
3. Add the salt.
4. Gradually bring the water to a simmer over medium heat.
5. Cover, and simmer until the potatoes are easily pierced with a fork, approximately 15 minutes.
6. Drain the potatoes, return them to the pot, and let them dry briefly over low heat until steam no longer rises.
7. Heat the butter in a saucepan over medium heat.
8. Add the potatoes. Gently toss to coat them evenly with butter and heat through.
9. Add the parsley and pepper, and season with salt to taste.
10. Serve immediately, or hold hot for service.

Recipe Categories

Vegetables

Chef's Notes

For a variation, substitute dill for the parsley.

Potentially Hazardous Foods

Cooked Potatoes

HACCP

Maintain at 135° F during service.

Nutrition

Calories	190
Protein	2 g
Fat	10 g
Carbohydrates	23 g
Sodium	257 mg
Cholesterol	5 mg

16 Fats and Oils

In This Chapter, You Will . . .

- Identify types of fats and oils
- Understand the nutritional value of fats and oils
- Learn how to use fats and oils when preparing food

Why You Need to Know This

Despite all the bad press that fats receive, they play an important role in making food taste good, giving foods a specific texture, and even making some foods stay fresh longer. They also provide energy and help our bodies absorb nutrients. The key is to understand the different types of fats that are available, and how to use them to increase the flavor of foods without increasing our waistlines and clogging our arteries. There are two basic types of fats: solid fats and liquid fats. As a class, name as many solid fats as you can. Then name liquid fats. Discuss the difference between the two.

Types of Fats and Oils

The main difference between fats and oils is their **phase**—or state—at room temperature. Fat that is solid at room temperature is called **fat**. Fat that is liquid at room temperature is called **oil**.

Solid fats commonly used in cooking include the following:

- **Butter:** Fat from cream which is separated by churning. Most people like the way butter tastes, and therefore, like foods made with butter.
- **Lard:** Fat separated from the fatty tissues of hogs. It is soft and bland in flavor.
- **Shortening:** Hydrogenated vegetable oil. It lacks flavor.
- **Margarine:** Hydrogenated vegetable oil that also contains water, whey, flavoring, and coloring. It lacks flavor.

Liquid fats commonly used in cooking include:

- **Neutral oils:** Made from canola, corn, and safflower oils. Neutral oils lack flavor.
- **Vegetable oil:** A blend of neutral oils. Vegetable oil lacks flavor.
- **Flavored oils:** Nut oils are made from nuts, such as walnuts, peanuts, sesame seeds, and almonds. Olive oil is made from olives. Flavored oils have distinctive flavors, depending on the source of the oil.

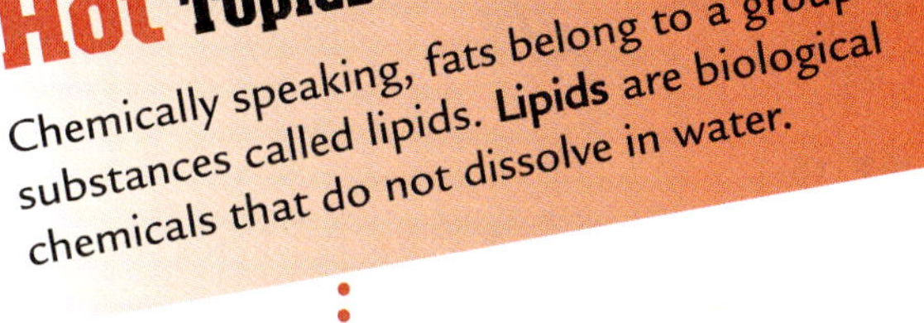

FIGURE 16-1
Visible fats are fats that you see and recognize, like the fat on the bacon and ham in this picture. **Invisible fats** are less obvious. They are hidden in foods, such as nuts. Can you think of other foods that contain invisible fats?

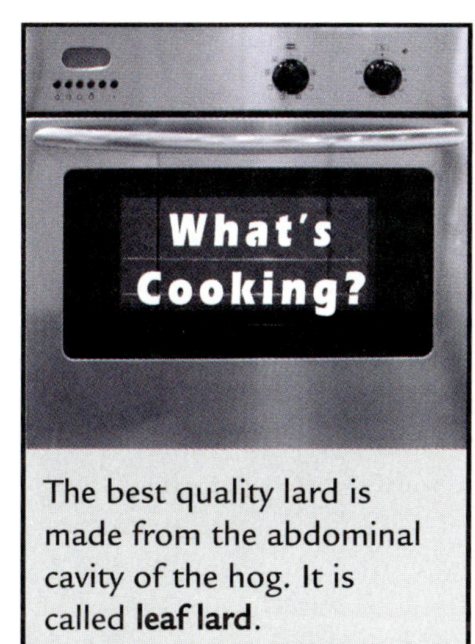

The best quality lard is made from the abdominal cavity of the hog. It is called **leaf lard**.

Forms of Fats and Oils

Fats like butter and some shortenings are usually sold in sticks, with four sticks in a box. Each stick equals 0.5 cup. The wrappers on the sticks often have marks indicating fractional amounts in tablespoons and cups to make it easy to measure the amount called for in a recipe.

FIGURE 16-2
Processed foods made with hydrogenated or partially hydrogenated vegetables oils are usually high in trans fatty acids. Why should you try to avoid eating too many foods containing trans fats?

Spreadable fats such as whipped butter, margarine, and other vegetable oil-based spreads come in tubs, or containers, of varying sizes. Oils are sold in a variety of sizes and containers. For convenience, most are sold in glass or plastic, though some high-quality olive oils are sold in cans.

Spray cans and bottles offer oil and oil products in a convenient form. Cooking sprays make it easy to lubricate pans, griddles, and grills. Pump bottles of salad oils and butter spreads let you control the amount of fat you spray on salads, vegetables, and other foods.

Nutritional Value of Fats and Oils

Fats are the richest source of energy, providing about 2.25 times as much energy as carbohydrates or proteins. Carbohydrates and protein both contain 4.1 calories per gram—about 120 calories per ounce—while fats have 9 calories per gram.

Some fats carry fat-soluble vitamins—A, D, E, and K—which are stable and not usually lost during cooking. The fats in fortified dairy products, such as milk and butter, are important sources of vitamins A and D. The fats in pure vegetable oils, nuts, and eggs contribute vitamin E.

Cholesterol is a fatty substance called sterol that is produced naturally by the body. Cholesterol in foods is called **dietary cholesterol**. It occurs only in animal food products, never in plant food products. Cholesterol in the blood is called **serum cholesterol**. When doctors check a person's cholesterol levels, they are trying to measure how much serum

Safe Eats

Unlike water-soluble vitamins, which are flushed out if they aren't used, fat soluble vitamins remain stored in the tissue of the body. Over time, an excess of fat soluble vitamins can cause hypervitaminosis, a condition that literally means too much vitamin in the body.

Specified as hypervitaminosis A (too much vitamin A), hypervitaminosis D (too much vitamin D), etc., the condition can be avoided simply by not taking an excess of fat soluble vitamin supplements.

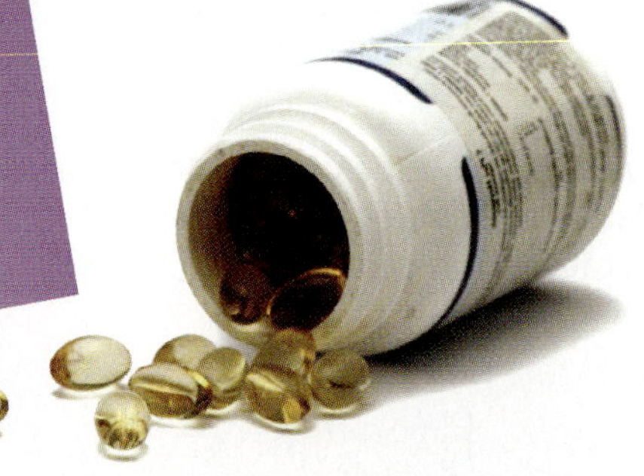

cholesterol is in the blood. To do this, they measure the levels of very-low-density lipoproteins (VLDL), low-density lipoproteins (LDL), and high-density lipoproteins (HDL) in the blood. VLDL and LDL are known as "bad cholesterol." Above a certain level, it could indicate a higher risk of heart disease. HDL is known as "good cholesterol." It clears cholesterol out of the circulatory system.

About Fatty Acids

Fats are made up of **fatty acids**, which are a chain of smaller units made up of atoms of carbon, hydrogen, and oxygen linked together. Your body needs fatty acids for growth, and to prevent skin diseases.

The one essential fatty acid your body can't make on its own is **linoleic acid**. You get it from foods—most vegetable oils, fish oil, meat, milk, and other dairy products. It is sometimes added to products such as margarine.

Fatty acids are grouped into three categories, according to their chemical structure: saturated, polyunsaturated, and monounsaturated. (For more information on fatty acids, check out the section on fats in Chapter 1.)

The difference between saturated and unsaturated fatty acids is that saturated fats contain all of the hydrogen they can hold. In other words, they are saturated with hydrogen. Monounsaturated fats are missing one hydrogen, and polyunsaturated fats are missing two or more hydrogen atoms.

Hot Topics

Margarine was developed as a healthier alternative to butter, until it was found that hydrogenation may create unhealthy trans fats, which can raise cholesterol. New alternatives claim to lower bad cholesterol without lowering the good.

Sterol spreads are made from vegetable oil with added **sterol esters**, a group of chemical compounds that occur naturally in plants.

Other spreads are made from natural saturates like palm fruit oil, blended with polyunsaturates like soy and canola oil.

FIGURE 16-3
Salmon, mackerel, dark-green leafy vegetables, walnuts and canola oil are all excellent sources of omega-3 fatty acids. Why should you try to eat foods containing omega-3 fatty acids?

Check the Label

Saturated fats Usually solid at room temperature. Most of them come from animal sources, except coconut and palm oil, which come from plants.

Polyunsaturated fats Liquid at room temperature. They come from plants.

Monounsaturated fats Liquid at room temperature. They come from plants, and are considered healthier than the other fats because they help balance cholesterol levels in the blood, which reduces the risk of heart disease.

Hydrogenated fats Polyunsaturated fats which have undergone a process called hydrogenation. Hydrogenation changes the fats from liquid to solid form and may produce trans fats, or trans fatty acids, a potentially harmful type of fat that has been linked to heart disease. For example, hydrogenated corn oil becomes margarine.

Omega-3 fatty acids A type of polyunsaturated fat that is considered particularly healthy. They are linked to a reduced risk of stroke and heart attack and to improved brain growth and development. They are found in some plants and all fish.

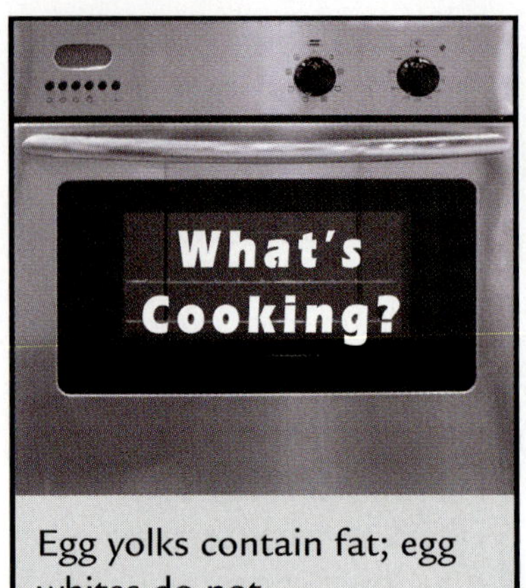

Egg yolks contain fat; egg whites do not.

The Fat Content of Foods

Fats are usually part of other foods; for example, milk and cheese contain butter fat. Some fats, like neutral oils, are in a pure form.

The richest sources of fat are meats, vegetable oils, vegetable shortenings, and nuts. In general, lean pork and beef contain about the same amount of fat, and fish has less fat than meat or poultry. Fruits—except for avocados—and vegetables contain very little fat. An average avocado has about 30 grams of fat.

Buying and Storing Fats and Oils

You select fats based on what you are cooking. Recipes will specify the type you need. One distinction is whether or not you want to taste the fat. When baking, you usually use butter, which adds a rich flavor. When sautéing or pan-frying, you might choose vegetable oil, which has no flavor.

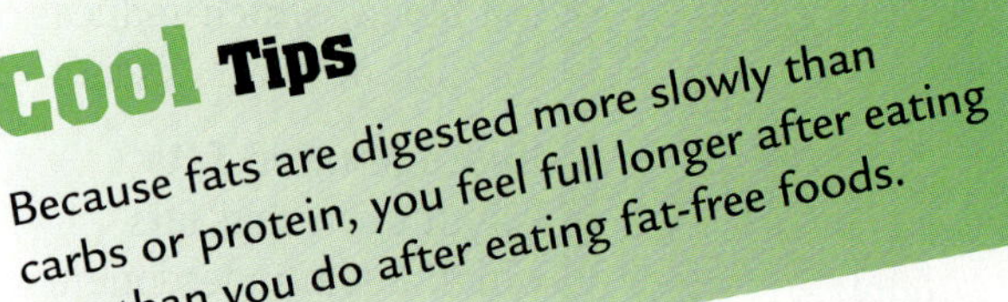

Cool Tips

Because fats are digested more slowly than carbs or protein, you feel full longer after eating fats than you do after eating fat-free foods.

In the United States, the quality of fats and oils is not regulated by the government, except for butter. The USDA assigns quality grades to butter based on aroma, flavor, and texture. Grade AA is the highest, then Grade A, followed by Grade B. You will rarely see Grade B butter in stores.

Vegetable oils and shortenings may be kept at room temperature, but animal fat products such as butter and margarine, should be stored in the refrigerator.

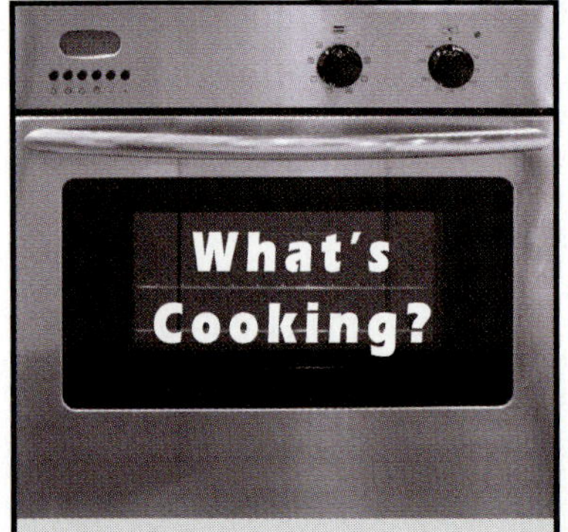

Butter is usually sold as either salted or unsalted. Salt acts as a preservative, giving salted butter a longer shelf life. It also affects the flavor. Most recipes call for unsalted butter, allowing you to control the amount of salt you add when cooking and baking.

FIGURE 16-4
Keep in mind that fats and oils will spoil and develop an unpleasant flavor if they are exposed to air for too long. They will also absorb the flavors of the foods around them. Keep them in airtight containers. What steps can you take to make sure fats and oils don't absorb other flavors?

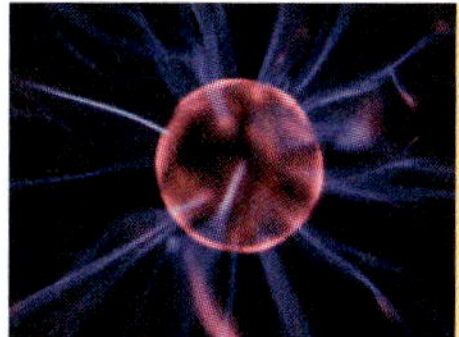

SCIENCE STUDY

Butter is actually made up of several different substances mixed together, including fat, milk proteins, and air. Use the following experiment to separate the substances:

1. Cut three tablespoons of unsalted butter into small pieces and put them in a pan.
2. Melt the butter over a low heat, observing it as it melts. Remove the pan from the heat when the butter starts to sizzle.

   You should see white foam floating on top of a clear yellow liquid, and white solid bits settling on the bottom. The liquid is the fat, the solids are the milk protein, and the foam is air trapped in the butter during processing.
3. Use a spoon to remove the foam and then carefully pour the liquid into a bowl. This liquid—with the milk proteins removed—is called clarified butter, drawn butter, or ghee.

Compare the flavor of the clarified butter with the flavor of regular butter.

Do you notice a difference? Why do you think they taste different? Can you think of any reasons why you might or might not want to use clarified butter for cooking?

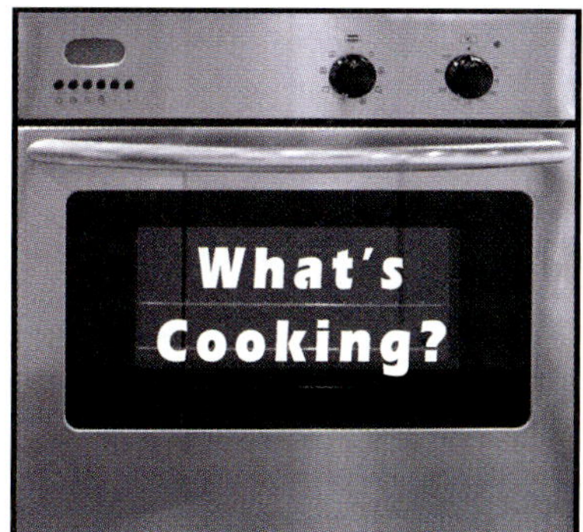

In the United States, most olive oil is classified as either virgin olive oil or just olive oil. Extra-virgin olive oil is a higher grade specified by the International Olive Oil Council.

Both virgin and extra-virgin olive oil have distinctive flavors, making them suitable for recipes when you want to taste the oil, such as salad dressings.

Olive oil lacks flavor, making it suitable for cooking when you do not want the fat to affect the taste. For example, it is good for sautéing or pan frying.

Preparing Foods with Fats and Oils

Fats bring a lot of utility to cooking. They add flavor, prevent foods from sticking, improve texture, and even help to keep some foods fresh longer. You can use them as a primary ingredient, as in salad dressings, or as one of many ingredients in a recipe, as in baked goods.

Most people like the flavor of butter. Use butter to season vegetables and sauces, or spread it on bread or other baked goods. As an ingredient, it impacts the texture as well as the flavor of baked goods. When baking, the more fat you add to a batter or dough, the softer it will be.

Hot Topics

Concern over the health risks of trans fatty acids have prompted government action. Cities and towns around the country–including New York City and Boston–banned the use of artificial trans fats in restaurants and other businesses that make freshly prepared food. The result is that businesses that once used hydrogenated fats now use healthier fats, such as canola oil.

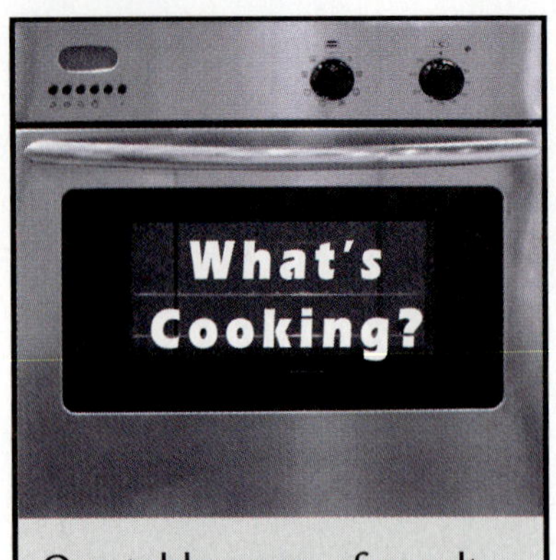

One tablespoon of unsalted butter contains the following:

100 calories (all from fat)
11.5 g of total fat
30 mg of cholesterol
3 mg of potassium (8% RDA of potassium)

One tablespoon of olive oil contains the following:

119 calories (all from fat)
14 g of total fat

Oil or meat fat, such as bacon grease, is often used to keep ingredients from sticking during sautéing, pan-frying, and stir-frying. Vegetable oil and non-virgin olive oil will not impact the flavor very much, but fats from meat or poultry will. The amount of fat you use depends on the cooking method:

- **Sautéing and stir-frying** are done in small amounts of fat. Vegetable oil, olive oil, and peanut oil are commonly used.
- **Pan-frying** is done in fat that is half the thickness of the food. Vegetable oil, or a combination of butter and vegetable oil, is commonly used.
- **Deep-frying** requires a large amount of fat—enough so that you can completely submerge whatever you are frying. Fats such as shortening, high quality lard or vegetable oil are used for deep-frying everything from chicken to vegetables. The best fats for deep-frying have a high **smoking point**, which is the temperature at which they start to smoke. When fats smoke, they break down chemically, producing an irritating odor and unpleasant flavor. Vegetable shortening and some oils—peanut oil, for instance, but not olive oil—have a high smoking point; butter and margarine do not.

Cool Tips

Add a pinch of salt to hot cooking oil to reduce splattering.

BY THE NUMBERS

If you are pan frying eggplant slices that are ½ inch thick, how much oil should you add to a 12-inch round pan?

Hint: The formula for calculating volume is area *x* depth, so find the area of the pan and multiply it by the depth of the oil. Multiply the result by .554 to convert it to fluid ounces.

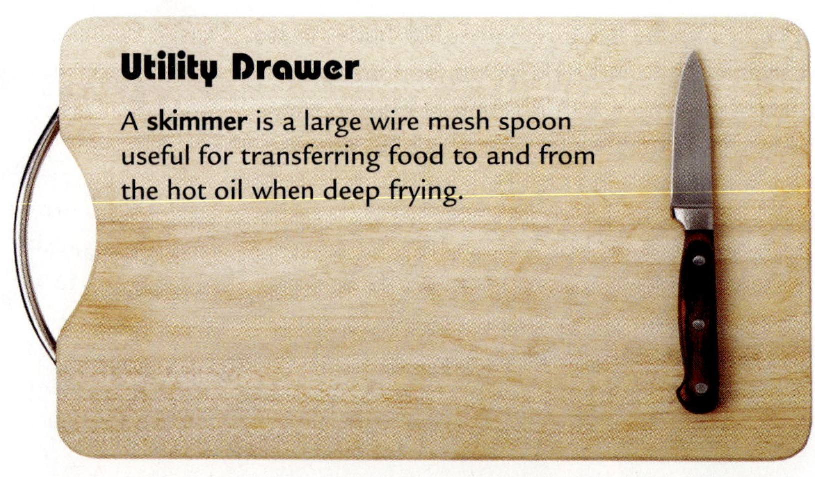

Utility Drawer

A **skimmer** is a large wire mesh spoon useful for transferring food to and from the hot oil when deep frying.

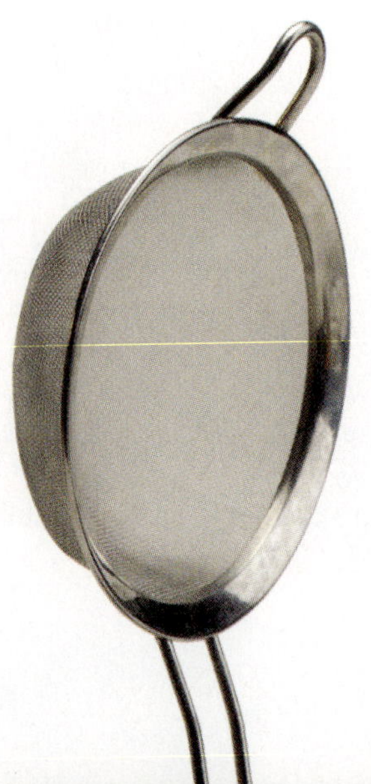

FIGURE 16-5
Fried foods develop a crisp, brown crust on the outside and absorb flavor from the fat used for frying. What are some of the health issues concerning deep-fried foods?

BASIC CULINARY SKILLS

Deep Frying

1. Heat fat, such as shortening or oil, in a deep fat fryer or a pot with tall sides. Heat to the temperature specified in your recipe, usually 350° to 375° F.
2. Blot food dry.
3. Coat food if your recipe calls for it.

4. Add food to the hot oil by using a frying basket or tongs. Fry in small batches, so the oil does not cool in temperature when the food is added.

5. Deep fry until food is an even golden brown and is fully cooked.
6. Remove food from the fat.

7. Drain excess oil. Blot food on a paper towel.

BASIC CULINARY SKILLS

Sautéing

1. Coat food with flour, if indicated in your recipe. Otherwise, blot the food dry and season it.

2. Heat the pan over direct heat.
3. Add oil or fat. Use only a small amount.

4. Add food to the hot pan. Do not crowd food.

5. Cook the first side. Do not disturb until the food is cooked halfway through.
6. Turn food once.

7. Complete cooking on the second side until properly cooked.

Utility Drawer

A **wok** is a large pan with high sloping sides, a round bottom, handles, and a lid. Originally associated with Chinese cooking, it is highly suited for stir-frying, but can also be used for deep-frying, steaming, and even roasting.

Case Study

Marissa is 15 years old. Her parents recently told her that they are moving to a new town, and that she will have to change schools. She is worried about fitting in. She thinks she will make friends more quickly if she loses a few pounds.

Marissa uses the Internet to look up diets that will help her lose weight fast. She finds one that suggests eating only cabbage soup, and another that says to eat only grapefruit. One diet guarantees she will lose 10 pounds in 10 days if she only eats in the morning. Another suggests she use over-the-counter diet pills.

Finally Marissa finds a diet that she thinks sounds sensible. It says she can eat whatever she wants, as long as it contains no fat. She prints the suggested menus and starts the diet immediately.

- Do you agree that the diet Marissa selected sounds sensible? Why or why not?
- What are some of the problems with the diets Marissa found online?

- Do you think Marissa should even be looking for a diet?
- How can Marissa find out if she really does need to lose weight, and if so, how to do so in a safe and healthy way?

Put It to Use

1. Develop a weekly menu for someone at risk of suffering a heart attack. Be specific about the types of fats and oils that should be used or avoided. Write a paragraph explaining your choices.

Put It to Use

❷ Make your own herb-infused olive oil. Select an herb, such as rosemary, garlic, basil, mint, or thyme, or a combination of these. You might want to add spices or other aromatics, too, such as lemon peel, garlic, or peppercorns. Wash and dry the fresh herbs (you can use dried herbs, if you want) and place them in a clean glass jar with a tight-fitting cover. Fill the jar with oil. Use about ¼ cup of oil for every two to three sprigs of herb. Place the jar in the refrigerator for 7 to 10 days. Test it as a dressing on salad or pasta, or as a dipping sauce for bread.

Write Now

What do you think about the government passing laws concerning the type of fats restaurants can use? Do you think it is necessary for the government to be involved, or do you think people should be able to make their own choices about the types of foods they want to eat? Write an essay stating your opinion. Support it with facts and real life examples.

Tech Connect 

Believe it or not, people collect used cooking oil and recycle it into biofuels to power cars. Use the Internet to learn about the pros and cons of recycling cooking oils. Try to find an example of a successful cooking oil recycling project in your area and report on it to your class.

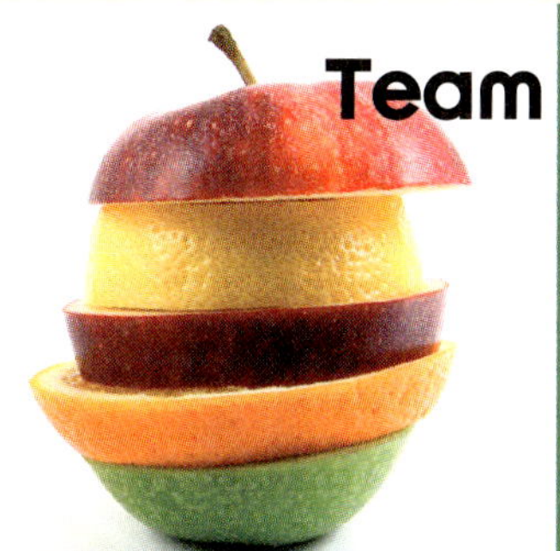

Team Players

In small groups, work together to create a brochure or flyer about the health benefits and risks of different types of oils and fats. Research the nutritional value of the items and incorporate the data into your document. Use pictures to illustrate your brochure/flyer. When the document is complete, print it. With permission, distribute it at a local elementary school, day care center, or senior center.

Put It Together

Match the explanation in column 1 with the term in column 2.

Column 1

- **a.** fat that is solid at room temperature
- **b.** fat that is liquid at room temperature
- **c.** biological chemicals that do not dissolve in water
- **d.** a process that changes the fats from a liquid into a solid
- **e.** lard made from the abdominal cavity of a hog
- **f.** chains of carbon, hydrogen, and oxygen atoms linked together
- **g.** an essential fatty acid the human body can't make on its own
- **h.** a fatty substance produced naturally by the body
- **i.** butter from which the milk proteins have been removed
- **j.** the temperature at which fats start to smoke
- **k.** a large pan with high sloping sides, a round bottom, handles, and a lid

Column 2

1. lipids
2. wok
3. smoking point
4. leaf lard
5. clarified butter
6. hydrogenation
7. oil
8. cholesterol
9. fat
10. linoleic acid
11. fatty acids

TRY IT!

Mayonnaise

Yield: 32 fl oz Serving Size: 2 oz

Ingredients

3 Egg yolks, pasteurized
1 fl oz White wine vinegar
1 Tbsp Water
2 tsp Dry mustard
24 fl oz Vegetable oil
1 tsp Salt
½ tsp Sugar
¼ tsp Pepper, white, ground
1 fl oz Lemon juice

Method

1. Combine the yolks, vinegar, water, and mustard in a bowl.
2. Mix well with a balloon whisk until the mixture is slightly foamy.
3. Gradually add the oil in a thin stream, constantly beating with the whip, until the oil is incorporated and the mayonnaise is smooth and thick.
4. Adjust the flavor with salt, sugar, pepper, and lemon juice to taste.
5. Refrigerate the mayonnaise immediately.

Recipe Categories

Garde Manger,
Dressings & Dips

Chef's Notes

Have all ingredients at the same temperature to prevent the sauce from breaking as you mix it.

Potentially Hazardous Foods

Eggs

HACCP

Store cold sauces below 41° F.

Nutrition

Calories	215
Protein	1 g
Fat	24 g
Carbohydrates	1 g
Sodium	84 mg
Cholesterol	23 mg

TRY IT!

Red-Wine Vinaigrette

Yield: 1 qt Serving Size: 1 oz

Ingredients

8 fl oz Red wine vinegar
2 tsp Mustard (optional)
2 Shallots, minced
24 fl oz Olive or canola oil
2 tsp Sugar (optional)
To taste Salt and pepper, freshly ground
3 Tbsp Chives, parsley, or tarragon, minced (optional)

Method

1. Combine the vinegar, mustard (if using), and shallots.
2. Gradually whisk in the oil.
3. Adjust seasoning with sugar (if using), salt, and pepper.
4. Add the fresh herbs if desired.

Recipe Categories

Garde Manger, Dressings & Dips

Chef's Notes

Add the minced fresh herbs just before serving for the best flavor and color.

Potentially Hazardous Foods

Raw Shallots

HACCP

None

Nutrition

Calories	190
Protein	0 g
Fat	21 g
Carbohydrates	0 g
Sodium	10 mg
Cholesterol	0 mg

TRY IT!

French Fried Potatoes

Yield: 10 Servings   Serving Size: 6 oz

Ingredients

8 Potatoes, baking, large (about 4 lb)
32 fl oz Vegetable oil
1½ tsp Salt

Method

1. Heat the oil to 300° F.
2. Peel the potatoes.
3. Cut into desired shape for your French fries.
4. Hold the French fries in cold water until ready to cook.
5. Rinse, drain, and dry thoroughly.
6. Add the potatoes, in batches.
7. Blanch until just tender, but not browned. (Time will vary depending on the size of the cuts.)
8. Drain well.
9. Transfer to sheet pans lined with paper towels.
10. Refrigerate until service. (Scale into portions if desired.)
11. Just before service, reheat the oil to 375° F.
12. Fry the potatoes, in batches, until they are golden brown and cooked through.
13. Drain well and season with the salt.
14. Serve immediately.

Recipe Categories

Vegetables

Chef's Notes

Sweet potatoes can be substituted for the baking potatoes.

Potentially Hazardous Foods

Cooked Potatoes

HACCP

Maintain at 135° F during service.

Nutrition

Calories	522
Protein	6 g
Fat	27 g
Carbohydrates	67 g
Sodium	330 mg
Cholesterol	0 mg

17 Dairy and Eggs

In This Chapter, You Will . . .

- Identify types of dairy products and eggs
- Understand the nutritional value of dairy products and eggs
- Learn how to prepare dairy products and eggs

Why YOU Need to Know This

Dairy products and eggs have been a staple of the human diet for hundreds of years. We eat them alone (a glass of milk, or a boiled egg), with other foods (milk in cereal or an omelet), and we use them as ingredients in a huge number of recipes. They are relatively inexpensive, easy to use, and readily available. Even if you have no other ingredients on hand, you can create a satisfying meal from eggs, milk, and cheese. Do you drink whole milk, or milk with less fat, such as 1% or fat-free? Survey the class to see how many students use whole, fat-free, 1%, or 2% milk. Some might not use milk at all, and some might use a milk substitute, such as soy or rice milk. Make a graph or chart showing the results. Discuss the differences between the types of milk, and why one might be better than another.

Types of Dairy Products

Dairy products come in a variety of forms, with a variety of flavors and textures. Each form is made up of water, solid particles, and fat. The fat in dairy is called **milkfat**, or **butterfat**. Dairy products are labeled according to the percent of milkfat they contain.

Dairy products all start with milk. Most milk comes from cows, but other animals produce dairy milk, too, including goats, sheep, camels, buffalo, and yaks. Milk can be served as a beverage or used as an ingredient. Cream, butter, yogurt, cheese, and ice cream are all dairy products made from milk. The USDA and FDA regulate milk production.

Milk is graded based on its bacterial count. Grade A has the lowest bacterial count and is the only milk sold in retail outlets. Grades B and C have a higher bacteria count, but are still considered safe and wholesome. They are usually used for processed products or making cheese.

FIGURE 17-1
The Holstein (above) is a common breed of dairy cattle. Have you ever seen a Holstein cow? Where?

Milk Processing

To insure safety, milk and milk products are **pasteurized**, which means they have been heated to destroy harmful organisms. There are two methods of pasteurization:

- **Flash:** Raw milk is brought to 160° F for a minimum of 15 seconds.
- **Holding:** Raw milk is brought to a temperature of at least 143° F for 30 minutes and then cooled rapidly.

Milk is **homogenized**, which means it is forced through fine screens to break up the milkfat into particles so small that they will remain evenly distributed throughout the milk. If milk is not homogenized, the fat separates from the milk and rises to the top.

Hot Topics

A small but growing number of people believe that drinking **raw milk**—which is unpasteurized—from grass-fed cows is healthier than drinking pasteurized milk. For safety reasons, however, the FDA advises consumers to use only pasteurized milk and dairy products.

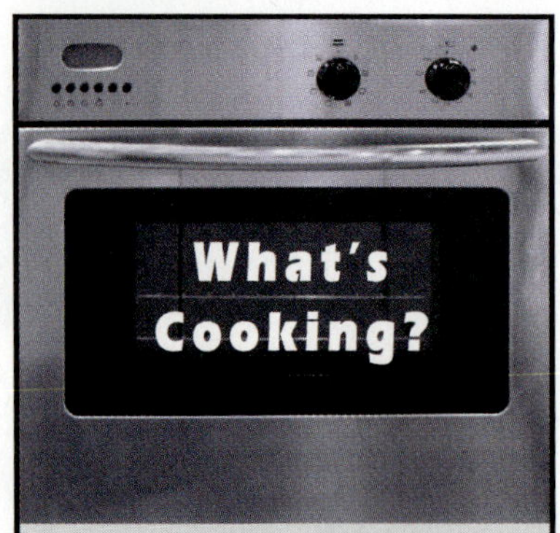

Crème fraîche is similar in flavor and texture to sour cream, but because it contains so much milkfat, it will not curdle in hot soups and sauces the way sour cream and yogurt can. Crème fraîche is French, literally meaning *fresh cream*.

FIGURE 17-2
Whipped cream is cream with at least 30% milkfat which has been beaten with air. The fat droplets in the cream form little pockets, causing the volume of the cream to double. What foods do you like to top with whipped cream?

Many farms give their cows artificial hormones to boost milk production. The government does not require labeling of dairy products that contain artificial hormones, but some producers voluntarily label products that *do not* contain the hormones.

Cream and Cultured Dairy Products

Cream is the milkfat layer skimmed off the top of milk before homogenization. It is thicker, richer tasting, and contains between 18 and 30% milkfat.

Cultured dairy products are made by adding a specific type of beneficial bacteria, called **probiotic** bacteria, to milk or cream. As the bacteria grow, they thicken the milk or cream and give it a tart taste. Buttermilk, sour cream, and yogurt are examples of cultured dairy products.

Probiotics are beneficial and occur naturally in some cultured dairy products, such as yogurt. They help to maintain the natural balance of organisms in the intestines, thereby promoting intestinal and digestive health. Probiotics are also available as a dietary supplement. Some food manufacturers have started adding probiotic bacteria to products such as yogurt. Scientists are still studying the health benefits of eating foods enriched with probiotics.

Storing Milk

Fresh milk is highly perishable. It should be stored in the refrigerator, in a tightly covered container. Pasteurization kills pathogenic bacteria, such as salmonella, but it doesn't stop the growth of spoilage bacteria, which causes food to go bad and develop unpleasant odors, tastes, and textures. Store dairy products at less than 40° F and discard them five days after the "Sell by" date stamped on the packaging.

Shelf-stable milk can be stored without refrigeration in unopened packages or tightly sealed containers for up to six months. Products include canned milk, powdered milk, and **UHT milk**. UHT, which stands for Ultra High Temperature, means the milk is pasteurized at a minimum of 275° F.

FIGURE 17-3
Milk and other dairy products have a "Sell by" date stamped somewhere on the packaging. If stored properly at less than 40° F, milk should stay fresh for four to five days past the marked date. Do you think it is safe to drink milk more than five days past the "Sell by" date? Why or why not?

Cheese

The process for making cheese involves separating the **curds**—chunks of cheese solids—from the **whey**—the clear liquid that contains some of the water-soluble substances in milk—and then pressing the curds in a mold.

To make the curds, a starter of lactic-acid–producing bacteria is added to warm milk. Once the milk turns slightly acid, **rennet**, an enzyme which causes the milk to curdle, is added. The curd is cut to separate the whey, mixed with salt, and packed into cheesecloth lined hoops for pressing. It is dried for several days, coated with hot paraffin (wax) to prevent moisture loss, and then placed in a ventilated room at a controlled temperature of about 50° F.

FIGURE 17-4
Making cheese. What step of the cheese-making process do you think is shown in this picture?

There are seven basic types of cheese, categorized based on texture, taste, appearance, and aging. All seven may be made using milk from cows, sheep, goats, or other animals.

- **Fresh cheeses** are moist, soft cheeses that have not ripened or aged significantly. Examples include cottage cheese, cream cheese, farmer cheese, mascarpone, fresh mozzarella, feta, and fresh ricotta.
- **Soft, rind-ripened cheeses** are soft cheeses that have been ripened by being exposed to a spray or dusting of friendly mold. They are aged until the **rind**—or surface—develops a soft downy consistency. Examples include Brie, Camembert, and Pont l'Eveque.

Check the Label

Whole milk Contains at least 3% milkfat.

Low-fat 1% milk Contains 1% milkfat.

Low-fat 2% milk Contains 2% milkfat.

Skim, or nonfat milk Contains less than 0.1% milkfat.

Dry, or powdered milk Milk from which all water is removed. It may be made from whole or skim milk.

Evaporated milk Milk that has been heated in a vacuum to remove 60% of its water. May be made from whole, low-fat, or skim milk. Milkfat content ranges from 0.5% all the way up to 8%.

Condensed milk Sweetened evaporated milk.

Half-and-half Equal parts milk and cream. Contains between 10.5 and 18% milkfat.

Buttermilk Thickened, cultured, tangy nonfat or low-fat milk.

Sour cream Thickened, cultured sweet cream. Contains between 16 and 22% milkfat.

Yogurt Thickened, cultured milk. Can be whole, low-fat, or nonfat. Available flavored or plain.

FIGURE 17-5
Soft-rind-ripened cheese. Many cheeses are named after the city or region in which they are produced. Some names, such as Roquefort, are protected by law, and cannot be used for cheeses made in a different location. Some names, such as Brie and Camembert, are not protected. Why might it might be important to protect the name of a cheese?

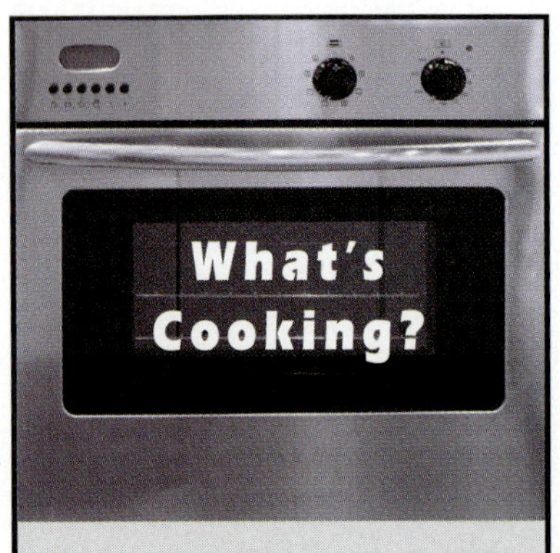

Raw milk cheeses are made from non-pasteurized milk. They are typically made by small cheese producers using traditional methods.

- **Semi-soft cheeses** are more solid than soft cheeses and retain their shape. There are three types:
  - **Rind-ripened** are semi-soft cheeses whose rinds are washed with a liquid such as grape juice, beer, brandy, wine, cider, or olive oil. The washing produces beneficial bacteria that penetrates and flavors the cheese. Examples are Muenster and Port-Salut.
  - **Dry rind** are cheese in which the rind is permitted to harden naturally through exposure to air. The rind becomes firm, but the interior of the cheese remains soft. Examples are Bel Paese, Monterey Jack, Morbier, and Havarti.
  - **Waxed-rind** cheeses are encased in a coating of wax during ripening. The interior remains consistently soft. Examples are Edam and Fontina.
- **Blue-vein cheeses** include Roquefort, Gorgonzola, Stilton, and Maytag Blue. They are made by injecting needles into the cheese to form holes in which mold spores multiply. The cheese is salted and ripened in a cave.
- **Hard cheeses** have a drier texture than semi-soft cheeses and a firmer consistency. They slice and grate easily. Cheddar, Gruyere, Colby, Jarlsberg, Provolone, and Manchego are examples of hard cheeses.

Cool Tips

Most cheeses develop a stronger flavor the longer they age.

FIGURE 17-6
The blue veins in a blue cheese are actually types of a beneficial mold. How do you feel about eating mold in cheese?

- **Grating cheeses** are solid, dry cheeses that have a grainy consistency, making them ideal for grating. Examples are Parmigiano, Pecorino, and Sapsago.
- **Processed cheeses** are made from one or more cheeses that are finely ground, mixed together with non-dairy ingredients, heated, and poured into a mold. American cheese is an example.

Safe Eats

Soft cheeses, such as Feta, Brie, Camembert, blue-veined, and Mexican-style cheeses made from unpasteurized milk can become contaminated with *Listeria monocytogenes*, a harmful bacterium. Pregnant women, the elderly, and people with weakened immune systems should avoid these products.

Buying and Storing Cheese

Before you buy cheese, look at the label to find out exactly what it is, what ingredients it contains, where it was made, and when it was made. Many cheeses and cheese products have a "Sell by" date stamped on the packaging. Like milk, you shouldn't buy cheese after this date, and you should throw it away five days after the date.

Examine the rind and the interior. The colors should be natural, and the texture should be whole, without cracks or damage. It is also a good idea to taste cheese before you buy it. Most cheese shops and distributors will give you the opportunity to try a cheese to make sure it suits your needs.

The interior of cheese will begin to deteriorate as soon as it is exposed to air. Cheeses should always be tightly wrapped in waxed paper or butcher paper or in a covered container, and stored in the refrigerator. Unwrapped cheese may get hard, dry, and brittle, or grow mold. Also, unwrapped cheese can absorb the flavor from other foods, such as onions, and other foods might absorb the flavor of the cheese.

When properly stored, soft cheese will last about two weeks, semi-soft cheeses about three weeks, and hard cheeses about a month.

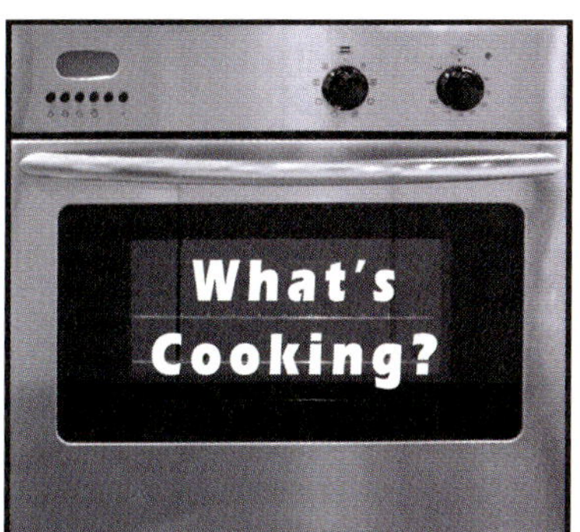

Milk substitutes are milk-like products that do not actually contain any milk or dairy products. They are used by people who cannot tolerate milk for health reasons—such as lactose intolerance—or who choose to not eat animal products—such as vegans.

Common milk substitutes include soy milk and rice milk. Less common products are almond milk, coconut milk, and oat milk. Note, though, that milk substitutes might not contain the calcium and protein that are found in milk, unless they are specially formulated or enriched.

Nutritional Value of Dairy

Milk and dairy products are nutrient-rich foods. They offer an abundant source of calcium and contain a high level of phosphorous. Dairy products are a source of high-quality protein and contain the carbohydrate milk sugar, called **lactose**. Some people are **lactose-intolerant**, which means their bodies do not process lactose efficiently. The lactose remains undigested in their intestines, producing pain, gas, and diarrhea.

Milk and milk products are fortified with vitamin D. They also contain vitamin A. However, they are not a significant source of either.

Milk is very low in iron and vitamin C, and is low in niacin. It is high in fat, but low fat or nonfat forms are available. The fat content of cheese and other milk products varies, depending on the type of milk used during processing.

Milk is the best food source of vitamin B_2—riboflavin. Light destroys the vitamin B_2 in milk, so buy milk in opaque containers, and keep it away from light.

Egg whites (albumen) are made of protein and water. They are liquid-like when raw, but turn white and hard when cooked.

Shell color depends on the breed and type of poultry that laid the egg.

Egg yolks contain protein, fat, and a natural emulsifier called **lecithin**. Like whites, yolks harden as they cook.

FIGURE 17-7
Eggs have three main parts: shell, white, and yolk. Do you think this egg is fresh? Why or why not?

Eggs

Eggs have unique properties that make them one of the most versatile foods. In the United States, chicken eggs are the most common type, but eggs from other types of poultry, including ducks, geese, ostrich, and even quail are also used.

The USDA sets standards for eggs, but grading is done on a voluntary basis. Graders look for freshness by examining the shells, the firmness of the white, and whether the yolk sits in the center of the white or slides to one side. They also assign standard size classifications based on weight.

The weight classes of eggs include:

- **Peewee:** 1.25 ounces
- **Small:** 1.5 ounces
- **Medium:** 1.75 ounces
- **Large:** 2 ounces
- **Extra Large:** 2.25 ounces
- **Jumbo:** 2.5 ounces

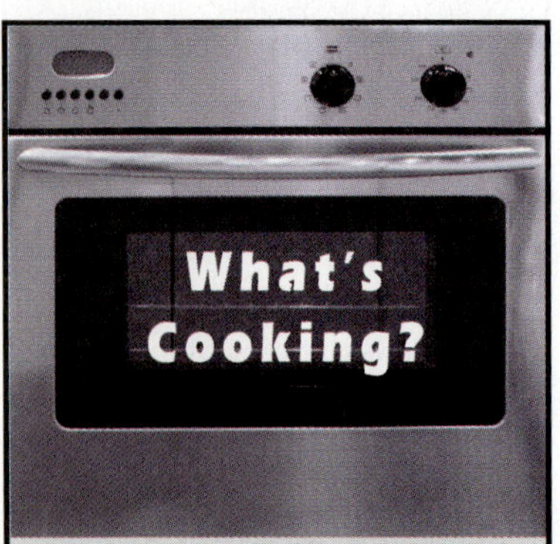

Egg whites actually have two parts—the thick white and the thin white. Fresh eggs contain more thick white than older eggs. Another indication that an egg might not be fresh is the position of the yolk. A fresh egg yolk sits in the center of the white. An older yolk is flatter.

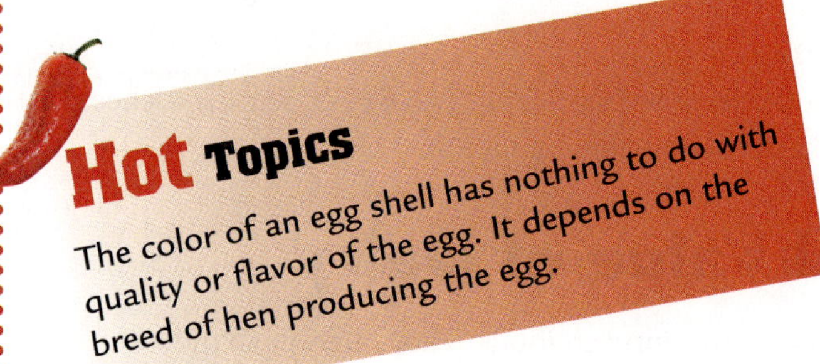

FIGURE 17-8
What type of bird layed these eggs? They are quail eggs. Why do suppose we don't normally see anything other than chicken eggs in the grocery store?

Buying and Storing Eggs

Consider the following factors when selecting eggs:

- **Use:** Think about how you plan to cook or use the eggs, the number of people you are cooking for, and how soon you will be doing the cooking.
- **Grade:** The quality you select depends on the use. For poached eggs, you want the highest quality—Grade AA. For scrambled eggs, or baking, Grade A may be fine.
- **Size:** Size also depends on use. For eggs on their own—poached, fried, scrambled, etc.—you can choose any size you want. Most recipes, however, are written for large eggs.
- **Cost:** A number of things influence the cost of eggs, including size, color, and grade. In addition, eggs from organic or free-range poultry will cost more.

FIGURE 17-9
This egg has a cracked shell. Why is it risky to use an egg that has a cracked shell?

Cool Tips

Because egg shells are porous, which means they let air and moisture in and out, they can absorb odors. Keep shell eggs in a covered container or away from smelly foods, such as onions or blue cheese.

Nutritional Value of Eggs

Like milk, eggs are a nutrient-rich food. Although most nutrients are contained in the yolk, the nutrient values are usually given for the entire egg.

Eggs contain high-quality protein. The yolk is an excellent source of vitamins A and D, a good source of riboflavin, and a fair source of thiamin. Egg yolks are also an important source of iron and phosphorous. Yolks also contain all of the fat in the egg, and the cholesterol. The white, by contrast, contains protein and some riboflavin.

If you overcook hardboiled eggs, or leave them sitting around a bit before use, you may see a green ring around the yolk. It is a chemical reaction caused by hydrogen in the egg white bonding with sulfur in the egg yolk. It is not harmful nutritionally, but may affect the flavor, and certainly affects the appearance.

Check the Label

Egg Grades:

Grade AA Denotes the freshest quality. The whites are compact and the yolks sit high in the center of the whites. The yolks are less likely to break when cooking.

Grade A Less fresh than Grade AA, these eggs have a slightly runny white. The yolks break more easily than the yolks of Grade AA eggs.

Grade B These eggs have a runny white and a flat off-centered yolk. They are best used as an ingredient in products like baked goods, or to make liquid, frozen, or dried egg products.

Check the Label

Shell eggs Fresh eggs, in their shells, sold in cartons or cases. Always open the package and check the condition. Do not buy any that have broken or cracked shells. Store shell eggs in the refrigerator in their original containers.

Bulk eggs Eggs out of their shells and sold in cartons or tubs. They may be whole (both whites and yolks), whole with added yolks, yolks only, or whites only. Bulk eggs are pasteurized to kill bacteria and other pathogens. Store bulk eggs in the refrigerator. Store frozen bulk eggs at 0° F, and allow them to thaw in the refrigerator before use.

Dried eggs Eggs that have been processed to remove all of the water. They are also called powdered eggs. Dried eggs may be stored on shelves in a cool, dry area.

Egg substitutes Made for use by people who cannot eat egg yolks. They are made from egg whites or soy-based products and have a similar color, texture, and flavor to an egg. Store egg substitutes according to the instructions on the packaging.

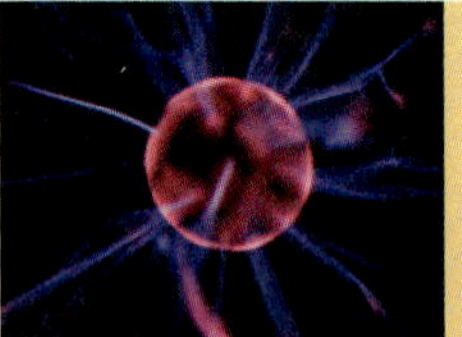

SCIENCE STUDY

Why do egg whites harden when cooked? One of the reasons eggs are so versatile is that egg proteins change when you heat them.

When you heat an egg, it coagulates, which means the chemical bonds in the proteins break down, straighten out, and form new, stronger bonds with other proteins. Eventually, they become a solid mass.

Egg whites are only proteins and water and coagulate at a lower temperature than yolks, which also contain fat and other compounds that have stronger bonds, and require more heat to break down.

You can watch the process in action: Fry an egg over a low heat and observe the changes in the egg white. How long before you see the white start to harden? How long before it is completely hard? What about the yolk?

Cooking with Dairy

As an ingredient, milk and other dairy products contribute nutrients, flavor, texture, and consistency. Milk is used primarily as a beverage or as an ingredient. As a beverage, it can be enjoyed cold or warm, plain or flavored. It is also often added to cereal.

Milk is used as an ingredient in savory and sweet dishes, and in recipes ranging from main courses to desserts. Some foods commonly prepared with milk or cream include cream soups, puddings, custards, cakes, and sauces.

FIGURE 17-10
Although milk can be drunk as a beverage at any meal, it is commonly served at breakfast or at snack time to accompany baked goods. What foods do you think go well with milk?

Because milk contains a lot of protein, it can form a thin film when it is heated. To avoid the film, heat the milk or milk mixture in a covered container, stir it often during cooking, or beat it to produce a layer of foam, which will block the film from forming.

If the film develops, it will cause pressure to build up beneath it, which will result in the mixture boiling over the edges of the pan.

There are two other problems that may occur when cooking with milk:

- Milk can **scorch**, or burn, when cooking. The best prevention is to use a low temperature.
- Milk can curdle when it comes in contact with an acid, such as tomatoes. To prevent curdling, thicken the milk or the acid with starch before combining, or use a low temperature for cooking.

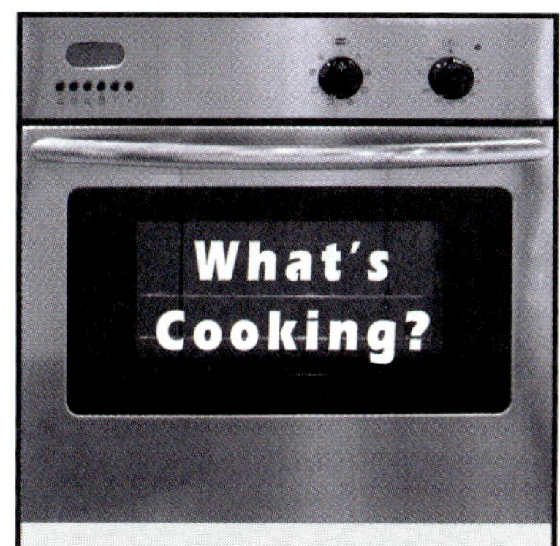

Scalded milk is milk heated to just below the boiling point. Before pasteurization, scalding was used to kill pathogens. Now, scalded milk is still used as an ingredient in some recipes, such as bread, hot cocoa, or when chefs want to infuse a flavor, such as vanilla, into the milk.

Cooking with Cheese

Cheese can be eaten alone, combined with other foods, or used as an ingredient. It may be used as an appetizer—such as plain on crackers or cooked into a dip or spread. It adds flavor when melted on vegetables, used as a topping on a sandwich or hamburger, or cooked into a sauce. Fresh cheese may be served at room temperature as part of the appetizer course, following a meal, or as a separate course called the cheese course.

Although cheese is often used in cooking, heat alters its unique flavor. Like milk, you should use low heat when cooking cheese because high heat affects the texture and consistency. The proteins will coagulate, causing it to become tough and rubbery.

Cool Tips

Evaporated milk can be diluted with an equal amount of water and used as a substitute for fresh whole milk. Likewise, dry milk can be reconstituted with water and used as a substitute for fresh milk in most recipes.

FIGURE 17-11
Fresh cheeses are often served on a flat platter called a cheese board, garnished with fruit or vegetables. Why do you think it is important to leave plenty of room between cheeses on the board?

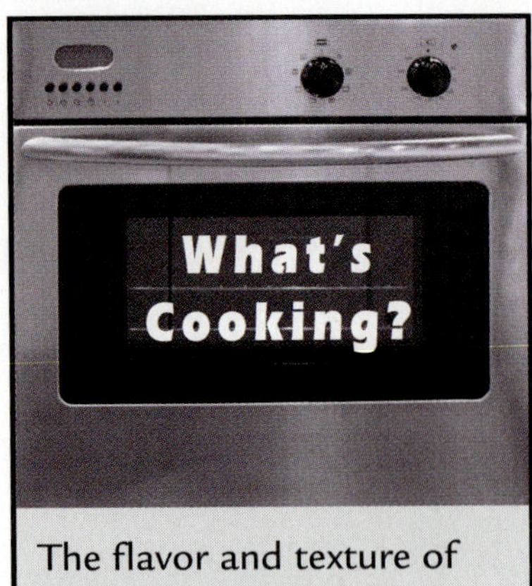

The flavor and texture of cheese combines well with fruit, fruit desserts, bread, crackers, and cured meats such as salami.

The three main ways in which cheese is used in cooking are:

- **As an ingredient in a dish.** Semi-soft cheeses are well suited for integrating in a dish because they don't leach excess water the way fresh cheeses can. Shredded cheese melts more easily and evenly than sliced cheese.
- **In a sauce.** Cheese can add body and flavor to sauces. For best results, stir the cheese into the sauce at the last minute.
- **As a topping or garnish.** Cheese can complement or offset the flavors and textures of other ingredients. It may be melted on top of a baked dish, such as a vegetable or pasta, or sprinkled on a finished dish, cold or hot.

FIGURE 17-12
Fondue is made of melted cheese and other ingredients. Typically, diners sit around the fondue pot and dip in bread and vegetables on skewers. Why would the texture of melted cheese be an important consideration for this dish? What type of cheese do you think would work well in fondue?

Preparing Foods with Eggs

Eggs may possibly be the most versatile food you will ever use. They may be prepared using a wide variety of cooking methods, used in baking and pastry work, and in the preparation of sauces and soups. They act as a thickening agent, and as a binder to keep other ingredients from falling apart.

Eggs may be cooked in their shells by simmering them in water. Simply bring the water to a simmer and then immerse the eggs for the specified amount of time. The shell protects the egg as it cooks. You can cook eggs in their shells to varying degrees:

- **Coddled eggs** are simmered for about 30 seconds—just long enough so that the whites are warm and thickened and semi-opaque, and the yolk is soft and very runny. They are usually used as an ingredient in a dish such as a Caesar salad.

- **Soft-cooked eggs** are simmered for 3–4 minutes. The whites are barely set and very moist. The yolk is hot, but still liquid. They are served in the shell in an egg cup. They may be called soft-boiled, or "a four-minute egg."
- **Medium-cooked eggs** are simmered for 5–7 minutes. They have fully set whites and yolks that have thickened. They are served in the shell in an egg cup.
- **Hard-cooked eggs** are simmered for 14–15 minutes. They have whites that are completely set and firm and yolks that are fully cooked and break apart easily when pushed. They may be served hot as part of a meal such as breakfast, or used in deviled eggs or egg salad. They may be cut or chopped to use as a garnish for a salad or vegetable dish.

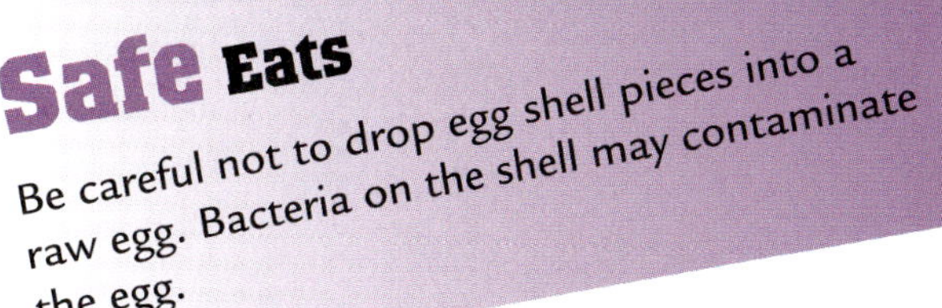

Other cooking methods for eggs require that they be removed from their shells:

- **Poached eggs** are removed from the shell and cooked in hot water, broth, or a sauce until the white is set, but still tender, and the yolk is slightly thickened. They may be cooked to any degree of doneness, and should be served immediately.
- **Fried eggs** are made by cooking a very fresh egg quickly in hot oil or butter. The white should be tender and fully cooked and the yolk should not break until the diner breaks it with a fork.
- **Scrambled eggs** are made by mixing the whites and yolks and stirring them as they cook in a sauté pan or double boiler over low to medium heat. Beat in water or milk before cooking to make the eggs fluffier and moister.
- **French omelets** are also called rolled omelets. You start the same way you start scrambled eggs, but you stir the eggs as they cook to keep them tender enough to roll or fold. As you stir, curds form on the bottom, and the heat of the pan sets the eggs on the bottom to make a smooth skin. Fillings and garnishes are folded into the eggs, stuffed in the middle or placed on top. The omelet is rolled out of the pan on to a heated plate.

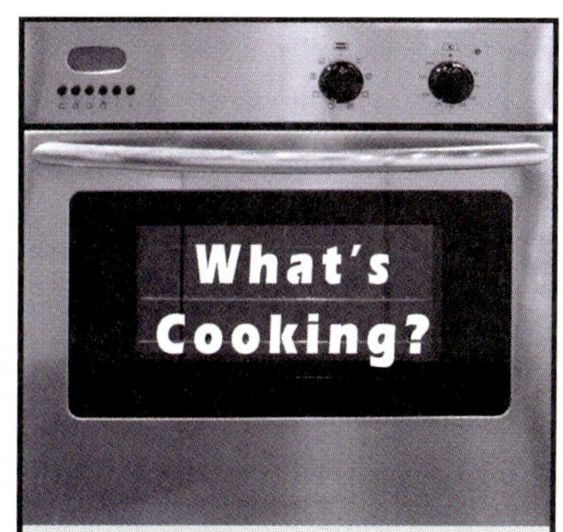

When you cook an egg in its shell, it is important to keep the water at a steady, gentle, simmer. Boiling shakes up the egg and can crack the shell. The whites may become overcooked and tough, and the yolk may be grainy instead of creamy.

FIGURE 17-13
Fried eggs may be cooked sunny-side up, which means cooked on one side only; or over easy, which means flipped over part way through. Which of the eggs in the picture is sunny-side up? How can you tell?

FIGURE 17-14
A **frittata** is a round, open-face omelet made by pouring beaten eggs into a preheated pan. When the bottom and sides are set, you finish the cooking in the oven. At what point do you think you should add fillings and garnishes to a frittata? Why?

- **American omelets** are also called flat omelets. Instead of stirring constantly, you push the eggs away from the bottom and sides from time to time. The result is larger curds, deep wrinkles, and a bit of browning on the outer layer. The fillings and garnishes may be folded into the eggs, stuffed in the middle, or placed on top. The omelet is folded in half on the plate.
- **Shirred eggs** are made by cracking a whole egg into a cup and then baking it in the oven at a moderate temperature. They are typically covered with a little cream and topped with breadcrumbs before baking. They have firm whites and soft yolks.
- A **quiche** is a baked egg dish that combines eggs with cream or milk to make a custard. The custard is poured into a crust and baked until it is fully cooked. Most quiche recipes include a filling such as cheese, sautéed potatoes or onions, vegetables, or ham.
- A **soufflé** is a light, puffed egg dish baked in a ceramic dish.

Utility Drawer

Use a nonstick pan for cooking eggs. They are easy to use and to clean and require very little fat. Some pans are made specifically for cooking omelets, but any round pan will do.

What's Cooking?

Egg white foam is used in many dishes, such as soufflés, meringues, and angel food cakes. It is made by beating air into egg whites until they form a stable foam. The air in the foam expands when heated, causing the egg mixture or batter to expand in volume until the egg white is heated enough to coagulate. The result is a light porous texture.

Fiction	Fact
Eggs cause heart attacks.	A 2007 study by the Egg Nutrition Center in the U.S. confimed that eating one or more eggs a day did *not* increase the risk of heart attack or stroke. The British Nutrition Foundation confirmed the same finding. In fact, many countries with high egg consumption are notable for low rates of heart disease according to a study at Wake Forest University.

BASIC CULINARY SKILLS

Poaching Eggs

1. Bring a pot of water to 165° F. Add vinegar if desired.
2. Crack each egg into a cup.
3. Slide the eggs into the water.

4. Poach the eggs until the white is set—approximately 3–4 minutes.

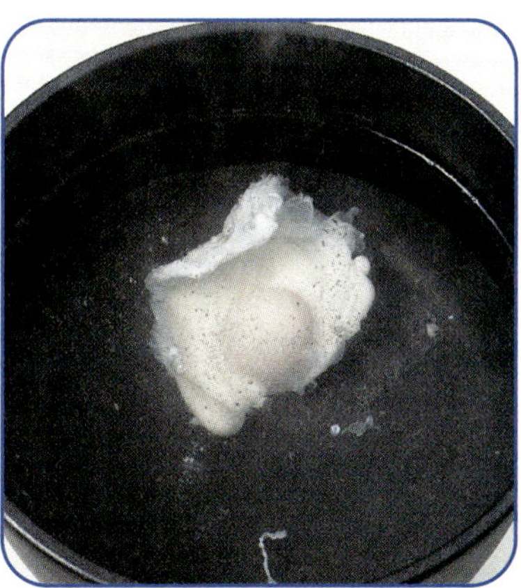

5. Cook to the desired doneness.
6. Lift the eggs from the water with a slotted spoon or spatula.

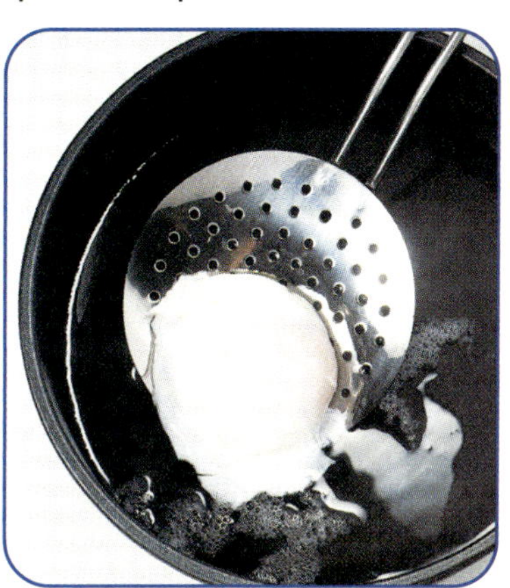

7. Blot the eggs on a paper towel.
8. Serve immediately.

Hot Topics

Beverages made from dairy products can be a nutritious and delicious treat. **Smoothies** usually combine yogurt and fruit juice; **milkshakes** blend milk, flavoring, and ice cream. An **egg cream** is a classic beverage that contains neither cream, nor eggs. It combines milk, syrup, and **seltzer** (carbonated water).

BASIC CULINARY SKILLS

Frying Eggs

1. Heat the pan and the cooking fat over moderate heat.
2. Break eggs into a cup, then slide them into the hot fat. Season if desired.
3. Baste the eggs if desired.

4. Cook to desired doneness.

5. For over eggs, flip them to cook on the second side.
6. Serve immediately.

BASIC CULINARY SKILLS

Scrambling Eggs

1. Break eggs into a bowl. Add water or milk if desired.
2. Season to taste with salt and pepper.
3. Beat the eggs with a fork until evenly blended.

4. Heat the pan and the cooking fat over moderate heat.
5. Add the eggs and stir frequently. Cook over lower heat until they are soft and creamy.

6. Serve immediately.

Case Study

After winning the volleyball match against a big rival, the team had a party at the captain's house. In all of the excitement, Jacqui ate an ice cream sundae.

About an hour after eating, Jacqui started feeling terrible stomach cramps, embarrassing gas, and diarrhea. She had to leave the celebration before anyone else.

Jacqui's teammates were angry when she showed up for practice the next day. They worried that they might catch whatever had made her sick.

She explained that she is lactose intolerant, which means she can only eat a small amount of fresh milk products at a time without having stomach problems. The ice cream had been more than her body could handle.

- What could Jacqui have done differently at the party?
- Were her teammates right to worry that they might get sick, too?
- What other foods does Jacqui need to avoid?

Put It to Use

1. You can add vinegar to egg-poaching water to encourage the white to set. Try poaching an egg without vinegar. Time how long it takes the white to set. Then, add a small amount of vinegar to the poaching water and poach another egg. See if the white sets more quickly. Is the time difference important? Be sure to use only about 2 tablespoons of vinegar per gallon of water, or the eggs will pick up the sour flavor of the vinegar.

Put It to Use

❷ Compare the whipping qualities of different types of cream. Use whipping cream, heavy cream, light cream, and half and half to see which whips more easily, and creates a better result. Discuss the results with the class.

Put It to Use

❸ Imagine that you are hosting a breakfast. One guest is lactose intolerant, and another is on a doctor prescribed low-fat, low-cholesterol diet. Plan a menu for the breakfast that everyone will be able to enjoy.

Write Now

In the United States, yogurt is usually eaten plain or with fruit. In other parts of the world, it is a main ingredient in many recipes, including soups, sauces, and salad dressings. Select a country that uses yogurt in a traditional recipe. Write a cookbook page for the recipe, including the recipe itself, and information such as its history, cultural significance (if any), and use.

Tech Connect

Do cows contribute to global warming? Many scientists say they do. Cows produce a lot of methane, a so-called greenhouse gas. It becomes an issue of sustainability, because global warming could result in less rain, which leads to less feed, which could impact the dairy business. Use the Internet to learn about how cows affect the environment, and how scientists and dairy farmers are working to address the problem. Present your findings to the class.

Team Players

Divide the class into teams of four. Two teams should research the pros and cons of one controversial issue regarding dairy products and farming, such as the use of artificial hormones, the safety of milk from cloned cows, the benefits of raw milk, the importance of sustainable dairy farming, etc. After completing the research, each team should select either pro or con, and then conduct a debate in front of the class. Classmates who watch the debate can select a winner.

Put *It* Together

Match the explanation in column 1 with the term in column 2.

Column 1

- **a.** fat in dairy products
- **b.** a black and white dairy cow
- **c.** the process of heating raw milk to 160° F for a minimum of 15 seconds
- **d.** the process of heating raw milk to at least 143° F for 30 minutes
- **e.** the process of forcing milk through fine screens to break up the milkfat into small particles
- **f.** chunks of cheese solids
- **g.** the clear liquid that contains some of the water-soluble substances in milk
- **h.** milk sugar
- **i.** egg whites
- **j.** a dish made primarily of melted cheese
- **k.** a round, open-face omelet

Column 2

1. flash pasteurization
2. lactose
3. albumen
4. Holstein
5. fondue
6. milkfat
7. whey
8. frittata
9. curds
10. homogenization
11. holding pasteurization

TRY IT!

Hard-Cooked Eggs

Yield: 10 Servings Serving Size: 2 Eggs

Ingredients

20 Eggs

Method

1. Place the eggs in a pot.
2. Fill the pot with enough cold water to cover the eggs by 2 inches.
3. Bring the water to a boil and immediately lower the temperature to a simmer. Begin timing the cooking at this point.
4. Simmer small eggs for 12 minutes, medium eggs for 13 minutes, and large eggs for 14 to 15 minutes.
5. Cool the eggs quickly in cool water.

Recipe Categories

Breakfast Cookery, Eggs and Dairy

Chef's Notes

Peel the eggs as soon as possible after cooking.

Potentially Hazardous Foods

Eggs

HACCP

Cook to at least 145° F for at least 15 seconds.

Nutrition

Calories	149
Protein	13 g
Fat	10 g
Carbohydrates	1 g
Sodium	132 mg
Cholesterol	425 mg

TRY IT!

French Toast

Yield: 10 Servings Serving Size: 3 Slices

Ingredients

30 slices of bread, such as Challah or brioche (¼ to ½ inch thick)
32 fl oz Milk
8 Eggs
2 oz (¼ cup + 1½ tsp) Sugar
Pinch Cinnamon, ground
Pinch Nutmeg, ground
To taste Salt
3 fl oz Butter, melted, or vegetable oil

Method

1. Lay the slices of bread in a single layer on sheet pans and allow to dry overnight.
2. Combine the milk, eggs, sugar, cinnamon, nutmeg, and salt.
3. Mix well with a whisk until smooth.
4. Refrigerate until needed.
5. Heat a skillet or nonstick pan over medium heat.
6. Brush with a small amount of butter or vegetable oil.
7. Dip the bread slices into the batter, coating them evenly.
8. Cook the slices on one side until evenly browned.
9. Turn the bread and brown on the other side.
10. Serve at once on heated plates.

Recipe Categories

Breakfast Cookery, Breakfast Foods

Chef's Notes

To dry the bread slices, you could also place them in a 200° F oven for 1 hour.

Potentially Hazardous Foods

Eggs, Dairy

HACCP

Cook to at least 145° F for at least 15 seconds.

Nutrition

Calories	277
Protein	10 g
Fat	11 g
Carbohydrates	34 g
Sodium	403 mg
Cholesterol	132 mg

TRY IT!

Tuna Melt

Yield: 10 Servings Serving Size: 1 Sandwich (8 oz)

Ingredients

1 lb, 8 oz Albacore tuna, canned and drained
10 fl oz Mayonnaise
4 oz (3/4 cup) Celery, minced
4 oz (2/3 cup) Onions, minced
To taste Garlic powder (optional)
2 tsp Worcestershire sauce, or as needed
To taste Salt and white pepper, freshly ground
To taste Dry mustard
20 English muffins, sliced in half crosswise
10 oz Cheddar cheese, sliced
4 oz (1/2 cup) Butter, as needed

Method

1. Flake the tuna and place it in a large bowl.
2. Add the mayonnaise, celery, and onions. Mix well.
3. Season with garlic powder (if using), Worcestershire sauce, salt, pepper, and mustard to taste.
4. Place 4 oz tuna salad on a slice of muffin.
5. Top with cheese and another slice of muffin.
6. Brush the outside of the muffins lightly with butter.
7. Griddle over medium heat until golden brown on both sides.

Recipe Categories

Sandwiches, Appetizers, & Hors d' Oeuvres

Chef's Notes

Serve garnished with salad greens.

Potentially Hazardous Foods

Mayonnaise, Fish

HACCP

Keep ingredients chilled to below 41° F until ready to prepare.

Nutrition

Calories	756
Protein	33 g
Fa	50 g
Carbohydrates	40 g
Sodium	1,032 mg
Cholesterol	89 mg

18 Fruits and Nuts

In This Chapter, You Will . . .

- Identify types of fruits and nuts
- Understand the nutritional value of fruits and nuts
- Learn how to prepare foods with fruits and nuts

Why YOU Need to Know This

With a vast array of colors, flavors, shapes, and textures, fruits and nuts are a fresh, beautiful, and healthy addition to almost any meal. Fruit is loaded with vitamins, minerals, fiber, and antioxidants, which help protect against cancer, heart disease, and Type 2 diabetes. Nuts contain protein, good fat, and nutrients. You can eat fruits and nuts fresh, or use them as a garnish or ingredient in both sweet and savory dishes. Once it was only possible to have fresh fruit and nuts in season, but modern transportation, packaging, refrigeration, and farming technologies make it possible to have them fresh year-round. Nutritionists often suggest eating a rainbow of colors as a way to get a variety of nutrients. Divide the class into four teams and assign each team the color red, green, yellow/orange, or blue/purple. See how many fruits of that color the teams can list in three minutes.

Types of Fruits and Nuts

Botanically speaking, **fruit** is the reproductive organ of a flowering plant. In culinary terms, fruit is usually any sweet, edible, plant product that grows from a **seed**. A seed is the part of a plant that can grow into a new plant. Every fruit contains at least one seed, and some fruits contain hundreds of seeds.

Like fruits, nuts have a different definition botanically than they do in cooking terms. In culinary terms, any large, oily kernel that has a shell and can be used as food is considered a **nut**. Botanically, nuts are the inseparable seed and fruit from a plant. All nuts are seeds, but not all seeds are nuts. Nuts are the seed and the fruit, which cannot be separated. Seeds can be removed from fruit. Table 18-1 lists the most commonly used nuts.

FIGURE 18-1
Peanuts are not technically a nut. They are a seed and grow underground in a pod. How many forms of peanuts can you think of? Which do you think is the most common form?

TABLE 18-1: MOST COMMON NUTS

Picture	Description
	**Chestnut:** Large and round with a high starch content. Chestnuts must be cooked. They are used in both sweet and savory dishes.
	**Hazelnut:** Small and nearly round, with a rich, sweet taste. Commonly used in desserts because it complements chocolate and coffee.
	**Pecan:** Golden brown kernel within a smooth, thin, hard, tan shell. It has a high fat content and a nutty, sweet taste. Often used in sweet dishes and desserts.
	**Almond:** Pale tan with a pitted, woody shell. Bitter almonds must be cooked. Sweet almonds can be used raw or cooked.
	**Pine nut:** Small, cream-colored seeds from a Mediterranean pine tree. They have a rich flavor and a high fat content. They are used in both sweet and savory dishes.
	**Walnut:** Hard wrinkled shell enclosing a nut with two tender sections. Oily, with a mild, sweet flavor, they are used in sweet and savory dishes and eaten as snacks. Different varieties are available.
	**Pistachio:** Pale green in color, it is available whole, shelled, and unshelled. The hard tab shell is sometimes died red. It is used in many desserts, and eaten as a snack.
	**Cashew:** Kidney shaped, sweet, and buttery, with a high fat content. Cashews are always sold shelled, because the skin contains irritating oils similar to those in poison ivy.
	**Macadamia:** Nearly round, with a rich, buttery flavor and a high fat content. They have an extremely hard shell, so are almost always sold shelled

FIGURE 18-2
When you cut an apple in half you can see all of its parts. What parts of the apple are edible? What parts are not edible?

Types of Fruits

Despite the amazing variety, each fruit shares a few common characteristics:

- Every fruit has a **stem end**, which is the place where the fruit was attached to the tree, bush, or vine on which it grew.
- Every fruit has a **blossom end**, which is opposite the stem.
- Fruit has a skin which protects the flesh while the fruit grows and ripens.

Fruits also have differences:

- The number and location of the seeds varies. For example, melons have a lot of seeds located in a center pocket. Fruit such as apples and pears have only a few seeds, located in the core, or center part of the fruit. Some fruits, such as plums, peaches, cherries, and apricots, have a single seed protected by a hard pit, which is sometimes called the stone. Some berries have seeds on the outside of the fruit.
- Fruits come in almost every color of the rainbow and in a multitude of shapes.
- You can eat the skin of some fruit, such as apples, but not others, such as bananas.

Fruits can be organized into the following groups:

- **Pomes** have smooth skin and an enlarged fleshy area surrounding a core. This group includes apples and pears. They grow on trees, have sweet, cream-colored flesh and a core of multiple seeds. Pomes come in many colors. Apples are available in shades of red, green, and yellow. Pears can range from mottled brown to pale green to deep red. Pomes should be eaten when they are firm, with good color. There should be no bruises or soft spots, and the fruit should feel heavy for its size.
- **Berries** grow on bushes, usually low to the ground. They are fragile, with lots of pulp and juice. Common berries are blueberries, strawberries, blackberries, and raspberries. Less common are gooseberries, boysenberries, currants, and cranberries. Berries should have good color, with no bruising or mold. If the packaging is stained with juice, the fruit was probably damaged during transportation. Keep berries as dry as possible until you are ready to use them. Once one berry in the package develops mold, they will all go bad quickly.

FIGURE 18-3
Cranberries are firm and tart. They last a long time and can be frozen. How are cranberries different from other types of berries?

Check the Label

Types of oranges:

Loose skinned oranges have an easy-to-peel skin. Examples include tangerines, tangelos, and clementines.

Sweet oranges, such as navel oranges, are large and easy to eat, and have few, if any, seeds.

Juicing oranges are sweet,have thin skin, lots of juice, and many seeds.

Bitter oranges have heavy rinds and a bitter taste. They are used to make marmalade.

Blood oranges have crimson colored flesh. They are smaller than regular oranges. Sometimes the rind also has a reddish color.

FIGURE 18-4
Grapes are juicy berries that grow in **clusters** on vines or shrubs. How are grapes different from other types of berries?

- **Citrus fruits** grow on trees in warm climates, and have a firm rind and pulpy flesh. Citrus fruits have a bright skin that contains its **essential oils**—the oils that give the fruit its distinct smell and flavor. Just below the outer skin is the pith, which is a white, bitter, indigestible layer. Common types include oranges, grapefruits, lemons, and limes, as well as kumquats, citrons, tangelos, and ugli fruit. Citrus is a good source of vitamin C. Look for good color and aroma, no soft or bruised portions, and no signs of mold.
- **Stone fruit**, or **drupes**, contain a hard pit covering a central seed or kernel. Peaches, cherries, apricots, plums, and nectarines are stone fruit. Peaches and apricots have a fuzzy skin which can be eaten fresh, but is usually removed before cooking.
- **Melons** come in several varieties, differing in size, taste, color, and skin texture. They have a hard outer skin—called the rind—and juicy flesh. Examples include cantaloupe, watermelon, casaba, and honeydew. Good quality melons should be firm and heavy for their size and have a sweet aroma.
- **Tropical** and **exotic** fruit are grown in warm climates. Common examples include bananas, pineapples, avocados, dates, and figs. As more become available in the U.S., a wider variety of tropical and exotic fruits are gaining popularity, including mangos, plantain, papaya, and pomegranates.

Cool Tips

Just because fruit reaches its full size does not mean it is ripe. **Ripe** fruit has developed the brightest color and deepest flavor, sweetness, and aroma. Most fruit starts out green, and becomes more colorful as it ripens. The bright, color is a signal that the fruit is ready to eat.

FIGURE 18-5
Rhubarb is not technically a fruit, but is used like a fruit in cooking. It has red celery-like stalks, is very tart and is usually combined with sugar or other sweet ingredients. Why do you think rhubarb is treated like a fruit in cooking?

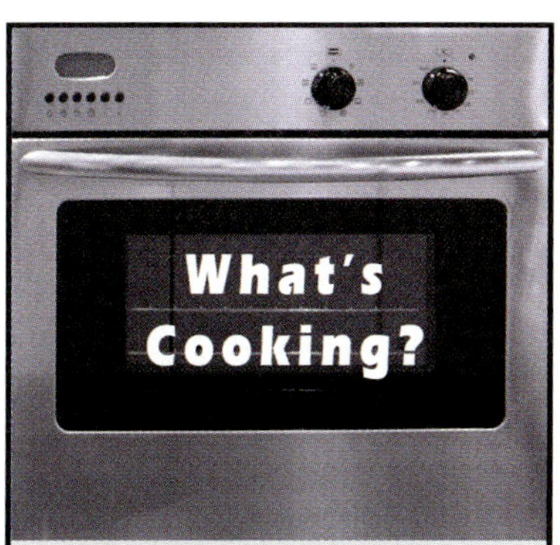

Many foods that are botanically classified as fruits are treated as vegetables in cooking. Usually, it is because they are not sweet. The list includes squash, cucumbers, tomatoes, corn, eggplant, and bell peppers.

Forms of Fruit

FIGURE 18-6
Watermelon has a thick green rind and a juicy red interior. Do you associate watermelon with a particular holiday, season, or event?

Fruit is available fresh or processed. Some fresh fruit is sold in individual pieces—apples and oranges—and some is sold in bunches—bananas and grapes. Some fruit is packaged before sale. For example, berries are usually sold in small containers. Fresh fruit is usually sold whole, but may be cut and packaged.

Processed fruit may be dried, frozen, or canned. Dried, frozen, and canned fruit last longer than fresh fruit because they are processed to control or kill the spoilage bacteria and enzymes.

- Dried fruit is fruit from which most of the water is removed. It has an extremely long shelf life and still contains most of its flavor and sweetness. Fruit that is often sold dried includes raisins, which are dried grapes, prunes, which are dried plums, figs, apricots, and dates. Dried fruit has basically the same nutritional value as the fruit before processing.
- Frozen fruit may be individually quick frozen (IQF), which means the fruit is frozen whole or in slices or chunks, without any added sugar or syrup. It may be sold as a puree or a paste. Some frozen fruit is packed in syrup. Frozen fruit keeps most of its flavor, color, and nutritional value.
- Canned fruit is usually available packed in syrup or in juice. The fruit may be whole, peeled, sliced, or cut, and is usually cooked before canning. If the fruit is not cooked before canning, it retains its nutritional value. However, cooking can destroy some nutrients. Also, liquid and sweeteners are sometimes added during canning, which will affect the nutritional value, flavor, and color of the fruit.

Hot Topics

The *local food movement* is a growing trend that encourages consumers and restaurants to use locally grown food. The idea is that locally grown food is not only fresher, but is also more environmentally friendly than food that must be transported over long distances. Energy is saved, transportation-related pollution is minimized, and small family farms stay in business.

Buying and Storing Fruit

The quality of fresh fruit usually depends on whether you are buying it during its growing season, and how close you live to where it grows. Most fruit has a relatively short growing season, and some fruit is highly perishable, lasting only a few days even under ideal storage conditions.

Although fruit is relatively inexpensive, prices will vary based on availability. Fruit in season is usually cheaper than fruit that is out of season. Prices may also be affected by growing conditions; for example, a freeze in Florida will impact the price of oranges and grapefruit.

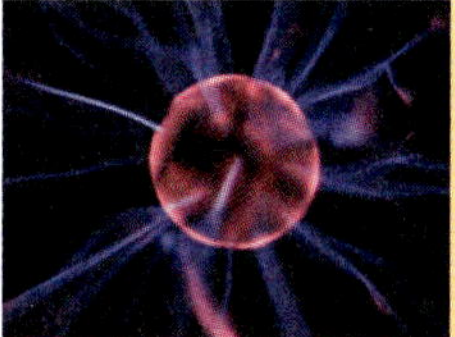

SCIENCE STUDY

When a plant releases the hormone ethylene, it causes the starch in the flesh of the fruit to convert to sugar. The more starch that is converted to sugar, the riper it is.

There's an experiment that will test this. You'll have to order 2% potassium iodide and 2% iodine in order to make an iodine solution. Iodine binds to starch.

1. Start by labeling eight, resealable plastic bags with the numbers 1–8. Bags 1–4 will be the control group. Bags 5–8 will be the test group.
2. Place one unripe pear in each of the control bags. Seal each bag. Place one unripe pear and one banana in each of the test bags. Seal each bag.
3. To make the iodine stain, dissolve 10 g potassium iodide (KI) in 10 ml of water, stir in 2.5 g iodine (I), and dilute the solution with distilled water to make 1.1 liters.
4. Pour the iodine stain into the bottom of a shallow glass or plastic tray, so that it fills the tray about half a centimeter deep.
5. Cut the first pear in half and set the fruit into the tray, with the cut surface in the stain. Allow the fruit to absorb the stain for one minute.
6. Remove the fruit and rinse the face with water. Record your observations of the color of the fruit. Repeat the procedure for the other pears, adding more stain to the tray, as needed.

Is the iodine stain darker on the pears in the control group or the test group? How does the presence of the banana affect the ripening process?

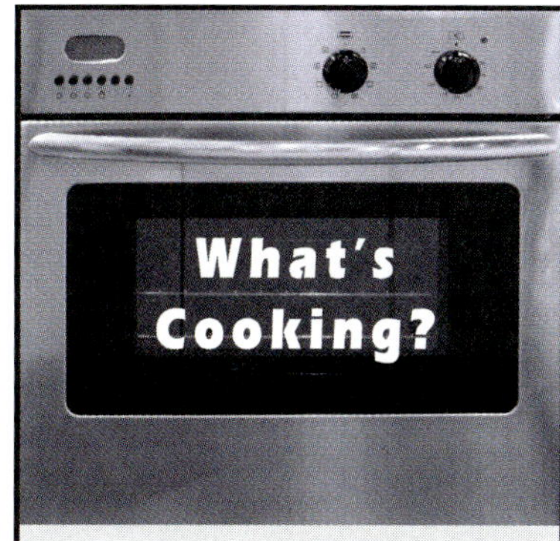

There are two types of stone fruit. **Clingstone** fruit has flesh that clings tightly to the pit, making it difficult to cut the flesh away cleanly. **Freestone** fruit has flesh that separates easily from the pit. Peaches and nectarines come in both clingstone and freestone varieties.

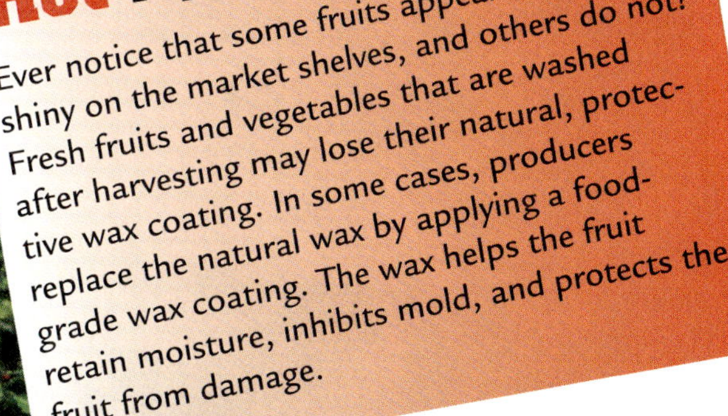

Hot Topics

Ever notice that some fruits appear bright and shiny on the market shelves, and others do not? Fresh fruits and vegetables that are washed after harvesting may lose their natural, protective wax coating. In some cases, producers replace the natural wax by applying a food-grade wax coating. The wax helps the fruit retain moisture, inhibits mold, and protects the fruit from damage.

FIGURE 18-7
For the best quality, purchase fresh fruit directly from the farm. Why do you think fruit purchased from a farm might be better than fruit purchased at a supermarket?

FIGURE 18-8
Bananas are picked mature, but still green. Why do you think bananas are picked before they ripen on the tree?

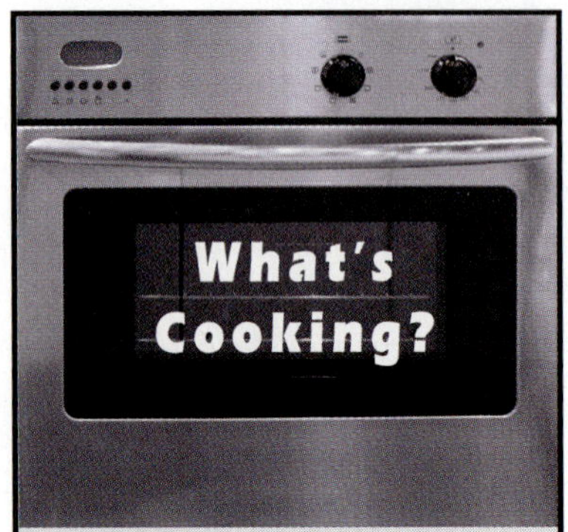

To preserve the quality of fruit, store it properly.

Keep unripe fruit at room temperature to continue ripening. Keep ripe fruit refrigerated to slow down the ripening process. Keep all fruit dry to prevent mold. Ideally, you should purchase fruit just before peak ripeness, and use it as soon as possible.

Ripe fruit tastes better and is more nutritious than unripe fruit. Some fruit must ripen before it is picked, like apples and peaches. Other fruit is picked after **maturing**—reaching its full size, but before fully ripe.

Fruit will become overripe if it is not used at peak ripeness. Overripe fruit becomes soft. It may become black, take on a strong smell, and attract insects. Eventually, it will rot.

Fruit and nuts are graded by the USDA's Agricultural Marketing Service (AMS), but grading has nothing to do with flavor, and everything to do with appearance. Judging is based on size, shape, weight, color, and the presence or absence of defects, such as splits in the skin. There are four basic grading levels for fresh fruit and nuts:

- Fancy is premium quality
- No. 1 is good quality, but not quite as good as Fancy
- No. 2 is medium quality
- No. 3 is standard quality

There are three grading levels for frozen fruit:

- Grade A is equivalent to Fancy
- Grade B is above average quality—also known as Choice
- Grade C is medium quality—also known as Standard

Processed fruit lasts longer than fresh fruit. Store frozen fruit in the freezer until you are ready to use it. Keep canned and dried fruit in a cool, dry location. Once you open the package of dried fruit, close it tightly or transfer the fruit to a container with a tight-fitting lid.

Cool Tips

When it's winter in the United States, it is summer in the southern hemisphere. That's why you are likely to find summer fruits imported from South American countries, such as Brazil, Peru, and Argentina in your local markets during December, January, and February.

Some fruit, such as apples and avocados, oxidize as soon as they are cut and exposed to air. Oxidation causes the flesh to turn brown. To keep fruit from oxidizing, cover it with **acidulated** water, which is water to which acid is added.

Make acidulated water by adding 3 tablespoons of an acidic citric juice such lemon, orange, or lime juice to one quart of water.

Buying and Storing Nuts

Nuts come in the shell or **shelled**, which means without the shell. They are available uncooked (raw), roasted, or blanched. Shelled nuts are available whole, halved, sliced, slivered, or chopped.

When you are buying nuts in the shell, choose clean nuts that have no splits, cracks, stains, mold, or holes. Look for shelled nuts that are plump and fairly uniform in color and size. Limp, rubbery, dark, or shriveled nuts are probably stale. Before buying nuts in the shell, feel free to give them a shake. **Nutmeats**—the edible part of the nut—that rattle in their shells are usually stale.

Always store nuts in a cool, dry, dark storage area. The shelf-life varies depending on the form and packaging:

- Vacuum-packed nuts will last forever if the package is not opened.
- In the shell, nuts may last for up to six months.
- Raw, unshelled nuts may last up to three months.
- Roasted nuts start to lose their quality after about a month.
- Sliced or chopped nuts have the shortest shelf life, usually no more than three to four weeks. They will keep longer in a sealed container in the freezer.
- Pastes such as almond paste or **tahini**—a paste made from sesame seeds—will keep in an unopened container for several months.
- Some butters or pastes can be kept in the refrigerator or freezer for up to three months.

FIGURE 18-9
Slivered almonds are often used as a topping for sweet and savory dishes. How should you store slivered almonds to keep them fresh?

Cool Tips

Nuts are among the most common food allergens. Some estimates claim close to 3 million Americans are allergic to nuts. The government requires labels on packaged foods containing common allergens, but not on fresh cooked food at institutions, including restaurants. There is a growing trend among food providers to list ingredients on menus and signs so that people who have allergies can make informed decisions about what to order and eat.

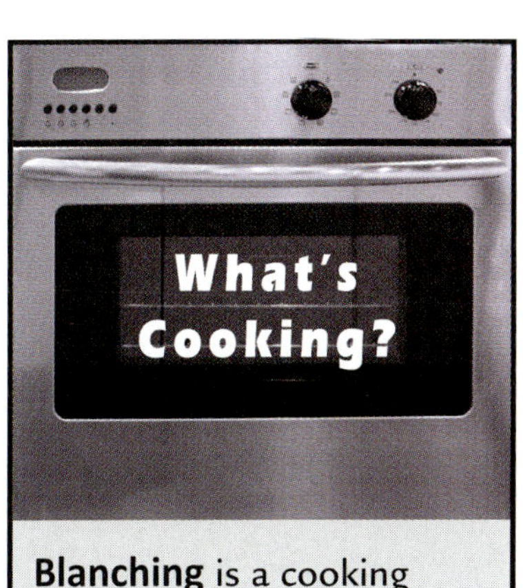

Blanching is a cooking method in which food such as nuts, fruits, or vegetables are cooked quickly in boiling water and then rapidly cooled.

Nutritional Value of Fruits and Nuts

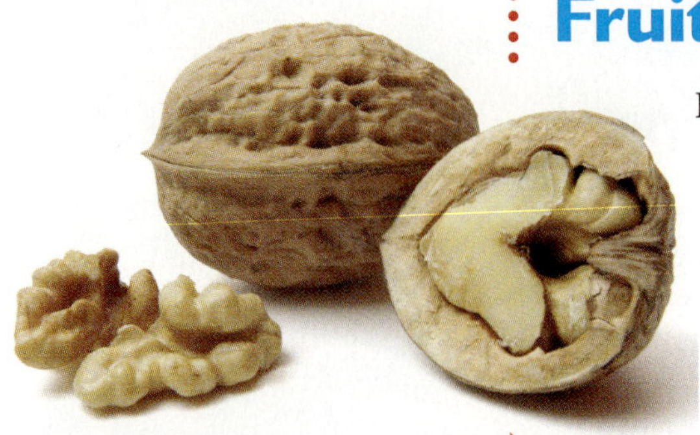

FIGURE 18-10
Walnuts are rich in omega-3 fatty acids. Why do you think nuts are part of the Meat and Beans group on the Food Pyramid?

Fruits and nuts are highly nutritious. Fruits have their own space on the food pyramid, and nuts are part of the Meat and Beans group.

Fruits

Fruits are an excellent and delicious source of vital nutrients. According to the food pyramid, you should eat about 1.5 cups of fruit a day.

Fruits are naturally low in calories, have little or no fats, and are high in essential minerals and vitamins, especially vitamins A and C. They also supply significant dietary fiber. In fact, the high water content of fruit, combined with the high fiber content, makes us feel full and satisfied. Of course, not all fruits contain the same vitamins and minerals, or the same amounts, and raw fruit is usually more nutritious than fruit that has been cooked or processed.

Hot Topics

Water is critical to good health.

- It dissolves vitamins and minerals so they can travel through the bloodstream.
- It cushions joints, organs, and sensitive tissues.
- It helps regulate temperature.
- It helps give cells their shape and stability.

People must replenish water daily, either by drinking it directly or by eating foods that contain water. Fruits are generally 80–95% water, so eating fruit is a sweet, healthy way to contribute to your daily intake of water.

FIGURE 18-11
Apricots are a deep golden yellow when ripe. What can you tell about their nutritional value from looking at their color?

Fiction	Fact
Peeling the skin off fruit before eating it is safer and healthier.	In many fruits, the nutrients are stored in the layer just below the skin. In order to consume the most nutrients, you should eat the skin. Of course, you cannot eat the skin of all fruits—think of bananas—and in some cases, the skin may contain dirt and chemicals.

Nuts

Most nuts are highly nutritious. They have a high fat content, which makes them an excellent energy source. They are part of the Meat and Beans group on the food pyramid, because of their high protein content. They contain many vitamins and minerals.

Even though nuts are generally high in fat, it's mostly monounsaturated. That means nuts can help lower LDL—bad cholesterol, without lowering HDL—good cholesterol.

Table 18-2 lists the nutritional values for a 1-oz serving of many of the most commonly eaten nuts.

TABLE 18-2: NUTRITIONAL VALUES OF COMMON NUTS

Nut	Nutritional Value of 1 oz
Almonds	6 g protein 160 calories 9 g monounsaturated fat Loaded with vitamin E and magnesium
Brazil nuts	4 g protein 190 calories 7 g monounsaturated fat Packed with selenium and phosphorus
Cashews	4 g protein 160 calories 8 g monounsaturated fat Rich in selenium, magnesium, phosphorus, and iron
Hazelnuts	4 g protein 180 calories 3 g monounsaturated fat Large amounts of vitamin E
Peanuts	7 g protein 170 calories 7 g monounsaturated fat Good source of vitamin B_3, vitamin E, zinc, potassium, and vitamin B_6
Pecans	3 g protein 200 calories 12 g monounsaturated fat Packed with vitamin B_1 and zinc
Pistachios	6 g protein 160 calories 7 g monounsaturated fat Full of phosphorus
Walnuts	4 g protein 190 calories 2.5 g monounsaturated fat Rich in omega-3 fatty acids

Check the Label

Vitamin C Most fruits: citrus fruits, cantaloupes, and strawberries provide the most Vitamin C

Vitamin A Yellow fruits—or those ranging from dark yellow to red—can help the body make vitamin A, because they contain beta carotene: examples include melons, peaches, and apricots

Vitamin E Avocados

Niacin Peaches, dates, avocados

B Vitamins Most fruits, although not in significant amounts

Iron and Calcium Oranges, strawberries, and cantaloupes

Potassium Bananas, peaches, dates, raisins, watermelon

Magnesium Bananas, dates, oranges, raisins, watermelon

FIGURE 18-12
Chestnuts are the only low-fat nut with 1 gram fat and 70 calories in 1 oz of dried or roasted nuts. Nut fat is mostly monounsaturated. Does that mean everyone should eat lots of nuts, all the time?

Preparing Foods with Fruits and Nuts

FIGURE 18-13
According to the FDA, pecans are one of the best sources of natural antioxidants. Why are antioxidants an important part of a healthy diet?

Nuts are often roasted or toasted to bring out their flavor, but they are also enjoyed raw. Because the flavor of nuts varies significantly, different types of nuts combine well with different foods. You can use them in both sweet and savory recipes, as an ingredient or as a topping. Be sure to remove the shells, first.

- Almonds, for example, go well with many different foods, including chicken, rice, fish, garlic, chocolate, vanilla, dates, raisins, cream, apricots, peaches, lemons, broccoli, spinach, berries, and yogurt.
- Walnuts, although closely related to almonds, go well with apples, pears, blue cheese, brie, celery, dates, figs, chocolate, goat cheese, spinach, artichokes, bananas, and pumpkin.
- Cashews are frequently used in cookies and candy.
- Peanuts are common in Asian and Indian cooking.
- Chestnuts are often used in soups and sauces.

Most ripe fruits can be eaten raw. Simply wash and serve them for a nutritious, delicious snack or dessert. Many can be served whole, but may be easier to eat and make a more appealing presentation if you cut them into wedges, slices, chunks, or cubes.

Wash fruit before use. Bacteria and other harmful organisms can live on the skin. Also, the fruit may have come in contact with chemicals, dirt, and insects while growing, being harvested, and during transportation. Wash the fruit under cold running water, and use a brush to scrub the skin.

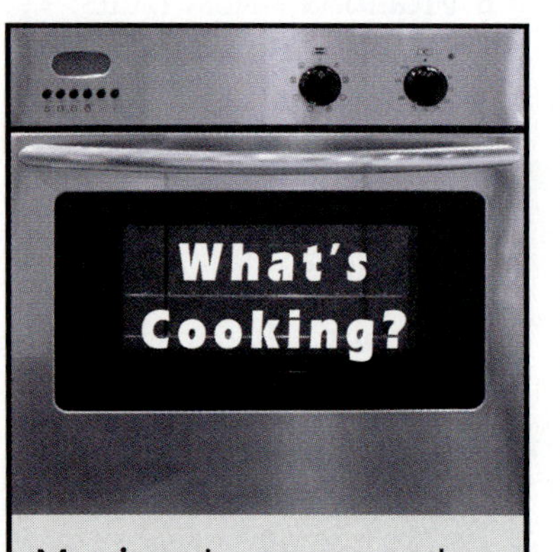

Marzipan is a paste made from almonds, sugar, and water. It can be colored with food coloring, and molded into shapes or used in place of frosting on cakes and other desserts.

FIGURE 18-14
Peanut butter may be smooth or crunchy. What do you think is the difference between the two?

Cool Tips

A variety of nuts are used in Middle Eastern cooking. You will find almonds, walnuts, and pistachios in many dishes.

BASIC CULINARY SKILLS

Toasting Nuts, Seeds, or Spices

1. Shell nuts or seeds.
2. Add to a dry, hot sauté pan.
3. Stir constantly.
4. Toast until aromatic and slightly brown. Do not overcook. Overcooked nuts and seeds become bitter.
5. Transfer to a cool bowl.

FIGURE 18-15
The **radura** is the international symbol for irradiation.

Irradiation is a process in which fruit and other food products are exposed to a source of ionizing radiation, such as electrons or gamma rays. This process kills bacteria and insects, inhibits further ripening of the fruit, and increases shelf-life.

Some people, however, think it may also destroy nutrients such as vitamin E and vitamin C, and mislead consumers into believing the food is safer than it really is.

Cool Tips

The colorful part of the peel on citrus fruit can be grated or cut into thin strips to produce bright, aromatic citrus zest. It makes an excellent seasoning or garnish. The pith, however, is very bitter and should not be eaten.

The FDA requires that irradiated foods include labeling with either the statement "treated with radiation" or "treated by irradiation" and the international symbol for irradiation. However, only whole fruit must be labeled; processed fruit or fruit as an ingredient in other foods does not have to be labeled.

Preparation Skills

To get fruit ready to eat, you may have to remove the skin, core, seeds, stones, and stems.

- To peel fruit, use a peeler or paring knife.
- To remove heavy rinds from melons or pineapples, use a chef's knife. Cut between the rind and the flesh, leaving as little flesh as possible attached to the rind.
- To remove a core from an apple or pear, cut the fruit in half from the stem to the blossom end, and use a melon baller to scoop out the core.

FIGURE 18-16
Use a peeler or paring knife to peel an apple. What other inedible parts of the fruit should you remove?

Utility Drawer

A **reamer** is a small tool with a ridged cone in the center, used to extract juice from citrus fruit. It fits over the top of a container, such as a bowl or measuring cup. You place half the fruit—skin up—over the cone, then twist and press down. The pulp and seeds remain on top of the reamer, while the juice drips through into the container. To make juice from other types of fruit, you should use an electric juice extractor.

- To remove seeds from melons, cut the fruit in half and scoop out the seeds and membranes with a serving spoon.
- To remove seeds from citrus fruit, use the top of a paring knife.
- To remove the stone from cherries, use a cherry pitter.
- To pit plums, peaches, and nectarines, use a paring knife to cut around the fruit, through the skin and flesh, up to the pit. Hold the piece of fruit with both hands and twist the halves in opposite directions.
- To remove the stems of strawberries, use the tip of a paring knife. Cut around the stem, angling your knife toward the center of the berry, to remove just the top and the white part around it.
- To prepare dried fruit for use in a recipe, you may have to rehydrate or soften it. Put it in a bowl, cover it with warm or hot liquid, and let it sit until it is ready to use.

Safe Eats

Wash fruit before using it, and use standard sanitary practices to avoid cross-contamination from fruit to the cutting board, knives, or your hands. Be particularly careful with melons—especially cantaloupes. Scrub the rind with a sanitized brush under running water before cutting.

FIGURE 18-17
Different types of melons create a colorful combination. Can you think of dishes or presentations for which you might use melon balls?

Cooking Fruit

So many fruits, so many recipes. You can cook fruit to create an appetizer, a side dish, or a dessert. You can add it as an ingredient in sweet and savory dishes. You can bake it in a pie, or add it to soup. You can squeeze it and use the juice as a flavoring, or drink it plain. The variations are nearly endless.

You can use a variety of moist and dry heat cooking methods to cook fruit. Use dry heat methods, such as grilling and broiling, sautéing, frying, and baking. Use moist heat methods, such as poaching and stewing.

The cooking method determines the texture of the cooked fruit. Some methods cause the fruit to break apart, and others allow it to keep its shape. Most moist heat methods cause the fruit to break apart; however, if the fruit is cooked in sugar, it will retain its shape.

FIGURE 18-18
Fruit dipped in chocolate is a special treat. What fruits do you think pair well with chocolate?

Serving Fruit

Fresh fruit can be served with any meal or as a snack. Serve fruit at room temperature to make sure it has the best flavor. Other ways to include fruit in meals include the following:

- Use fresh fruit to top cereal or yogurt.
- Fresh fruit, fruit syrups, fruit purees, or other cooked fruit may be used as a topping for dishes, such as French toast or waffles.
- Fruit jam or jelly may be spread on toast, sandwiches, or baked goods.
- Fruit may be an ingredient in baked goods, including muffins, pies, cakes, breads, quick bread, cookies, and scones.
- Stewed fruit may be served as a dessert.
- Fruit may be used as a garnish for many dishes, including desserts, omelets, or cheese boards.
- Fruit sauces may be served as a side dish.
- Baked fruit, such as apples or pears, are often used as desserts.
- Fruit fritters may be served at breakfast or for a snack. A fritter is food coated in batter and deep fried.
- Fruit makes an excellent flavoring for ice cream and sorbet.
- Fruit plates and fruit salads are popular, and may be used at any meal. Fruit cocktails include a mixture of fruit and sweeteners, such as syrup.
- Squeeze fruit to extract the juice for drinking, blending, or using as a flavoring.

Hot Topics

When you use your imagination and creative skills, you can turn your fruits and other foods into edible works of art. For example, create shapes such a spirals or fans from apples, pears, or other fruit to use as a garnish or table decoration. You can blend fruit purees of different flavors and colors, which you can squirt directly on a plate or dish. You can carve melons into bowls or baskets and fill them with other fruits or foods.

BASIC CULINARY SKILLS

Poaching Fruit

1. Prepare the fruit by trimming, peeling, and cutting, as necessary.
2. Simmer the liquid, along with any flavoring ingredients called for in your recipe.

3. Place the fruit into the liquid. Add more liquid, if necessary, to barely cover the fruit.

4. Simmer over low to moderate heat. Allow the liquid to come up to a simmer temperature of 170° F.
5. Poach the fruit until tender and flavorful.
6. Cool the fruit in the cooking liquid, drain, and serve or store.

BASIC CULINARY SKILLS

Puréeing Fruit

1. Prepare the fruit as necessary.
2. Poach the fruit, if your recipe calls for it, with sweeteners and flavorings.
3. Place the fruit in a food processor or blender, or press fruit through a strainer, to purée to the desired consistency.

4. Adjust the flavoring as necessary.
5. Serve or store the purée.

Case Study

On a typical day, 17-year old Nate eats a bagel with cream cheese for breakfast, a peanut butter and jelly sandwich on whole wheat bread for lunch, and a burger with fries for dinner. He rarely drinks soda or eats sweets, gets plenty of exercise, and is of average weight.

Nate's family teases him about only eating food that is flat, brown, and round. His mom keeps fresh fruit in the house, and always serves a vegetable and salad with dinner, but Nate avoids them all. He seems turned off by color and texture in food. "I feel fine," he says. "Don't bug me about it."

Do you think Nate's diet is fine?

- Are there any problems that might arise from only eating food that is flat, brown, and round?
- If Nate really won't eat food with color or texture, what would you recommend he do to improve his diet?

Put It to Use

1. You can quick freeze fresh berries so you have them available to use whenever you need them. Place the berries in a single layer on a sheet tray and freeze them, uncovered, until they are solid. Transfer the frozen berries to freezer containers and keep frozen until you need them. Use them frozen, or allow them to thaw in the fridge before use.

Put It to Use

❷ Visit a local fruit market. Look for fruit that you think has been waxed. Look for fruit that has an irradiation label. Compare the prices of fruit that is in season, and fruit that is out of season. Take note of where the fruit comes from—is it local? Grown in the United States? Is it imported? How does the season and country of origin impact the price?

Write Now

Imagine you own a small fruit farm. A lot of your customers have started buying their fruit at the supermarket, instead of from you. Write an article or press release about your farm, explaining what you grow, and why people should shop there.

Tech Connect 

If you know the seasons when fruits are available, you can plan menus accordingly. Use the Internet to uncover information about when fruits are in season, and where. What type of fresh fruits would you have available for a wedding in June in New England? How about for a brunch in Florida in January?

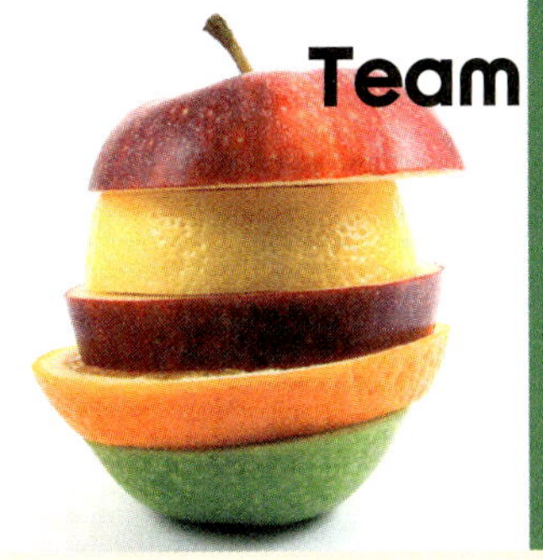

Team Players

In teams of three or four, create beautiful, edible fruit sculptures. Work together to plan and design the sculpture. Consider color, texture, shape, and flavor. Assign tasks to each member of the team. When the sculpture is complete, present it for evaluation. Explain your design and how you selected and prepared the fruit. Your classmates, or students from a different class, can judge the sculptures and select a winner.

Put It Together

Match the explanation in column 1 with the term in column 2.

Column 1

a. any sweet, edible, plant product that grows from a seed
b. a pale green nut
c. the place on the fruit where it was attached to the tree, bush, or vine
d. fruits that contain a hard pit covering a central seed or kernel
e a hormone that causes the starch in fruit to convert to sugar
f. water to which acid is added
g. nuts that have been removed from their shells
h. a cooking method in which food is cooked quickly in boiling water, and then cooled rapidly
i. nutmeats
j. a paste made from sesame seeds
k. a process in which fruit is exposed to electrons or gamma rays

Column 2

1. shelled nuts
2. blanching
3. irradiation
4. tahini
5. pistachio
6. the edible part of a nut
7. ethylene
8. drupes
9. stem end
10. fruit
11. aciduated water

TRY IT!

Tropical Fruit Salad

Yield: 10 Servings   Serving Size: 3½ oz

Ingredients

6 oz (1 cup) Mango, diced
6 oz (1 cup) Pineapple, diced
6 oz (1 cup) Melon, diced (or melon balls)
6 oz (1 cup) Papaya, diced
3 fl oz Orange juice
6 oz (1 cup) Bananas, sliced
2 oz (¾ cup) Coconut, shredded, unsweetened

Method

1. Toss the mango, pineapple, melon, and papaya together with the orange juice.
2. Keep chilled until service time.
3. Arrange the fruit salad on chilled plates and top with bananas. Top with coconut and serve at once.

Recipe Categories

Garde Manger, Salads

Chef's Notes

Cut all fruit as close as possible to service time.

Add guava, passion fruit, star fruit, or other tropical fruits in addition to, or as a substitute for, those listed above.

Potentially Hazardous Foods

Cut Fruits

HACCP

Store cold salads below 41° F.

Nutrition

Calories	70
Protein	1 g
Fat	2 g
Carbohydrates	12 g
Sodium	5 mg
Cholesterol	0 mg

TRY IT!

Applesauce

Yield: 4 cups Serving Size: ½ cup

Ingredients

3 lb (9½ cups) Apples, peeled, cored, and sliced
4 fl oz Apple juice or water
2 Tbsp Sugar (plus more as needed)
⅛ tsp Salt
1 tsp Cinnamon, ground (optional)

Method

1. Combine the apples, apple juice or water, sugar, and salt in a heavy-gauge saucepan.
2. Simmer over low heat until the apples are tender and starting to fall apart, about 10 minutes.
3. Purée the apples and any liquid in the pan by pushing them through a food mill or puréeing in a food processor.
4. Taste and adjust the seasoning with cinnamon (if using) or additional sugar.
5. Use warm or cool.

Recipe Categories

Fruit

Chef's Notes

You can make applesauce that is very smooth or more textured.

Potentially Hazardous Foods

None

HACCP

Cool to below 41° F within 4 hours (1-stage cooling method) or within 6 hours (2-stage cooling method).

Nutrition

Calories	79
Protein	0 g
Fat	0 g
Carbohydrates	21 g
Sodium	3 mg
Cholesterol	0 mg

TRY IT!

Broiled Pineapple

Yield: 8 Serving Size: 4 oz

Ingredients

1 Pineapple, fresh (about 3½ to 4 lb), trimmed, cored, and sliced in half lengthwise
1½ Tbsp Dark brown sugar, packed
1 cup Pineapple juice
⅓ cup Coconut, shredded

Method

1. Preheat broiler.
2. Slice halves of pineapple into halves. You will have 4 slices.
3. Line a 1-qt shallow baking dish with foil.
4. Combine the brown sugar and pineapple juice in a small bowl.
5. Stir the ingredients until the sugar is dissolved.
6. Arrange the pineapple slices in a single layer in the baking dish.
7. Drizzle evenly with the rum mixture.
8. Broil the pineapple 5 inches from the heat until the slices are lightly browned, about 5 to 8 minutes.
9. Sprinkle with the coconut.
10. Continue broiling until the coconut is lightly browned, about 3 to 4 minutes.
11. Plate and drizzle the pineapple slices with the juices from the dish.
12. Serve immediately.

Recipe Categories

Fruit

Chef's Notes

Cut a half pineapple into two slices, each about 2 oz. A serving is two of the 2-oz portions.

Potentially Hazardous Foods

None

HACCP

None

Nutrition

Calories	120
Protein	1 g
Fat	0 g
Carbohydrates	20 g
Sodium	0 mg
Cholesterol	0 mg

TRY IT!

Orange-Blueberry-Banana Smoothie

Yield: 2 cups Serving Size: 2 cups

Ingredients

1 cup Vanilla low-fat yogurt
1 cup Blueberries, washed and patted dry
½ Banana, peeled
½ cup Orange juice
½ cup Ice

Method

1. Combine all ingredients in a blender.
2. Process on high speed until smooth.
3. Serve immediately.

Recipe Categories

Beverages, Breakfast Foods, Fruit

Chef's Notes

You may substitute other fruit in place of the blueberries or strawberries.

You may substitute different juice in place of the orange juice.

Frozen fruit can be used successfully in this recipe. Use them directly from the freezer without thawing.

Potentially Hazardous Foods

Dairy, Blueberries

HACCP

Keep cold ingredients chilled below 41° F.

Nutrition

Calories	380
Protein	13 g
Fat	4 g
Carbohydrates	72 g
Sodium	9 mg
Cholesterol	15 mg

19 Vegetables

In This Chapter, You Will . . .

- Identify types of vegetables
- Understand the nutritional value of vegetables
- Learn how to prepare vegetables
- Learn about vegetarian diets

Why YOU Need to Know This

Vegetables are versatile, colorful, and available in many types and varieties all over the world. You can eat many of them raw, or with minimal cooking, so they are easy to prepare and serve; and they contain both nutrition and flavor. As more people become aware of the importance of including vegetables in a healthy diet, it is useful to know how to create beautiful and delicious vegetable dishes as a main meal or as an accompaniment to any meal.

Pass examples or pictures of uncommon types of vegetables around the classroom. As a class, see how many of them you can identify. Discuss how you might use them in a dish, or as a meal.

Types of Vegetables

FIGURE 19-1
Vegetables add color, texture, flavor, and nutrition to a meal. Can you identify the vegetables shown here?

Vegetables are edible parts of plants. Depending on the plant, we might eat the roots, stems, leaves, flowers, fruits, tubers, or seeds. Some vegetables can be eaten raw, and some must be cooked.

Most vegetables are grouped according to the part of the plant they come from. For example, carrots, radishes, and turnips are roots; peas, beans, and okra are pods and seeds.

Vegetables may be served at any meal, including breakfast. Traditionally, they have been served as a side dish to accompany a main dish, or used as an ingredient. As more people look for healthy, vegetarian alternatives, however, more main dishes are created using vegetables alone.

Vegetables also make excellent garnishes and embellishments for tables, plates, and dishes. They are colorful and come in a variety of textures and shapes. Like fruits, you can carve and slice them into shapes such as fans and flowers, or blend a puree to add color and flavor to a dish.

- **Cabbages:** The cabbage family provides a wide range of vegetables, including cabbages, Brussels sprouts, broccoli, and cauliflower. There are many varieties of cabbages. Red and green cabbage should be heavy for their size and have tightly packed leaves. Other varieties, including Savoy cabbage, bok choy, and Napa cabbage, have looser leaves.
- **Gourds:** The gourd family includes cucumbers, eggplant, and the many varieties of summer and winter squash. Summer squash includes zucchini, yellow

Hot Topics

In cooking, certain vegetables may be grouped by flavor—either strong, such as cabbage and onions; or mild, such as lettuce. They may also be classified based on water content. Vegetables such as celery or broccoli, that have a high water content, are called **juicy**, or **succulent.** Vegetables such as corn and potatoes that have less water and contain more carbohydrates are called **starchy**.

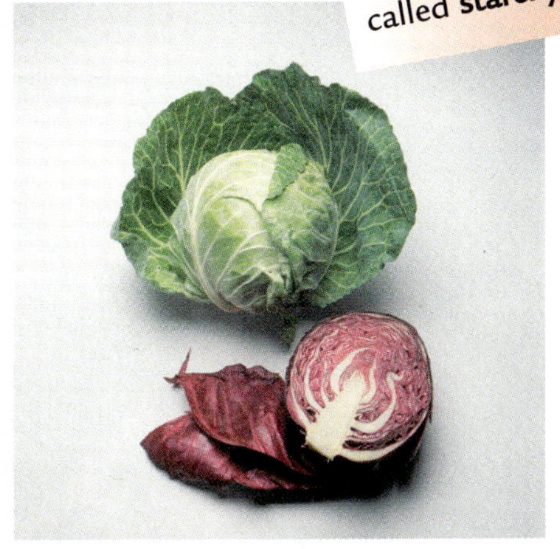

FIGURE 19-2
Bok choy (left) and **red cabbage** (right) are both members of the cabbage family. What differences can you see between the two?

Check the Label

Types of salad greens:

Mild greens Iceberg lettuce, Boston lettuce, Bibb lettuce, Romaine lettuce, and baby spinach

Spicy greens Arugula, mustard leaves, and watercress

Bitter greens Radicchio, Belgian endive, dandelion, escarole, chicory, and frisee

squash, and pattypan squash. They are picked when immature because that is when they are the most tender. Typically, you can eat all parts of these vegetables. Winter squash includes acorn, butternut, and delicata varieties. They have thick, inedible rinds, and large seeds that must be removed before serving. The seeds of some winter squash, such as pumpkins, may be roasted and served as snacks. The flesh is usually yellow.

- **Leafy Greens:** In addition to salad greens, the leafy greens family includes green vegetables used for cooking such as spinach, Swiss chard, turnip greens, collards, and kale. Cooking greens are often sautéed, steamed, or braised.
- **Mushrooms:** There are many types of mushrooms, in a variety of sizes, shapes, colors, and flavors. Cultivated mushroom varieties include white mushrooms, Portobello, cremini, shiitake, and oyster mushrooms. Wild varieties—which may, in fact, be cultivated—include porcini, chanterelles, morels, and truffles.
- **Onions:** The onion family includes garlic, shallots, green—or fresh—onions, and dry—or cured—onions. Leeks, scallions, and chives are green onions. Pearl onions, white onions, shallots, garlic, yellow onions, and red onions are dry onions.
- **Peppers:** The two basic types of peppers are sweet peppers and chiles. Sweet peppers are sometimes called bell peppers because of their shape. They all start out green, and then change colors as they ripen. Varieties of chiles include Anaheim, poblano, jalapeno, cayenne, Scotch bonnet, and habanero. Chiles contain **capsaicin**, a compound that makes them hot, or spicy. The capsaicin is strongest in the seeds. There is actually a scale for measuring the hotness of chiles. The Scoville Scale measures the heat based on how much capsaicin the chile contains. A bell pepper has a zero rating—it contains no capsaicin. A habanero can rate as high as 350,000 units.

Safe Eats

The capsaicin in chiles can burn and irritate. Although it is not permanently harmful, it will hurt if it comes in contact with your eyes, lips, or other sensitive areas. Some people are more sensitive than others.

Wear gloves, wash all cutting surfaces and knives immediately, and wash your hands well with soap and water after handling chiles.

FIGURE 19-3
Mushrooms are usually grown in dirt. You should wash them carefully just before use. Why do you think you should wait until you are ready to use them before you wash them?

- **Pods and Seeds:** Pod and seed vegetables include peas, beans, bean sprouts, corn, and okra. They are usually picked and eaten when they are young because that is when they are most sweet and tender. Some have pods which you can eat, including sugar snap peas, snow peas, green beans, and wax beans. Some have inedible pods; these include green peas, fava beans, and lima beans.

FIGURE 19-4
Fiddleheads are new shoots of an edible fern plant. When they first break through the soil in early spring, they look like the curled spiral handle of a fiddle or violin. Have you ever seen fiddleheads growing? Have you ever eaten a fiddlehead?

- **Root Vegetables:** Root vegetables include beets, carrots, parsnips, radishes, and turnips. They grow underground and are rich in sugars, starches, vitamins, and minerals.
- **Shoots and Stalks:** Artichokes, asparagus, celery, fennel, and fiddleheads are examples of shoot and stalk vegetables.
- **Tomatoes:** Although really a fruit, tomatoes are used like vegetables in cooking. They come in hundreds of varieties. They all have juicy flesh, edible seeds, and smooth, shiny skin.

Cool Tips

Some salad greens have greater nutritive value than others. For example, spinach has more nutrients than iceberg lettuce. One clue is the color—the deeper the green, the more nutrients there are.

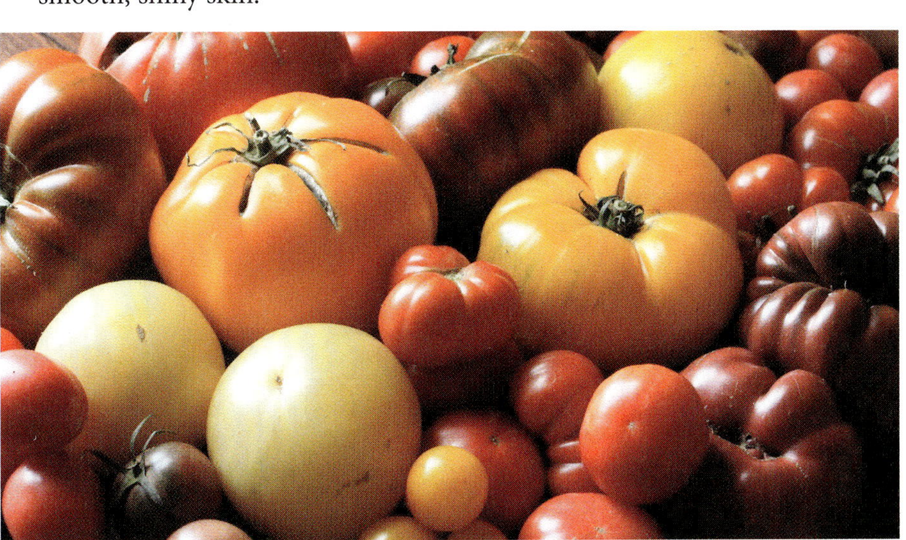

FIGURE 19-5
Heirloom fruits and vegetables are varieties that existed many years ago, before produce was grown for mass-market consumption. The seeds of the plants have been collected and used to grow new plants that are not modified in any way. The results are plants that are true to the original varieties in color, size, texture, and flavor. **Heirloom tomatoes** come in all shapes and sizes. What might some benefits be of serving heirloom tomatoes?

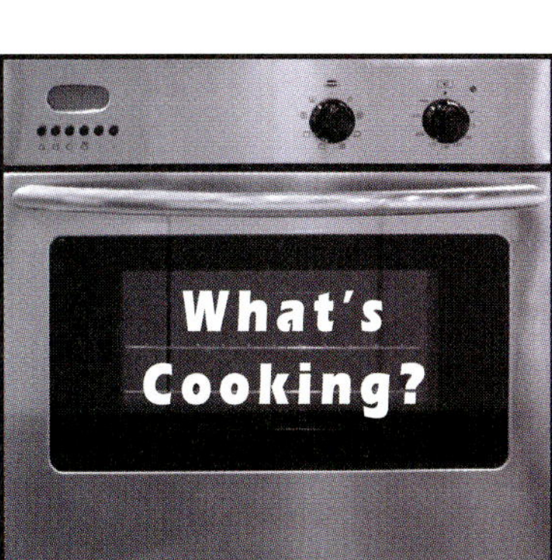

Some foods that we think of as vegetables are really fruits. Recall that a fruit is the reproductive organ of a flowering plant. In culinary terms, fruit is usually any sweet, edible, plant product that grows from a seed. Tomatoes and cucumbers fall into this category. We think of them as vegetables because we use them like vegetables when we cook.

■ **Tubers:** The fleshy portions of certain plants that usually grow underground are called tubers. Potatoes are the most common type of tubers.

Buying Vegetables

Because of modern transportation and refrigeration methods, fresh vegetables are usually available in all areas throughout the year. The cost and quality may vary, however, depending on where they are grown, and how far they have to travel.

Hot Topics

Food miles refer to the amount of energy used and pollution generated by shipping food. By calculating food miles, you can place a value on the environmental impact of shipping food from its source to the consumer. Groups advocating sustainable farming and support of local farms theorize that the fewer miles food travels, the less gas and other fossil fuels are consumed, and the better it is for the environment.

For example, asparagus grown in Argentina in January is likely to cost more than asparagus grown locally—for example, in Wisconsin—in May or June.

Fresh vegetables are ready to eat or cook when purchased. They are sold by weight and count, as well as in boxes, crates, and bags. Some fresh vegetables are peeled, trimmed, and cut before they are packaged and sold. Others are sold whole.

Once picked, the natural sugars in most vegetables start to convert into starch. This makes them less sweet as they age. That's one reason fresh picked vegetables taste better.

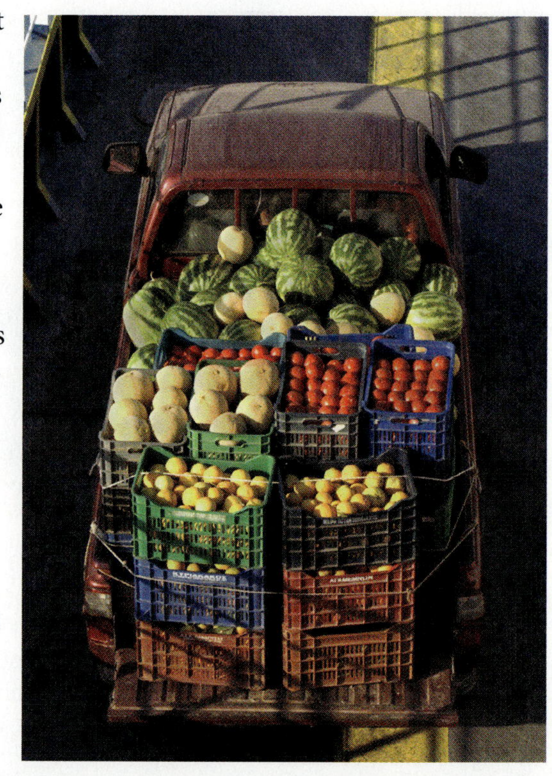

FIGURE 19-6
Fresh vegetables may be shipped in crates or boxes. What do you think a chef should look for before accepting a delivery of fresh vegetables?

Check the Label

Types of tubers:

High-starch/low-moisture potatoes Russet (or Idaho) potatoes. They are dry and granular after cooking and are suitable for baking, pureeing, and mashing. Their low moisture content makes them less likely to splatter or absorb grease during frying. They are also good for scalloped or other casserole-style potato dishes.

Low-starch/high-moisture potatoes Red-skinned potatoes, yellow potatoes, all-purpose potatoes, boiling potatoes, and heirloom varieties, such as purple potatoes, and fingerlings. They hold their shape even after they are cooked. Good for boiling, steaming, sautéing, roasting, and braising or stewing.

Yams and sweet potatoes Often confused, yams and sweet potatoes are different species of plants. Sweet potatoes have tapered ends, deep orange flesh, dense texture, sweet flavor, and thin, smooth skin. They can be handled like low-starch/high-moisture potatoes. Yams are starchier, dryer, and less sweet. They should be handled like high-starch/low moisture potatoes.

Fresh vegetables should be crisp, firm, and sound. They should have good, bright color and be free from bruising, rot, and mold.

Fresh vegetables are graded for quality by the USDA's Agricultural Marketing Service. Like fruit, the grade is based on the color, shape, and size, but not on flavor.

When fresh vegetables are not available, you can purchase them frozen, canned, or dried. The best method to use when trying to assess the quality of frozen, canned, or dried vegetables is to read the label and to familiarize yourself with different brands.

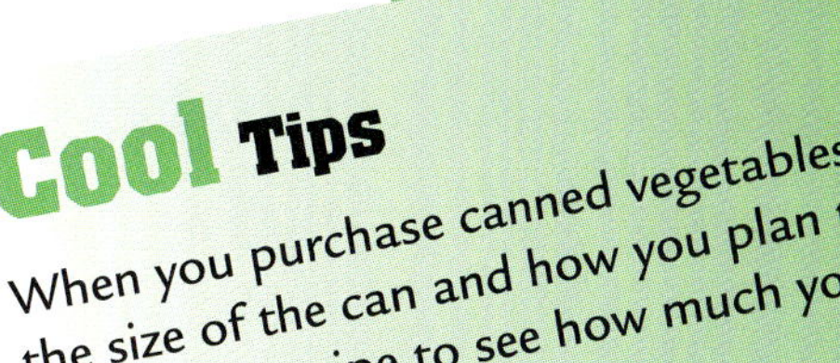

Cool Tips

When you purchase canned vegetables, consider the size of the can and how you plan to use it. Read your recipe to see how much you need before selecting a size.

The U.S. government sets standards for canned foods, and the USDA sets quality grades for canned vegetables. The labeling is voluntary, however, so you might see a grade on the can label, or you might not.

- Grade A, or Fancy, is the top grade
- Grade B, or Extra Standard, is a notch below Fancy
- Grade C, or Standard, and Grade D, Substandard, are unlikely to appear on cans

There are some vegetables that should not be stored in the refrigerator:

- Store whole tomatoes at room temperature.
- Store avocados at room temperature until they ripen.
- Store all tubers in dry storage with good ventilation, away from direct light, heat, and moisture. Separate the dry onions from other items to avoid transferring flavors.
- Store winter squash in a cool, dark place.

Storing Fresh Vegetables

Proper storage insures that vegetables will remain fresh and flavorful until you are ready to use them. Most fresh vegetables, and any vegetables that have been trimmed, peeled, or cut, are highly perishable. They should be kept in the refrigerator, wrapped loosely to keep them from getting too wet.

Vegetables will lose vitamin C if they are not stored properly; and the color, flavor and texture will deteriorate. Most vegetables taste and look their best right when they are harvested.

Storing Canned, Frozen, and Dried Vegetables

Store canned vegetables at room temperature or in a cool, dry location. Some may have a Use by date on the can or label. You should discard the can after the Use by date.

Frozen vegetables should be kept in the freezer at 0° F or below. If they thaw, you should use them immediately. Do not thaw and then refreeze the vegetables, because the quality will deteriorate.

FIGURE 19-7
Most vegetables are available canned or frozen. What are some benefits and drawbacks of using canned or frozen vegetables instead of fresh?

FIGURE 19-8
Carrots are rich in carotene, which your body can process into vitamin A. How can you identify vegetables that have a lot of carotene?

Dried and dehydrated vegetables may be stored at room temperature in their original packaging. Once the package is opened, they should be kept in tightly sealed containers to prevent insects from getting into them.

Nutritional Value of Vegetables

Vegetables are highly nutritious. Like fruits, they have their own space on the food pyramid. The recommended daily allowance is 2.5 cups of vegetables a day.

Most vegetables are naturally low in calories, have little or no fats, and are high in essential minerals and vitamins. Most vegetables contain only small amounts of incomplete protein. Different types of vegetables provide different nutrients.

- Leafy greens and deep yellow vegetables are excellent sources of beta-carotene, which becomes vitamin A. The deeper the yellow and the darker the green, the higher the content of beta-carotene. Spinach, broccoli, carrots, and sweet potatoes are all excellent sources of beta-carotene.
- Leafy green vegetables are also a good source of vitamin C. So are broccoli, green peppers, tomatoes, and raw cabbage.
- Seed vegetables such as lima beans and peas are good sources of vitamin B.
- Leafy green vegetables are an excellent source of calcium, iron, and folic acid.

Fiction	Fact
You must eat veggies raw in order to get all the nutritional value.	Scientists say too much heat and water cause veggies to lose nutrients during cooking. You can minimize nutrient loss by cooking veggies quickly, and with little or no water. That means you should use steaming or stir-frying instead of boiling.

Safe Eats

Fresh vegetables can harbor bacteria and pathogens that cause illness. Follow standard sanitary practices to minimize risk:

- Wash all vegetables with cool tap water immediately before eating or using.
- Cut away damaged areas before eating.
- Wash hands, utensils, and working surfaces frequently, with hot soapy water, and sanitize them after working with fresh vegetables.

- Most vegetables are low in carbohydrates. Exceptions include tubers and seeds, which are a good source of starch.
- The skin and pulp of most vegetables provides cellulose—or dietary fiber.

FIGURE 19-9
Vegetables are usually served raw in salads. What vegetables do you like in a salad?

Preparing Vegetables

Some vegetables are served raw, while others are cooked. When cooked right, vegetables are beautiful, delicious, and nutritious. When cooked wrong, they are drab, lifeless, and lose their nutrients. The goal is to preserve color, texture, flavor, and nutrients.

The first step in preparing vegetables is always to clean them properly. You can then cut and trim them as needed.

Wash vegetables thoroughly before use in cool running water, even if you plan to cook them. Washing removes surface dirt, bacteria, insects, and chemicals.

Pay particular attention to leafy vegetables, to those that have stalks, and those that have roots. They almost always trap dirt between the leaves and the stalk, and along the root.

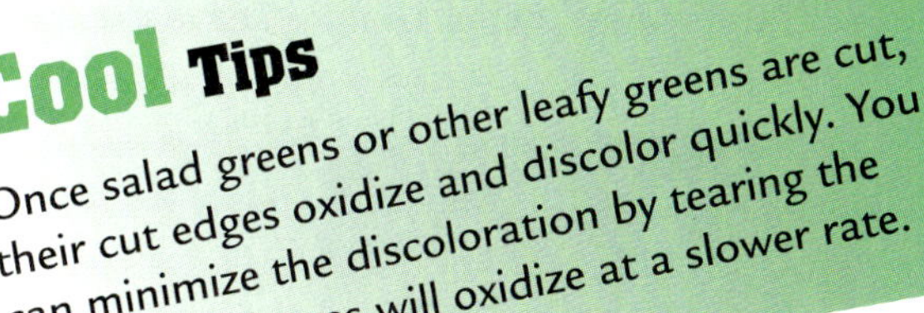

Preparation Skills

- Some vegetables are peeled before cooking. Use a swivel-bladed peeler, a paring knife, or a chef's knife.
- Remove woody stems from mushrooms, asparagus, artichokes, and broccoli.
- Peel onions and garlic by removing the dried outer layers of skin to reveal the moist inner layers.

BY THE NUMBERS

Does the way you chop a vegetable affect the volume?

Start with three carrots weighing 4 ounces each. Small chop—or dice—the first, medium chop the second, and large chop the third. Measure and record the volume of each, using a measuring cup. Discuss the results as a class.

FIGURE 19-10
Onions are usually diced, chopped, or minced before cooking. Can you think of any vegetables that are usually cooked whole?

BASIC CULINARY SKILLS

Trimming and Dicing Onions

1. Cut away a thin slice from the stem and root ends of the bulb with a paring knife, making a flat surface on both ends.
2. Pull away the peel by catching it between your thumb and the flat side of your blade.

3. Trim away any brown spots from the underlying layers.
4. Cut the onion in half from the root end to the stem end. Lay half the onion, cut side down, on the cutting board.
5. Make evenly spaced cuts, running lengthwise, with the top of a chef's knife. Leave the root end intact.

6. Make two or three horizontal cuts parallel to the work surface, from the stem end to the root end, but do not cut all the way through.

7. Make even, crosswise cuts, working from stem end up to the root, cutting through all layers of the onion.

FIGURE 19-11
Tomato concassé is a technique used to prepare fresh tomatoes for cooking. It involves peeling, seeding, and dicing the tomato. What types of recipes do you think tomato concassé would be used for? Can you think of any alternatives to this technique that might be used instead?

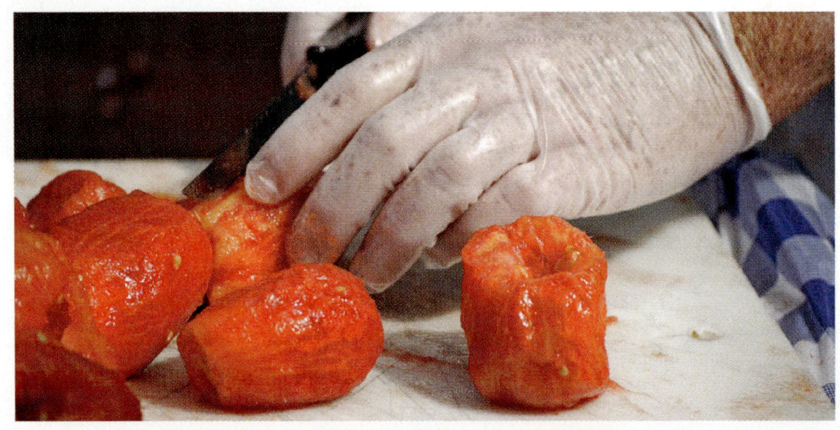

BASIC CULINARY SKILLS

Trimming and Mincing Garlic

1. Separate garlic cloves by wrapping an entire head of garlic in a towel and pressing down on the top.

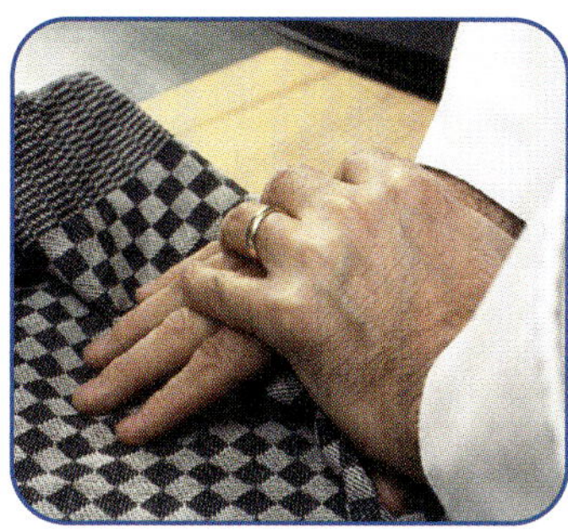

2. Loosen the skin from each clove by placing a clove on the cutting board, placing the flat side of the blade on top, and hitting the blade with fist or the heel of your hand.

3. Peel off the skin and remove the root end and any brown spots. If the clove has sprouted, split it in half and remove the sprout.

4. Crush the cloves by laying them in the cutting board and using the same technique as for loosening the skin, but this time apply more force.

5. Mince the cloves with the blade of your chef's knife using a rocking motion. To mash the garlic, hold the knife blade nearly flat against the garlic and press down.

Check the Label

Dice Cut into uniform pieces—usually ⅛ to ¼ inch on all sides

Cube Cut into uniform pieces—usually ½ inch on all sides

Chop Cut into irregular pieces; may be fine, medium, or coarse

Mince Chop into tiny irregular pieces

Slice Cut into flat, thin pieces

Julienne Cut into long, thin pieces—usually 2 inches long and ⅛ inch wide

Cooking Techniques

Cooked vegetables are served at any meal and sometimes used as an ingredient in other dishes. Almost all cooking methods can be used with vegetables. Most vegetables should be served immediately after cooking.

- Boiling and steaming are common cooking methods for many types of vegetables. The amount of water and the cooking time should be kept to a minimum in order to preserve nutrients.
- Some vegetables, such as potatoes and squash, can be baked in their own skin. Baking preserves water-soluble vitamins that might be lost using moist-heat cooking methods.
- Roasted vegetables have a deep flavor which results from cooking in the dry environment of the oven. It is often used for squashes, yams, eggplant, potatoes, and beets.
- Frying and pan-frying are common cooking methods for vegetables. French fries are deep fried potatoes; **vegetable tempura** is an Asian dish of deep fried vegetables. Eggplant slices are often pan-fried. When you pan-fry vegetables, you usually apply a bread crumb coating, or use flour or batter.
- Stir-frying and sautéing are both good methods for cooking high moisture vegetables. Stir-frying is commonly used for vegetables in Asian cooking. The veggies are usually sliced thin before cooking. Start vegetables that require a longer cooking time first. For example, start with red peppers and cabbage stalks, followed by snow peas, scallions, and cabbage greens.
- You can broil or grill many types of vegetables. For best results, brush them with oil before cooking.
- Stewing and braising lets vegetables cook in their own juices. Some recipes call for one type of veggie; others for a combination. The veggies should be fork tender, or in some cases, meltingly soft.

Properly steamed and boiled vegetables have vivid colors and identifiable, fresh flavors. Most should be simmered, not cooked at a full boil, and drained completely before serving.

To preserve the color of bright-green vegetables during boiling or steaming, leave the cover off for the first few minutes.

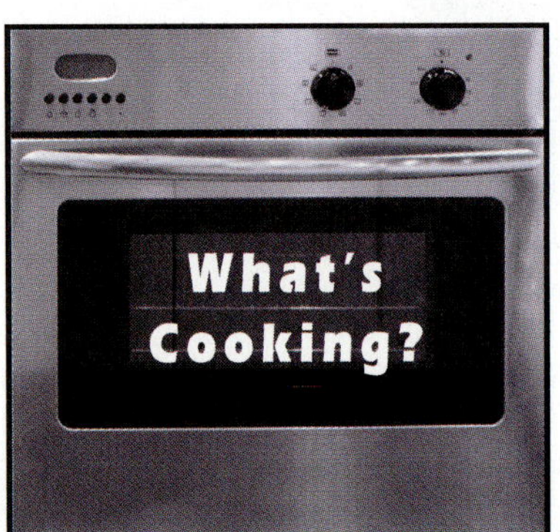

Crudités are raw vegetables cut into bite-sized pieces and served with a dip, such as a vinaigrette. Typically, the crudités are arranged on a platter around the dip, and served as an appetizer or accompaniment to a meal. Common vegetables used as crudités are carrots, celery, bell peppers, cauliflower, and broccoli.

FIGURE 19-12
French fried potatoes are deep fried. Can you think of other ways to prepare potatoes?

When you cook vegetables, there are many considerations that determine "doneness." The recipe, the vegetable's characteristics, and personal preference all factor into how long you should cook the food.

Be careful not to overcook vegetables. Too much heat and water cause vegetables to lose their nutrients, their color, and their flavor. Overcooked veggies also tend to be mushy.

Safe Eats

Before roasting or baking vegetables whole, you should pierce the skin to allow steam to escape. Otherwise, they might burst in the oven, which, in addition to ruining the dish, makes a colossal mess.

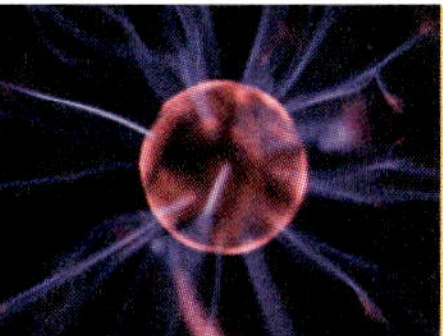

SCIENCE STUDY

Vegetables become soft during cooking because the cellulose structure of the plant softens in the presence of water and heat.

You can test this by cooking samples of the same vegetable for different amounts of time.

Assemble four bunches of broccoli of the same weight.

Steam the first batch for two minutes, the second batch for five minutes, the third batch for eight minutes, and the fourth batch for twelve minutes.

Which batch is softest? Which is hardest? How is the color affected? How is the flavor affected?

Check the Label

Levels of doneness:

Blanched Vegetables are blanched for 30 seconds to 1 minute. It is appropriate for vegetables served cold, or for those that will complete cooking in a separate process.

Parcooked/parboiled Vegetables are cooked to partial doneness to prepare them for grilling, sautéing, or stewing.

Tender-crisp Vegetables are cooked until they can be bitten into easily, but still offer a slight resistance and a sense of texture.

Fully cooked Vegetables are quite tender, although they should retain their shape and color.

FIGURE 19-13
Vegetables such as carrots and tomatoes can be juiced to create beverages. Have you ever tasted carrot juice? Would you like to?

BASIC CULINARY SKILLS

Making Vegetable Braises or Stews

1. Cook the aromatic vegetables in a cooking fat, beginning with members of the onion family.

2. Add the remaining ingredients as specified by your recipe, stirring as necessary.

3. Adjust seasoning and consistency of the dish as needed.
4. Stew or braise the vegetables until flavorful, fully cooked, and fork tender.

5. Serve immediately, or hold for future use.

BASIC CULINARY SKILLS

Deep Frying Vegetables

1. Heat the oil in a deep fryer or kettle.
2. Add the vegetables to the hot oil, using a basket or tongs.

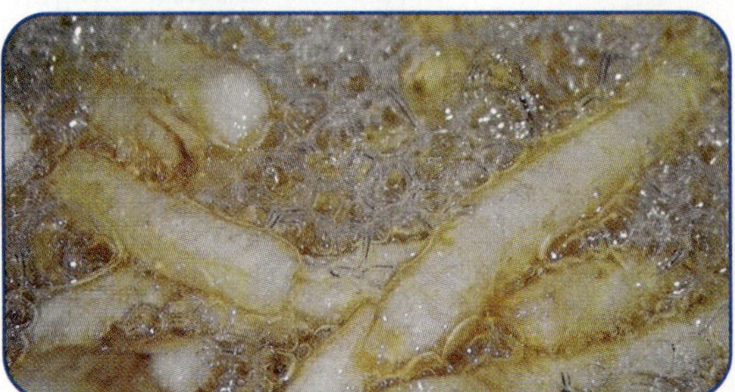

3. Fry the vegetables until fully cooked.

BASIC CULINARY SKILLS

Baking a Potato

1. Preheat oven to 425° F.
2. Scrub potato under cold water and blot dry.
3. Season outside with oil, salt, and pepper.
4. Pierce skin in one or more places.
5. Bake for 1 hour, or until done. To test for doneness, insert a fork into the potato. The skin should be crisp and the interior should be tender.

Vegetarian Diets

A **vegetarian** is someone who does not eat meat, poultry, or fish. Most vegetarians eat fruit, vegetables, legumes, grains, seeds, and nuts. Many vegetarians eat eggs and/or dairy products, but avoid hidden animal products such as beef and chicken stocks, lard, and gelatin.

Some people become vegetarians for personal health reasons, spiritual or religious reasons, or for environmental, economic, and world hunger concerns. Others feel compassion for animals or believe in nonviolence. Still others simply prefer a vegetarian diet.

Because a healthy vegetarian diet is high in fiber and low in fat, most vegetarians are at a lower risk for developing heart disease, certain cancers, diabetes, obesity, and high blood pressure. Even so, a vegetarian diet can be unhealthy and high in fat if it includes excessive amounts of fatty snack foods, fried foods, whole milk dairy products, and eggs.

It is possible to eat a healthy, well-balanced vegetarian diet, but it takes some effort. Vegetarians must take care to make sure they are consuming enough protein, vitamin B_{12}, vitamin D, calcium, phosphorous, and iron. If they cannot consume these vitamins and nutrients, they may have to take a supplement in order to avoid health problems.

There are different types of vegetarians.

- Most vegetarians in the U.S., are ovo-lactovegetarians. **Ovo-lactovegetarians** exclude meat, poultry, and fish, but eat dairy products and eggs.
- **Lactovegetarians** exclude meat, poultry, fish, and eggs, but eat dairy products.
- A **vegan** is someone who excludes all animal products from his or her diet, including dairy products, eggs, and sometimes even honey.

FIGURE 19-14
Ovo-lactovegetarians exclude meat, poultry, and fish, but eat dairy products and eggs. How is this diet different from a vegan diet?

Case Study

Mr. Ross, a high school health teacher in Louisiana, is concerned about the number of students who don't eat the recommended daily 2.5 cups of vegetables.

He met with the school principal and proposed planting a community garden on the school grounds. He hopes it will increase the students' awareness and interest in vegetables, and help promote healthier eating habits.

The principal approved the plan, and Mr. Ross started a gardening club. By the end of the year, the club had 15 members. They were harvesting beans and peas, and were planning to keep the garden growing during the summer, so they could enjoy tomatoes, cucumbers, and peppers.

- Why do you think Mr. Ross is concerned about students not eating enough vegetables?
- Do you think a community garden and a gardening club is a good way to raise awareness and interest in vegetables?
- Do you think the project was successful? Why or why not?
- What problems might you encounter if you tried to do this at your school?

Put It to Use

1. You are going hiking with two friends, and you are responsible for bringing lunch. One friend is a vegan, and the other has celiac disease. Plan a lunch menu that you can bring on the hike that will meet everyone's requirements.

Put It to Use

❷ Traditional recipes and menus usually make use of vegetables that are available locally. Select a country and explore the vegetables that are popular or used in traditional dishes. Select a recipe and write a paragraph comparing it to a recipe that is common in the United States, that uses the same or similar vegetables. Compare the cooking methods, amounts, other ingredients, and nutritional value. If possible prepare the recipes and share them with your class, or collect them into a class cookbook.

Write Now

Do you think the way vegetables are grown, processed, and transported affects the environment? Conduct research to learn more about food miles, and how producing and transporting vegetables might impact the environment. Then, write a letter to the editor of your local newspaper, or to a government representative expressing your opinion on the subject. Include facts and real world examples to support your opinion. Share your letter with the class.

Tech Connect 

A raw food diet is a diet based on unprocessed and uncooked plant foods, such as fresh fruit and vegetables, sprouts, seeds, nuts, grains, beans, dried fruit, and seaweed. Use the Internet to learn about raw food diets. How do they differ from vegan diets? What are the benefits and risks? Prepare an oral report on the subject to deliver to your class.

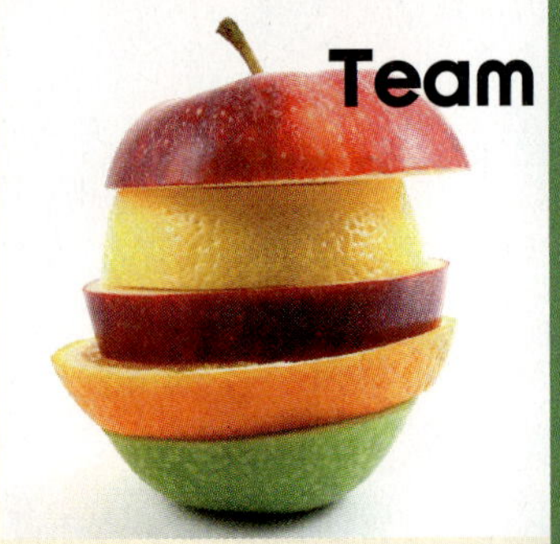

Team Players

The recipe and cooking method you will use helps to determine the type of vegetable you should select. In small teams, randomly select a cooking method/recipe for potatoes. For example, one team might select French Fries; one might select mashed; one might select baked, etc. In teams, visit a supermarket and select the type of potato you think is best for your cooking method/recipe. If you cannot get to the market, find pictures of the type you think is best. Write a paragraph explaining your decision. Discuss your selection as a class with the other teams.

Put It Together

Match the explanation in column 1 with the term in column 2.

Column 1

- **a.** a compound that makes peppers hot
- **b.** vegetable varieties that existed many year ago, and are still grown today
- **c.** the amount of energy used and pollution generated by shipping food
- **d.** cut into uniform pieces of up to 0.25 inches on all sides
- **e.** cut into uniform pieces of about 0.5 inches on all sides
- **f.** chop into tiny irregular pieces
- **g.** cut into pieces 2.0 inches long and 1/8 inch wide
- **h.** someone who excludes all animal products from his or her diet
- **i.** an Asian dish of deep fried vegetables
- **j.** the doneness of vegetables that are cooked until they can be bitten, but offer slight resistance and a sense of texture
- **k.** raw vegetables cut into bite-sized pieces and served with a dip

Column 2

1. julienne
2. crudites
3. vegetable tempura
4. vegan
5. capsaicin
6. food miles
7. cube
8. tender-crisp
9. dice
10. mince
11. heirloom vegetables

TRY IT!

Ratatouille

Yield: 10 Servings   Serving Size: 1 cup

Ingredients

2 fl oz Olive oil
12 oz (3 cups) Onion, medium dice
¾ oz (2 Tbsp) Garlic, minced
1 oz (2 Tbsp) Tomato paste
4 oz (1 cup) Green pepper, medium dice
14 oz (5 cups) Eggplant, medium dice
10 oz (2 cups) Zucchini, medium dice
5 oz (2 cups) Mushrooms, quartered or sliced
7 oz (1 cup) Tomato concassé, medium dice
6 fl oz Chicken or vegetable stock
1 oz (¾ cup) Herbs, fresh, chopped
To taste Salt and black pepper, freshly ground

Method

1. In a large pot, heat the oil over medium heat.
2. Add the onions.
3. Sauté until the onions are translucent, about 4 to 5 minutes.
4. Add the garlic.
5. Sauté until the aroma is apparent, about 1 minute.
6. Turn the heat to medium-low.
7. Add the tomato paste.
8. Cook until paste completely coats the onions and a deeper color is developed, about 1 to 2 minutes.
9. Add the vegetables. Cook each vegetable until it softens before adding the next.
10. Add the stock.
11. Turn the heat to low, allowing the vegetables to stew.
12. Stew until vegetables are tender and flavorful.
13. Adjust the seasoning with fresh herbs, salt, and pepper.
14. Serve immediately or cool and hold for later service

Recipe Categories

Vegetables

Chef's Notes

Add the vegetables in the following sequence: peppers, eggplant, zucchini, mushrooms, and tomatoes.

When stewing, the vegetables should be moist, but not soupy.

If you don't have time to make the tomato concassé, use canned chopped tomatoes.

Potentially Hazardous Foods

Stock

HACCP

Maintain at 135° F during service.

Cool to below 41° F within 4 hours (1-stage cooling method) or within 6 hours (2-stage cooling method).

Nutrition

Calories	115
Protein	4 g
Fat	2 g
Carbohydrates	21 g
Sodium	91 mg
Cholesterol	0 mg

TRY IT!

Glazed Carrots

Yield: 10 Servings Serving Size: 3½ oz

Ingredients

2 oz (¼ cup) Butter, unsalted
2½ lb (8 cups) Carrots, oblique-cut
1⅓ oz (3 Tbsp) Sugar
¼ tsp Salt
⅛ tsp White pepper, freshly ground
10 fl oz Chicken or vegetable stock, hot

Method

1. Melt the butter in a large sauté pan.
2. Add the carrots.
3. Cover the pan.
4. Lightly cook the carrots over medium-low heat for about 2 to 3 minutes.
5. Add the sugar, salt, pepper, and stock.
6. Bring the stock to a simmer.
7. Cover the pan tightly.
8. Cook over low heat until the carrots are almost tender, about 5 minutes.
9. Remove the cover.
10. Continue to simmer until the cooking liquid reduces to a glaze and the carrots are tender, about 2 to 3 minutes.
11. Adjust seasoning with salt and pepper.

Recipe Categories

Vegetables

Chef's Notes

At Step 10, an additional sweetener, such as maple syrup or honey, may be added.

Potentially Hazardous Foods

Stocks

HACCP

Maintain at 135° F during service.

Nutrition

Calories	122
Protein	1 g
Fat	5 g
Carbohydrates	21 g
Sodium	164 mg
Cholesterol	12 mg

TRY IT!

Potatoes au Gratin

Yield: 10 Servings Serving Size: 1 cup

Ingredients

7 Chef's potatoes, large (about 3½ lb)
16 fl oz Heavy cream
8 fl oz Milk
1 tsp Garlic, minced
½ tsp Salt
¼ tsp Black pepper, freshly ground
3 Tbsp Butter, unsalted
5 oz (1¼ cups) Cheddar cheese, grated

Method

1. Peel the potatoes.
2. Slice them very thin (1/16 inch thick) by hand or on a mandolin. Reserve.
3. Combine the cream, milk, garlic, salt, and pepper.
4. Bring the mixture to a simmer.
5. Rub the butter in an even layer on the bottom and sides of a baking dish (10 × 12 inches).
6. Combine the potatoes and the cream mixture.
7. Place the mixture in the buttered pan.
8. Top with grated cheese.
9. Cover the pan with aluminum foil.
10. Bake the potatoes (in a hot-water bath, if desired) at 350° F until nearly tender, about 50 minutes.
11. Uncover and continue to bake until the potatoes are creamy and the cheese is golden brown, another 20 minutes.
12. Remove the potatoes from the oven and let them rest 10 to 15 minutes before slicing into portions.
13. Serve immediately, or hold hot for service.

Recipe Categories

Vegetables

Chef's Notes

Gruyere cheese could be substituted for the cheddar cheese.

Potentially Hazardous Foods

Cooked Potatoes, Dairy, Eggs

HACCP

Cook to an internal temperature of 145° F or higher.

Maintain at 135° F during service.

Cool to below 41° F within 4 hours (1-stage cooling method) or within 6 hours (2-stage cooling method).

Nutrition

Calories	211
Protein	7 g
Fat	9 g
Carbohydrates	26 g
Sodium	821 mg
Cholesterol	29 mg

20 Soups and Casseroles

In This Chapter, You Will . . .

- Identify types of soups and casseroles
- Understand the nutritional value of soups and casseroles
- Learn how to prepare soups and casseroles

Why YOU Need to Know This

Soups and casseroles are hearty foods that combine nutrition, flavor, and convenience in one pot. Because they are made from combinations of ingredients, they can stretch limited supplies to feed more people. They can even be made from leftovers from another meal. Soups and casseroles save time, too. When properly stored, they can be used again; or you can freeze them for another time.

In the folk tale, *Stone Soup*, a beggar convinces an unfriendly town to make a pot of soup. They try to shoo him away, claiming they are poor and have no food to share. He puts a stone in a pot of water, and asks each townsperson to add one vegetable. By the end, they have a hearty soup that feeds them all. What lessons can you learn from the story—both morally and culinary? Discuss how cooking can be a community activity that brings people together.

Types of Soups and Casseroles

Soups and casseroles are made by combining similar or complementary ingredients in one pot or dish. **Soups** are liquid foods served in a bowl and eaten with a spoon. A **casserole** is actually a type of ovenproof dish, as well as the food cooked and served in that dish. Both soups and casseroles are considered "**comfort food**," which means they are basic, familiar foods that give diners a sense of warmth and wellbeing. "Comfort food" may help cheer up people who are lonely or sad.

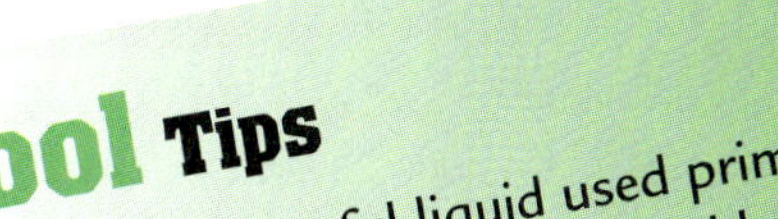

A **stock** is a flavorful liquid used primarily to prepare soups, sauces, stews, and braises. Stocks are made by simmering the basic ingredients of bones, seafood shells, or vegetables with aromatics such as *mirepoix*—a combination of vegetables—spices, and herbs, in a liquid such as water.

Soups

Soups may be served at the beginning of a meal, or for the meal itself. There are two basic types of soup:

- **Clear soups** are richly flavored, aromatic, very clear liquids. Ingredients used to flavor the soup are strained out before serving. The more care you take when cooking a broth, the clearer it will be.
- **Hearty** soups are thick and contain ingredients such as vegetables, meat, grains, or pasta. The main flavoring ingredients remain in the soup. There is more variety in hearty soups than in broths.
    - **Cream** soup is thick, with a velvety smooth texture made by combining a broth or stock with the main flavoring ingredients, and finishing it with a cream or cream variant, such as milk or sour cream.

FIGURE 20-1
Soups let you feature seasonal foods, such as pumpkin. Why might a restaurant want to feature seasonal foods?

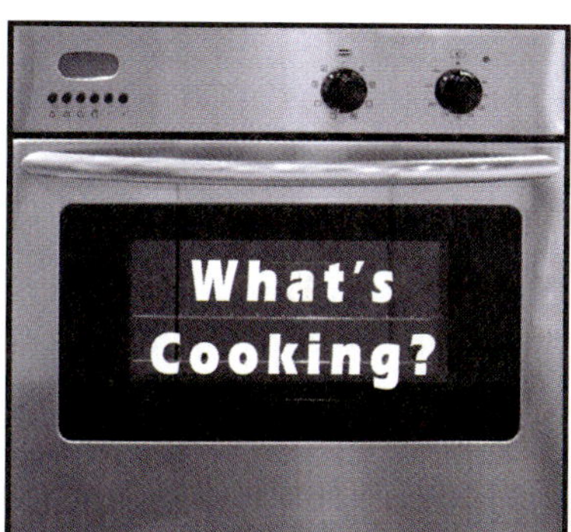

Some clear soups are **broths**, which are clear and thin, but have a small amount of flavorful fat. Another type of clear soup is **consommé**, which is a very clear soup, more refined than broth,and completely fat-free.

Bouillon is the French word for broth. Bouillon is sold dried in cubes or granules which can be combined with water to make instant broth.

- A **purée** soup is made by simmering a starchy ingredient, such as beans or potatoes, along with other ingredients in a liquid until they are tender enough to puree.
- A **bisque** is a soup in which the main flavoring is seafood, such as lobster or shrimp. After simmering, the bisque is pureed, strained, and finished with cream.

If you do not have time to make a soup from scratch, you can buy soups in cans, dehydrated, or frozen. Some canned soup can be very high in sodium, high in fat, contain preservatives, and have very little nutritional value. However, they are also a low-cost, convenient option for lunch.

Some canned soups are **condensed**, which means they are prepared and packaged with a minimum of water. These soups are very thick and you must add water to the soup when you heat it. Others are ready-to-eat, which means you just heat them and serve.

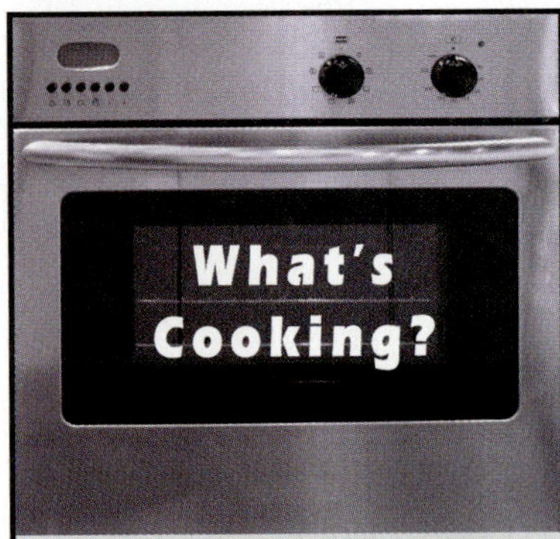

Soups can often be made from material trimmed from other foods, which reduces the overall food cost in the kitchen. **Trim** is the amount of scraps you produce when cutting something. For example, fat cut off meat, bones removed from chicken, skin peeled off vegetables, etc.

FIGURE 20-2
Some clear soups are served with a garnish, such as fresh herbs, diced cooked meats, or cooked noodles. Which of these soups is the hearty soup, and which is the clear soup? How can you tell?

Hot Topics

Many soups are identified by their country of origin. For example:

- **Minestrone** is a hearty vegetable-based soup from Italy.
- **Borscht** is beet-based soup from Russia or Poland.
- **Mulligatawny**, from India, combines a variety of ingredients such as meat, curry, rice, and coconut.
- **Menudo** is a hearty, spicy Mexican soup made with tripe, calf's feet, chiles, hominy and seasonings.

Casseroles

A casserole is a baked main dish or side dish made with a sauce and a combination of foods similar to those used to make soups. Casseroles come in many types and are made from almost every ingredient you can think of. They might include meat, poultry, vegetables, legumes, grains, and pastas. The ingredients might be pre-cooked, raw, or left over from another meal. They might be blended together or layered in the dish before baking.

FIGURE 20-3
Casseroles can be made from a variety of fresh and left over ingredients. Why are casseroles considered a convenient way to serve a complete meal?

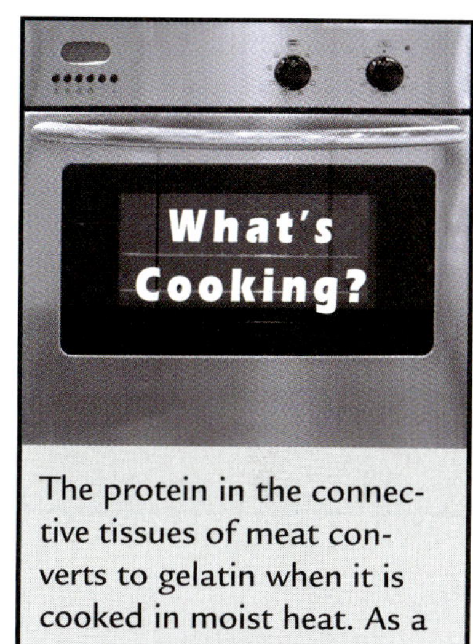

The protein in the connective tissues of meat converts to gelatin when it is cooked in moist heat. As a result, broth or soup in which meats are cooked will gel when cooled.

Nutritional Value of Soups and Casseroles

The nutritional value of soups and casseroles is dependent on the ingredients. Therefore, the healthiest soup is rich in nutritious ingredients, such as vegetables, fish, and lean meat or chicken.

- Creamed soups, bisques, and chowders are typically higher in fat and calories than clear or non-creamed soups.
- Clear soups contain only the dissolved flavoring substances from the ingredients. They have very little nutritional value, but are generally very low in calories.
- Hearty soups and casseroles have some nutritional value from the meat, vegetables, grains and other ingredients that they contain.
- Casseroles that contain cream are higher in fat and calories than those that do not contain cream.

Cool Tips

You can minimize unnecessary calories and fats in soups and casseroles by removing fat from the ingredients, and using low-fat or nonfat products whenever possible. For example, you can substitute low-fat milk for cream.

FIGURE 20-4
A garnish of fresh vegetables contributes nutritional value to soups. What other benefits do garnishes bring to a soup or casserole?

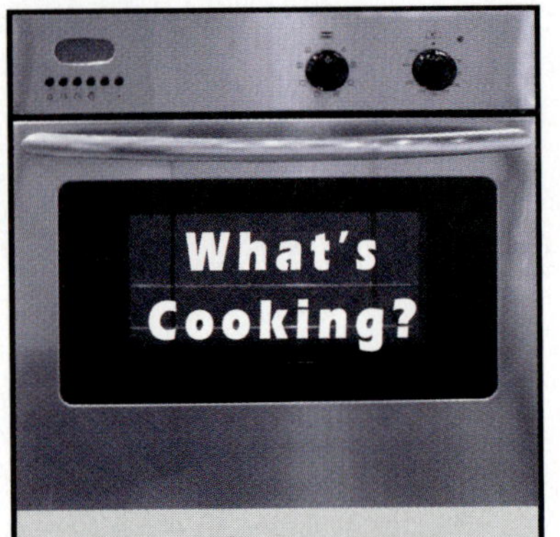

Cooking method does not affect the nutritional value of soups and casseroles. It's true that the nutrients from the ingredients leach into the cooking liquid, but you are going to eat the liquid, anyway.

Preparing Soups and Casseroles

Soups and casseroles can feature a wide variety of ingredients, alone or in combination. You could easily make a different soup or casserole every day of the year.

Preparing Soups

- When making a broth, select flavorful ingredients and keep the broth at a slow, even simmer to extract the most flavor. Carefully skim off the impurities that might make the broth cloudy.
- When you make a consommé, you use a clarification to make sure the soup is clear and fat-free. A **clarification** is a set of ingredients designed to attract impurities in consommé, and bind them together into a blob—called a **raft**—which can then be removed. It also adds flavor and color to the soup.

  A common clarification includes egg whites, finely chopped vegetables and herbs, an acid such as tomatoes or lemon juice, and possibly ground meat, fish, or poultry.
- When you make a hearty soup, the order in which you add ingredients has an impact on the flavor and texture of the soup. Usually, the aromatic ingredients, such as onions and garlic are added first. The remaining ingredients are added starting with the ones that take the longest to cook.

Fiction	Fact
All soups are best served hot.	Some soups are meant to be served hot, and others should be served cold. Soup should be served at the temperature that brings out the full flavor, color, and texture. Hot soups should be held and served at a minimum of 165° F, and cold soups should be served at room temperature or colder. Examples of cold soups include Cucumber Soup, Gazpacho, and Vichyssoise (cold potato leek soup).

Hot Topics

The word *soup* comes from the word *sop*, which originally meant bread soaked in liquid. Soup is still often served with bread or crackers, and some recipes—such as French onion soup—still call for bread to be cooked in the broth.

Reheating Soups

You can keep soup refrigerated for up to two days and reheat it for serving. Reheat soups over direct heat in a heavy-gauge pot, being careful to stir the soup frequently so it does not scorch.

FIGURE 20-5
To reheat thick or creamy soups, pour a thin layer of water or broth into the pot before adding the cold soup, then warm the pot over low heat until it is softened and warmed through. Then, raise the heat and simmer until done. What do you think is the purpose of the layer of water or broth?

Garnishing Soups

A garnish adds an extra dimension of flavor, texture, or color to a soup. Garnishes should be added immediately before serving.

- Use a garnish such as herbs or grated citrus zest, to add flavor and freshness.
- Use a garnish such as diced meats, grated cheese, or noodles, to add substance.
- Use a garnish such as a dollop of cold sour cream or a crunchy **crouton** (a small cube of toasted bread), to add texture and temperature contrast.

Cool Tips

If a soup recipe calls for water, substitute stock instead. The result will be a richer, more flavorful soup.

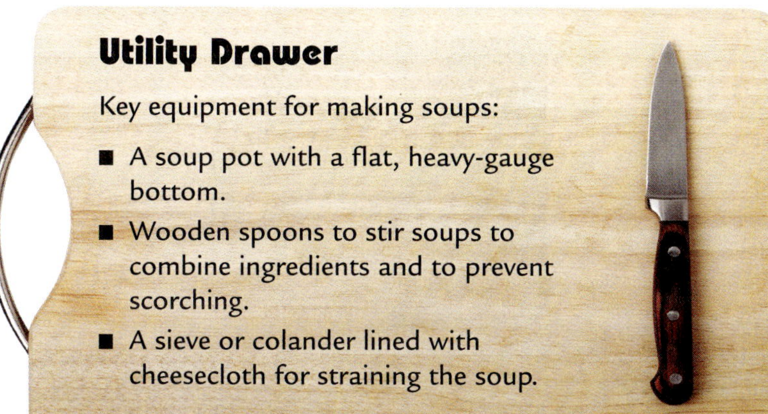

Utility Drawer

Key equipment for making soups:

- A soup pot with a flat, heavy-gauge bottom.
- Wooden spoons to stir soups to combine ingredients and to prevent scorching.
- A sieve or colander lined with cheesecloth for straining the soup.

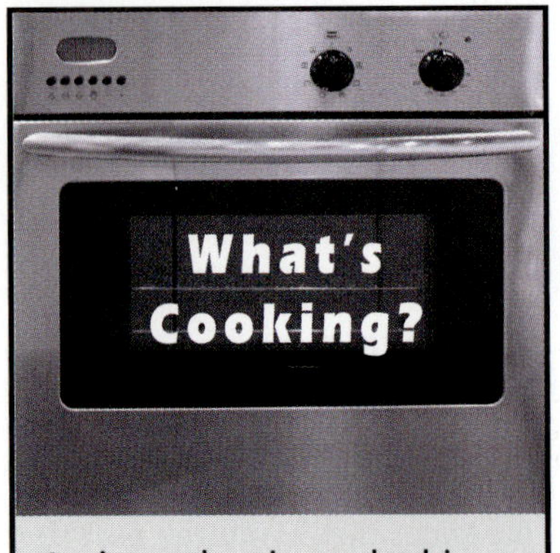

An item that is cooked in the soup is not a garnish. Only items added to the soup after it is cooked are considered a garnish.

FIGURE 20-6
Wonton soup garnished with wontons (dumplings), noodles, and cilantro. Can you think of other traditional soup and garnish combinations?

BASIC CULINARY SKILLS

Broth

1. Prepare ingredients according to your recipe.
2. Add cold liquid to cover the main ingredient.

3. Add remaining ingredients.

4. Bring to a simmer.
5. Maintain a slow, gentle simmer. Skim to remove any foam that rises to the surface.
6. Taste the broth from time to time. Adjust the seasonings as necessary.
7. Add aromatics, such as a bouquet garni or sachet d'epices, during the final 30 to 45 minutes.

8. Strain the broth.

BASIC CULINARY SKILLS

Consommé

1. Prepare the clarification ingredients.
2. Blend the clarification ingredients together. They must be kept very cold.

3. Add cold stock or broth to the cold clarification in a soup pot.

4. Bring to a simmer slowly, stirring occasionally to keep the clarification from sticking and scorching.
5. Stop stirring the consommé when the clarification ingredients start to form a large soft mass—the raft.

6. Break a small opening in the raft.
7. Lower the heat. Very small bubbles should rise to the surface through the hole in the raft.
8. Baste the raft while the consommé simmers by gently ladling the consommé over the top of the raft.

9. Simmer the consommé for the recommended time.
10. Ladle the consommé out of the pot. Do not break the raft apart.
11. Strain the consommé through a very fine sieve, cheesecloth, or filter.

12. Skim or blot any fat on the surface of the consommé.

FIGURE 20-7
Strawberry soup. Fruit soups are refreshing, and are usually served as an appetizer on a hot summer day. Do you think fruit soups are served hot or cold?

BASIC CULINARY SKILLS

Hearty Soup

1. Prepare ingredients according to your recipe.
2. Sauté aromatic ingredients, such as mirepoix, mushrooms, onions, bacon, and garlic.

3. Add a flavorful liquid.
4. Add roux or thickening agent, if called for by your recipe.
5. Add the remaining ingredients in a sequence that ensures they will be cooked to the correct point of doneness.

6. Simmer gently, stirring frequently.
7. Skim the surface to remove foam or fat.
8. Taste the soup as it cooks, adjusting seasonings and consistency as necessary.
9. Continue simmering until all the ingredients are fully cooked, very tender, and very flavorful.
10. If necessary, remove the sachet d'epice or bouquet garni and discard.
11. Purée the soup, if your recipe calls for it.

12. Add finishing ingredients or garnishes, if your recipe calls for it.

BY THE NUMBERS

A recipe for 1 gallon of brown beef stock calls for 8 pounds of roasted bones, 1 pound of mirepoix, 6 ounces of tomatoes, and 6 quarts of water. How much of each of these ingredients would you need to make 2.5 gallons of brown beef stock?

Hint: Multiply each ingredient by 2.5.

Preparing Casseroles

Like soups, casseroles are made from a combination of ingredients. Read your recipe carefully to determine whether the ingredients should be cooked first, or not. Usually, grains and pastas are precooked. Also, if ingredients are layered, the order of the layering affects the flavor and texture of the dish.

Casseroles include a sauce to blend the flavors of the other ingredients. Sometimes the sauce is simply stock or broth, sometimes it is a red sauce, with a tomato base, but often it's a white sauce made with milk or cream, which gives the casserole a smooth, creamy texture.

Most casseroles can be frozen for future use. However, some ingredients freeze better than others, and freezing may affect the texture of the dish when it is reheated. For example, some starch-thickened sauces become watery when they thaw.

You can prepare and freeze the main part of your casserole, and then add the starch when you reheat it. Likewise, you might want to go light on spices such as pepper, onion, garlic, and cloves before freezing, and then add them during reheating.

If you plan to freeze a casserole, undercook it the first time. Otherwise, it will be overdone when you reheat it.

Safe Eats

According to the U.S.D.A., it would take an 8-inch stockpot of soup 24 hours to cool to a safe temperature in your refrigerator. To avoid bacteria growth, you should cool soups and casseroles thoroughly before refrigerating, and then refrigerate for at least two hours before freezing.

For quick cooling, transfer soup to shallow containers, cover loosely, and refrigerate immediately. Cover tightly once the soup has cooled.

To quickly cool a casserole, put the dish in very cold or ice water, with the water up nearly to the lip of the dish.

Do not let soups or casseroles sit at room temperature for more than two hours.

FIGURE 20-8
A casserole dish has a cover and rounded sides. It is designed to keep food from drying out or burning while developing a crust that seals in the flavor of the food. Why do you think it is important to cook a casserole in a casserole dish?

Case Study

Sasha's mother recently accepted a job where she has to work the evening shift three days a week. On those days, Sasha is responsible for preparing dinner for herself and her younger sister.

Sasha's mom doesn't want to worry about Sasha having a cooking accident when she is not around. She stocked the freezer with pizza and hot dogs. She bought boxes of macaroni and cheese dinners, and keeps milk, cereal, and eggs in the house at all times. She thinks this should be enough to get the girls through the nights when she is not home.

After two weeks Sasha's sister starts to complain. Their mother gets angry. She tells them it is only three days a week, and that they should just get used to it.

- Do you think the girls are right to complain?
- Do you think the meals their mother left are healthy and well-balanced?
- Can you think of ways she might be able to make a wider variety of foods available to the girls, without risk of an accident?

Put It to Use

1. Using a cookbook or recipe guide, look up recipes for 10 different soups. Categorize them based on whether they are clear or hearty. If they are hearty, classify them as cream, purée, or bisque. Select one and make a shopping list for the ingredients.

Put It to Use

❷ Make a chart showing the nutritional comparison between different brands and types of canned chicken broths. You can get the information at the market, or online. Compare calories, fat, and sodium, as well as vitamins and nutrients. Select the one you would recommend for use, and write a paragraph explaining your selection.

Write Now

You have an elderly relative who has lost her appetite. You are worried that she might be depressed. You think a casserole might tempt her to eat. Find a casserole recipe that you think would appeal to someone who is elderly or ailing. Write a descriptive essay or advertisement promoting the casserole and explaining why you think it is the right food to tempt your relative. Write out the recipe on a piece of paper or recipe card and attach it to your essay.

Tech Connect 

Use the Internet to look up the recipe for a soup or casserole traditionally served in a foreign country. Learn what you can about the history of the dish and any traditions or holidays associated with the dish. Write a recipe card for the dish, including notes and tips. Present the recipe to your class.

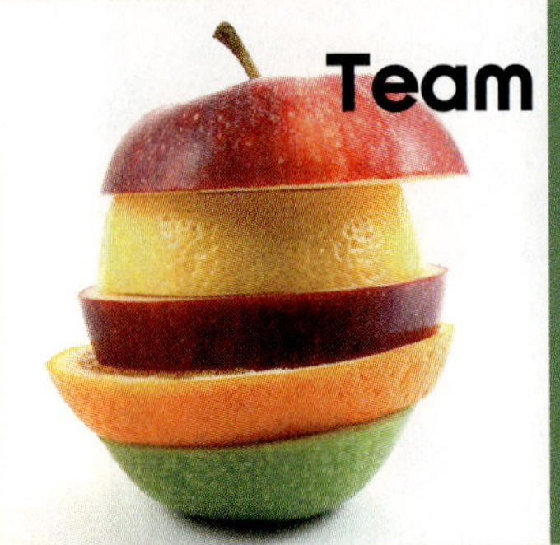

Team Players

Divide into teams of four or five. Each team should randomly select eight to ten ingredients from the meat, poultry, fish, vegetables, fruits, grains, pastas, dairy, spices, and herbs categories. In 10 minutes, work as a team to create a recipe for a soup or casserole using your selected ingredients. Present your recipe to the class, explaining the combination of flavors, how you would cook it, and how you would serve it.

Put It Together

Match the explanation in column 1 with the term in column 2.

Column 1

- **a.** liquid food served in a bowl and eaten with a spoon
- **b.** a flavorful liquid used primarily to prepare soups, sauces, stews, and braises
- **c.** the French word for broth
- **d.** a very clear soup that is completely fat-free
- **e.** a beet-based soup from Russia or Poland
- **f.** a dish made with a sauce and a combination of ingredients, cooked for a long time in an oven
- **g.** the amount of scraps produced when cutting food items
- **h.** a set of ingredients designed to attract and bind impurities in consommé
- **i.** the blob of ingredients created by a clarification in consommé
- **j.** a small cube of toasted bread
- **k.** food that is basic and familiar and gives diners a sense of warmth and wellbeing

Column 2

1. stock
2. casserole
3. comfort food
4. clarification
5. trim
6. crouton
7. raft
8. consommé
9. soup
10. Borscht
11. bouillon

TRY IT!

Court Bouillon

Yield: 1 gallon Serving Size: 8 fl oz

Ingredients

5 qt Cold water
8 fl oz White wine vinegar
To taste Salt
12 oz Carrots, sliced
1 lb Onions, sliced
1 pinch Thyme, dried
3 Bay leaves
12 Parsley stems
½ tsp Peppercorns

Method

1. Combine all the ingredients, except the peppercorns, in a large stockpot.
2. Simmer for 50 minutes.
3. Add the peppercorns.
4. Simmer for 10 minutes more.
5. Serve the stock, or cool and hold for later use.

Recipe Categories

Stocks, Sauces, Soups

Chef's Notes

Different types of vinegars (including flavored vinegars) can add subtle variations to the flavor of the court bouillon.

Potentially Hazardous Foods

None

HACCP

Cool to below 41° F within 4 hours (1-stage cooling method) or within 6 hours (2-stage cooling method).

Refrigerate at 41° F, or below.

Reheat to 165° F.

Nutrition

Calories	20
Protein	0 g
Fat	0 g
Carbohydrates	0 g
Sodium	4 mg
Cholesterol	0 mg

TRY IT!

Chicken Consommé

Yield: 6 cups Serving Size: 8 fl oz.

Ingredients

1 lb Chicken parts, including bones
1 large Onion
3 stalks Celery, including some leaves
1 large Carrot
1½ tsp Salt
3 whole Cloves
6¼ cups Water, cold
1 Egg

Method

1. Quarter the onion. Scrub and chop celery and carrot into 1-inch chunks.
2. Rinse the chicken pieces in cold water.
3. Place the chicken in large stockpot and cover with 6 cups of water.
4. Add onion, celery, carrot, salt, and cloves.
5. Bring to a boil.
6. Skim the surface as necessary.
7. Reduce the heat, cover, and simmer for 1 hour.
8. Remove the chicken and vegetables.
9. Strain the stock through a sieve or colander lined with rinsed cheesecloth.
10. Skim the surface as necessary.
11. To clarify the stock, continue with the following steps.
12. Separate the egg white from the egg yolk. Crush and reserve the shell.
13. In a small bowl, combine ¼ cup cold water, egg white, and crushed eggshell to create the clarification.
14. Add the clarification to the strained stock, and bring to a boil.
15. Remove from the heat and let stand for 5 minutes.
16. Strain again through a sieve lined with cheesecloth.

Recipe Categories

Stocks, Sauces, Soups

Chef's Notes

For a deeper flavor and color, roast the chicken in a 350° F oven until it is golden before preparing the soup.

Potentially Hazardous Foods

Poultry, Egg

HACCP

Cool to below 41° F within 4 hours (1-stage cooling method) or within 6 hours (2-stage cooling method.)

Refrigerate at 41° F, or below.

Reheat to 165° F.

Nutrition

Calories	200
Protein	15.5 g
Fat	13 g
Carbohydrates	4.5 g
Sodium	678 mg
Cholesterol	94 mg

TRY IT!

Tuna Noodle Casserole

Yield: 6 servings Serving Size: 8 fl oz.

Ingredients

2 quarts Water
1 (8 ounce) package Wide egg noodles
2 Tbsp Butter, unsalted
2 Tbsp All-purpose flour
1 tsp Salt
1 cup Low-fat milk
1 cup Sharp Cheddar cheese, shredded
1 (6 ounce) can Albacore tuna, canned and drained
1 (15 ounce) can Peas, drained

Method

1. Preheat oven to 350° F.
2. Coat a 2 quart casserole dish with cooking spray or oil.
3. Bring 2 quarts of water to a boil.
4. Add noodles and boil until done, according to directions on the package, and then drain well.
5. In a medium saucepan, melt the butter.
6. Add the flour and salt; stir until the ingredients are evenly combined.
7. Add the milk and whisk until the sauce begins to thicken.
8. Add the cheese and continue to whisk until the ingredients are well blended.
9. Stir in the tuna, peas, and noodles.
10. Spread the mixture into the prepared casserole dish.
11. Bake in the preheated oven for 30 minutes.
12. Remove and serve immediately.

Recipe Categories

Fish, Casseroles

Chef's Notes

For variety, use a combination of Monterey Jack and cheddar cheeses.

Those concerned with sodium content may choose to omit the salt.

Potentially Hazardous Foods

Fish, cheese, milk, egg noodles, butter

HACCP

Cool to below 41° F within 4 hours (1-stage cooling method) or within 6 hours (2-stage cooling method.)

Refrigerate at 41° F, or below.

Reheat to 165° F.

Nutrition

Calories	349
Protein	21.3 g
Fat	14 g
Carbohydrates	34 g
Sodium	742 mg
Cholesterol	77 mg

21 Breads

In This Chapter, You Will . . .

- Identify types of quick breads and yeast breads
- Understand the nutritional value of quick breads and yeast breads
- Learn how to prepare quick breads and yeast breads

Why YOU Need to Know This

Archaeological evidence shows that people have been baking bread since time began. In its simplest form, it can be made from the most basic of ingredients to create an inexpensive, filling, and nutritious dish. It can also be enhanced using different types of flour, flavoring, fat, leavening, spices, herbs, fruits, and vegetables to create a beautiful, aromatic, and delicious treat. Breads are used to accompany and enhance meals, and also as a base for some main dishes, such as pizza and sandwiches. As a class, discuss the types of bread that are associated with different countries around the world—such as baguettes in France, tortillas in Mexico, and pitas in Egypt. How many have you tasted? What are the differences?

Types of Yeast Breads and Quick Breads

Yeast breads are breads made from dough that is leavened with **yeast**—a tiny single-celled edible fungus that requires moisture, warmth, and food in the form of sugar in order to grow.

There are three basic types of dough used to make yeast breads:

- **Lean dough**, which is also called hard dough, is the most basic. It is made from flour, yeast, salt, and water, with very little or no sugar and fat. Spices, herbs, dried nuts, and fruit are sometimes added to lean dough. Examples include pizza crust, Italian-style bread, and French baguettes. Whole wheat, rye, pumpernickel, and sourdough breads are variations of lean dough.
- **Soft dough**, which is also called medium dough, is lean dough with between 6 and 9% sugar and fat added. The fat and sugar help make the soft dough tender when it is baked, and give it a soft crust. Examples include Pullman bread and soft rolls.
- **Enriched dough**, which is also called sweet rich dough, is lean dough enriched with butter, oil, sugar, eggs, or milk products. It has fat and sugar amounts up to 25%, and the finished product is almost cake-like. The added ingredients slow down the yeast activity so that enriched dough takes longer to rise than lean or soft dough. The eggs and butter also create a soft crust and a golden color. Examples include cinnamon buns, challah, stolen, and brioche.

FIGURE 21-1
Breads made from lean dough include French baguettes, pumpernickel loaves, and hard rolls. Do you think you eat more bread made from lean dough, soft dough, or enriched dough?

Be careful not to confuse **enriched dough** with doughs made using **enriched flour**. Enriched flour has vitamins and minerals added to it, and can be used to make any type of dough or batter. See Chapter 11 for more about enriched flour.

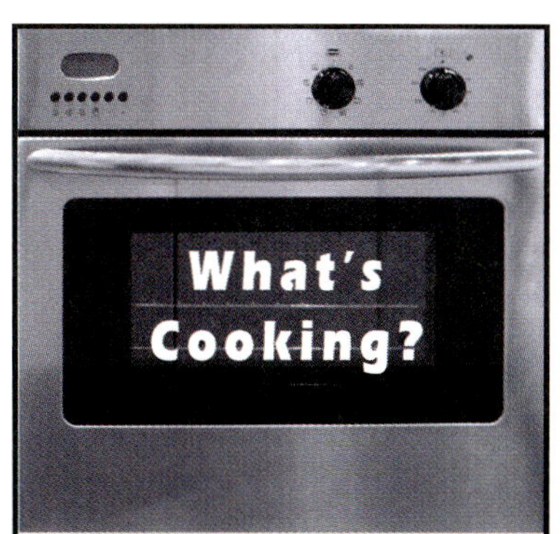

A **leavener** is anything that increases the volume of a dough or batter by adding air or other gas. Bakers rely on three basic types of leaveners:

- Organic leaveners, such as yeast
- Chemical leaveners, such as baking powder or baking soda
- Physical leaveners, such as steam or air

Check the Label

Bread shapes:

Pan loaves A pan loaf is made by pressing dough into a mold or a pan. This is the most common shape for packaged breads.

Flat breads Flat breads are common in many countries. They may be single-layered, like a Mexican tortilla, or puff into a double layer during cooking, like Greek pita. Other types of flat breads include injera in Ethiopia, focaccia in Italy, and naan in India.

Free-form loaves A free-form loaf is shaped by hand into an oval, round, or other shape. Boule—the French word for ball—is the term for any round, free-form loaf of bread.

Braided loaves Braided loaves are made from three ropes of dough, which are braided together into one loaf. Challah is often made in a braided loaf.

Baguettes Long narrow loaves are called baguettes. Typically used for French bread, they have a skinny, cylindrical shape that results in a crispy, golden crust and a soft interior.

Quick breads are made from batter than is leavened with fast-acting leavening agents, such as baking powder or baking soda and acid, rather than with yeast. The batter does not have to rise, so it can be baked immediately. Quick breads may be sweet or savory, depending on the ingredients.

Although all types of quick breads are made from the same basic ingredients, they may be categorized based on the thickness of the batter:

- **Pour batters** have a thin consistency and can be poured from the mixing bowl. They may contain equal amounts of liquid and flour, or slightly less liquid than flour. For example, if you use one cup of flour, you would use 1 or ¾ of a cup of liquid. Examples of quick breads made from a pour batter include waffles, pancakes, and popovers.
- **Drop batters** are thicker than pour batters and must be scraped from the mixing bowl into baking pans. They usually contain about twice as much flour as liquid. For example, if you use 1 cup of flour, you would use ½ cup of liquid. Examples include muffins, coffee cakes, and banana bread.
- **Soft dough** is thick enough to roll and shape by hand. It contains about one-third as much liquid as flour. For example, if you use 1 cup of flour, you would use ⅓ cup of liquid. Examples include biscuits, doughnuts, and scones.

Hot Topics

Bakers and pastry chefs often call their recipes formulas, and use percentages to indicate how each ingredient compares to the total amount of flour. Using percentages makes it easier for them to accurately increase or decrease the recipe. For example, if you have a recipe that calls for 2 pounds of flour and 1 pound of sugar, the formula is 100% flour and 50% sugar.

FIGURE 21-2
Muffins are a type of quick bread made with a drop batter. Why do you think it is necessary to fill muffin tins only part way before baking?

Buying and Storing Breads

Bakeries and most supermarkets sell fresh baked bread of all types. They will usually slice it for you upon request. You can also buy loaves of packaged bread. Packaged bread usually contains preservatives to increase its shelf life. It may also contain other ingredients such as soy and corn syrup; read the label so you know what you are buying.

Safe Eats

Discard bread that shows even the slightest hint of mold. Some types of mold are harmless, but some are highly toxic.

Use the Sell By date on bread only as a guideline. The freshness of the product is greatly affected by the environment, including humidity and temperature.

Packaged bread will stay fresh longer than fresh bread, due to the preservatives. Most packaged bread has a Sell By date. You should not buy the product after the Sell By date, and you should use it within five days of the Sell By date. Bread that is not fresh will become stale and hard, and lose its flavor. It may also grow mold, a multi-celled type of fungus that may be harmful when eaten.

You should store all bread at room temperature in a container designed to retain moisture. You can freeze bread in an airtight container for two to three months.

FIGURE 21-3
You can purchase many types of packaged bread in a supermarket. What are some of the benefits of buying packaged bread? What are some of the disadvantages?

Fiction	Fact
Bread stays fresh longer if it is stored in the refrigerator.	Staleness is caused when starch molecules crystallize—a process called **retrogradation.** Retrogradation occurs faster in colder temperatures, so bread actually gets stale faster when it is stored in the fridge. However, storing bread in the fridge does help slow the growth of mold.

The Nutritional Value of Quick Breads and Yeast Breads

Bread falls into the Grains category of MyPyramid. It is an important source of energy. The nutritional value of both yeast breads and quick breads depends on the ingredients that you use.

Flour is the ingredient used in the largest amount for all types of breads. Breads made with whole grain flours are rich in B vitamins and minerals, such as iron and phosphorous. They are full of starch which supports energy, and fiber which promotes healthy digestion. If the bread is made using enriched flour, it will provide thiamin, riboflavin, niacin, folic acid and iron, but less fiber. (Refer to Chapter 11, Grains, Pasta, and Legumes, for more information on types of grain and flour.)

When milk and eggs are used in a bread dough or batter, they increase the mineral and vitamin content of the bread and provide some complete protein. (For information on the nutritional value of milk and eggs, refer to Chapter 17, Dairy and Eggs.) When fat and sugar are used in a bread dough or batter, they affect the tenderness and flavor of the bread and increase the caloric value yet add little other nutritional value. However, if fruit or vegetables are added to dough or batter, they will enhance the nutritive value of the finished product.

FIGURE 21-4
Sugar and butter add calories, but few nutrients to bread. Which type of bread do you think is typically more nutritious—yeast bread or quick bread?

Utility Drawer

A **bread box** is a tightly sealed container designed for storing bread. It traps moisture, which prevents the bread from going stale. An added benefit is it keeps out insects and mice.

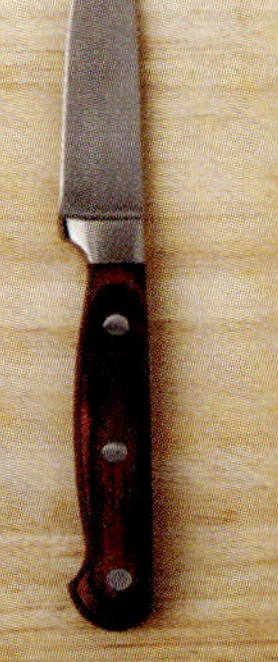

Hot Topics

You can cut down the fat content of quick breads and muffins by substituting applesauce or fruit purée in place of half the fat—butter, oil, or shortening—called for in the recipe. The fruit fibers hold moisture and the natural sugars promote browning.

Preparing Yeast Breads

All types of yeast breads combine the same basic ingredients:

- Flour (usually wheat)
- Yeast
- Liquid (usually water, milk, or a combination of the two)

Some types of yeast bread may also include eggs, salt, sugar, and shortening, such as butter or oil.

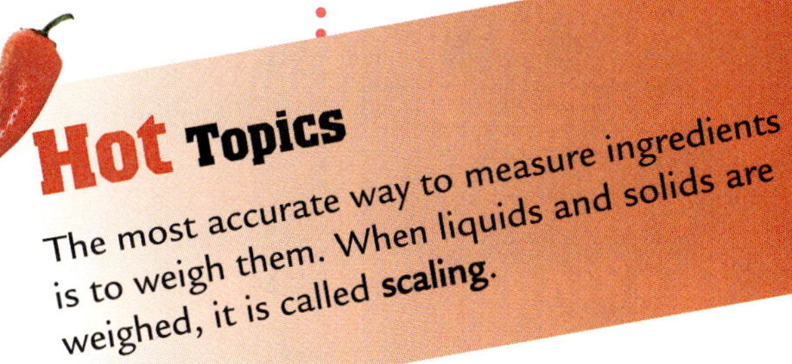

Hot Topics

The most accurate way to measure ingredients is to weigh them. When liquids and solids are weighed, it is called **scaling**.

FIGURE 21-5
During fermentation, yeast bread dough doubles or triples in size. Do you think dough will rise more in a cool place or a warm place? Why?

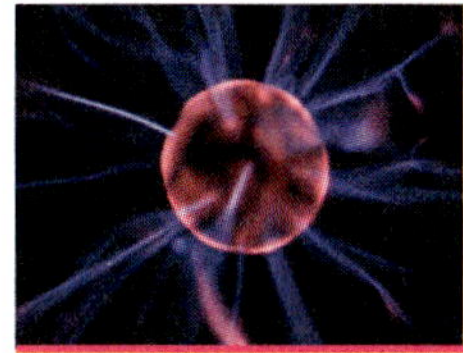

SCIENCE STUDY

Temperature plays an important role in baking, particularly baking yeast breads. The temperature of ingredients affects the temperature of the dough, which in turn affects the way the yeast reacts and ferments.

You can test the way temperature affects yeast using a simple experiment. You will need:

- 2 empty, clean 1-litre soda bottles
- 2 rubber bands
- teaspoon
- all-purpose flour
- water, room temperature
- 2 latex balloons
- glass measuring cup, 1-cup capacity
- tablespoon
- granulated sugar
- 2 packages of active dry yeast (¼ ounces each)

1. Pour one ¼ ounce package of Active Dry Yeast, 1 teaspoon sugar, 2 tablespoons all-purpose flour and 1 cup of room temperature water in each soda bottle.
2. Place one bottle in a container filled with hot—not boiling—water.
3. Place the other bottle in a container filled with ice water.
4. Secure a balloon on top of each bottle with a rubber band.
5. Keep both containers at a constant temperature—you may have to add ice and/or hot water.
6. Record your observations.

What happens to the balloons? What does that prove about the affect of temperature on yeast?

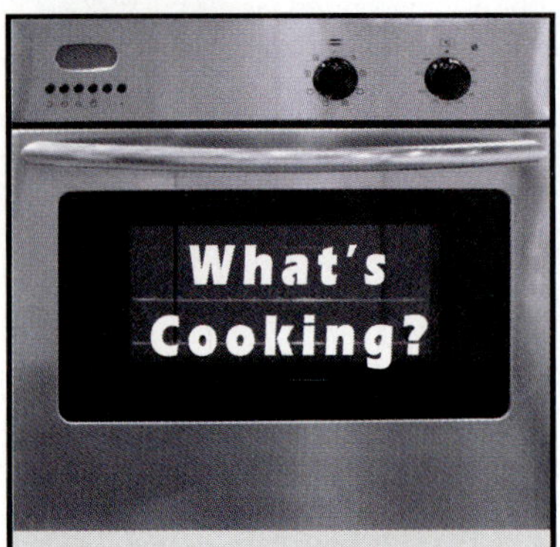

During fermentation, the yeast in the dough grows, producing carbon dioxide and acid as by-products. The dough doubles or triples in size. The process is known as **proofing**.

Making a yeast bread is a lengthy and precise process:

- You must use the correct proportion of ingredients. Baking is a science, and relies heavily on chemistry. In fact, recipes for yeast dough are often formulas, with the basic ingredients listed as percentages, or parts, based on the weight of the flour.
- You must combine the ingredients in the proper order, according to the recipe.
- You must knead the dough in order to develop the gluten. Dough that is properly kneaded is shiny and elastic. You can test your dough using the so-called gluten window test. Pinch off a piece, pull it, and hold it up to the light. It should be stretchy, should not tear, and thin enough so that some light can come through.
- The dough must sit in a warm, moist place for a fermentation period.
- After fermentation, you must fold and push down the dough.
- You must shape the dough correctly, and allow it to finish rising.
- You must bake the dough at the correct temperature for the correct amount of time.
- You must allow the bread to cool sufficiently to insure that all excess moisture has evaporated.

If you perform all of these steps correctly, you will be rewarded with beautiful, fragrant, delicious bread.

You may be able to purchase frozen or refrigerated yeast bread dough. For example, many markets sell refrigerated pizza dough, and frozen dough for rolls and baguettes. These products will take less time to prepare than homemade yeast breads. They are also more fool-proof, because the ingredients are already measured and mixed.

FIGURE 21-6
Pizza dough is a yeast dough, even though it is flat. What is your favorite type of pizza?

Check the Label

Types of yeast:

Active dry yeast Commonly used in bread recipes, it is dormant, which means it must be rehydrated in warm liquid before use.

Instant yeast Works about twice as fast as active dry yeast. It does not need to be rehydrated before use. It is also called bread yeast, instant yeast, or rapid-rise yeast.

Fresh yeast Works faster and longer than active dry yeast, but it is very perishable and loses potency a few weeks after it is packed.

FIGURE 21-7
This dough is fermenting. What do you see in the dough that indicates the yeast is producing carbon dioxide?

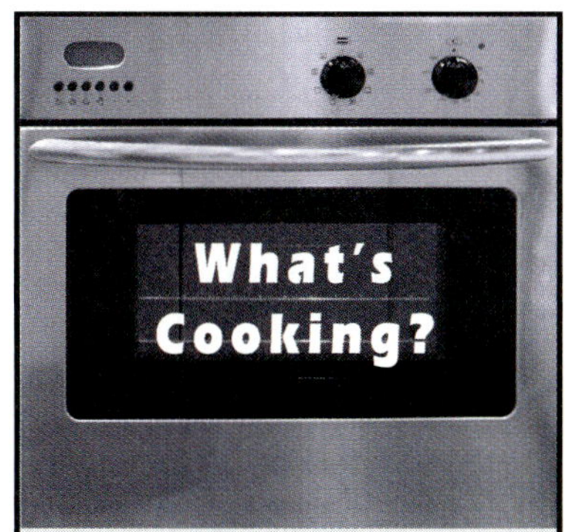

A **Panini** is a hot pressed sandwich that originated in Italy. It is made by toasting the sandwich on a heavy, two-sided cooking press that compresses and grills the outside of the bread until the sandwich and filling is heated all the way through. The Cuban version of a Panini is called a cubano.

Serving Yeast Breads

Yeast breads are a staple food served at every meal, and as a snack between meals. Restaurants often place a basket with a variety of breads on the table as soon as guests are seated, or serve bread with butter or a dipping sauce before the main meal. Bread is used as the base for many dishes, including pizza, meat rolls, and turnovers.

One of the most popular uses of bread is for sandwiches. Sandwiches combine bread, a spread such as mustard or mayonnaise, and a filling. While once considered a quick, portable lunch, sandwiches are now served at every meal.

Sandwiches can be made on a wide variety of breads, with an almost infinite choice of filling. Slices from standard loaves of bread such as white, wheat, rye, and sourdough are typical, but rolls, focaccia, pita, tortillas, and bagels are also used.

Cool Tips

Yeast loses its ability to grow and reproduce over time. Once opened, both instant yeast and active dry yeast lose about 10% of their potency per month. They will keep for 3 months in the refrigerator, or 6 months in the freezer. Unopened, it will keep for about one year past its expiration date, longer in the freezer.

Check the Label

Types of sandwiches:

Closed sandwich The classic type of sandwich, with filling enclosed between two slices of bread, roll, or bun.

Open-faced sandwich One slice of bread topped with filling.

Grilled sandwich A hot sandwich made by cooking one side and then the other directly on a heat source, such as a griddle.

Pressed sandwich A hot sandwich made in a two-sided cooking press that toasts both sides of the sandwich at the same time.

Wrap A sandwich made by rolling the filling up inside a single flatbread, such as a tortilla.

Pita pocket A sandwich made by opening one end of a piece of pita and stuffing the bread with filling.

Club sandwich A double-decker sandwich made with three slices of bread.

Finger sandwich A small sandwich that can be picked up and eaten in one or two bites.

FIGURE 21-8
A hamburger on a bun is a sandwich, usually served for lunch or dinner. What type of sandwich might you serve for breakfast?

Preparing Quick Breads

Quick breads also combine the basic ingredients of flour, leavening, salt, fat, and liquid. They frequently use eggs and sugar, as well as add-ins and flavorings such as fruit, vegetables, and chocolate.

Although the process of making a quick bread is much shorter than the process of making a yeast bread, it is still important to use the proper proportion of ingredients, and to combine the ingredients using the appropriate method.

FIGURE 21-9
Some bread recipes call for sifting dry ingredients together. Why do you think sifting might be important to baking success?

BASIC CULINARY SKILLS

Straight Dough-Mixing Method

1. Scale the ingredients. Be sure to weigh the ingredients precisely.
2. Hydrate the yeast by combining it with water to activate it.

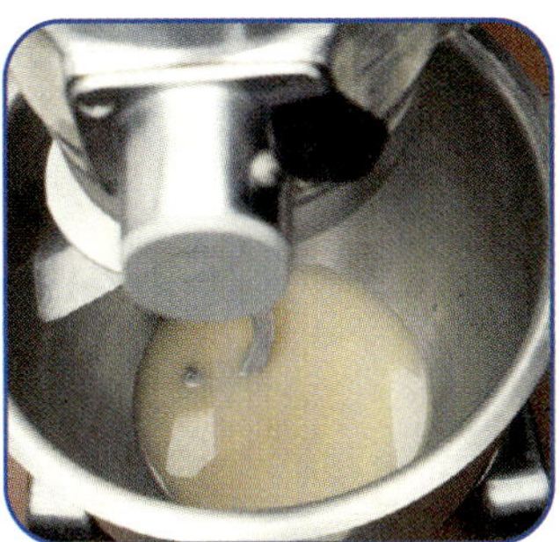

3. Add other ingredients all at once, with the mixer on a slow speed.
4. Increase the mixer speed to medium, until the dough begins to catch on the dough hook. Properly kneaded dough is satiny and forms a ball in the mixing bowl.

5. Fermentation allows the dough to double or triple in size.

6. Fold and push the dough down to release the carbon dioxide.

7. Shape and bake as directed by the recipe.

BASIC CULINARY SKILLS

Pizza Dough

1. Hydrate the yeast in water.
2. Add the flour and salt and mix at a low speed until the dough is evenly moistened.
3. Mix and knead the dough by hand on a floured surface or on medium speed in a mixer until the dough is very smooth and springy to the touch.
4. Transfer the dough to an oiled bowl, oil the surface lightly, and cover.
5. When the dough has doubled in size and retains an imprint when pressed with a gloved fingertip, remove it from the bowl.
6. Fold the dough over on itself in several places.
7. Shape into pizza crust.

It is important not to overmix the batter for quick breads. Overmixing causes too much gluten to form; then the batter will not rise correctly. The result will be heavy, funny shaped loaves or muffins, dotted with air holes.

The two basic methods of mixing a quick bread batter are the well method and the creaming method.

- **Well method:** Blend liquids in one bowl and sift dry ingredients in a separate bowl. Make a depression—a well—in the middle of the dry ingredients, pour in the liquids, and mix minimally.
- **Creaming method:** Cream together the sugar and fat to incorporate air; then beat in the eggs one at a time. Finish by alternately adding the sifted dry ingredients and any liquid ingredients.

You should grease the baking pans before adding the batter, in order to promote browning and make the finished product easier to remove. (Refer to Chapter 22, Desserts, for more information on baking skills.)

Quick breads are often available frozen, refrigerated, and as dry mixes. For the dry mixes, you usually add fat, liquid, and eggs, and then bake the item according to the directions on the packaging. The frozen and refrigerated products are usually ready-to-bake.

Although these items may be convenient, they may actually cost more than homemade, and—in the case of the dry mixes—may take as long to prepare as homemade.

Hot Topics

You can enhance quick breads by adding a wide variety of ingredients, called **stir-ins**. Common stir-ins include shredded vegetables, fresh fruits, nuts, whole grains, meat, cheese, and chocolate. For example, banana bread is a quick bread often served for breakfast. Corn bread with green chilies, onions, ground beef, and cheese is a Mexican-style main dish. Proportionally, a good rule of thumb is to use one cup of additions per one cup of flour.

FIGURE 21-10
You can use dried fruit, such as cranberries, as a stir-in to enhance quick bread. What types of stir-ins do you like in muffins and quick breads?

Serving Quick Breads

Quick breads may be served at any meal, for a snack, or for dessert. Although they are typically associated with breakfast, many restaurants include savory quick breads, such as biscuits or muffins flavored with cheese or herbs, in bread baskets placed on the table with dinner or lunch.

Quick breads are delicious served warm from the oven, but they may also be served cold, or toasted and topped with butter, jam, or other spread.

Some quick breads are used as the base for dessert. For example, biscuits are layered with fruit and whipped cream to create shortcake; Belgian waffles are thick, sweet waffles topped with fruit or ice cream.

FIGURE 21-11
Biscuits and scones are small, individual-sized quick breads. Biscuits usually have little or no sugar, while scones are sweet, and may have fruit or nuts added to the dough. How would you serve biscuits and scones?

BY THE NUMBERS

You have a yeast bread recipe that calls for 3 tsps, of active dry yeast, but you only have instant yeast in your pantry. How much instant yeast should you use?

Hint: To substitute instant yeast for active dry yeast, use 0.67 times the weight of the active dry yeast for the instant yeast. For example, 1 tsp of active dry yeast = ⅔ tsp of instant yeast.

BASIC CULINARY SKILLS

Creaming Method for Blueberry Muffins

1. Sift flour and baking powder together.
2. Cream butter, sugar, and salt together with a paddle attachment in the bowl of an electric mixer.

3. Add eggs, one at a time, at low speed. Scrape the sides of the bowl each time.
4. Alternate adding milk with dry ingredients, one-third each time
5. Remove from mixer.
6. Fold in blueberries with a rubber spatula.

7. Portion equal amounts of batter into greased muffin tins.
8. Bake at 400° F for 20 minutes, or until lightly browned on top.
9. Remove muffins from tins. Serve warm.

Case Study

Jake is on his school's debate team. They have an all day event scheduled on Saturday from 8:00 a.m. until 5:00 p.m. Each team member is responsible for bringing food for the team, and Jake is assigned to bring the bread for sandwiches.

Jake buys two loaves of packaged whole wheat bread on Friday. It snows overnight and the event is postponed until the following week. Jake leaves the bread in his locker, and brings it to the debate. When he takes it out at lunch time, it is hard around the edges, and has dark green spots.

- What happened to the two loaves of bread?
- Should the team eat the bread anyway?
- What steps could Jake have taken to keep the bread from getting hard and spotty?

Put It to Use

1. Develop a lunch menu of sandwich choices for a whole week. Specify the type of bread and filling. Indicate whether the sandwich should be served hot or cold. Include variety, and provide options for those on special diets, including gluten-free, peanut-free, low-fat, and low-sodium. Use a computer program to create a menu document that you can print and distribute to your classmates.

Put It to Use

❷ Compare the costs of different types and forms of bread. If possible, visit a bakery and supermarket, and write down the prices of fresh baked and packaged products. (If you cannot visit a bakery or market, use a supermarket advertising circular or the Internet.) You can also check the prices of frozen and refrigerated breads, and dry mixes. Make a comparison chart showing your findings.

Our Breads

Whole grain bread can refer to the same as "wholemeal bread", or to white bread with added whole grains to increase its fibre content (i.e. as in 60% whole grain bread). The inner, soft part of bread is known to bakers and other culinary professionals as the crumb, which is not to be confused with small bits of bread that often fall off, called crumbs. The outer hard portion of bread is called the crust.

A croissant is a buttery flaky pastry, named for its distinctive crescent shape. It is also sometimes called a crescent or crescent roll. Croissants are made of a leavened variant of puff pastry by layering yeast dough with butter and rolling and folding a few times in succession, then rolling.

A popover is a light, hollow roll made from an egg batter similar to that used in making Yorkshire pudding. The name "popover" comes from the fact that the batter swells or "pops" over the top of the muffin tin while baking. They can also be baked in individual custard cups.

Write Now

Imagine you own a bakery specializing in bread. You want to develop a glossary that explains the different breads you offer, so customers can read it and learn about your products. Select four types of bread—two yeast breads and two quick breads—and write descriptive paragraphs about each. Include adjectives and adverbs that describe the color, texture, aroma, and taste of each. Describe the ingredients and the processes used to create them. Use a word processing or desktop publishing program to format the paragraphs. Add pictures to illustrate the breads, if possible. Print the documents, and share them with the class.

Tech Connect 

Use the Internet to research the difference between baking powder and baking soda. Make a chart comparing the two. Include the chemical composition, other ingredients, and how they react. You can also include other uses for each of them. Discover whether you can substitute one for the other. Write a paragraph explaining your findings and share it with the class.

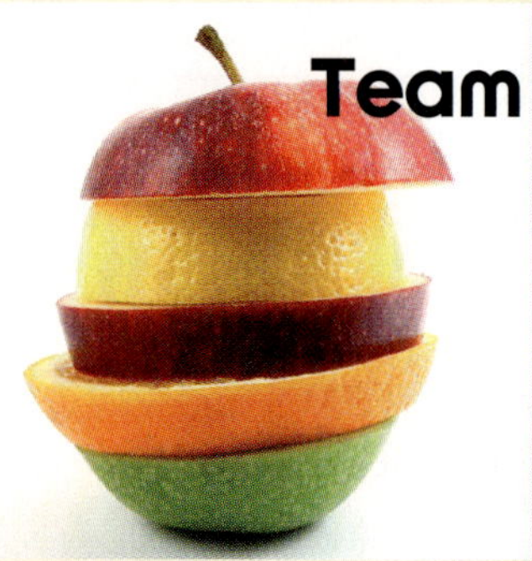

Team Players

Divide the class into teams of three or four students. Each team should use cookbooks or the Internet to locate a yeast bread recipe that is not written as a formula. The teams will share their recipe with the other teams. Each team will then work together to convert all of the class recipes to baking formulas. They will write the formulas neatly on recipe cards, and compare the results with the other teams. The team that correctly converts the most recipes to formulas is the winner. If possible, each team should select a formula and prepare it. Publish the recipes in the school paper, or print them for a class cookbook. Include photographs of the finished dish.

Put It Together

Match the explanation in column 1 with the term in column 2.

Column 1

a. a tiny, single-celled organism that require moisture, warmth, and food in order to grow
b. lean dough with between 6 and 9% sugar and fat added
c. Ethiopian flat bread
d. a round, free-form loaf of bread
e. a batter that contains about twice as much flour as liquid
f. the most accurate way to measure ingredients
g. the process during which yeast changes sugar into carbon dioxide and alcohol
h. a double-decker sandwich made with three slices of bread
i. a hot pressed sandwich that originated in Italy
j. ingredients added to quick breads to enhance and modify flavor and texture
k. a dessert made from biscuits layered with fruit and whipped cream

Column 2

1. drop batter
2. stir-ins
3. shortcake
4. club sandwich
5. scaling
6. injera
7. soft dough
8. panini
9. boule
10. yeast
11. proofing

TRY IT!

Buttermilk Pancakes

Yield: 12 Pancakes Serving Size: 3 Pancakes

Ingredients

1½ cups All-purpose flour
2 Tbsp Sugar
½ tsp Baking powder
¼ tsp Baking soda
¼ tsp Salt
1¾ cups Buttermilk
2 Eggs, large
3 Tbsp Butter, melted and cooled slightly
As needed Butter, melted, or cooking oil (for oiling the pan)
As needed Butter, syrup, honey, or fruit purees (for serving)

Method

1. Sift the flour, sugar, baking powder, baking soda, and salt into a mixing bowl.
2. In a separate bowl, blend the buttermilk, eggs, and butter.
3. Add the buttermilk mixture to the flour mixture.
4. Stir by hand just until the batter is evenly moistened.
5. Heat a large nonstick skillet or griddle over medium-high heat.
6. Oil the skillet or griddle lightly by brushing with melted butter or spraying with cooking oil.
7. Drop the pancake batter into the hot skillet or griddle by large spoonfuls (about ¼ cup).
8. Leave about 2 inches between the pancakes to allow them to spread and make turning easier.
9. Cook on the first side until small bubbles appear on the upper surface of the pancake and the edges are set, about 2 minutes.
10. Use an offset spatula or a palette knife to turn the pancakes.
11. Finish cooking on the second side, another 2 to 3 minutes.
12. Adjust the temperature beneath the griddle to produce a good brown color.
13. Serve the pancakes at once with butter, syrup, honey, fruit purees, or other toppings, as desired.

Recipe Categories

Breakfast Cookery, Breakfast Foods

Chef's Notes

If you are using a griddle, be sure to keep it cleaned as you work.

Use a ladle for consistently sized pancakes.

Potentially Hazardous Foods

Eggs, Dairy

HACCP

Cook to at least 145° F for at least 15 seconds.

Nutrition

Calories	293
Protein	8 g
Fat	13 g
Carbohydrates	37 g
Sodium	662 mg
Cholesterol	84 mg

TRY IT!

Club Sandwich

Yield: 10 Servings   Serving Size: 1 Sandwich (10 oz)

Ingredients

30 White Pullman bread slices, toasted
6 fl oz Mayonnaise
10 Red leaf lettuce leaves
1 lb, 4 oz Turkey, thinly sliced
1 lb, 4 oz Ham, thinly sliced
20 Tomato slices
20 Bacon strips, cooked and cut in half

Method

1. For each sandwich, spread 1 tsp mayonnaise on one slice of toast.
2. Layer a lettuce leaf on the toast.
3. Place 2 oz each of turkey and ham on top of the lettuce leaf.
4. Spread ½ tsp mayonnaise on both sides of another slice of toast.
5. Place the toast on top of the ham.
6. Top the toast with the remaining lettuce leaf, 2 slices of tomato, and 2 slices of bacon (4 halves).
7. Spread 1 tsp mayonnaise on one more slice of toast.
8. Place the toast on the sandwich, mayonnaise side down.
9. Secure the sandwich with sandwich picks.
10. Cut the sandwich into quarters.
11. Serve immediately.

Recipe Categories

Sandwiches, Appetizers & Hors d' Oeuvres

Chef's Notes

For club sandwich variations, try different combinations of meats and cheese.

Potentially Hazardous Foods

Mayonnaise, Meats

HACCP

Keep ingredients chilled to below 41° F until ready to prepare.

Nutrition

Calories	780
Protein	21 g
Fat	42 g
Carbohydrates	58 g
Sodium	857 mg
Cholesterol	125 mg

TRY IT!

Pizza Dough

Yield: Ten 8-inch Pizzas   Serving Size: One 8-inch Pizza

Ingredients

2 Tbsp Flour, all-purpose
1½ cups Warm water (105° F)
1 Tbsp Honey
1 pkg Dry yeast
4 to 5 cups Flour, bread
½ tsp Salt

Method

1. In a large bowl, whisk together the all-purpose flour, warm water, honey, and yeast.
2. Allow the yeast to proof for 15 minutes or until there is visible growth.
3. Use a wooden spoon to stir in 2 cups of the bread flour and the salt.
4. Gradually add the remaining flour, as necessary, to make a stiff, elastic dough.
5. Turn the dough out onto a lightly floured board.
6. Knead the dough by hand until it is smooth and elastic, about 12 minutes.
7. Let the dough rise until it is nearly doubled in size.
8. Punch the dough down and divide into 10 equal pieces.
9. Roll the dough into balls and allow them to rest for 20 to 30 minutes.
10. Roll each ball into a disk about 8 inches in diameter.

Recipe Categories

Yeasted Breads

Chef's Notes

The dough may be mixed in a standing mixer with a dough hook. Kneading time would be about 5 minutes.

Stack the disks you don't want to cook right away between parchment paper or aluminum foil sprinkled with cornmeal. You can freeze them for up to 2 months.

Potentially Hazardous Foods

None

HACCP

None

Nutrition

Calories	175
Protein	5 g
Fat	2 g
Carbohydrates	33 g
Sodium	390 mg
Cholesterol	0 mg

TRY IT!

Soft Rolls

Yield: 12 Dozen Rolls Serving Size: 1 Roll

Ingredients

2½ lb Water
6 oz Yeast
5½ lb Bread flour
4 oz Milk powder
8 oz Sugar
2 oz Salt
8 oz Butter
8 oz Eggs

Method

1. Combine the water and the yeast. Stir to mix.
2. Place all other ingredients in a bowl.
3. Mix on low speed for 2 minutes.
4. Increase to medium speed and mix for 10 minutes.
5. Allow to bulk ferment for 1 hour.
6. Scale into 2-lb, 8-oz pieces for rolls.
7. Scale into 1-lb, 2-oz pieces for loaves.
8. Bench rest for 15 to 30 minutes.
9. Shape into desired items. Apply egg wash.
10. Pan proof.
11. Bake in a 375° F oven for 20 minutes or until rich golden brown on top.

Recipe Categories

Yeasted Breads

Chef's Notes

Work sequentially, starting with the first piece of dough you divided and rounded.

Potentially Hazardous Foods

Eggs

HACCP

Refrigerate at 41° F or below.

Nutrition

Calories	120
Protein	3 g
Fat	5 g
Carbohydrates	16 g
Sodium	200 mg
Cholesterol	10 mg

TRY IT!

Zucchini Bread

Yield: 2 Loaves Serving Size: 1 Slice

Ingredients

3½ cups Flour
1 tsp Salt
2 tsp Baking powder
½ tsp Baking soda
½ tsp Cinnamon, ground
½ tsp Nutmeg, ground
¼ tsp Cloves, ground
2 Zucchini, large, grated (2½ cups)
1 cup Sugar
4 Eggs
½ cup Vegetable oil
1 cup Pecans/walnuts, chopped coarse and toasted

Method

1. Preheat the oven to 350° F.
2. Grease and flour two loaf pans.
3. Sift together the dry ingredients: flour, salt, baking powder, baking oda, cinnamon, nutmeg, and cloves.
4. Combine the zucchini, sugar, eggs, and oil in a large bowl. Mix well.
5. Stir the sifted dry ingredients into the zucchini mixture until the dry ingredients are blended into the batter.
6. Fold in the nuts.
7. Transfer the batter into the prepared loaf pans.
8. Bake the bread in the preheated oven until fully baked, about 50 to 55 minutes.
9. Remove the bread from the pans.
10. Cool on racks.

Recipe Categories

Quick Breads

Chef's Notes

The breads are fully baked when the edges are browned and starting to pull away from the pan. The bread should spring back when lightly pressed with your fingertip.

To toast the nuts, toss them in a skillet over high heat until they give off a good aroma and are just starting to brown. Immediately transfer them to a bowl to prevent them from burning.

Potentially Hazardous Foods

Eggs

HACCP

Refrigerate at 41° F or below.

Nutrition

Calories	159
Protein	2 g
Fat	8 g
Carbohydrates	22 g
Sodium	57 mg
Cholesterol	0 mg

TRY IT!

Corn Muffins

Yield: 16 Muffins Serving Size: 1 Muffin

Ingredients

As needed Cooking spray
2 cups Cornmeal
⅔ cup Flour, bread
2 tsp Baking powder
½ tsp Baking soda
¼ cup Sugar
1 tsp Salt
3 Eggs
2 cups Milk or buttermilk
⅓ cup Vegetable oil

Method

1. Preheat the oven to 350° F.
2. Line muffin tins with paper liners or spray them lightly with cooking spray.
3. Mix together the dry ingredients: cornmeal, bread flour, baking powder, baking soda, sugar, and salt.
4. Stir together the eggs, milk (or buttermilk), and oil until blended.
5. Add the wet ingredients to the dry ingredients. Combine until just mixed.
6. Pour the batter into the prepared muffin tins.
7. Bake at 375° F for 18 to 20 minutes or until the surface is golden brown and springs back when lightly pressed with a fingertip.
8. Cool the muffins on a cooling rack.
9. Serve while still quite warm.

Recipe Categories

Muffins and Quick Breads

Chef's Notes

There are many regional types of cornbread. Southerners tend to use very little sugar in their cornbread and often specify white cornmeal. In other parts of the country, yellow cornmeal is preferred, and the amount of sugar is increased.

This bread does not keep well, but any left over can be made into breadcrumbs and used as stuffing or frozen to use later on.

Potentially Hazardous Foods

Eggs, Dairy

HACCP

Refrigerate at 41° F or below.

Nutrition

Calories	220
Protein	4 g
Fat	9 g
Carbohydrates	32 g
Sodium	320 mg
Cholesterol	35 mg

22 Desserts

In This Chapter, You Will . . .

- Identify types of desserts
- Understand the nutritional value of desserts
- Learn how to prepare desserts

Why YOU Need to Know This

Finishing strong is not just important in sports, but also in food. Dessert is the last course you serve and eat. It should signal a clear end to the meal by providing a contrast both in flavor and appearance from the previous courses. You want to leave the meal with positive thoughts and a good taste in your mouth. Nutritional value is not usually the first thing people associate with dessert. Although many desserts are high in fat and calories, you can certainly provide healthy and delicious alternatives. In general, the only rules are that dessert should be fun to eat, taste remarkable, and create a lasting impression. As a class, list your favorite desserts and discuss what makes them memorable and special.

Types of Desserts

Dessert is the last course of a meal. The focus is usually on sweet tastes, but may also include savory foods and flavors, as well. For example, apple pie with cheddar cheese is a traditional dessert combination.

Desserts range from fresh fruit, basic drop cookies and a scoop of sorbet, to elaborate multi-layered cakes that take days to assemble and decorate. Some are baked; some are chilled or frozen; and some are simply assembled and served at room temperature.

Common baked or cooked desserts include:

- **Cookies** are small, flat, baked desserts. They come in many shapes, sizes, and flavors.
- **Cakes** are baked desserts made from some combination of flour, sugar, and eggs. Most cakes also include fat or oil, chemical leavening such as baking powder or baking soda, flavoring, and a liquid.
- **Pies** are baked desserts made by lining a dish with dough to create a pie shell, and filling the shell with fruit, cream, or custard. They may be topped with crust or crumbs, or left uncovered.

  Pies are actually a type of **pastry**, which is a baked crust made from a dough that is rich in fat. Other common pastries include cream puffs, puff pastry, croissants, and sweet rolls. Most pastries are usually made from four basic ingredients—flour, fat, salt, and liquid.

The scope and variety of desserts extends well beyond cookies, cakes, pies, mousse, and custard. For example, there are many types of puddings, gelatins, and tarts. Some desserts are made by combining different elements, such as a trifle, which has layers of sponge cake, custard or pudding, fruit or jelly, and whipped cream. Fresh fruit and cheese are often served for dessert, and, of course, there are confections such as chocolates and other candies. You can learn more about desserts at your library or on the Internet.

Cool Tips

In England and some other countries, cookies are called biscuits.

FIGURE 22-1

(Top) **A spritz cookie** is a type of drop cookie made by piping dough through a pastry bag into decorative shapes. Can you think of other kinds of piped cookies?

(Bottom) **Biscotti** are a common type of **twice-baked cookie**. They are made of dough formed into a large log shape and baked, then cut into slices and baked again. Twice-baked cookies are popular in many countries around the world. How do you think baking the cookie twice affects its texture?

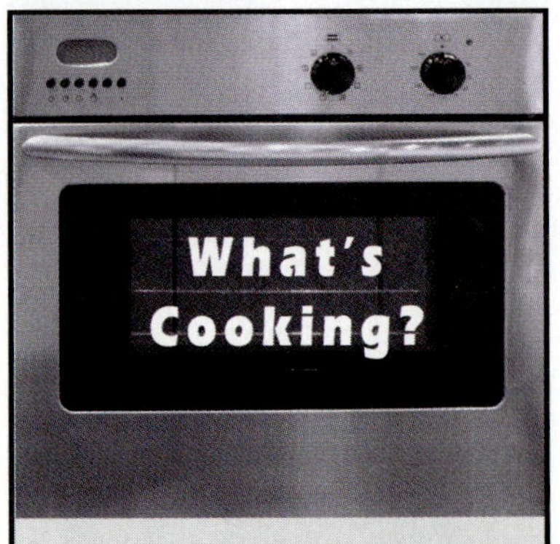

Mousse is usually chilled, so that it holds its fluffiness. If it is frozen, it is called a **parfait** or **frozen soufflé**.

Check the Label

Types of cookies:

Drop cookies The most common cookie, made from a firm dough that holds its shape on a pan. They are generally high in fat. Examples: chocolate chip and oatmeal. Variations: icebox (refrigerator) cookies, piped, and stenciled cookies.

Bar cookies Made from a soft batter that is spread into a pan. Once baked, they are cut into individual cookies. Examples: lemon bars and brownies. Variations: twice-baked cookies.

Rolled cookies Also called **cut-out cookies**, made of stiff dough that is rolled flat and then cut into decorative shapes, sometimes using cookie cutters. Examples: gingerbread men and holiday butter cookies. Variations: molded cookies.

FIGURE 22-2
Mousse, like custard, contains eggs and dairy products. How do you think you should store these items in order to keep them fresh and safe to eat?

- **Mousse** is a rich, fluffy dessert. It is made by adding air to a blend of a flavored base—such as chocolate or fruit puree—egg foam, gelatin, and whipped cream.
- **Custard** is a liquid thickened with eggs. For desserts, the liquid is usually milk. Some custards are baked, some are stirred; and some are boiled.
- **Pudding** is similar to custard, but is thickened with a starch, such as tapioca or flour.

FIGURE 22-3
Crème Brulee is a custard topped with a crust of caramelized sugar. Can you describe the contrast between the crust on crème brulee and the custard?

Common desserts that do not have to be cooked or baked before serving include:

- Fruit in all its forms is a popular, refreshing, and delicious dessert. It may be served on its own, or to complement other foods, such as cheese. It is frequently provided as a healthy, nutritious alternative to sweet desserts which tend to be high in fat and calories. (Refer to Chapter 18, Fruit and Nuts, for more information on fruit.)
- Cheese is often served as a separate course after the meal or with dessert. (Refer to Chapter 17, Dairy and Eggs, for more information on cheese.)
- Ice cream, sorbet, and other frozen desserts.

  Americans eat an average of 2.7 gallons of ice cream a year, as well as soft-serve (or **frozen) custard**, sorbet, sherbet, frozen yogurt, and shaved ice. But they are not alone in their love of frozen desserts. Examples of international frozen treats include:

  - **Gelato:** A dense, flavorful version of ice cream, enjoyed in Italy, along with sorbetto—sorbet—and granite, which is sorbet that is frozen in a way that leaves it in crunchy, crystalline flakes.
  - **Glacé:** The French version of ice cream, although it is more similar to frozen custard.
  - **Kulfi:** A frozen treat made in India. It combines milk, sugar, water and flavoring. Often, the flavorings are savory, such as pistachio or saffron, rather than sweet.

FIGURE 22-4
Fruit sorbets are often garnished with fresh fruit and mint. Can you think of a time when you might use sorbet as a garnish for something else?

Check the Label

Types of cakes:

Butter cakes Made with solid fats—although it may not actually be butter. They may be called **shortened cakes** because they include shortening.

Chiffon cakes Made with oil instead of a solid fat. They usually include a large number of eggs which gives them airy lightness.

Foam cakes Also called **sponge cakes**, these are made without fat, oil or shortening of any kind. Air is beaten into the egg whites to develop a light, fluffy texture.

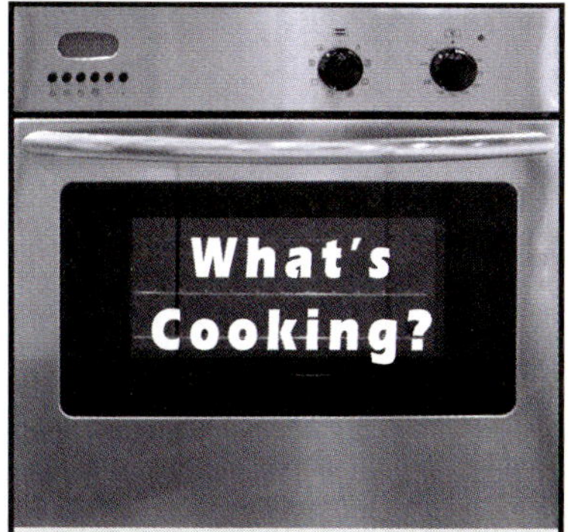

Frosting, or **icing**, is a concentrated sugar mixture used to top cakes, cookies, and other pastries. Some frostings are cooked like candy, similar to fudge or syrup. Other frostings are uncooked, made by combining powdered sugar and butter with a liquid such as milk, cream, or fruit juice.

Beverages and Dessert

Beverages such as coffee and tea are usually served after dessert. Some people prefer to have them both at the same time, however, so that the beverage can complement or balance the richness of the dessert. Milk is commonly served with chocolate desserts and baked goods, because the flavors blend well together.

FIGURE 22-5
Why do you suppose so many people like to have coffee or tea with dessert?

In some cases, beverages may served as the dessert itself—a frozen drink such as a milkshake or ice cream soda, for example.

Buying and Storing Desserts

Fresh desserts are usually available at supermarkets and specialty shops, such as bakeries and ice cream parlors. Most are made using methods similar to those you would use to make the item at home. The difference is that they are made in larger quantities.

You can buy some desserts, such as ready-made pastry shells and different types of dough, in the refrigerator section, and they may also be available frozen.

Packaged desserts come in boxes or bags, or may also be frozen.

- Many packaged desserts, such as cookies and ice cream, contain preservatives to extend their shelf life.
- Some desserts are sold frozen so they last longer. You simply defrost the item and serve.
- You can also purchase dry mixes to bake cakes, cookies, and other desserts.

The way you store a dessert depends on the type of dessert it is.

- Before storing baked desserts, such as cookies, let them cool completely on wire racks. Then, store them in airtight containers at room temperature.
- Desserts containing fruit, custard, or mousse should be covered and stored in the refrigerator. In general, they should be used within three days.
- Frozen desserts, such as ice cream, should be stored in the freezer until just prior to use. You may have to let them thaw a bit before serving, so you can scoop or cut the appropriate serving size.

Hot Topics

You may find premade mixes convenient, and in some cases, they even cost less than baking an item from scratch. Usually, you add fat, liquid, and eggs, and then bake the item for the specified amount of time. The quality varies from product to product, so you might have to test a few varieties in order to find one that you like.

BASIC CULINARY SKILLS

Creaming Method of Mixing

1. Prepare the pans and preheat the oven, as directed by your recipe.
2. Sift the dry ingredients.
3. Place the sugar and fat into the bowl of a mixer.
4. Cream the ingredients on medium speed until the mixture is smooth, light, and fluffy. Use a paddle attachment if possible.

5. Add the eggs one at a time, beating well after each addition.
6. Scrape down the sides of the bowl as necessary.

7. Add the dry ingredients alternating with the liquid ingredients, mixing on low speed, until just blended.
8. If called for by your recipe, stir in garnishes until evenly combined. Do not overmix.

Mixing

You must use the proper mixing method in order to be certain that there is an appropriate amount of air in the batter, and that the ingredients are combined correctly.

- **Creaming method:** Most cakes and cookies, and some pastries, use the creaming method of mixing. Creaming blends the sugar and fat thoroughly until light and fluffy, ensuring that all ingrediants are completely incorporated.
- **Warm foaming method:** Sponge cakes and some mousses are mixed using the warm foaming method, which relies on a warm egg-foam base to develop a light, airy texture.
- **Cold foaming method:** Angel food cakes and some mousses are mixed using the cold foaming method, which relies on a meringue-base in order to incorporate as much air as possible into the batter.
- **Rubbed dough method:** Flakey pie crusts and some other pastries are made using the rubbed dough method, in which you cut the fat into the dry ingredients before adding the liquid. The small pea-sized pieces of fat melt during baking, leaving the crust with a flakey texture.

BASIC CULINARY SKILLS

Warm Foaming Method of Mixing

1. Prepare the pans and preheat the oven, as directed by your recipe.
2. Sift the dry ingredients.
3. Combine eggs and sugar in a bowl, over a pan of simmering water.
4. Stir the mixture constantly until it reaches 110° F.

5. Remove the mixture from the heat and whip on high speed for 5 minutes.

6. Whip the mixture on medium speed for 10 to 15 minutes.
7. Fold in the dry ingredients in several batches.

8. Combine a small portion of the batter with the fat, and then fold it into the rest of the batter.

9. Pour the batter into the prepared pans and bake as directed.

10. Cool as directed.

Hot Topics

Creative chefs experiment with sweet and savory flavors. Chocolates, traditionally made with sweet tastes such as caramel and cherry, are now flavored with spices like curry or wasabi. Ice cream may feature avocado or lavender, or combine anise with black pepper, or jalapeno with coffee.

BASIC CULINARY SKILLS

Rubbed Method Pie Dough

1. Chill the shortening or butter until very firm.
2. Chill the liquid as specified by your recipe.
3. Sift the dry ingredients.
4. Cut the fat into cubes, ½-inches square.
5. Place the fat and the dry ingredients together in a bowl.
6. Break the fat into pea-sized pieces by rubbing the mixture with your fingers in the bowl. Alternatively, use two knives and cut the mixture until the fat is pea-sized.

7. Add the liquid and mix carefully until the dough comes together and forms a ball. Don't overmix.
8. Wrap the dough in plastic wrap and chill for several hours before using.

Baking

Successful baking depends mostly on temperature and time.

- Always preheat the oven to the temperature specified by your recipe. A preheated oven helps make sure cookies, cakes, and pastries cook evenly.
- Arrange the pans so that air can circulate around them.
- Avoid opening the oven door during baking, as it will cause the temperature to fall.
- Fill the pans accurately. Too much batter will overflow; too little batter will result in uneven browning.
- Cookies should be spaced so that they have room to spread without running into each other.
- Remove cookies from the pan promptly. They will continue to bake for a bit, and will harden as they cool.
- Use a cooling rack to cool all baked goods, so the air can circulate all around the items.

FIGURE 22-10
Bar and other types of cookies should be baked on a cookie sheet. Why do you think the shape of a cookie sheet is important?

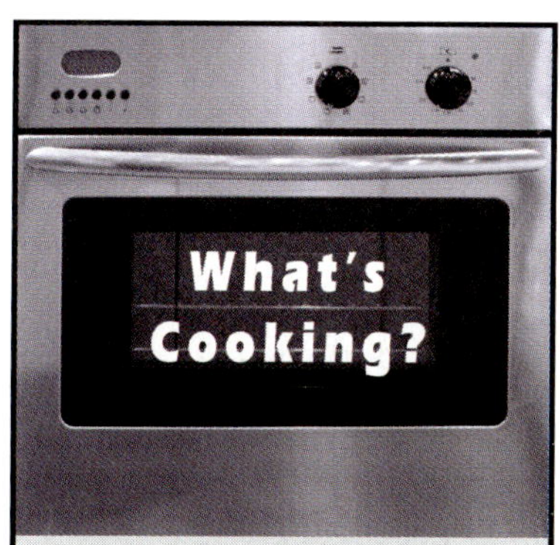

Custards can be baked, stirred, or boiled:

Stirred custards are cooked slowly on the stove. Examples include sauces, such as Crème Anglaise and bases for ice creams and mousse.

Baked custards are baked in a water bath. Examples include crème brulee, crème caramel, and custard pies.

Boiled custards are cooked on the stove. Examples include pastry cream and cream pie fillings.

FIGURE 22-11
Use cream and fruit between layers to build a tasty and beautiful cake. How might you finish the outside of this cake?

A dessert is done when it has achieved the proper texture and consistency. For some, like ice cream, this means being completely frozen. For custards, it means being thick and creamy. For cakes, it means being baked through.

When baking cakes, there are three ways to check for doneness:

- Look for even browning, and a slight gap along the sides where the cake has pulled away from the pan.
- Gently touch the center of the cake with your fingertip. It should not leave an impression.
- Insert a toothpick into the center of the cake and pull it out. If it comes out clean, with no moist batter stuck to it, the cake is done.

In a restaurant, there are many chefs responsible for different parts of the meal. A **pastry chef** is the top person in the pastry section. He or she is highly skilled in making desserts, pastries and many other baked goods, and may have assistants to help. Typical responsibilities include ordering, costing, and menu planning, developing new recipes, and, of course, preparing the desserts.

BASIC CULINARY SKILLS

Frosting a Two-Layer Cake

1. Let the cake cool completely.
2. Place the first layer upside down on the plate.
3. Place the appropriate amount of frosting in the center of the layer, and spread it evenly to the edge.

4. Place the top layer, bottom side down, on top of the frosted layer.

5. Use a spatula to spread the frosting around the sides of both layers.

6. Place frosting on the top of the cake. Use the spatula to spread the frosting evenly to the edge.

FIGURE 22-12
Finishing refers to the way you assemble, frost, or garnish a dessert. What do you think is the purpose of decorating a cake?

Cool Tips

You must allow baked goods to cool completely before you apply frosting, or the frosting will melt.

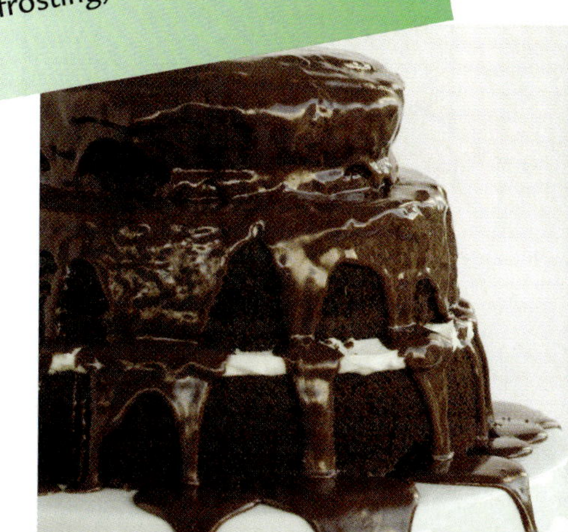

BASIC CULINARY SKILLS

Rolling Pie Dough

1. Allow dough to properly chill and rest, and then gently press it into a flat cylinder.
2. Wipe your work surface with a damp sponge, and then spread plastic wrap or wax paper. The moisture will help keep the paper from sliding.
3. Sprinkle the paper with flour.
4. Place the dough in the center of the paper and cover it with a second piece of paper.
5. Begin rolling from the center of the dough toward the edge. Do not roll back and forth, but rather lift the rolling pin at the outer edge and place it back in the center. Repeat rolling from the center out in all directions to form a circle.
6. Continue until the dough is ⅛-inch thick and 1 inch larger than the pie plate.

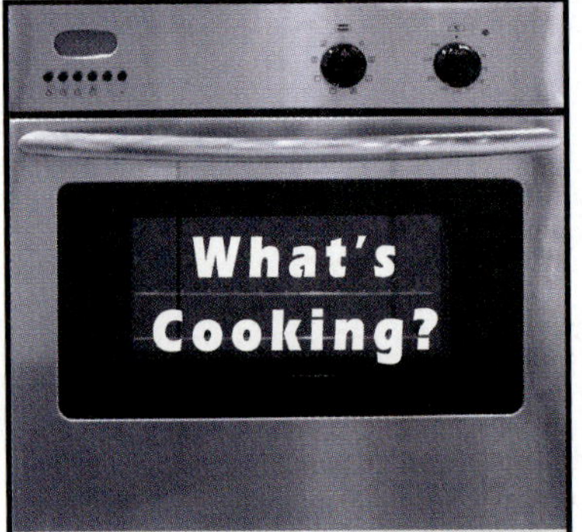

Some pie crusts are **blind-baked**, which means they are pre-baked before filling. Blind-baked crusts are usually used when you don't want to bake the filling. To keep a blind-baked crust from bubbling, line it with parchment paper and fill it with weights during baking.

FIGURE 22-13
A lattice topped pie has strips of dough evenly spaced and woven together across the top of the filling. How do you think you can seal the edges of the strips to keep them from coming apart?

FIGURE 22-14
Streusel is a crumb topping made by mixing flour, sugar, other dry ingredients, such as nuts or oatmeal, with butter. It's often used to top fruit pies, cobblers, or crisps. Why might you use a streusel topping instead of a top crust?

BASIC CULINARY SKILLS

Assembling a Pie

1. Roll the dough to the correct size.
2. Place the pie plate to one side of the rolled dough.
3. Remove the top piece of paper. Grasp one edge of the bottom paper and gently lift it over and across the pie plate, centering the dough over the plate with the paper side up.
4. Peel off the top paper.
5. Gently press the dough into the bottom and sides of the plate without stretching or tearing the dough.

For a one-crust pie:

1. Trim the overhanging pastry so that it is ½ inch larger than the plate.

2. Fold the extra pastry back and under to form a high edge, and then flute the edge by pinching it with your fingertips.

3. Fill and bake as directed.

For a two-crust pie:

1. Trim the overhanging pastry at the edge of the plate, as in the one-crust directions.
2. Fill the pie shell as directed, as in the one-crust directions.
3. Moisten the flutted edge of the bottom pastry with water.
4. Cover the pie with the top crust, and trim the crust so that it is ½ inch larger than the plate.
5. Fold the extra edge of the top pastry under the edge of the bottom pastry. Seal the edge by pressing with your fingertips on the edge of the pan. Flute the edge.
6. Cut several slits near the center of the top pastry to allow steam to escape during baking.

FIGURE 22-15
Flute the edges of a pie to make it look more attractive. Why is it important to serve desserts that look attractive?

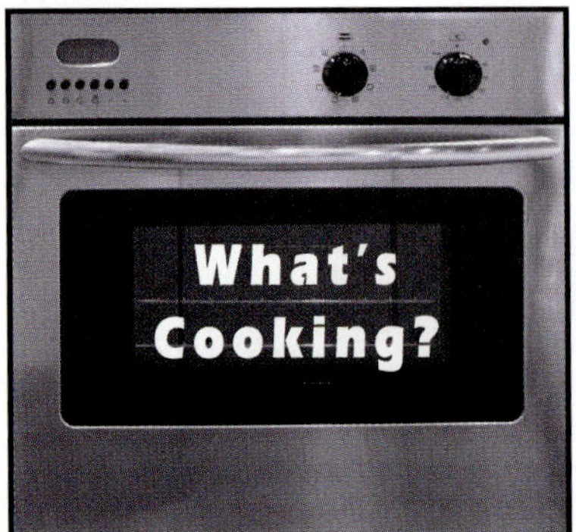

Chocolate is the main ingredient or flavoring in many desserts. Chocolate is made from cleaned cocoa kernels, called **nibs**, which are milled into a thick paste, called **unsweetened chocolate, chocolate liquor**, or **baker's chocolate**. This paste is then made into the three basic types of chocolate used for cooking:

- Dark chocolate is made by adding cocoa butter, sugar, vanilla, and other flavorings to chocolate liquor. Bittersweet dark chocolate must contain at least 35% chocolate liquor. Semisweet dark chocolate can contain from 15 to 30%.
- Milk chocolate is a combination of milk powder and dark chocolate. It must contain at least 12% milk solids and 10% chocolate liquor.
- White chocolate isn't really chocolate at all, but a combination of cocoa butter, sugar, milk powder, and flavorings.

If you bake at high altitudes, you'll soon learn that—starting at about 2,000 feet above sea level—you'll have adjust your ingredients, cooking times, and temperatures in order to achieve the desired result. Here are some general rules for baking at higher altitudes:

- Water boils at a lower temperature at high altitudes, which means it takes longer to cook foods in or over liquid; custards will take longer to set; dense moist cake batter may resist setting or crust over on top before the interior gets hot enough to set; and pie crust can over-brown on top before the fruit inside bakes through.
- Water evaporates faster at high altitudes, which will change the ratio of liquids to solids, leaving a higher concentration of sugar and fat in the batter or dough. This can cause the batter or dough to set too slowly, have a coarse texture, or collapse.
- Finally, leavening gases expand faster at high altitudes, which can cause baked goods to rise too fast, and then collapse.

There's no strict formula for adjusting your recipes and baking methods. Some ingredients will give directions for use at high altitudes. Otherwise, you might have to resort to trial and error.

Fiction	Fact
Chocolate is full of empty calories and has no nutritional value at all.	Some studies claim that dark chocolate contains antioxidants, which have been shown to reduce the risk and severity of some types of heart disease.

Case Study

Kareem and Jannelle plan to bake a birthday cake for their grandmother. They have time to bake together on Saturday morning, and on Sunday morning. The party is Sunday afternoon.

It is their grandmother's 60th birthday, so they want the cake to be special. They know she loves strawberries and chocolate, so they would like to include those flavors in the cake. She also likes citrus flavors, like oranges and lemon.

They have both baked cakes before, but they usually use a store-bought mix. This time, they plan to make the cake from scratch.

- What advice can you give Kareem and Jannelle as they plan the cake?
- What type of cake do you think they should bake?
- How do you think they could include all of their grandmother's favorite flavors?
- What steps can they take to make sure the cake is special?

Put It to Use

1. Assemble the ingredients for baking a chocolate cake from scratch, and for baking one using a store-bought mix. Compare the ingredients, the nutritional value, and the cost. If possible, bake them both, and compare the taste and texture, and the ease of preparation. Select the one you think is the best value, and explain why.

Put It to Use

❷ Plan the menu for a dessert buffet that must serve 25 people. Think about the different types of dessert you would want to include, and the flavors and styles of each. Make sure you include beverages, and options for people on special diets. Draw a picture of how you would set up the buffet.

Write Now

Imagine that you run a restaurant and have just decided to start making desserts in-house, instead of buying them from someone else. Think about the employee(s) you will need to help you achieve this goal, and then write a job description for the position(s). Explain the education and experience required for the job, describe the working environment, and list the responsibilities. If there is time, pair up with a classmate and act out the process of interviewing candidates for the position.

Tech Connect 

Use the Internet to research different types of cake pans, cookie sheets, and other baking tools. Compare the costs and the qualities. Make a list of the items you think are most important to have in a well-equipped baking kitchen.

Team Players

As a class, organize a cake auction to raise money for a charity in your community. Set a date and distribute announcements inviting people to attend. You might want to invite another class to participate as well. Then, work in teams of three or four to design, bake, and decorate the cakes for the auction. You should design a cake that will be visually appealing to many people, so they will bid up the price. Be creative and artistic, by using the elements and principles of design. For example, consider the elements of line, shape, texture, form and color, as well as the principles such as balance, proportion, harmony, and unity. Work together to select the type of cake, size, and decoration, and to plan a schedule so the cake will be complete in time for the auction. Assign tasks for baking, assembling and finishing the cake. Display the cakes—or photographs of the cakes—before the auction so people can see them. Be prepared to explain how you made the cakes, what ingredients you used, and the nutritional value. Auction the cakes, and donate the proceeds to the charity.

Put *It* Together

Match the explanation in column 1 with the term in column 2.

Column 1

- **a.** a rich fluffy dessert made by adding air to a blend of a flavored base, egg foam, gelatin, and whipped cream
- **b.** a custard topped with a crust of caramelized sugar
- **c.** cookies made from a soft batter that is spread into a pan before baking
- **d.** cookies made from a firm dough or batter that holds its shape on a sheet pan
- **e.** cakes made with solid fats
- **f.** cakes made without fat, oil, or shortening of any kind
- **g.** a dense, flavorful version of ice cream that originated in Italy
- **h.** a frozen treat made in India that combines milk, sugar, water, and flavoring
- **i.** a concentrated sugar mixture used to top cakes, cookies, and other pastries
- **j.** a type of pan that has a fastener on the side that you release in order to lift the pan off the dessert
- **k.** a crumb topping made from butter, flour, sugar, and other dry ingredients

Column 2

1. gelato
2. sponge cakes
3. frosting
4. kulfi
5. springform
6. mousse
7. drop cookies
8. streusel
9. crème brulee
10. butter cakes
11. bar cookies

TRY IT!

Chocolate Mousse

Yield: 10 Servings Serving Size: 5 oz

Ingredients

10 oz Bittersweet chocolate
1½ oz Butter
5 Eggs, separated
1 fl oz Water
2 oz Sugar
8 fl oz Heavy cream, whipped

Method

1. Combine the chocolate and the butter.
2. Melt over a hot-water bath.
3. Combine the egg yolks with half the water and half the sugar.
4. Whisk the egg yolk mixture over a hot-water bath to 145° F for 15 seconds.
5. Remove from the heat.
6. Whip until cool.
7. Using a large rubber spatula, fold the egg whites into the egg yolks.
8. Fold the butter-chocolate mixture into the egg mixture.
9. Fold in the whipped cream.

Recipe Categories

Desserts

Chef's Notes

You can add additional flavoring (such as espresso powder) to the mousse after you fold in the whipped cream.

Potentially Hazardous Foods

Eggs, Dairy

HACCP

Cook to an internal temperature of 145° F or higher.

Cool to below 41° F within 4 hours (1-stage cooling method) or within 6 hours (2-stage cooling method).

Refrigerate at 41° F or below.

Nutrition

Calories	279
Protein	6 g
Fat	21 g
Carbohydrates	15 g
Sodium	94 mg
Cholesterol	190 mg

TRY IT!

Oatmeal Raisin Cookies

Yield: 12 Dozen Cookies Serving Size: 2 Cookies

Ingredients

2¼ lb Flour, all-purpose
1 oz Baking soda
½ oz Cinnamon, ground
½ oz Salt
3 lb Butter, soft
1 lb, 3 oz Sugar
3½ lb Light brown sugar
10 Eggs
1 fl oz Vanilla extract
3 lb, 3 oz Rolled oats
1½ lb Raisins

Method

1. Line sheet pans with parchment.
2. Sift together the flour, baking soda, cinnamon, and salt.
3. Cream the butter and sugars on medium speed with a paddle attachment until the mixture is smooth and light in color, about 10 minutes. Scrape down the bowl periodically.
4. Blend the eggs and vanilla.
5. Add the egg-vanilla mixture to the butter-sugar mixture in three additions. Mix until fully incorporated after each addition and scrape down the bowl as needed.
6. On low speed, mix in the sifted dry ingredients and the oats and raisins until just incorporated.
7. Scale the dough into 2-oz portions.
8. Arrange on the prepared sheet pans in even rows.
9. Bake at 375° F until the cookies are light golden brown, about 12 minutes.
10. Allow to cool slightly on the pans.
11. Transfer to racks and cool completely.

Recipe Categories

Desserts, Cookies

Chef's Notes

Alternatively, the dough may be scaled into 2-lb units, shaped into logs 16 inches long, wrapped tightly in parchment paper, and refrigerated until firm enough to slice. Slice each log into 16 pieces and arrange on the prepared sheet pans in even rows.

Potentially Hazardous Foods

Eggs

HACCP

Cook to an internal temperature of 145° F or higher.

Cool to below 41° F within 4 hours (1-stage cooling method) or within 6 hours (2-stage cooling method).

Refrigerate at 41° F or below.

Nutrition

Calories	340
Protein	5 g
Fat	14 g
Carbohydrates	50 g
Sodium	270 mg
Cholesterol	45 mg

TRY IT!

Marbleized Pound Cake

Yield: 6 2-lb Loaves Serving Size: 2 oz

Ingredients

As needed Butter or spray oil (for preparing loaf pans)
3 lb, 4½ oz Flour, cake
1½ oz Baking powder
2 lb, 5½ oz Butter, soft
2 lb, 5½ oz Sugar
½ oz Salt
30 Eggs, beaten
12 oz Chocolate, bittersweet, melted and cooled

Method

1. Coat the loaf pans with a light film of fat, or use appropriate pan liners.
2. Sift together the flour and baking powder.
3. Cream together the butter, sugar, and salt on medium speed with the paddle attachment, until the mixture is smooth and light in color, about 5 minutes. Scrape down the bowl as needed.
4. Add the eggs to the butter and sugar mixture in three additions.
5. Add the sifted dry ingredients, mixing on low speed until just blended. Scrape down the bowl as needed.
6. Transfer one-third of the batter to a separate bowl.
7. Stir the melted chocolate into the batter in the separate bowl.
8. Fold the chocolate batter into the plain batter just enough to swirl the chocolate throughout. Do not blend evenly.
9. Scale 2 lb of batter into each prepared loaf pan.
10. Bake at 350 °F until a skewer inserted near the center of a cake comes out clean, about 50 minutes.
11. Cool the cakes in the pans for a few minutes before transferring them to racks to cool completely.

Recipe Categories

Desserts, Cakes

Chef's Notes

Use the tip of a sharp paring knife to cut through the top crust of the cakes after they have baked for 30 to 35 minutes. This creates a neat, attractive split in the top of the fully baked cakes.

Potentially Hazardous Foods

Eggs

HACCP

Refrigerate at 41° F or below.

Nutrition

Calories	220
Protein	4 g
Fat	11 g
Carbohydrates	30 g
Sodium	130 mg
Cholesterol	25 mg

TRY IT!

Angel Food Cake

Yield: 5 Tube Cakes Serving Size: 1 Slice

Ingredients

2½ lb Sugar
½ oz Cream of tartar
15½ oz Flour, cake
1½ tsp Salt
40 Egg whites
1 Tbsp Vanilla extract

Method

1. Sprinkle the insides of five 8-inch tube pans lightly with water.
2. Combine half the sugar with the cream of tartar.
3. Sift together the remaining half of the sugar with the flour and salt.
4. Whip the egg whites and vanilla to soft peaks, using the whip attachment on medium speed.
5. Gradually add the sugar and cream of tartar mixture to the egg whites, whipping on medium speed until medium peaks form.
6. Gently fold the sifted sugar and flour mixture into the egg whites until just incorporated.
7. Scale 15 oz of batter into each prepared tube pan.
8. Bake at 350° F until a cake springs back when lightly touched, about 35 minutes.
9. Invert each tube pan onto a funnel or long-necked bottle on a wire rack to cool.

Recipe Categories

Desserts, Cakes

Chef's Notes

Alternatively, invert a small ramekin on a wire rack and prop the cake pan upside down and at an angle on the ramekin. Allow the cakes to cool completely upside down. Carefully run a palette knife around the sides of each pan and around the center tube to release the cake. Shake the pan gently to invert the cake onto the wire rack.

Potentially Hazardous Foods

Eggs, Dairy

HACCP

Cook to an internal temperature of 145° F or higher.

Nutrition

Calories 138
Protein 3 g
Fat 0 g
Carbohydrates 31 g
Sodium 74 mg
Cholesterol 0 mg

TRY IT!

Pie Crust

Yield: 6 lb, 6 oz Serving Size: 1 oz

Ingredients

3 lb Flour, all purpose
½ oz Salt
2 lb Butter, cut into pieces and chilled
16 fl oz Water, cold

Method

1. Combine the flour and salt thoroughly.
2. Gently rub the butter into the flour, using your fingertips to form large flakes for an extremely flaky crust or until it looks like a coarse meal for a finer crumb.
3. Add the water all at once.
4. Mix until the dough just comes together. It should be moist enough to hold together when pressed into a ball.
5. Turn the dough out onto a floured work surface and shape into an even rectangle.
6. Wrap the dough with plastic and chill for 20 to 30 minutes.
7. The dough is ready to roll out now, or it may be held under refrigeration for up to 3 days or frozen for up to 6 weeks.
8. Scale the dough out as necessary, using about 1 oz of dough per 1 inch of pie pan diameter.
9. To roll out the dough, work on a floured surface and roll the dough into the desired shape and thickness, using smooth, even strokes.
10. Transfer the dough to a prepared pie or tart pan.
11. The shell is now ready to fill or bake blind.

Recipe Categories

Desserts

Chef's Notes

Thaw frozen dough under refrigeration before rolling it out.

Potentially Hazardous Foods

None

HACCP

Refrigerate at 41° F or below.

Nutrition

Calories	133
Protein	2 g
Fat	8 g
Carbohydrates	12 g
Sodium	115 mg
Cholesterol	16 mg

TRY IT!

Apple Pie

Yield: One 9-inch Pie   Serving Size: 1 Slice (4 oz)

Ingredients

1¼ lb Pie dough
1½ lb Golden delicious apples, peeled, cored, and sliced
5 oz Sugar
½ oz Tapioca starch
¾ oz Cornstarch
½ tsp Salt
½ tsp Nutmeg, ground
½ tsp Cinnamon, ground
1 Tbsp Lemon juice
1 oz Butter, melted

Method

1. Preheat the oven to 375° F.
2. Prepare the pie dough according to directions.
3. Divide the dough into 2 equal pieces.
4. Roll half of the dough ⅛-inch thick.
5. Line the pie pan with the rolled pie dough.
6. Reserve the other half, wrapped tightly under refrigeration.
7. Toss the apples with the remaining ingredients.
8. Fill the pie shell with the apple mixture.
9. Roll out the remaining dough ⅛-inch thick.
10. Place it over the filling.
11. Crimp the edges to seal.
12. Cut several vents in the top of the pie.
13. Bake at 375° F until the filling is bubbling, about 45 minutes to 1 hour.

Recipe Categories

Desserts

Chef's Notes

Serve warm or cool to room temperature before serving.

Potentially Hazardous Foods

None

HACCP

Cool to below 41° F within 4 hours (1-stage cooling method) or within 6 hours (2-stage cooling method).

Refrigerate at 41° F or below.

Nutrition

Calories	265
Protein	2 g
Fat	13 g
Carbohydrates	37 g
Sodium	211 mg
Cholesterol	16 mg

A Careers

Choosing a Career

When considering an occupation in culinary arts, it is important to focus on your interests, values, and abilities. Understanding yourself makes it is easier to select the right occupation. There are many different career opportunities in the culinary arts field, so learning how to use resources to research occupations will make it easier for you to choose a career.

Put It to Use

Use a word processing or spreadsheet program to make tables listing your interests, values, and abilities. Prioritize each list so that the most important item is at the top. You can refer to the list throughout your job search to help you identify a fulfilling career.

Interests

Your **interests** mark what you like to do and what you do not like to do. Recognizing your interests helps you to make a good career choice. If you discover that many of the tasks listed in an occupation are not interesting to you, reconsider your choice, and research careers that match your interests.

Values

A **value** is the importance that you place on various elements in your life. Money might be more important to you than leisure time. Working with people might be more important to you than what shift you work. Spending time with your family might be more important to you than earning a promotion. Knowing what you value avoids compromising the things that are most important to you. This also helps you prioritize your work-related values, such as job security, leisure time, wages, recognition, creativity, advancement, working environment, home life, responsibility, and management. Ask yourself the following questions.

- **Job security.** Is it important that you find a job immediately upon the completion of your training program? How important is job availability?
- **Leisure time.** Is it important for you to have extra time for leisure activities?
- **Wages.** Is an average wage acceptable if you like your work, or is a very high wage necessary?
- **Recognition.** Is it important that the job you choose is respected by the people in your community?
- **Creativity.** Do you like to come up with new ideas to solve problems, or do you prefer a job in which there is exactly one way to do things?
- **Advancement.** Do you want a career that provides opportunities for promotion?

- **Working environment.** Do you prefer to work indoors or outdoors?
- **Home life.** Do you want to work a daytime schedule (9 to 5) with some overtime and with weekends and holidays off, or are you willing to do shift work (all hours, any day of the week)?
- **Responsibility.** Do you want a job that requires you to make a number of decisions?
- **Management.** Do you want to be responsible for supervising the work of other people or for organizing many tasks at once?

All of these factors affect your job choice. Make a list of these work values in order of their importance to you. When you research an occupation, refer to your list so you do not choose a job that conflicts with many of your values.

Abilities

An **ability** is something you do well. You have many abilities. For example, you may work well with your hands, or you may be very good at mathematics. It is much more pleasant to work in an occupation that uses your abilities. If you choose an occupation that is too far below your ability level, you will be bored. If it is too far above your ability level, you will be frustrated. It is important to evaluate your abilities during your career search. List your abilities, and use the list when researching an occupation. Match your abilities to the job description.

Education

Once you have decided to pursue a career in the culinary industry, it is important to choose a quality education program that fits your interests. Culinary education and training involves formal education, certification, continuing education, professional development, and the establishment of a professional network.

- **Formal education.** A sound and thorough culinary education is the logical first step in the development of your culinary career. Employers look for applicants who have culinary degrees. There are more than 800 schools in the United States alone that offer some form of post-secondary culinary education. Some schools are dedicated to the culinary arts; others are part of a community college or university. Schools may offer programs that result in an associate or bachelor's degree. Master's programs with a strong emphasis on food, as well as degrees in related fields such as nutrition and food science, are also important to professional chefs. The best culinary schools incorporate lots of hands-on applications in their curriculum.
- **Apprenticeship.** This provides a way to achieve a formal education without attending culinary school. The apprenticeship program sponsored by the American Culinary Federation (ACF) combines on-the-job training with technical classroom instruction.

FIGURE A-1
Food service industry professionals must have specialized training in order to succeed. What courses can you take in high school to help you advance in the food service industry? Make an appointment with a career counselor to discuss your educational goals and make a plan for how you can achieve them.

- **Certification.** Certification provides a way to prove that you have met certain standards in the culinary field. A certification is recognition of your skill level. Typically, a certification program involves a specific level of experience in the field, course work, and passing a written and practical cooking examination. To maintain your certification, you will need to refresh your knowledge and provide documentation of continuing education and professional development.
- **Continuing education.** Once you have achieved your initial training, you must keep your skills current. The culinary professions are constantly evolving, and you will need to attend classes, workshops, and seminars to hone your skills in specialized areas and to keep up with new methods and styles of cooking.
- **Professional development.** As your career progresses, you should join professional organizations; read professional magazines, newsletters, and books; and participate in culinary competitions. Many culinary arts careers have professional associations which provide support, scholarships, continuing education, workshops, and other ongoing benefits. Following is a list of some of the associations, and links to their websites where you can find additional information.

  American Dietetic Association, www.eatright.org

  International Association of Culinary Professionals, www.iacp.com

  American Culinary Federation, www.acfchefs.org

  National Restaurant Association, www.restaurant.org

  American Hotel and Lodging Association, http://ahma.com
- **Establishing a professional network.** You will want to network with other professionals to gain insight, recommendations, or guidance. Perhaps you can find a mentor or coach to help you reach a new level in your career. You can also mentor others, when appropriate.

Put It to Use

Find out if your school hosts a chapter of a career and technical student organization such as FCCLA. Attend a meeting to learn more. If you are not already a member, join. If you are already a member, find a way to become more active. If there is no chapter, contact the state organization to learn how to start one.

Career and Technical Student Organizations

While in school, you can join a career and technical student organization. These organizations provide a valuable opportunity for students to explore career possibilities and to participate in activities, conferences, and competition. They provide leadership training, and give you access to information about jobs, internships, and scholarships.

Student organizations are an excellent way to gain practical experience and skills that employers look for when they are hiring. These organizations emphasize leadership and communications skills. They encourage members to become involved in meetings and learn how to manage people and events.

The Family, Career and Community Leaders of America (FCCLA) is an example of a student organization that you could join. Many schools have chapters of FCCLA, or you can contact the organization in your state to join. You can learn more about FCCLA on their Web site, at www.fcclainc.org.

Setting Your Goals

Goals are things you want to achieve over time. Short-term goals, such as doing well on a particular exam, can be achieved quickly. Long-term goals, such as becoming an executive chef or purchasing a home, will take time to achieve.

To start, identify your goals, rank them in order of importance, and classify them as long-term or short-term. Then you can look for schools and internship programs that offer training in your areas of interest. You can assess these programs on the basis of how long you would study, and how much it could cost. You can also look for jobs for which you are already qualified. These jobs can provide valuable experience that will help you learn more about yourself, your chosen field, and how you will be able to achieve your long-term goals. You can then develop a timeline to help you plan how you will work toward achieving your goals.

Preparing Your Resume

To start any job search, you need a resume. At first, this will be a functional resume—a summary of your job abilities. It tells prospective employers about your skills, education, work experience and personal interests. The main purpose of a resume is to get you an interview.

- **Contact information.** Start your resume with your name and contact information.
- **Objective.** Your objective describes the job you want. You can customize the objective depending on the job for which you are applying.
- **Education.** List your formal training, including degrees and certificates that you have earned. Include dates, the name of the school or institute, and the location.
- **Skills.** List the skills you want to highlight and give examples of your performance.
- **Work experience.** List your most recent job experiences first. Include the employer's name, the location, the job title, and the dates you worked.
- **Professional involvement and awards.** List special awards, honors, and accomplishments that highlight your contributions to the food service profession, for example, if you won a cooking contest.

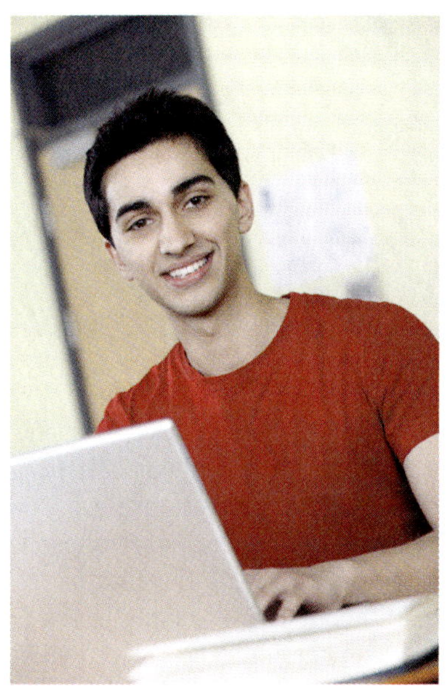

FIGURE A-2
You send a resume to prospective employers so they can quickly assess your qualifications. What are some things you should list on your resume? What are things you should omit? Use a word processing program to create a resume. Most have templates that make it easy. Ask a teacher or career counselor to look it over and make suggestions for improvements.

- **Personal interests.** List activities you can discuss during an interview to show that you are a well-rounded, interesting person.
- **References.** A reference is a person who will describe your character or job abilities. You should maintain a list of three or four people to offer as references. They should know you well in either a personal or professional capacity; they should be willing to discuss you with a potential employer; and they should be able to verify the information on your resume. Former employers are the best references. People who have supervised you in a volunteer organization are also good, as are teachers who know you well. Be sure to ask their permission before using them as a reference.

  Some references may provide a letter of reference that you can keep, copy, and give to prospective employers. In any case, you should maintain a reference list that includes the name, title, and contact information for each reference; you can hand this out to prospective employers.

Locating Job Opportunities

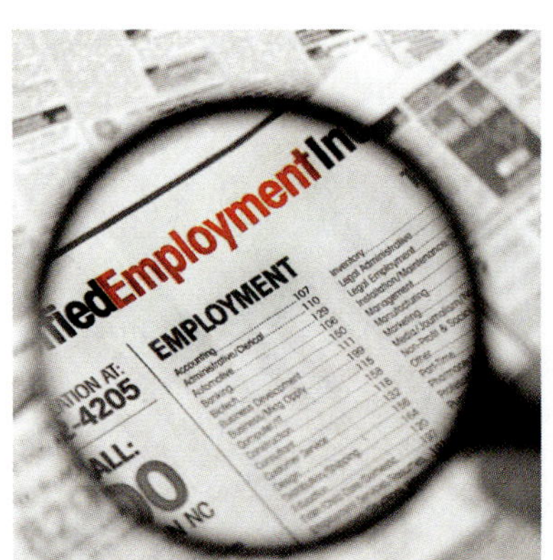

FIGURE A-3
Classified ads are just one of many ways to locate job opportunities. Why is it important to use every available resource to learn about potential jobs? Use the Internet to learn about jobs available in the food service industry. Select two or three and write classified ads for each one. Share them with your class.

There are many resources, including reference books and the Internet, that you can use to find potential jobs. Occupational research will help you learn about the tasks to be performed, the job outlook, the education required, and the working environment. It requires time and effort to research the occupations that interest you and to prepare for a specific career, but this will result in a greater chance of job satisfaction.

The *Dictionary of Occupational Titles* (http://www.occupationalinfo.org/) lists job titles, tasks, and duties for 20,000 occupations, and the *Occupational Outlook Handbook* (http://www.bls.gov/oco/) discusses the nature of the work, employment outlook, training and qualification requirements, earnings, and working conditions for a variety of occupations. *Work Briefs* by Science Research Associates, the Career Exploratory Kit, the *Encyclopedia of Careers*, and computer programs are also good resources.

Use available resources to prospect for opportunities.

- **General mailings and phone calls.** Send your resume to all companies that you are considering. Ask about job openings in your cover letter. Follow up with a phone call to request an informational meeting.
- **Apply in person.** Pounding the pavement is a legitimate job-search method in the hospitality industry. Dress professionally, bring along multiple copies of your resume, and be polite and respectful to everyone you meet.
- **Internet advertisements.** Look on company Web sites for employment opportunities. Many businesses post available positions online, and even let you apply online. There are also general employment sites that organize job opportunities according to category or field. Many of these sites let you post your resume online so potential employers can find you.

- **Classified ads.** Read your newspaper and check the online publications to find out what jobs are available locally. Check trade publications such as Nation's Restaurant News, as well.
- **Employment agencies.** Agencies are hired by the employers to screen job applicants. They work for employers, so you should act professionally at all times when talking to them or meeting them in person. You should never have to pay a fee; the employer will pay the agency if you are hired.
- **Networking.** One of the best ways to find a job is by networking. Your network includes your personal contacts—friends, family, teachers, acquaintances, etc. The more you talk about your goals and ambitions, the more people will know about what you are looking for. You should always be prepared to exchange contact information with people you meet, so you can get in touch with them to discuss opportunities. You can expand your network by becoming involved in clubs, activities, professional organizations and volunteer work.
- **Informational interviews.** An informational interview offers a chance to learn about a company or specific career from a professional on the inside. Some companies hold regular information sessions; others may respond if you request a meeting. It is best if you contact the specific person working in the position that interests you. Make it clear that you are not applying for a job, but are looking for information about the type of experience you need and the type of skills you should acquire.

Applying for a Job

Once you identify a job—or jobs—in which you are interested, you must begin the application process.

- **Job application.** Most employers will ask you to complete a job application form that requests your name, contact information, education, work experience, and references. The information you put on the application should be consistent with your resume. Make sure you have access to all of the relevant information so that you can complete the application accurately. For example, bring along a list of references, the addresses and telephone numbers of your previous employers, and the dates when you worked.
- **Cover letter.** A cover letter is a letter of introduction that you submit with your resume. It should identify the job for which you are applying and explain how you heard about the opening. If someone recommended you, be sure to include his or her name. You can also use your cover letter to highlight the skills and abilities that make you well-suited for the position.
- **Phone calls.** If a job advertisement asks you to respond by phone, write a cover letter and use it as a script for your conversation. Be sure to ask for an interview, and to thank the employer for speaking to you.

FIGURE A-4
You should submit a cover letter whenever you submit your resume. What type of information can you put in a cover letter that you might not be able to put in a resume? Write a cover letter you might send when applying for a job in a restaurant.

- **Interviewing and demonstrating.** An interview is a chance for you and the employer to get to know each other. It is important to make a positive impression, and to show the employer that you would be an enthusiastic and hard working employee. Dress professionally, and make sure you are well-groomed, clean, and neat. Exhibit positive personality traits, such as friendliness, honesty, and a willingness to learn.

  Bring your resume and reference list, and be ready to talk about yourself professionally and personally. Bring a pen and pad of paper to take notes during your conversation, and be prepared to ask questions about the company, the position, and even the application process; for example, you should ask when they will contact you again.

  You may be asked to prepare, or describe the preparation of, a dish. Practice in advance if possible, and remember to highlight safety and sanitation.

Put It to Use

Practice your interviewing skills with a classmate. One of you should be the interviewer, and the other should be the applicant. Dress and act professionally, and use excellent communication skills. Swap roles and repeat the process.

- **Follow-up**. Always write a thank-you note immediately after an interview. It should be short and polite, and thank the employer for his or her time. You can also take this opportunity to remind the employer of the skills you possess that match those needed for the job.

  If you call to follow up on the status of your application, be positive and appreciative. If they have hired someone else, ask them to keep your name and resume on file for future opportunities.

- **Career portfolio.** As you gain experience, assemble a portfolio that highlights your work and abilities. Bring this along on interviews to show prospective employers. A portfolio can include:
  - Diplomas and certificates.
  - Letters of reference.
  - Letters of commendation, awards, and prizes.
  - Written job reviews by employers.
  - Your special recipes.
  - Photos of your creations.
  - Reviews of restaurants where you have worked.
  - Articles or recipes you have published.

- **Organization.** Keep your job search materials organized. Maintain a master list that includes each of the prospects that you have identified. Record the company name and contact information, and the dates on which you deliver your resume or call them, as well as your next steps. Set up a real or virtual folder for each company, storing all of your correspondence and notes.

Jobs and Professions

There are hundreds of job opportunities in the food service industry. They can be divided into the following general categories:

- Chefs and cooks
- Banquets and institutions
- Caterers and private chefs
- Research and development
- Diet and nutrition
- Management
- Food communications

Chefs and Cooks

Chefs and cooks can determine the reputation and success of a restaurant. The primary goal is to please the customer. Chefs typically supervise cooks. Both must work independently and as a team, often under a great deal of pressure in tight quarters.

Executive Chef

The role of executive chef is a management position. They are responsible for a large kitchen, or even a chain of kitchens. The executive chef must have a diploma or certificate from a school or organization that can grant the title "Certified Executive Chef." Someone who earns this title may place the initials C.E.C. after his or her name.

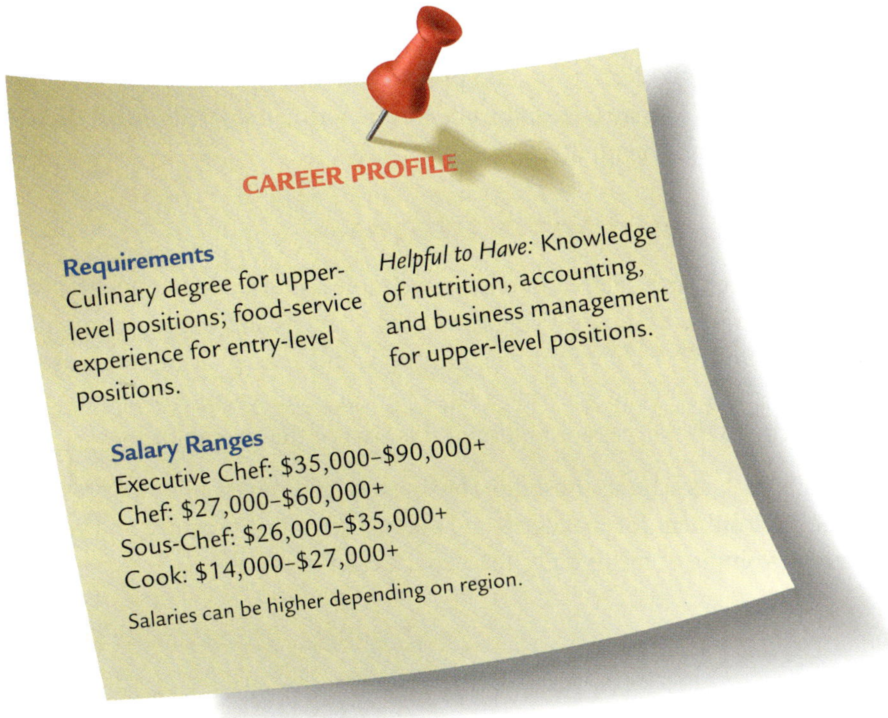

Put It to Use

Contact a local restaurant and ask them to describe the qualifications of an executive or head chef. If possible, talk directly with the person in that role. Ask about his or her experience and education. Write a paragraph summarizing the information, and share it with the class.

Head Chef

A head chef is also in a management position. Head chefs supervise professional cooks. If there is no executive chef, the head chef is in charge.

Sous-Chef

The sous-chef is the second in command, reporting to the chef who is in charge of the kitchen. This is the lowest managerial or administrative position in the kitchen.

Sauté Cook

Sauté cooks are responsible for sautéed items and their sauces. This is often considered the most glamorous of the various cooks' jobs. It demands experience, stamina, split-second timing, an excellent memory, and the ability to multi-task.

Grill Cook

Grill cooks are responsible for the preparation of all grilled or broiled items. Grill cooks have the same job requirements as sauté cooks.

Fish Cook

Fish cooks must know the various types of fish and shellfish and understand the anatomy of fish.

Apprentice or Prep Cook

Apprentices or prep cooks are the positions in which most people begin their culinary careers. The work is very important, though not highly skilled. Apprentices generally clean, trim, and prepare vegetables for stocks, soups, and salads. They also may be responsible for preparing the salads, salad dressing, and other simple menu items.

Banquets and Institutions

Banquet managers and chefs at hotels, convention centers, and banquet halls serve large numbers of diners, using their own kitchens and dining rooms. Many operations rely on conventions, meetings, and special events, such as weddings or receptions, to generate a large part of their revenue. The banquet chef is crucial to the success or failure of this part of the operation.

Institutions such as schools, hospitals, businesses, and nursing homes present a unique environment for managers and chefs. They must organize and prepare a high volume of meals daily, for a specific set of clients.

Banquet Manager and Chef

Banquet managers and chefs must develop menus that can be adjusted to meet the guests' requirements. Menus must be carefully priced so that customers perceive them as a value, while the business generates the greatest possible profit. These managers also manage staff, and organize many operational details, ranging from the color of the tablecloths to the timing of the service.

In a large facility, several events may take place on the same day, even at the same time. The advance preparation, cooking, and service requirements demand concentration, endurance, and skill. They work in public view and must have strong interpersonal skills to accommodate customer needs.

Institutional Chef

Institutional chefs—also called high-volume chefs—are responsible for cooking food appropriate for a specific institutional setting. Some institutions use cafeteria-style service, with multiple choices for each course, as in a school cafeteria. In others, they prepare meals for individuals with special dietary needs, such as in a nursing home or hospital. Management and organizational skills are critical for this position; there is little contact with the public.

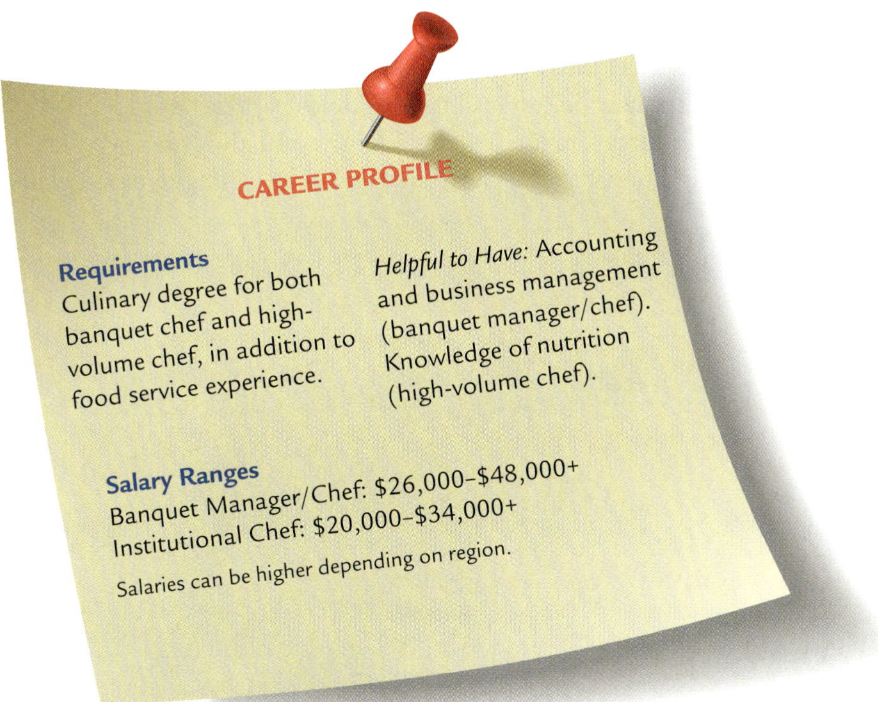

Put It to Use

Visit a banquet hall or institution (your own school cafeteria will do) to learn more about what they do. Ask if you can have a tour of the facility. If possible, meet with someone who works there and ask about his or her experience and education.

Put It to Use

Use an Internet search engine to look up job opportunities for caterers and private chefs. Use the information you find to write an advertisement for either a caterer or a private chef looking for work.

Caterers and Private Chefs

A caterer's job is similar to that of a banquet manager or chef. They develop menus, manage support staff, and organize details. The major difference is that caterers usually travel to different locations to prepare and serve the meal.

Private chefs may be hired by families or businesses to provide meals for individuals or small groups.

Caterer

Caterers must be creative chefs who can plan and price a pleasing menu *and* deal with clients. They must be experts in all aspects of food service, from preparation to presentation. Visualizing an event from beginning to end, they may be called on to provide dishes, cooking supplies, tableware, and paper goods as well as the ingredients required. A caterer may be responsible for ordering portable refrigerators and stoves, or for renting tables and chairs. People who choose catering as their vocation must be extremely organized and effective managers. They are often entrepreneurs who run their own business.

Private Chef

Private chefs oversee the operation of their employer's kitchen. They must be easygoing, versatile, and creative. Private chefs must develop menus that avoid repetition and meet the specific, and often changing, requirements of their employers. They may be responsible for dinner parties—from a small dinner for eight to a reception for 100. For a large event, they may need to hire and manage additional staff. Private chefs must be willing to have flexible schedules.

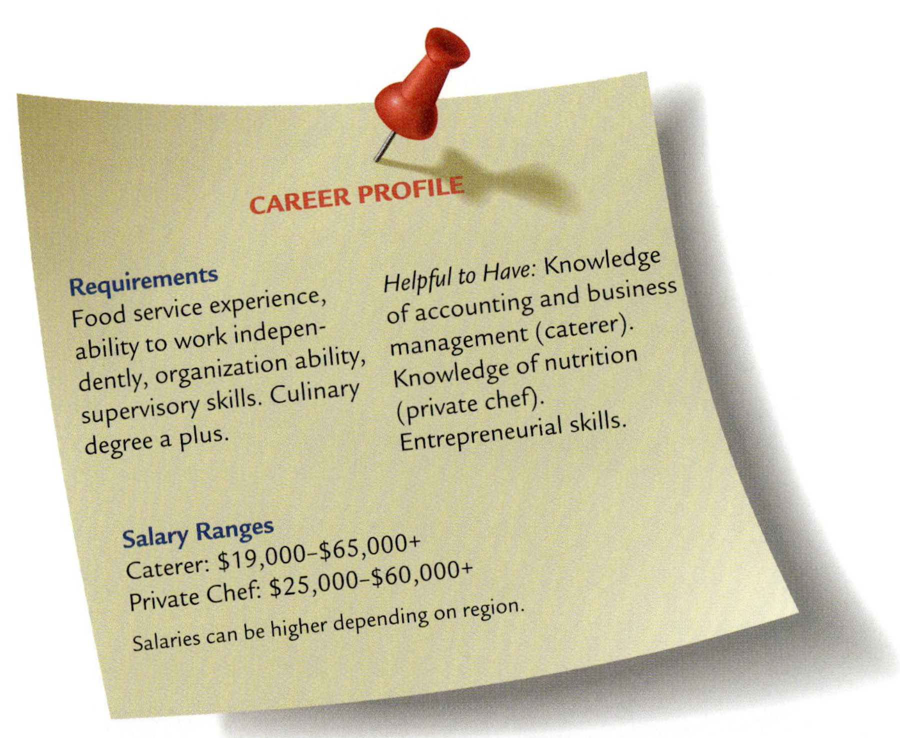

Research and Development

Companies that produce food products are always looking for new items to capture the interest of the buying public. Before a new product is introduced, it must undergo exhaustive evaluation, testing, and research.

In addition, before recipes can be published in books or on Web sites, or shown on television, they must undergo extensive testing. These recipes are checked by independent testers for accuracy.

Test Kitchen Researcher

Test kitchen researchers help major food companies, restaurant chains, television shows, and specialty food producers develop new products. Using their culinary skills in professional test kitchens, researchers analyze how the product responds when heated, refrigerated, stored on a shelf, or frozen. Test kitchen professionals must be organized, methodical, and scientific in their approach to their work.

Test Kitchen Recipe Developer

Test kitchen recipe developers work for food magazines, television shows, and Web sites. They must be able to develop recipes that will showcase the flavor, texture, color, or nutritional characteristics of a particular food. Recipe developers must research the specific food and present it in a way that will be useful for the home or professional cook. This work may offer an opportunity to break into the world of food styling or food photography.

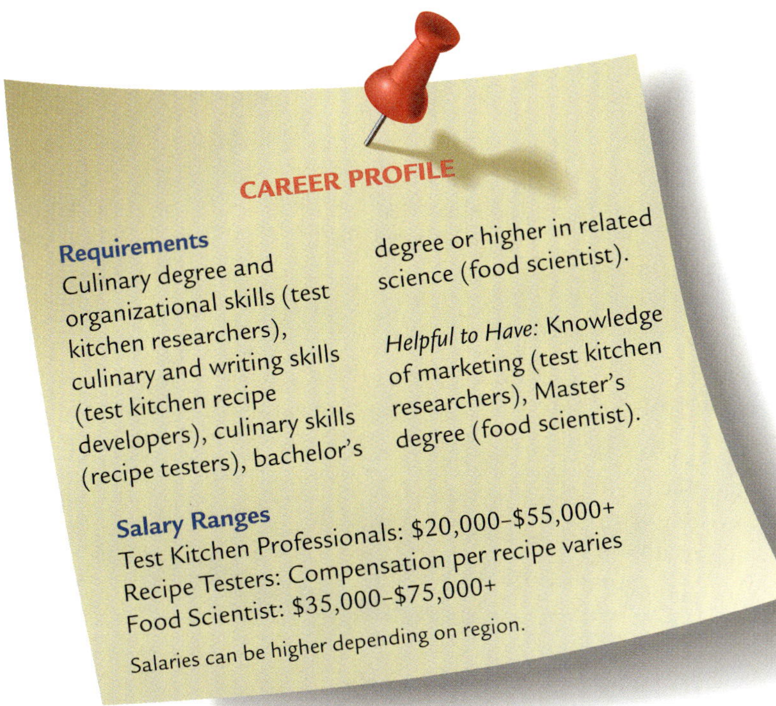

Recipe Tester

Recipe testers are hired by cookbook authors, Web site content developers, and television producers to check recipes before they are offered to the public. Each recipe must be tested to be sure it is easily followed, that all measurements are accurate, and that the end product is a dish that is appealing, tasty, and attractive. Recipe testers are generally paid a flat fee per recipe and are reimbursed for the food items purchased to prepare the recipes.

Food Scientist

Food scientists study the scientific properties of food and develop methods for its safe processing, preservation, packaging, storage, and transportation. They look for ways to improve the flavor appearance, and nutritional value of food, as well as make it more convenient to use. They might specialize in one particular area, such as canning or dehydration, or on a particular product, such as seafood or fruit. They must achieve a bachelor's degree in food technology or a relevant science and may need a graduate degree as well.

Diet and Nutrition

Dietitians work in places such as schools and hospitals, where they create nutrition programs and oversee the preparation and serving of meals. They educate people on healthy eating habits and conduct research to prevent and treat illnesses and encourage healthy living.

Put It to Use

Contact the American Dietetic Association and request information about education and jobs. Find out if they have any programs for students interested in careers in diet and nutrition. Make a poster about the organization, and display it in your classroom.

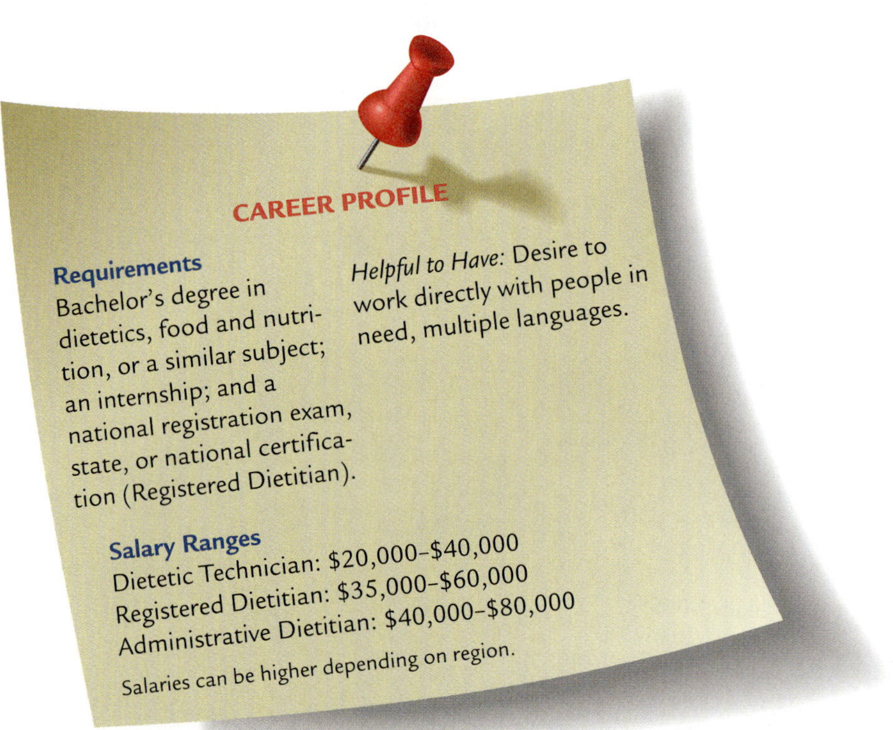

CAREER PROFILE

Requirements
Bachelor's degree in dietetics, food and nutrition, or a similar subject; an internship; and a national registration exam, state, or national certification (Registered Dietitian).

Helpful to Have: Desire to work directly with people in need, multiple languages.

Salary Ranges
Dietetic Technician: $20,000–$40,000
Registered Dietitian: $35,000–$60,000
Administrative Dietitian: $40,000–$80,000

Salaries can be higher depending on region.

Dietitian

Dietitians work in a variety of settings to provide nutritional care and counseling. Clinical dietitians work in institutions, such as medical centers and extended care facilities. They prepare menus and offer advice to ill patients. They may also manage the food service department.

Community dietitians work in public health facilities, such as health clinics. They provide education and develop personal nutrition plans. Opportunities are also available for nutritional consultants, who work independently with clients seeking counseling. Dietitians and nutritionists must work closely with individual clients who may be ill or have increased risk for disease. In addition to extensive knowledge of the nutritional value of food, and how diet affects health and wellness, they must have strong communication skills. They must also be willing to abide by a code of ethics established by the American Dietetic Association. A bachelor's degree and dietetic internship leading to certification as a Registered Dietitian (RD) are required, and for some positions, a graduate or master's degree is necessary as well.

Management

There are many types of managers in the culinary profession. Some, like executive chefs, manage large kitchens and all of their staff. Mid-level managers might be responsible for the servers in a restaurant, or act as assistants to higher management. Lower-level managers are needed for specific areas within a food service business.

Small business owners, such as caterers and restaurant owners, manage their own companies. The success or failure of their business depends on their decision-making skills.

Manager

Managers in a food service business usually serve in a specific function. For example, a large hotel restaurant might employ a purchasing agent or storeroom supervisor. This person would need a strong background in math and accounting. Managers typically need strong computer skills and specific management training relating to their position, such as safety and sanitation education.A manager must be able to organize work that is done by others. The ability to communicate effectively, including providing positive and negative feedback to workers, is a critical job skill. In some situations, being multilingual is an advantage.

Managers are sometimes required to work long hours.. They must be able to perform any job function required of their subordinates, in case they have to cover for someone unable to work.

Put It to Use

Invite the owner or manager of a local restaurant to visit your class to discuss his or her experience and education. Prepare questions and treat the visitor politely and with respect. Write an article about the visit for the school newspaper or Web site.

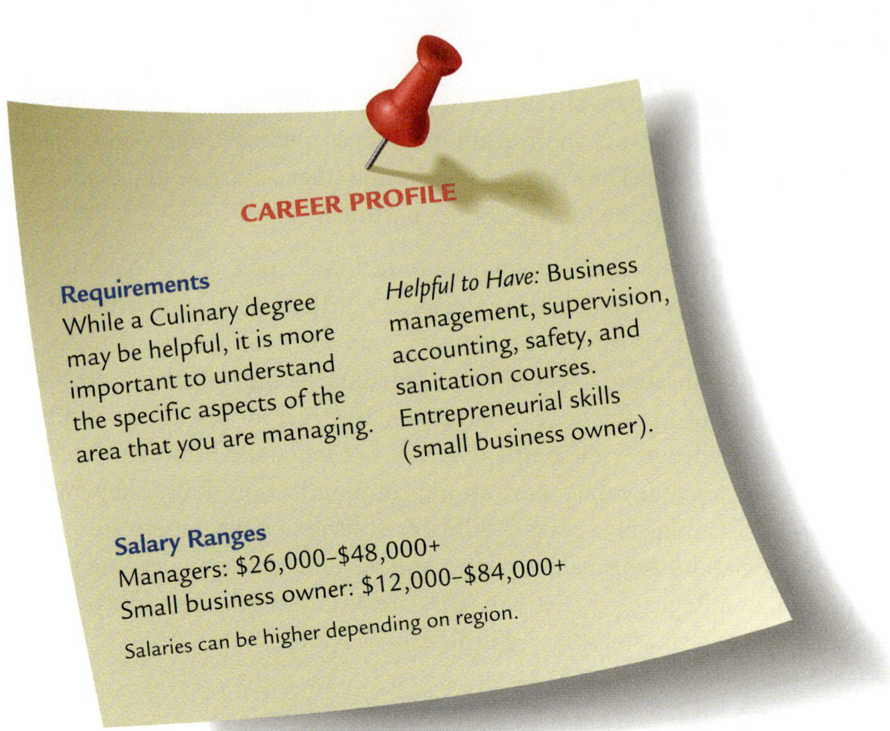

Small Business Owner

Small business owners must have a full complement of skills—to perform specific tasks, run a business, and manage others. They are usually risk-takers, with a take-charge, can-do attitude. These people are willing to work long hours and to make decisions regarding everything from where to order ingredients to whether or not to hire a worker. Among other skills, they need math and accounting skills (to run the financial aspects of the business), business skills, marketing and communication skills, computer skills, and culinary arts skills. Small business owners have to understand and implement all of the regulations governing the business, work well with others, and command respect.

Food Communications

Food communications is the coverage of culinary topics in the media, which includes newspapers, magazines, books, radio, television, and the Internet. Opportunities range from television chef to food writer, as well as the support staff that works behind the scenes to develop and produce the content.

Food Writer

Food writers are communicators with a strong, basic knowledge of food and cooking. Some food writers edit or write books. Others submit articles to magazines, newspapers, and Web sites. A food writer with good presentation skills might be featured on a local radio or television program, or broadcast on the Internet. Many have journalism, public speaking, or writing degrees in addition to culinary training.

Put It to Use

If you are interested in a career as a food writer or critic, read food articles in magazines or on Web sites. Then, write a story or review about a local restaurant.

If you are interested in being a food photographer or stylist, look for pictures and video of food. Then, try your hand at taking photographs or making a video that shows off food items or a meal.

Food Critic

Food critics are reviewers who understand what good food, good cooking, and good service is all about. They are able to discuss the style of a restaurant and trends in the restaurant business. Critics write or present reviews in magazines and newspapers, on television, the radio, or online. Sometimes they remain anonymous, so that they can visit a restaurant without receiving special treatment. Successful food critics rate establishments in an objective and straightforward manner, and present the information in an entertaining way.

Food Photographer

Food photographers make food look visually appealing in print or online. This takes creativity and patience. They are often hired by magazines, cookbook publishers, and marketing or advertising companies on an as-needed basis.

Food Stylist

Food stylists work with food photographers to prepare and place food so that it will look fresh and appealing. Their culinary knowledge is critical; they have to select the best product, apply the right technique, and present the item expertly and correctly.

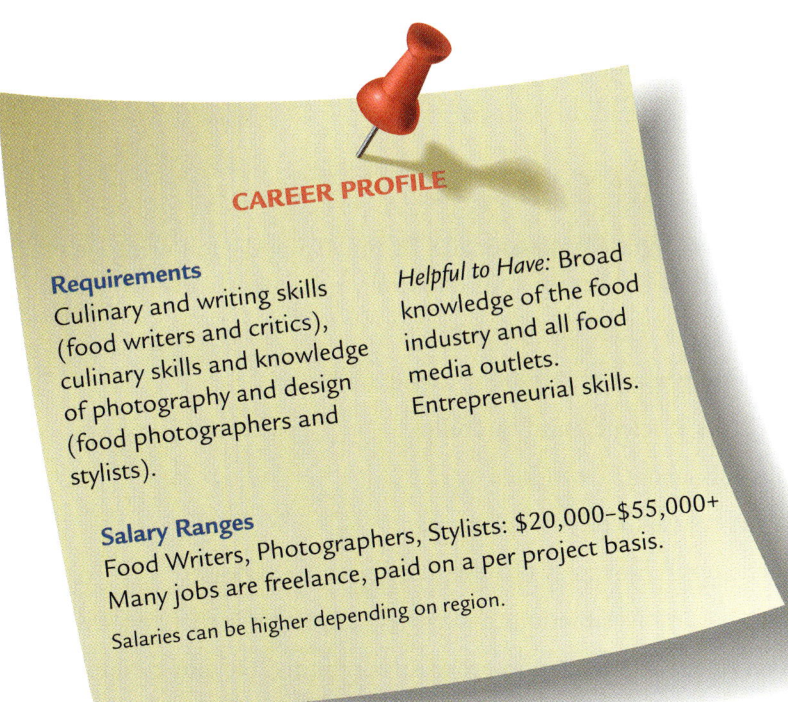

FIGURE A-5
It is important to career success to be professional in both appearance and behavior. What should you wear to a job interview? Make a presentation or poster that shows the Do's and Don'ts of professional appearance and behavior. Display it to your class.

Professional and Personal Characteristics

Once you are hired, you have a responsibility to your employer and your customers to maintain a professional appearance and attitude.

Standards of Appearance

A professional appearance makes a statement about your commitment. You should always be well-groomed, clean, and dressed in a professional manner. The following are recommendations for maintaining a well groomed, professional appearance.

- Dress according to your facility's dress code.
- Wear clean and appropriate shoes every day.
- Keep your hair clean.
- Follow rules of good hygiene:
- Brush your teeth at least twice a day.
- Floss daily.
- Use mouthwash or breath mints.
- Bathe daily.
- Use unscented deodorant.
- Wear your hair up and off your collar.

Standards of Behavior

You should always behave professionally. By practicing the following set of behaviors, you show respect for your employer, customers, and co-workers.

- Maintain a calm, courteous manner.
- Listen carefully when others are speaking with you.
- Perform tasks efficiently and carefully.
- Do not gossip about co-workers.
- Do not use coarse or offensive language.
- Do not engage in horseplay or other dangerous behaviors.
- Watch for hazardous situations and correct any hazards that you see.
- Follow all safety and sanitation procedures required by your facility.
- Uphold ethical principles, such as honesty, integrity, and fairness.

Personal Characteristics

In addition to professional behavior, workers who exhibit the following personal characteristics will make a positive impression on their employers, co-workers, and customers.

- **Honesty.** Be fair, truthful, and morally upright.
- **Dependability.** Be reliable or trustworthy.
- **Willingness to learn.** Have the openness to admit that you don't know the answer or that you can be helped to understand a situation more fully.
- **Patience.** Have the ability to put up with waiting, delay, or provocation without becoming annoyed or upset, or to act calmly when faced with difficulties.
- **Acceptance of criticism.** Have the ability to deal with disapproval or a suggestion that something can be improved.
- **Enthusiasm.** Show excited interest in or eagerness to do something.
- **Self-motivation.** Be energetic, ambitious and able to get things done without direction from others.
- **Tact.** Avoid giving offense; have an intuitive sense of what is right or appropriate.
- **Competence.** Have the ability to do something well, measured against a standard, especially ability acquired through experience or training.
- **Responsibility.** Accountability; the state of being accountable to somebody or for something.
- **Discretion.** Show tact; the good judgment and sensitivity needed to avoid embarrassing or upsetting others.
- **Teamwork.** Work cooperatively with other people, subordinating personal interests in order to achieve a common goal.

Communication

When you work in the field of culinary arts, you must be able to communicate with a wide range of people, including co-workers, employers, customers, and vendors. Understanding how people communicate will help you achieve your goals, and succeed in the workplace.

Verbal Communication

Verbal communication includes spoken and written messages.

- **Spoken messages.** When you speak to someone, you send a message. The tone of your voice, the language you use, and the message you send are interpreted by the receiver. Always speak clearly and concisely. This ensures that your message is understood.
- **Written messages.** You communicate frequently with the written word. You take and write orders. You might write notes in a recipe file and you might need to leave instructions for fellow workers. It is important to spell correctly, use proper grammar, and write in a clear and concise manner.

Put It to Use

With a classmate or a small group, practice nonverbal communication. Use gestures, facial expressions, and body language to communicate an emotion. Create a skit to demonstrate how miscommunication can lead to mistakes in a kitchen or other food service environment.

Nonverbal Communication

Communication also takes place in nonverbal ways. It is not necessary to speak in order to send a message. You send messages with your eye contact, facial expressions, gestures, and touch.

- **Eye contact.** Making eye contact with the person with whom you are communicating is important. Eye contact lets others know that you are paying attention. When you do not make eye contact, you send others a message that you are not interested or that you wish to avoid them.
- **Facial expressions.** A smile sends a different message than a frown does. It is possible to say something very kind and still send a message of anger with your eyes. Try to think of an instance when you knew that what was being said was not what was meant. How did you know? The expression on the sender's face probably sent you a different message.
- **Gestures.** Shrugging your shoulders, turning your back, and leaving the room while someone is talking to you certainly convey a lack of interest in the sender's message. You need not say, "I am not interested" because you have effectively sent that message through your gestures.
- **Touch.** Touch can convey great caring, warmth, concern, and tenderness. It can also convey anger, rejection, and distaste. Touch is a very important part of your communication. It is important that your nonverbal communication be supportive and positive.

Barriers to Communication

Recognizing barriers to communication allows you to become a better food service worker. The following are four major communication barriers:

- **Labeling.** Deciding the other person is mean, lazy, a complainer, or difficult causes a breakdown in communication. You do not pay attention to the message being sent. If you listen, you might find out the reason for the behavior.
- **Sensory impairment.** Deafness or blindness can be a communication barrier. Always evaluate the people you are communicating with to be certain that they do not have a sensory impairment.
- **Talking too fast.** It is especially important when you are working in a fast-paced environment to speak slowly. Communication can break down when the message is delivered too rapidly.
- **Cognitive impairment.** Cognitive impairment includes memory, perception, problem solving, emotional reaction, and idea formulation. These types of impairments might result from autism, brain injury, Parkinson's, Alzheimer's, or old age. Be careful to make sure your customer or co-worker understands what you are saying.

B Glossary

A

A la carte menu (AH LA CART) Menu on which each food item or beverage is priced and served separately.

Accident report Standard form used to report accidents to OSHA; must be filed within eight hours. Also used to describe an event related to a serious health problem or physical injury to a customer; should include the customer name, any server involved, the date, and the time it occurred.

Actual cost method Menu-pricing method in which the actual cost for the raw food, labor, other expenses, and profit are all added together to determine a menu price.

Aerating Adding air to food.

Aflatoxin (aff-la-TOX-in) Toxin produced by molds, sometimes found on legumes.

Aged (beef) Beef that undergoes a process that gives it a darker color, a more tender texture, and a fuller flavor.

Al dente (al DEN-tay) Italian expression meaning "to the tooth;" used to describe pasta that is cooked only until it gives a slight resistance when you bite it.

Albumen (al-BYOO-men) White part of an egg, composed of water and protein.

All-purpose flour Blend of "soft" (low protein) and "hard" (high protein) wheat; the most common type of flour used in the bakeshop.

Ambience Feeling or mood of the restaurant.

American buttercream Dense and rich buttercream icing that is just buttercreamed with powdered sugar.

American service Service style in which food is fully prepared and plated on individual serving plates in the kitchen, brought to the dining room, and served by the right hand at the right side of the guest, leaving the server's left hand free to carry other plates.

Amino acids (ah-MEEN-oh) Basic building blocks of protein, some of which our bodies make. Others, the essential amino acids, we must get from food.

Anadromous fish Fish that live part of their lives in saltwater and part in freshwater.

Antioxidants Substances, such as certain vitamins, that prevent tissue damage in the body.

Antipasto platter (an-tee-PAHS-toh) Assortment of cured meats (such as prosciutto and salami), cheeses, and pickled vegetables served on a platter.

Appetizer Dish that is served as the first course in a meal.

Appetizer salad Salad designed to whet the appetite before the main course.

Aromatic (AIR-o-mat-ic) Foods with especially strong smells.

Arson Fire that is intentionally set.

Assembly points Meeting points at a predetermined spot at a safe distance from the building where everyone should gather after a fire.

Assets Items a business owns, such as furnishings and appliances.

Assignment In terms of mise en place, the food for which you will be responsible.

Asymmetrical In terms of plate presentation, unequal numbers of items on either side of the plate.

Automated external defibrillator (AED) Device that shocks the heart into starting again.

Automatic systems Extinguishers, sprinklers, and alarms triggered by the heat of a fire.

B

Baba ghanoush (BAH-bah gha-NOOSH) Vegetable-based dip made from roasted eggplant that has been pureed and seasoned with olive oil, tahini, lemon juice, and garlic.

Back of the house In a restaurant, the kitchen area.

Back waiter Provides overall assistance to the front waiter; may deliver food and drinks to the front waiter, clear plates, and refill bread and water.

Bacteria Single-celled organism that can live in food or water and also on our skin or clothing; some are a potential biological hazard, capable of producing food-borne illness.

Baguette (bag-EHT) Long, narrow French bread with a crispy, golden brown crust and light, chewy crumb dotted with holes.

Bain marie (BANE ma-REE) Another name for a double boiler.

Baker's chocolate Chocolate that has no sugar added. Also called unsweetened chocolate or chocolate liquor.

Baking Dry heat method of cooking in which food is cooked by hot air trapped inside an oven. Baking typically means you are preparing smaller pieces of food than for roasting.

Baking stones Unglazed ceramic pieces used to line an oven rack; they help develop a crisp crust on breads and pizza by holding and transferring the oven's heat evenly.

Balance scale Scale typically used for weighing baking ingredients. Ingredients are placed on one side; weights are placed on the other side. When the sides balance, the ingredients weigh the same as the weights.

Balsamic vinegar (bahl-SAH-mek) Vinegar that takes as long as 15 to 20 years to ferment and age. It has a mellow, sweet-sour taste and syrupy consistency and is very expensive.

Bar cookies Cookies made from a soft batter that is spread into a pan before baking and then cut into individual cookies.

Barley Grain that looks like a doubled grain of rice.

Base price method Menu-pricing method that analyzes what customers want to spend per meal and then works backward to come up with menu items, their prices, and a built-in level of profit. Requires data about customers' eating habits in the restaurant over a long period.

Batch cooking Process of preparing a small amount of food several times during a service period so that a fresh supply of cooked items is always available.

Batonnet (bah-tow-NAY) Long, rectangular cut that is ¼-inch wide by ¼-inch thick and 2 to 2½ inches long.

Batter Coating option on foods made by blending a type of flour and a liquid. (8.1) Wet form of dough made from flour, oil or melted butter, eggs, milk or other liquids, salt, and usually baking powder. Used to create waffles, cakes, and other items.

Battuto (bah-TOOT-oh) Aromatic combination used in Italian soups, sauces, stews, and meat dishes; consists of cooking fat, garlic, onions, parsley, carrots, and celery.

Bavarian cream Aerated dessert similar to a mousse but without an egg foam.

Beans Legume that is longer than it is round.

Béchamel sauce (BAY-sha-mell) One of the grand sauces; a white sauce made by thickening milk with a white roux.

Belly bones Fish bones found along the thinner edge of the fillet.

Bench boxes Covered containers in which dough is bench proofed; a skin forms on the dough that holds in the carbon dioxide.

Bench proofing Brief resting period that allows gluten to relax after dough has been pre-shaped.

Bench scraper Tool with a rectangular steel blade, usually six inches wide, that is capped with a wooden or plastic handle. Used like a knife to cut soft ingredients such as butter or soft cheese, to lift and turn soft or wet dough, or to transfer ingredients from a work surface to a mixing bowl.

Bid Proposal from a supplier that states the price to be charged for an item.

Biga In baking, an Italian dough starter developed overnight or longer; can be wet or dry.

Bi-metallic-coil thermometer Thermometer that uses a metal coil in the probe to measure temperature. Oven-safe version can stay in food while cooking and gives a reading in 1 to 2 minutes. Instant-read version is not oven-safe and gives a reading in 15 to 20 seconds.

Biological hazards Living organisms such as bacteria, viruses, fungi, and parasites, which are a health risk.

Biscuits Small quick breads that have little or no sugar.

Bisque (BISK) Hearty soup made with shellfish (lobster, crayfish, or shrimp shells).

Bittersweet chocolate Dark chocolate with less sugar than semisweet chocolate.

Black pepper Dried, unripe berries of the pepper vine; used as a seasoning.

Blanching Moist heat method of cooking that involves cooking in a liquid or with steam just long enough to cook the outer portion of the food. The food is immediately placed in ice water to stop carryover cooking.

Blender Electrical mixing device used for combining ingredients by means of a rotating blade.

Blending Type of mixing in which the ingredients are chopped so the overall mixture has a uniform consistency.

Blind baked Pre-baked pie shell.

Blinis (BLEE-nees) Very thin Russian crêpes.

Blue-vein cheeses Cheeses in which needles are injected into the cheese to form holes in which mold spores multiply. The cheese is salted and ripened in a cave.

Boiling Moist heat method of cooking in which food is cooked at 212° F.

Bolster Point at the heel of a knife blade where the blade and handle come together.

Boning knife Knife about 6 inches long with a narrow blade, used to separate meat from the bone.

Bottom line Money left in a business after subtracting expenses from earnings.

Bouillon (bool-YOHN) French term for broth.

Boule (BOOL) Round loaf of white bread named for the French word for "ball."

Bouquet garni (boo-KAY GAR-nee) Combination of fresh herbs and other aromatic ingredients used to flavor dishes.

Box grater Hand tool that has four sides with various side holes; used for grating.

Boxed meat Meat that is fabricated to a specific point (such as primal, subprimal, or retail cuts) and then packed, boxed, and sold to food-service establishments.

Braising Combination cooking method in which food is first seared and then gently cooked in flavorful liquid. Braising usually indicates that the food is left whole or in large pieces.

Braising pan Pan with medium high walls and a lid to keep moisture in.

Bran Coating found on some kernels of grain, located just beneath the hull.

Brand Public image of a business, including the business name, logo, and sometimes a slogan.

Bread flour Flour that contains more protein than all-purpose flour; it is used in most yeast-bread recipes.

Brigade Group of workers, such as the staff of a restaurant, assigned a specific set of tasks.

Brioche (BREE-ohsh) French version of sweet bread with a knotted top made in individual molds with a fluted base; it can also be made into a round loaf or rolls.

Brochettes (BRO-shets) Small skewers containing grilled or broiled meat, fish, poultry, or vegetables. Brochettes are often served with a dipping sauce.

Broiler Cooking unit with a radiant heat source located above the food. Some units have adjustable racks that can be raised or lowered to control cooking speed.

Broiling Dry heat method of cooking that is similar to a grill except the heat source is above the food.

Broth Clear, thin soup made by simmering a combination of meat, fish, poultry, or vegetables in a liquid with aromatics.

Brown rice Rice with some or all of the bran still intact.

Brown sauce One of the grand sauces; a sauce with a rich brown color made from brown stock. (See Espagnol sauce, demi-glaçe, and jus de veau lié).

Brown stock Type of stock made from roasted animal or poultry bones. Brown stock has a deep reddish-brown color and a roasted meat flavor.

Brunch service Combination of a buffet-style breakfast and lunch.

Brunoise (brewn-WHAZ) Smallest dice cut, about ⅛-inch square. Means "to brown" in French.

Bruschetta (brew-SKEH-tahs) Type of open-faced sandwich served as an appetizer. It consists of toasted bread drizzled with olive oil and topped with tomatoes, olives, cheese, or other ingredients. See crostini.

Budget List of planned income and expenses.

Buffalo chopper Machine that holds food in a rotating bowl that passes under a hood where blades chop the food. Some units have hoppers or feed tubes and interchangeable disks for slicing and grating.

Buffet service Serving style practical for serving a large number of people a wide variety of dishes over a period of time; servers behind the buffet table may serve guests or guests may serve themselves.

Bulgur (BUHL-guhr) Grain cereal made from steamed, dried, and cracked wheat.

Bus person Person responsible for clearing and cleaning tables. Also called dining room attendant.

Business plan Written plan that a business owner develops to launch a new business, such as a restaurant. Includes a mission statement, goals that support the mission statement, sample menus, preliminary operating budgets, and staffing needs.

Butler service Serving style in which the server brings a platter to the table, provides a serving spoon and fork, and guests serve themselves.

Buttercream Icing made by aerating butter, shortening, or a combination of the two; used to decorate cakes and as a cake and pastry filling.

Butterfat Fat content of dairy products, measured by the weight of the fat compared to the total weight of the product. Same as milkfat.

Butterflied Split down the middle and then spread open, as with boneless meat.

C

Cafeteria service Self-service where diners choose their own foods from behind a counter or barrier. Servers dish out controlled portions from the other side and diners carry their dishes on trays to their tables.

Caffeine Chemical found in coffee, tea, chocolate, and sodas; it stimulates your body and mind.

Cajun trinity (CAGE-uhn) Aromatic combination used in Creole and Cajun cooking; consists of onion, celery, and green pepper.

Cake comb Triangular or rectangular piece of metal or plastic with serrated edges, used to create a decorative edge on iced cakes or to give texture to a chocolate coating.

Cake flour Flour that has less protein than bread or all-purpose flour; it is used in most cake recipes and many cookie and muffin recipes.

Calamari (cahl-ah-MAHR-ee) Another name for squid.

California menu Single menu listing breakfast, lunch, and dinner foods; it offers customers the freedom to order any item at any time of day.

Calories Measured units, derived from food, that provide energy.

Canadian bacon Leaner than regular bacon and similar to ham; it comes in a chunk ready for slicing.

Canapés (KAN-up-pays) Bite-sized pieces of bread or crackers with a savory topping, used as hors d'oeuvres.

Capsaicin (cap-SAY-ih-sin) Compound that gives a chile its heat; it is most potent on the white ribs inside the pepper.

Captain At fine dining restaurants, the person responsible for explaining the menu to guests and taking their orders; the captain is also responsible for the smooth running service in a specific group of tables.

Caramelize Change that takes place in food that contains sugar when it is heated. The surface of the food starts to turn brown.

Carbohydrates Energy sources for the body, made up of smaller units known as sugars.

Carcinogenic Causing cancer, such as a toxic chemical exposure.

Cardiopulmonary resuscitation (CPR) Technique used to restore a person's breathing and heartbeat.

Carpaccio (car-PAH-chee-oh) Raw beef, sliced very thinly and dressed with a sauce.

Carryover cooking Cooking that takes place in a food after it is removed from a source of heat.

Carver Person in charge of carving and serving meats or fish and their accompaniments. Also called trancheur.

Casserole Pan with medium high walls and a lid to keep moisture in.

Caviar Type of salted fish eggs; in France and the United States, only sturgeon eggs are classified as caviar.

Chafing (CHAYF-ing) dish Metal holding pan mounted above a heat source and used to keep food warm. The pan is usually contained within a larger unit that holds water. When the water is heated, the steam heats the food evenly.

Challah (HAL-la) Sweet, airy, braided bread made with a lot of eggs; it is a Jewish bread traditionally served on the Sabbath and on holidays.

Channel knife Tool used to cut grooves lengthwise in a vegetable such as a cucumber or carrot. A rondelle cut from the grooved vegetable has decorative edges that resemble a flower.

Cheese board Flat platter on which cheese is served.

Cheese cart Cart that is wheeled to the guests' table to give them an opportunity to choose cheeses of different kinds.

Chef de cuisine (CHEF duh KWEE-zine) Head chef who commands the kitchen, designs the menu, and oversees food costs. Also called executive chef.

Chef's knife All-purpose knife used for peeling, trimming, slicing, chopping, and dicing. Blade is usually 8 to 12 inches long. Also known as a French knife.

Chef's tasting Method of presenting appetizers; it is a sampler plate with an assortment of different appetizers. The portions are often only one bite, just enough to sample the various appetizers.

Chemical hazards Toxins such as metals, cleaning compounds, food additives, and fertilizer found in food and water.

Chemical leavener Baking powder or baking soda, which increases the volume of a batter by the addition of air or gas.

Chiffonade (shiff-en-ODD) Cut used for cutting herbs and leafy greens into fine shreds.

China Dishware designed to contain food, including plates, bowls, dishes, cups, saucers, and creamers.

Chlorine dioxide (KLOR-ene die-OX-ide) Chemical dough conditioner to facilitate handing of lean dough.

Chocolate liquor Chocolate that has no sugar added. Also called unsweetened chocolate or baker's chocolate.

Cholesterol (koh-LESS-ter-all) Fatty substance the body needs to perform various functions; it becomes a health risk when certain protein levels appear as elevated in the blood, indicating a possible build-up of cholesterol on the walls of arteries, reducing blood flow to the heart.

Choux paste Versatile pastry dough made from liquid, fat, flour, and eggs; used for both sweet and savory baked goods.

Chutney Sauce with a chunky texture, typically fruit- or vegetable-based and made with a sweet and sour flavoring. Served hot or cold.

Clarification Mixture of ingredients including ground meat, aromatic vegetables, and an acid such as tomatoes or lemon juice used to add flavor and clear a broth to make a consommé.

Clarified butter Butter with all water and particles removed, leaving pure fat for cooking at high temperatures.

Cleaver Cutting tool with a large, rectangular blade; available in a range of sizes and weights. Used for many of the same applications as a chef's knife.

Client base Group of customers who come to a restaurant.

Clingstone Describes fruit that clings tightly to its pit, making it difficult to cut the flesh away cleanly.

Closed sandwich Two pieces of bread with a filling between them.

Club sandwich Double-decker closed sandwich, made with three slices of bread (or toast) and traditionally filled with chicken or turkey, bacon, lettuce, and tomato.

Coating chocolate Chocolate made with vegetable fat instead of cocoa butter. Also called compound chocolate.

Cocktail sauce Dipping sauce of ketchup, horseradish, and possibly Tabasco sauce; used with shellfish.

Cocoa butter Cream-colored fat from cocoa beans; used in the chocolate-making process.

Cocoa powder Unsweetened chocolate with some of the fat removed and then ground into a powder.

Coddled eggs Eggs cooked in their shells for 30 seconds, leaving the whites warm and thickened and the yolks warm but still runny.

Colander Large, perforated stainless-steel or aluminum bowl used to strain or drain foods.

Cold food presentation Collection of cold foods that are presented in an artful manner, often in a buffet setting.

Cold storage area Kitchen area where walk-in refrigerators, reach-in refrigerators, and other large refrigeration equipment is located.

Combination steamer oven Oven powered by either gas or electricity. It can cook like a convection oven, a steamer, or both.

Complex carbohydrates Carbohydrates that contain long chains of many sugars; found in plant-based foods such as grains, legumes, and vegetables.

Composed salad Salad with any combination of ingredients (greens, vegetables, proteins, starches, fruits, or garnishes) that are arranged carefully and artfully on a plate or in a bowl.

Compote Dish of fresh or dried fruit that is slow-cooked in stewing liquid.

Compound butter Flavored butter made by blending aromatics or garnishes with softened butter, typically served with grilled meats.

Compound chocolate Chocolate made with vegetable fat instead of cocoa butter. Also called coating chocolate.

Condiments (CON-di-ments) Prepared mixtures that are used to season and flavor foods. Condiments are served on the side and added by the individual diner.

Conditioning the pan Process of letting the pan heat up before adding any oil or food, when sautéing.

Confectioner's sugar Sugar that has been ground into a fine, white, easily dissolvable powder.

Conical sieve Made of very fine mesh and shaped like a cone. Also called a chinois or a bouillon strainer. Used to strain or purée foods.

Consommé (KAHN-soh-may) Very clear broth made by simmering a broth or stock with a clarification. A consommé should be fat-free.

Continental breakfast Light breakfast of baked goods served with coffee, tea, and juice.

Continuous seating plan Seating plan that allows use of tables according to the flow of business.

Convection oven Oven with fans that force hot air to circulate around the food, cooking it evenly and quickly. Some convection ovens have the capacity to introduce moisture. Special features may include infrared and/or microwave oven functions.

Convection steamer Cooking unit that generates steam in a boiler and then pipes it to the cooking chamber, where it is vented over the food. It is continuously exhausted, so the door may be opened at any time without danger of scalding or burning.

Converted rice Rice that is parcooked before it is milled to shorten the cooking time.

Conveyer belt dishwasher Large piece of dishwashing equipment that can process a high volume of dishes as a continuous flow.

Cookware Utensils used for cooking, such as pots and pans.

Copycat method Simple menu-pricing method that involves going to a nearby restaurant that has the same menu items and copying their prices.

Corer Tool used to remove the core of an apple or pear in one long, round piece; can also be used to remove eyes from potatoes or the stem and core from tomatoes.

Corn syrup Thick, sweet syrup made from cornstarch. It is available light or dark.

Cornmeal Cereal made by grinding whole or processed kernels of corn.

Corrective action Steps a food service establishment takes to correct a problem or situation, such as food held too long at an unsafe temperature.

Corrosive Having the ability to irritate or even eat away other materials.

Cost control Keeping variable expenses in check, such as avoiding waste or conserving electricity.

Coulis (coo-LEE) Thick puréed sauce, usually made from vegetables or fruit.

Count Number of shrimp per pound.

Counter scale Countertop device used for weighing moderate size packages.

Counter service Alternative to table dining; guests sit at a counter, often on stools.

Countertop blender Blender with the motor and blades at the base and a glass, metal, or plastic container on top to hold ingredients. Also called a bar blender.

Countertop mixer Mixer used on top of a counter in small to moderate size kitchens. It can stand about 2 feet high and weigh 100 pounds.

Cover Complete place setting for one person; includes china, glassware, and flatware.

Cracked grain Coarsely ground or crushed grain kernel.

Cream soup Soup made with cream that is noticeably thick with a velvety smooth texture.

Creaming In baking, a mixing method in which fat and sugar are combined vigorously to incorporate air.

Crème anglaise Classic dessert sauce made with the stirred custard method; often used as a base for ice cream and mousses.

Crème fraîche (krehm fraysh) Cultured dairy product similar to sour cream but with more butterfat. French for "fresh cream."

Crêpe (KRAYP) Thin, French-style pancake made with very thin batter in a special crêpe pan; often folded or rolled and spread with a sweet mixture or filled with savory ingredients.

Crêpe pan Shallow skillet with very short, sloping sides; often has a nonstick coating.

Critical control point Specific time in the process of food handling when you can prevent, eliminate, or reduce a hazard.

Critical limits Measurements of time and temperature that indicate when a food is at risk and in need of a corrective action.

Croissant (kwah-SAHNT) Buttery-rich crescent-shaped yeast roll.

Cross cuts Large sections of a large drawn fish that has been cut into sections.

Cross-contamination Contamination of food that occurs when safe food comes in contact with biological, physical, or chemical hazards while it is being prepared, cooked, or served.

Crostini (kroh-STEE-nee) Type of open-faced sandwich served as an appetizer. It consists of toasted bread drizzled with olive oil and topped with tomatoes, olives, cheese, or other ingredients. See bruschetta.

Crouton (CREW-tahn) Small cube of bread that is toasted or fried until crisp and golden brown; a popular garnish for hearty soups.

Crown roast Roast prepared by tying a rib roast into a crown shape.

Crudités (kroo-deh-TAYS) Vegetables that have been cut into bite-size pieces.

Crumb topping Crumbly mixture of fat, sugar, and flour; often applied to muffins or quick breads.

Crustaceans (crus-TAY-shuns) Shellfish that have jointed exterior shells.

Cubano The Cuban version of a pressed sandwich.

Cube Large dice that is ¾ inch or greater.

Cultured Describes dairy products such as buttermilk, sour cream, and yogurt that are made by adding a specific type of beneficial bacteria to milk or cream to achieve a desired texture, taste, and aroma.

Cured foods Foods that are preserved by drying, salting, pickling, or smoking. Examples are ham, bacon, and salted anchovies.

Custard cup Baking dish that is round and straight-edged; comes in various sizes.

Custard Liquid, such as milk or cream, thickened with egg and then baked.

Cut-out cookies Cookies made of stiff dough that is rolled flat and then cut into decorative shapes, often using cookie cutters.

Cyclical menu Menu that is written for a certain period of time and then repeats itself. Some cyclical menus change four times a year, according to the seasons. Some change every week.

D

Daily values Daily requirements for nutrients, as established by the FDA; amounts are listed on nutrition labels as a metric weight and also as a percent value and are based on a 2,000-calorie diet.

Dark chocolate Bittersweet or semisweet chocolate; it is less sweet than milk chocolate.

Deadline In terms of the mise en place timeline, your completion time; when the dish you are preparing must be ready to serve.

Deck ovens Ovens stacked like shelves, one above the other, like pizza ovens. Food is placed directly on the deck instead of on a wire rack.

Deep frying Dry heat method of cooking in which foods are cooked in hot oil that completely covers the food.

Deep poaching Moist cooking method in which food is cooked in enough liquid to completely cover it.

Deep-fat fryer Freestanding or countertop unit that holds frying oil in a stainless-steel reservoir. A heating element, controlled by a thermostat, raises the oil to the desired temperature and maintains it. Stainless-steel wire baskets are used to lower foods into the hot oil and lift them out.

Demi-glaçe (DEM-ee-glahs) Type of brown sauce; made by simmering equal amounts of Espagnol sauce and brown veal stock until the sauce is intensely flavored and thick enough to coat foods.

Denaturing Altering the chemical structure of a protein, as with acid.

Denominator Bottom number in a fraction. (The top number is the numerator).

Depurated (DEP-yew-rate-ed) Shellfish that have been placed in tanks of fresh water to purge them of their impurities and sand.

Derivative sauces Sauces that use a grand sauce as the main ingredient. The grand sauce is combined with other seasonings or garnishes for a specific flavor, color, or texture.

Dessert salad Salad served as dessert often features fruits and nuts.

Deveining Process of removing the vein in a shrimp.

Diagonal cut Variation of a rondelle, cutting diagonally instead of straight down, to expose a greater surface area of the vegetable.

Dice Cut that produces a cube-shaped piece of food.

Dietary Guidelines Developed by the USDA as a method for helping people create a healthy and well-balanced diet; emphasis is on reducing risk for major diseases through diet and physical activity.

Dining room attendant Person responsible for clearing and cleaning tables. Also called bus person.

Dining room manager Person running the dining room portion of the restaurant; also responsible for training service personnel, working with the chef on the menu, arranging seating, and taking reservations. Also called maître d'hôtel or maître d'.

Dip Sauce or condiment served with raw vegetables, crackers, bread, potato chips, or other snack food.

Direct contamination Contamination of food caused by improperly storing, cooking, or serving food that causes the biological hazards in the food itself.

Disjointing Cutting poultry into halves, quarters, or eighths, before or after cooking.

Double boiler Cookware that is actually a pair of nesting pots. The bottom pot is filled with water and heated, providing steady, even heat for the top pot.

Double-panning Using two stacked sheet pans to gently heat the bottom of cookies and avoid over-browning.

Double-strength sanitizing solution Mixture of water and a sanitizer that contains twice the recommended amount of sanitizer suggested for normal use; used on food contact surfaces such as knives, meat slicers, and cutting boards during food preparation.

Double-strength stock Stock that is simmered long enough to cook away half of the water.

Dough divider Baking equipment that cuts a quantity of dough into equal pieces so they can be shaped in rolls.

Dough sheeter Baking equipment used to roll large batches of dough out into sheets.

Dough starter In baking, a dough mixture that starts the fermentation process before the final mixing of all the ingredients. The longer fermentation gives the dough time to develop more gluten strength and depth of flavor. Also called a pre-ferment.

Drawn butter Butter that is melted to use as a sauce for shellfish.

Drawn fish Whole fish that has the stomach removed.

Drop cookies Cookies made from a firm dough or batter that is dropped onto a sheet pan.

Drum sieve (SIV) Screen stretched on an aluminum or wood frame; used to sift dry ingredients or purée very soft foods.

Dry aging Process of storing meat by hanging it in a climate-controlled area to make it more tender and flavorful.

Dry cured Method of preserving food by rubbing it with salt and seasonings.

Dry goods Foods such as flour, tea, sugar, rice, or pasta.

Dry sautéing Another name for pan broiling. Dry heat method of cooking very much like sautéing except no fat is used.

Dry storage area Kitchen area where goods such as flour, dry pasta, canned goods, and supplies are stored on shelves at room temperature.

Du jour menu (DOO ZHOOR) Menu that lists food served only on that particular day. "Du jour" means "of the day" in French.

Dumplings Type of pasta; made from dough that is soft enough to drop into boiling water.

Dupes Duplicates of guest checks, passed from the dining room to the back of the house.

Dutch processed Process used for cocoa powder to make it less acidic.

E

Earnings Money coming into a restaurant; also known as income or sales.

Éclair Long, straight pastry filled with cream and glazed on top.

Effective criticism Criticism that not only points out what went wrong or where things could be better, but also indicates how you can improve.

Egg wash Mixture of egg and water or milk; brushed on the tops of breads and pastries to give a glossy sheen.

Emulsifier (e-MULL-si-fy-er) Ingredient added to an emulsion that makes an emulsion permanent.

Emulsion (e-MULL-shon) Mixture of two ingredients that would otherwise not combine; an emulsion has a uniform consistency.

En papillote (ahn pap-ee-YOTE) Fish or shellfish baked in a wrapped package, usually made of parchment paper. Often the fish or shellfish is wrapped with aromatics and vegetables.

Endosperm Largest part of a grain; it contains the food necessary to support a new plant and is made up almost entirely of carbohydrates, or starch.

English service Service style for special groups or private dinners in which the table is fully preset, food is delivered on platters to the dining room, serving dishes are placed on the table or on a table nearby, and a server serves the food to the guests.

Enriched dough Lean dough that has added butter, oil, sugar, eggs, or milk products. Also called sweet rich dough.

Entrée (AHN-tray) The main course of a meal.

Entremetier (ehn-tray-mee-tee-AY) Chef responsible for hot appetizers, pasta courses, and vegetable dishes. Also called soup and vegetables station chef.

Environmental Protection Agency (EPA) Federal agency that plays a part in regulating workplace safety along with OSHA by requiring food service operations to track any chemicals that pose a risk to health.

Espagnol sauce (ess-pan-YOLL) Type of brown sauce; made by thickening a brown veal stock with a roux.

Essential oils Quickly evaporating oils that occur in plants and their fruit and that give the plant or fruit its characteristic odor and/or flavor.

Ethylene (EH-thih-leen) Gas that accelerates the ripening and rotting process in fruit and vegetables.

Evacuation routes Escape routes that give everyone in the building at least two ways to get out of the building.

Executive chef Head chef who commands the kitchen, designs the menu, and oversees food costs. Also called chef de cuisine.

Expediter (ex-PED-eye-ter) Person who accepts orders from the dining room, relays them to the various station chefs, and reviews the dishes before service to make sure they are correct.

Expenses Money being spent by a restaurant.

Extra-virgin olive oil Finest grade of olive oil, produced by pressing olives once without using any heat. It has a fruity, grassy, or peppery taste with a pale yellow to bright green color and a very low acid content.

Extruded Act of pushing material through an opening. Pasta machines extrude dough to make special shapes, such as elbow macaroni, spaghetti, and penne.

F

Fabrication (of meat) Additional butchering done by a restaurant to break down a subprimal cut of meat into portion-sized cuts of meat.

Factor method One of the oldest, simplest methods for pricing menu items; it involves multiplying the raw food cost by an established pricing factor.

Family service Table service in which food is placed on the table in serving dishes and guests help themselves.

Farinaceous (fare-eh-NAY-shus) Rich in starch.

Farm-raised fish Fish raised in ponds or in penned waters.

Fat-soluble vitamins Vitamins A, D, E, and K; they dissolve in fat; are stored in body fat, cannot be easily flushed out once ingested, and so should not be taken in excess.

Fatty acids Small units, made of carbon, hydrogen, and oxygen atoms linked together, that are contained in a fat such as olive oil or butter.

FDA Food Code Set of recommendations for safe food handling, provided by the Federal Department of Agriculture, that may be adopted (all or in part) by local governments as law.

Feedback Review of one's work; it could come from a co-worker, a boss, or a customer.

Fermentation Chemical reaction that is triggered when hydrated yeast is mixed with food; it makes dough rise until double or triple in size.

Fermiere (FARM-ee-air) Rustic cut that produces ⅛- to ½-inch pieces.

Fettuccini (feht-too-CHEE-nee) Flat, ribbon-style pasta; may be fresh or dried.

Fillet Boneless piece of fish.

Filleting knife Knife with flexible blade; used for filleting fish.

Finger food Hors d' oeuvres that are served on a napkin and eaten with the fingers.

Finger sandwich Simple, small sandwich usually made with firm, thinly sliced pullman loaves. Can be made both as closed sandwiches and as open-faced sandwiches. Also called tea sandwich.

Fire detectors Devices that warn you about a fire so you can get out of a building safely; the two basic types of fire detectors are smoke detectors and heat detectors.

Fire emergency plan Established plan of action in case of a fire.

Fire extinguishers Handheld devices used to put out a small fire; specific types of extinguishers are designed to handle specific types of fires.

Fish fumet (foo-MAY) Type of stock made from fish bones that are sweated until they change color and release some of their moisture.

Fish poacher Long, narrow, metal pan with a perforated rack used to raise or lower the fish so it doesn't break apart.

Fish station chef Chef responsible for preparing and cooking fish and seafood in a restaurant. Also called poissonier.

Fixed cost Business expense that is the same from one month to the next, such as rent.

Fixed menu Menu that offers the same items every day.

Fixed seating plan Seating plan that uses set, staggered meal times (such as 6 p.m., 8 p.m., and 10 p.m.), enabling the kitchen to work at a steady, reliable pace.

Flat fish Fish with both eyes on the same side of their heads.

Flat omelet Round open-face omelet that is cooked in a pan and then often baked to produce a dense product that is cut in wedges.

Flattop range Cooking unit with a thick solid plate of cast iron or steel set over the heat source; provides an indirect, less intense heat than an open burner. Pots and pans are set directly on a flattop, which is ideal for items that require long, slow cooking.

Flatware Utensils used at the table or for serving; includes knives, forks, and spoons.

Flavor Taste, aroma, texture, sound, and appearance of a food.

Flight of cheeses Offering a number of different cheeses at the same time.

Floor scale Device at floor level at a receiving area, used for weighing bulky and heavy packages.

Flow of food Route food takes from the time a kitchen receives it to the time it is served to the customer.

Fluting Giving pie crusts a decorative edge; it is done by squeezing the dough between your fingers or using a special tool.

Foccacia (foh-KAH-chee-ah) Large, flat Italian bread, traditionally flavored with olive oil and herbs.

Fonds de cuisine (FAHND du kwee-ZEEN) French term for stocks; translates as "foundations of cuisine."

Food chopper Machine that holds food in a rotating bowl that passes under a hood where blades chop the food. Some units have hoppers or feed tubes and interchangeable disks for slicing and grating. Also called a buffalo chopper.

Food Guide Pyramid Tool developed by the USDA to help people find a balance between food and physical activity; it is based on choosing foods from five basic food groups.

Food mill Tool used to strain and purée at the same time. Has a flat, curving blade that is rotated over a disk by a hand-operated crank to purée foods.

Food processor Machine used to grind, mix, blend, crush, and knead foods; it houses the motor separate from the bowl, blades, and lid.

Foodborne illness Illness that results from eating contaminated foods.

Food-safety audit Inspection of a food service establishment by a representative of the local health department.

Food-safety system System of precautionary steps that take into account all the ways food can be exposed to biological, chemical, or physical hazards.

Forced food method Menu-pricing method that is determined by the market; the choices your customers actually make in a restaurant. It takes into account loss and spoilage and assumes that food that is at a high risk of loss or spoilage should have a higher price. It also includes volume in the calculation. The lower the volume, the higher the price (and vice versa).

Forequarter Front quarter of an animal carcass, such as beef.

Foresaddle Front portion of the saddle of an animal carcass, such as veal.

Forged blade Knife blade made from a single piece of heated metal; it is dropped into a mold and then the metal is cut free and hammered into the correct shape.

Fork tender Describes foods that are fully cooked and allow a knife or fork to slide all the way into the food easily.

Formulas Term that bakers and pastry chefs often use for recipes; it points out the importance of accuracy in all aspects of baking.

Free-form loaf Loaf of bread that is not pressed into a mold or a pan but is shaped by hand into an oval, a round ball, or another shape.

Freestanding mixer Mixer that sits on the floor and is typically used in commercial bakeries; it can stand about 5 feet high and weigh 3,000 pounds.

Freestone Describes fruit that has flesh that separates easily from its pit.

French buttercream Yellow buttercream icing made by adding sugar and butter to whipped egg yolks.

French service Elaborate style of service based on serving a meal in three courses: the first course, or entrée, the second course, and the dessert.

French toast Piece of bread dipped in a mixture of milk and eggs, fried until golden brown on both sides, and then served with syrup, fruit, or other toppings.

Frenching Technique of scraping clean the bones for roasts or chops before they are cooked.

Fresh cheeses Moist, soft cheeses that typically have not ripened or significantly aged.

Freshwater fish Fish that live in freshwater ponds, lakes, rivers, and streams.

Frittata (free-TAH-ta) Type of flat omelet made by pouring eggs mixed with other ingredients into a pan, cooking the mixture, and then finishing it in a hot oven.

Fritters Small deep-fried pieces made by dipping food items, such as fruit, in a batter or other coating and then frying.

Front of the house In a restaurant, the dining room area.

Front waiter Person second in line of responsibility after the captain. Helps the captain take orders, makes sure tables are set properly for each course and that food is delivered properly to the correct tables.

Frozen soufflé Frozen mousse, also called a parfait.

Fully cooked Describes food that is cooked all the way through or to the doneness requested by a customer.

Fungi Single-celled or multi-celled organisms (plural of fungus). May be beneficial; such as a mold used to produce cheese; may be a biological hazard, such as a fungus that causes a foodborne illness.

G

Game General term for meat of wild mammals and birds.

Ganache Emulsion made from chocolate and a liquid, typically heavy cream.

Garde manger (GAHRD mohn-ZHAY) Person or persons responsible for cold food preparation.

Gaufrette (go-FRET) Cut typically made by a mandoline. Means "waffle" in French.

Gauge (GAGE) Thickness of the material of which cookware is made.

Gelatin Protein processed from the bones, skin, and connective tissue of animals; it is used as a gelling agent to thicken and stabilize foams or liquids.

General safety audit Review of the level of safety in an establishment.

Germ Smallest part of a grain; the germ can produce a new plant and contains most of the grain's oils and many vitamins and minerals.

German buttercream Rich buttercream icing made from adding butter to pastry cream.

Giblet bag Small bag in the cavity of a whole bird; includes the liver, stomach, and neck of the bird.

Gizzard Stomach of a bird, such as a chicken gizzard.

Glassware Glass containers for liquids, including water glasses, wine glasses, champagne goblets, cocktail or liquor glasses, beer mugs, pitchers, and carafes.

Glaze Stock that is simmered long enough to produce an intense flavor and a very syrupy consistency.

Glazed fish Whole fish that has been dipped in water and then frozen several times to build up a layer of ice.

Glucose (GLOO-kohs) What our bodies turn carbohydrates into to use as fuel for warmth and for muscle, brain, and nervous system function.

Gluten Network of long, stretchy strands that trap the carbon dioxide given off by yeast when kneading dough.

Gluten window test Test to check the strength of the gluten in dough—pinch off a piece of dough to see if it stretches without tearing and if it is thin enough for some light to come through.

Gnocchi (NYOH-kee) Italian dumpling.

Goujonette (goo-zhohn-NET) Straight cut of fish, sometimes called a fish finger, usually about the width of a thumb.

Grain (of meat) Direction that the fibers in the meat are running.

Grains Seeds of cereal grasses.

Grand sauces Five basic sauces. They are: brown sauce, velouté sauce, béchamel sauce, tomato sauce, and Hollandaise. Also called mother sauces and leading sauces.

Granité Frozen dessert made from a flavored water base; it has large ice crystals, similar to shaved ice.

Granton edge Knife edge that has a series of ovals ground along the edge of the blade to prevent moist foods from sticking to the blade while slicing.

Granulated sugar Ordinary white sugar that is refined from sugar cane or sugar beets.

Gratin dish Shallow baking dish made of ceramic, enameled cast iron, or enameled steel.

Grating cheeses Solid, dry cheeses that have a grainy consistency; they are grated or shaved on food rather than cut into slices because of their crumbly texture.

Griddle Thick cast iron or steel plate used as a cooking surface heated from below. Foods are cooked directly on this surface, which is usually designed with edges to contain foods and a drain to collect used oil and waste.

Grill Cooking unit with a rack over a radiant heat source. Grills that burn wood or charcoal require special ventilation. Restaurant units use gas or electric heat sources.

Grill station chef Chef responsible for all the grilled items made in a restaurant. Also called grillardin.

Grillardin (gree-yar-DAHN) Chef responsible for all the grilled items made in a restaurant. Also called grill station chef.

Grilled sandwich Sandwich that is assembled, the outside surface of the bread spread with butter, and then cooked directly on a heat source, usually a griddle. Also known as a griddled sandwich.

Grilling Dry heat method of cooking that uses a grill to cook food, with the heat source below the grill.

Grit Degree of coarseness or fineness of a sharpening stone or steel.

Grits Type of cornmeal made from yellow or white corn.

Gross profit method Menu-pricing method that determines a specific amount of money that should be made from each customer who comes into the restaurant.

Grosse pièce (GROHSS pee-YES) Method of serving a main item in a presentation or buffet in which a large part of the main item is left unsliced.

Guacamole (gwo-kah-MOH-lee) Mexican dip made from mashed avocado, seasoned with lime or lemon juice, tomatoes, cilantro, onions, and chiles.

HACCP Hazard Analysis Critical Control Plan; a system for maintaining food safety; often pronounced "HAS-sup."

Hanging scale Device at a receiving area, used for weighing large items that can be lifted on a hook, such as a side of beef.

Hard cheeses Cheeses with a drier texture than semi-soft cheeses and a firmer consistency. They slice and grate easily.

Hard dough Basic yeast dough made with the bare essentials—flour, yeast, salt, and water. Also called lean dough.

Hare Type of larger rabbit; it is usually wild.

Hash Breakfast mixture of pan-fried chopped meat (typically corned beef), potatoes, and seasonings.

Hash browns Finely chopped or grated potatoes, pressed down in a pan or on a griddle to brown on one side and then flipped to brown on the other side.

Haunch Hindquarters of a game animal, such as a deer, consisting of the leg and the loin.

Hazard analysis Review of the ways foods may become unsafe during handling, preparation, and service.

Hazard Communication Program Part of an effective safety program; it includes several important documents that can be used as evidence that reasonable care was taken if someone is injured.

Hazard Communication Standard (HCS) Also known as Right-to-Know or HAZCOM; a health regulation that makes sure an employer tells all employees about any chemical hazards present on the job.

Headed and gutted fish Drawn fish with the head cut off.

Heat lamp Light with a special bulb placed directly above an area where food is held.

Heat transfer How efficiently heat passes from cookware to the food inside it.

Heel Widest, thickest point of a knife blade, closest to the handle; used for cutting tasks that require some force, such as cutting hard vegetables, bones, and shells.

Heimlich maneuver Emergency procedure performed to remove an obstruction from the throat of a choking victim.

Heirloom plant Variety of fruit or vegetable that existed many years ago, before produce was grown for mass-market consumption. Grown from heirloom seeds that have been saved by farmers.

Herbs de Provence (AIRBS duh proVAWNS) Dried herb mixture associated with France's Provence region. Includes basil, thyme, marjoram, rosemary, sage, fennel seeds, and lavender.

Herbs Leaves, stems, and roots of various plants; used either fresh or in a dried form to flavor dishes.

Hero sandwich Large, closed sandwich that uses a long thin loaf of bread (often called a hero loaf). Hero sandwiches (or heroes) are known by different names in different parts of the country. Also called submarines, grinders, po' boys, or hoagies.

High-sodium food Food that has a strong salty taste and that contains a significant amount of sodium.

Hindquarter Rear quarter of an animal carcass.

Hindsaddle Back portion of the saddle of an animal carcass, such as veal.

Holding cabinet Metal container on wheels that can hold large quantities of food or plates on trays, ready to serve.

Hollandaise sauce (HOLL-uhn-daze) One of the grand sauces; made by blending melted or clarified butter into slightly cooked egg yolks.

Hollowware Large objects, decorative or utilitarian, including silver platters, candlesticks, large tea or coffee pots, sauceboats, fondue sets, and cake stands.

Home fries Sliced potatoes that are pan-fried, often with chopped peppers and onions.

Hominy (HOM-uh-nee) Kernel of corn that is processed in lime to remove the hull and make the grain easier to cook and digest.

Hominy grits Meal made from hominy.

Homogenized Process that evenly distributes and emulsifies the fat particles in milk.

Honing Straightening a knife's edge on a whetstone or a steel to sharpen the knife.

Hood systems Fire protection systems installed in the ventilation hood over ranges, griddles, broilers, and deep fat fryers; instead of water, they release chemicals, carbon dioxide, or gases that can smother and put out a fire.

Hors d' oeuvre (or-DERV) Small, savory, flavorful dish, usually consumed in one or two bites. Means "outside the meal" in French.

Hors d' oeuvres varies (or-DERV van-REEZ) Method of presenting hors d'oeuvres. A variety plate for one person with a combination of hors d'oeuvres on it, usually fewer than ten small offerings.

Hot cross bun Sweet yeast bun with an icing cross drizzled on top; they originated in England and were traditionally served on Good Friday and Easter.

Hot plate Device with an electrical heating element typically used to warm coffee and water.

Hotel pan Stainless-steel or plastic container used for cooking, holding, and storage. Available in standard sizes that fit in steam tables and other serving equipment.

Hull Outer layer of a grain; provides a protective coating.

Hummus (HOOM-uhs) Popular Middle-Eastern spread made from mashed or pureed chickpeas that are seasoned and served with pita bread, chips, or raw vegetables.

Husk Loose or firmly attached wrapper on the grain as it grows.

Hydrogenation Process that changes a liquid polyunsaturated fat, such as corn oil, into a solid fat, such as margarine.

I

Icebox cookies Drop cookies made from a batter that is formed into a cylinder, chilled, and then sliced and baked.

Immersion (ih-MER-zhuhn) blender Long handheld machine that houses a motor at one end and a blade on the other end; it is placed directly in the container of the food to be blended.

Income Money coming into a restaurant; also known as earnings or sales.

Income statement Record of earnings (or income), losses, and profits of a business. Also called a profit and loss statement or a P&L.

Individually quick frozen (IQF) Describes fruit that has been frozen whole or in slices or chunks, without any added sugar or syrup.

Infrared thermometer Device for reading the surface temperature of food without touching the food by measuring invisible infrared radiation. Also used in kitchens to check holding temperatures.

Injera (in-JEER-ah) Spongy, sourdough-tasting flatbread used as a utensil to scoop up meat and vegetables in Ethiopia.

Insoluble fiber Fiber that does not dissolve in water. Also called roughage, it cleans our digestive tracts, assisting in waste elimination.

Instant oats Rolled oats that have been partially cooked and then dried before being rolled again.

Inventory List of all the assets in a restaurant, usually organized by category.

Iodized salt Table salt to which iodine has been added.

Italian buttercream White buttercream icing made by adding butter to meringue.

J

Job description Duties to be performed and responsibilities involved for a particular job, as well as the level of education and training needed for the job.

Julienne (JU-lee-ehn) Long, rectangular cut measuring ⅛-inch wide × ⅛-inch thick and 1 to 2 inches in length.

Jus de veau lié (JHOO duh voh lee-AY) Type of brown sauce; made by simmering brown stock with flavorings and aromatics and, in some cases, additional bones or meat trimmings.

K

Kaiser roll (KIGH-zer) Large, round, crusty roll used for sandwiches. Also known as a hard roll or a Vienna roll.

Keel bone Bone that joins the two halves of a breast of poultry.

Kernel Whole seed of a grain before it is crushed or milled.

Kitchen shears Scissors used for kitchen chores such as cutting string and butcher's twine, trimming artichoke leaves, cutting grapes into clusters, and trimming herbs.

Knead Work dough by hand or in a mixer to distribute ingredients.

Kosher salt (KOH-shure) Salt made without any additives; sold in coarse or fine grain styles. Typically flakier than table salt.

Kuchen Sweet, yeast-raised cake filled with fruit or cheese; originally from Germany.

Kugelhopf (KOO-guhl-hof) Light yeast cake, filled with candied fruit, nuts and raisins, traditionally baked in a fluted ring mold; a tradition in Austria as well as Poland, Alsace, and Germany.

L

Ladle Tool with a bowl (1 to 16 oz. capacity) and a long handle for reaching to the bottom of a deep pot.

Lamb Tender meat produced by young, domesticated sheep.

Laminated yeast dough Yeast dough that has fat rolled and folded into it, creating alternating layers of fat and dough, adding flavor and flakiness to the finished product. Also called rolled-in yeast dough.

Lasagna (luh-ZAHN-yuh) Layered baked pasta dish.

Lattices Strips of pie dough laid across the top of the filling to create a cross-hatch effect. (

Lean dough Basic yeast dough made with the bare essentials—flour, yeast, salt and water. Also called hard dough.

Leavener Baking ingredient that increases the volume of a dough or batter by adding air or other gas.

Lecithin Fatty substance in egg yolk, liver, and legumes; it acts as a natural emulsifier.

Legume (LEG-yoom) Plant that has a double-seamed pod containing a single row of seeds. Some varieties have edible seeds, some have edible seeds and pods. Seeds are often removed from the pod and dried.

Lentils Legume shaped like a round disk.

Liabilities Losses that occur when a restaurant business uses up assets without making a profit.

Liaison (lee-AY-zohn) Mixture of egg yolks and cream used to lightly thicken and enrich a sauce.

Limited menu Menu that offers a limited range of choices to the customer, such as four sandwiches, two soups, and a salad for lunch.

Line chef Chef responsible for a particular food. Also called station chef.

Liquid-filled thermometer Thermometer that has either a glass or metal stem filled with a colored liquid. Designed to stay in the food while cooking. Specialized versions measure high temperatures used in candy and jelly making as well as deep fat frying.

Liquor (shellfish) Natural juices of a shellfish.

Lo mein (low mane) Asian-style noodle, often purchased fresh.

Loaf pans Rectangular pans used for simple cakes and quick breads.

Logo Drawing, picture, or other symbol that identifies your restaurant. It should be instantly recognizable.

Low boy Undercounter reach-in refrigerator unit for storing a small amount of ingredients within easy reach at a workstation.

Lox Cured salmon.

Lozenge (LOZ-enj) Diamond-shaped cut measuring ½-inch long, ½-inch wide, and ¼-inch thick.

M

Macaroni Common name used to refer to pasta in general.

Maillard reaction Color change seen in food that contains protein when it is heated. The food turns brown.

Main-course salad Salad that is the main course.

Maître d' (MAY-truh DEE) Person running the dining room portion of the restaurant; also responsible for training service personnel, working with the chef on the menu, arranging seating, and taking reservations. Also called maître d'hôtel or dining room manager.

Maître d' hotel (MAY-truh doh-TELL) Person running the dining room portion of the restaurant; also responsible for training service personnel, working with the chef on the menu, arranging seating, and taking reservations. Also called maître d' or dining room manager.

Mandoline Tabletop device used for making slices of various thicknesses.

Marbling Amount of fat present in lean meat.

Marinade (MAHR-i-nahd) Combination of citrus juice, oil, aromatic ingredients, and other flavoring components.

Market quote Statement from a supplier of a product's selling price and an indication of the length of time that the price will be effective.

Market research Information collected to find out what customers like or dislike.

Masa harina (MAH-sah ah-REE-nah) Cornmeal made from posole.

Material Safety Data Sheet (MSDS) A product identification sheet, provided by a chemical manufacturer or supplier, that describes the specific hazards posed by a chemical.

Matignon (mah-tee-YOHN) Aromatic combination of onions, carrots, celery, and ham used to flavor dishes.

Maturation Process of reaching the full potential size and weight, as with fruit.

Mayonnaise (MAY-oh-nayz) Thick, creamy emulsion of oil and egg yolks.

Meal Grains that are milled into fine particles by rolling the grain between steel drums or stone wheels.

Meat grinder Freestanding machine, or an attachment for a mixer, that grinds meat dropped through a feed tube. The meat is pushed through the machine, cut by blades, and forced out.

Meat slicer Slicing tool with a circular blade on a horizontally titled frame across which food is passed by means of a carriage.

Medium dough Lean dough with some sugar and fat added. Pullman loaves, the soft sliced bread used for sandwich making, are made from medium dough. Also called soft dough.

Melon baller Tool used to scoop smooth balls from melons, cheese, and butter.

Menu List of food and drink choices available in a restaurant.

Meringue Mixture of stiffly beaten egg whites and sugar.

Mesclun (MEHS-kluhn) French-style salad mix that often includes baby red romaine, endive, mâche, radicchio, and arugala.

Metric system Standard international system of measurements. Volume measurements are milliliter (ml) and liter (l). Metric weight measurements include the milligram (mg), gram (g), and kilogram (kg).

Micro plane General-purpose tool used for grating food, such as grating the skin of a lemon to produce fine lemon zest.

Milk chocolate Chocolate made with milk powder; it is sweeter than dark chocolate.

Milkfat Fat content of dairy products, measured by the weight of the fat compared to the total weight of the product. Same as butterfat.

Milling Cutting, crushing, rolling, or grinding grain; part of the processing of grain.

Mirepoix Combination of vegetables used as an aromatic flavoring ingredient in many dishes. Common varieties are standard, white, Cajun trinity, matignon, and battuto.

Mise en place (MEEZ uhn PLAHS) French term meaning to gather all the raw ingredients required and have all the equipment and tools necessary to carry out a culinary operation at a workstation.

Mission statement Goal statement of an organization such as a restaurant.

Mixer Machine consisting of a bowl and mixing tool for combining ingredients, primarily for batters and doughs.

Mixing Process of combining ingredients so they are evenly spread throughout the mixture.

Modified à la carte menu Menu on which appetizers and desserts will be priced and served separately. Often the main course includes a soup or salad, a starch, a vegetable, and possibly a beverage. Found in family-style restaurants.

Modified straight-dough mixing method Baking method in which ingredients are added in steps, providing better distribution for fat and sugar. Useful for enriched dough.

Molded cookies Made with stiff dough that is shaped by hand; it can also be stamped, pressed, or piped into carved molds.

Mollusks Shellfish that have soft bodies and no skeletons.

Monosodium glutamate (MSG) Used in much the same way as salt. Provides the umami taste rather than the salty taste and is often associated with Chinese or Japanese food. Enhances the meaty or brothy flavor in meat, poultry, fish, and vegetables. The source of MSG is seaweed.

Monounsaturated fats Fats that come from plants and are liquid at room temperature. Considered healthy fats, they help balance cholesterol levels in the blood.

Mousse Aerated dessert made with a flavored base, gelatin, egg foam, and whipped cream.

Muesli (MYOOS-lee) Swiss version of granola; a mixture of cereal (such as oats and wheat), dried fruit, nuts, bran, and sugar; eaten with milk or yogurt.

Mutton Meat from sheep that is over 16 months old; it is tougher than lamb and has a strong, gamey taste.

Mutual supplementation Combining foods, such as rice and beans, to create a complete protein with all the essential amino acids.

N

Nappé (nap-AY) French term used to describe a sauce that has been properly thickened (thick enough to lightly coat foods).

Neutral stock Another name for white beef stock; it has a very mild flavor and a light color.

Nibs Cleaned cocoa kernels, removed from their hard outer shell.

Nonverbal feedback Form of feedback that is not spoken.

No-reservation policy Restaurant policy of not accepting reservations but instead serving customers on a first-come-first-served basis.

Numerator Top number in a fraction. (The bottom number is the denominator).

Nutrients Parts of food our bodies use.

Nutritional balance Providing enough calories to meet energy needs and enough specific nutrients to promote health.

Nutritive value All the benefits a food might have for our bodies; nutritional value.

Nuts Dried fruit of a tree.

O

Oat groats Whole grain of the oat, with the hull removed.

Oatmeal Coarsely ground oats; cooked as a hot cereal or used in baking.

Obesity (oh-BEE-city) Condition of being dangerously overweight and prone to health risks. (

Oblique (ob-LEEK) cut Cut for vegetables where sides are neither parallel nor perpendicular but cut on an angle, with the vegetable rolled after each cut. Used for long, cylindrical vegetables such as parsnips, carrots, and celery.

Obstructed airway maneuver (Heimlich maneuver) Emergency procedure performed to remove the obstruction from the throat of a choking victim.

Occupation Safety and Health Administration (OSHA) Federal agency that is charged with keeping the workplace safe.

Offal (AH-full) Organs and other portions of an animal, including the liver, heart, kidneys, and tongue. Also known as variety meat.

Omega-3 fatty acids Type of polyunsaturated fat found in some plants and in all fish; they are linked to reducing the risk of stroke and heart attack and improving brain growth and development.

Omelet Blended eggs cooked in a sauté pan with or without other ingredients. It can be folded, rolled, or finished in an oven and served flat.

Omelet pan Shallow skillet with very short, sloping sides; often has a nonstick coating.

On the half shell Method of serving shellfish in which they are opened and served on one of their shells.

One-stage cooling method Safely cooling foods to below 41° F within four hours to avoid foodborne illness.

Opaque (o-PAKE) Indicates that light will not travel through an object.

Open-burner range Electric or gas-fueled cooking unit with a set of adjustable open burners. Pots and pans are set directly on an electric element or on a grid over a gas flame.

Open-faced sandwich Sandwich made with one slice of bread and topped with ingredients.

Organic leavener Yeast, a living organism, which is used to increase the volume of dough.

Orientation Period of time during which new employees learn about the business and their roles in it.

Oven spring In baking, last stage of rising that determines volume; it occurs in the oven when gluten strands expand rapidly and trap steam, allowing full size to be reached.

Over egg Fried egg cooked and then flipped once during frying.

Ovo-lacto vegetarian Person who does not eat meat, poultry, and fish but does eat eggs and dairy products.

P

P&L Record of earnings (or income), losses, and profits of a business. Also called a profit and loss statement or an income statement.

Palette knife Tool with a long, flexible blade and a rounded end; used for turning cooked or grilled foods and spreading fillings or glazes. Sometimes used in baking. Also called a straight spatula.

Pan broiling Dry heat method of cooking very much like sautéing, except no fat is used.

Pan frying Dry heat method of cooking in which food is cooked in hot oil in a pan.

Pan loaf Loaf of bread made by pressing dough into a mold or a pan.

Pan proofing Allowing dough to rise one last time outside of the oven before baking.

Pan-dressed fish Fish with the fins removed, and sometimes the head and tail cut off; usually small enough to fit easily in a pan and make a single serving.

Panini Italian version of a pressed sandwich.

Parasites Multi-celled organisms that can cause illness when eaten; roundworms are an example. Potential biological hazards.

Parboiled Moist heat method of cooking in which food is cooked at 212° F.

Parchment paper Grease-resistant, nonstick, heatproof paper; often used to line pans.

Parcooked grain Grain that is processed by partially cooking it.

Parcooked Stands for "partially cooked." Method of cooking in which the food is not cooked fully.

Parfait Frozen mousse, also called a frozen soufflé.

Paring knife Small knife with 2- to 4-inch blade; used mainly for trimming and peeling fruits and vegetables.

Parisienne (pah-REE-see-ehn) **scoop** Melon baller with a scoop at each end, one larger than the other.

Par-stock list Listing of the quantity of supplies you need to have on hand in a restaurant to make every item on the menu.

Pasta Italian for dough; used to describe the category of starchy foods made from shaped dough that includes flour and liquid. Typically cooked in boiling or simmering water.

Pasteurized Heated at high temperature to kill harmful bacteria.

Pastry bag Cone-shaped bag with two open ends. Fill the bag with dough or whipped cream, apply a decorative tip to the pointed end, and then squeeze the bag to add fillings to pastries, make delicate cookies, and apply decorative finishes to cakes and pastries.

Pastry blender Tool with a crescent-shaped loop of thin wires attached to a handle; it is used to mix fat into flour when you make a pastry dough.

Pastry brush Brush used to apply egg wash and to butter pans and muffin tins; made of soft, flexible nylon or unbleached hog bristles.

Pastry chef Chef responsible for making pastry and other desserts. Also called pâtissier.

Pastry wheel Cutting tool with a round blade mounted on a handle; roll the blade over pastry dough to make a single, clean cut. Blade may be straight or scalloped to make a decorative edge.

Pâté (pah-TAY) Well-seasoned ground meat, fish, poultry, or vegetables mixture that has been baked. Usually served cold.

Pâte fermentée (pah-TAY fer-mahn-TAY) In baking, piece of dough saved from one batch and added at the end of mixing to the next batch. French term for "old dough."

Pâté mold Oven cookware that is deep, rectangular, and made of metal, sometimes with hinged sides.

Pathogen Disease-producing organism, such as bacteria, viruses, parasites, or fungi.

Pâtissier (pah-tee-SYAY) Chef responsible for making pastry and other desserts. Also called pastry chef.

Paupiette (pah-pee-YET) Thin fish fillet that is rolled up before it is cooked.

Paysanne (pahy-SAHN) Rustic type of cut that produces ½-inch square by⅛-inch thick pieces. Means "peasant" in French.

Pearl barley Barley that has been polished to remove the bran.

Pearl grain Grain that has the bran completely removed.

Peas Legume that is round.

Pectin Substance naturally found in certain fruits; it is used to thicken a liquid.

Peel Large, flat wooden or metal paddle used to slide bread onto baking stones and to retrieve loaves when they are done.

Performance evaluation Meeting at which a manager and an employee talk about whether the employee has met expectations, based on the job description.

Perishable goods Foods, such as meats and milk, that must be properly wrapped or kept cold until they can be stored in a refrigerator or freezer.

Pest management Approach to controlling and eliminating rodents, insects, and other pests from the kitchen by keeping the kitchen clean; maintaining the building, especially doors, windows, roof, and drains; covering garbage; and using pesticides when necessary.

Physical hazards Object that falls into food and can cause injury or illness.

Physical inventory Counting the actual assets in a restaurant, such as the number of cans or boxes on a shelf.

Physical leavener Steam or air incorporated into a batter, causing it to increase in volume.

Pickup In baking, the first stage of mixing ingredients.

Pierogi (peer-OH-gee) Polish half-moon-shape dumplings with a sweet or savory filling.

Pilaf (PEE-lahf) Rice dish made by cooking grain in a little oil or butter before a measured amount of liquid is added.

Pin bones Fish bones found in the middle of the fillet.

Piped cookies Drop cookies made of soft dough that can be piped through a pastry bag to form decorative shapes.

Pita bread (PEE-tah) Flat round or oval Middle Eastern bread; also known as pocket bread. When cut in half, each half forms a pocket that can be filled as a sandwich.

Pith The white, bitter, and indigestible layer that is just below the outer skin of a citrus fruit.

Planetary mixer Mixer with a stationary bowl and a mixing tool that moves within it, like a planet orbiting the sun. Three standard attachments are a paddle, a whip, and a dough hook.

Plate cover Metal cover placed over a plate of food to keep the food warm on its way to the customer.

Plate presentation The way you put food into a dish or on a plate.

Platform scale Device on a platform of a receiving area, used for weighing bulky and heavy packages.

Platter service Serving style typically used for banquets in which completely prepared hot food is delivered from the kitchen in large platters to a table and then served to guests without plating at a side table. Also called Russian service.

Poaching Moist heat method of cooking in which food is cooked at 160° F to 170° F.

Poissonier (pwah-sawng-YAY) Chef responsible for preparing and cooking fish and seafood in a restaurant. Also called fish station chef.

Polenta (poh-LEHN-tah) Italian cornmeal porridge; also Italian cornmeal.

Polished grain Grain with the bran completely removed.

Polyunsaturated fats Fats that come from plants; they are liquid at room temperature.

Poolish (poo-LEESH) In baking, a wet dough starter with a consistency like pancake batter.

Portable refrigeration cart Movable refrigerator units for temporary refrigeration or off-site catering.

Portion control Controlling the quantity of particular foods by using appropriately sized servings.

Portion scale Scale that measures the weight of a small amount of food or ingredient (typically a portion). Can be reset to zero to allow for the weight of a container or weigh more than one ingredient at a time.

Portion Serving size for one person, expressed in pieces, weight, or volume.

Portioning food Serving the correct amount of a particular food.

Posole (poh-SOH-leh) Whole kernel of corn, with the germ and bran still intact. It is soaked in a solution of lime and water to make the hull softer and easier to digest.

Potentially hazardous foods Foods that, because of conditions or the nature of the food itself, provide a friendly environment for the rapid growth of pathogens.

Poultry Refers to any domesticated bird used for human consumption.

Pre-ferment In baking, a dough mixture that starts the fermentation process before the final mixing of all the ingredients, giving the dough time to develop more gluten strength and depth of flavor. Also called a dough starter.

Prep chef Chef responsible for washing and peeling vegetables and fruits, cutting meat, and preparing any other ingredients that will be used by other chefs.

Presentation side The most attractive side of a food item.

Pressed sandwich Sandwich that is toasted on a heavy, two-sided cooking press that compresses and grills it until it are hot and heated through on the inside.

Pressure steamer Cooking unit that heats water under pressure in a sealed compartment, allowing it to reach temperatures above the boiling point. Cooking time is controlled by automatic timers that open the exhaust valves, releasing steam pressure so the unit can be opened safely.

Pricing factor Factor by which a raw food cost is multiplied to arrive at the price of a menu item.

Pricing system comparison chart Aids in making pricing decisions; it shows a comparison of the prices from various pricing methods, two competitor's prices, the final decision for the menu price, and your value judgments.

Primal cuts Cuts made to saddles or quarters of meat that meet uniform standards for beef, veal, pork, and lamb.

Prime cost method Used in cafeteria operations; it prices menu items and also calculates the cost of preparing the menu item.

Prime cost Raw food cost plus the direct cost of labor involved in preparing a menu item.

Prix fixe menu (PREE FEKS) Menu that offers a complete meal, often including a beverage, for a specific price, allowing a diner to choose one selection from each course. Similar to table d'hôte menu.

Processed cheese Cheese made from one or more cheeses that have been finely ground, mixed together with other non-dairy ingredients, heated, and poured into a mold.

Processed grain Grain prepared to use as foods.

Producer Person or business selling items to a restaurant. Also called vendor, purveyor, or supplier.

Product specifications Description of a product, including its size, quality, grade, packaging, color, weight, or count.

Profit and loss statement Record of earnings (or income), losses, and profits of a business. Also called a P&L or an income statement.

Profit Earnings (money coming into a restaurant) minus expenses (money spent by a restaurant).

Promotion Extra effort taken, in addition to advertising, to make a restaurant business known and get people interested in coming to it.

Proofer Special box used in baking that holds dough as it rises. Some models have thermostats to control heat and are able to generate steam.

Protein Nutrient our bodies need to grow and to replace worn out tissues and cells; it comes from foods such as meat, fish, eggs, milk, and legumes.

Psychological factors Factors that take into account how a customer perceives a specific menu item. Customers may psychologically associate high-end menu items, such as lobster, caviar, or truffles, with a higher price.

PUFI mark Mark from the United States Department of Commerce that indicates a facility has passed a Type 1 inspection.

Pullman loaf Long loaf of bread that is baked in a rectangular pan with a lid. A slice from a Pullman loaf is square on all sides.

Purée (pyur-AY) To process food until it has a soft, smooth consistency. Very fine paste made by cooking a flavorful ingredient until it is very soft and then straining it or using a food processor or blender to chop it very fine.

Purée soup Hearty soup made by simmering a starchy ingredient such as dried beans or potatoes along with additional vegetables, meats, or aromatics in a broth or other liquid and then puréeing it to the appropriate texture.

Purveyor Person or business selling items to a restaurant. Also called vendor, supplier, or producer.

Q

Quarter fillets Common name for the four fillets cut from a flat fish.

Quarters (of meat) Four pieces of an animal carcass that are made by dividing two sides.

Quiche (KEESH) Baked egg dish made by blending eggs with cream or milk and other ingredients and baking in a pie shell.

Quick bread Type of bread that is quick to make because baking soda or baking powder, instead of yeast, is used for leavening, resulting in a ready-to-use batter rather than a dough that needs fermentation time.

Quick-cooking oats Oats that are parcooked before they are cut and rolled into flakes.

Quinoa (KEEN-wah) Grain originally grown in South America; it has a round kernel and becomes fluffy and light when you cook it.

R

Radiant heat Heat transferred by rays that come from a glowing, or red hot, heat source such as burning coals, flames, or a hot electric element.

Raft Name for a clarification that has cooked enough to form a mass and rise to the surface of a simmering consommé. (See clarification.)

Ramekin (RAM-I-kin) Baking dish that is round and straight-edged; comes in various sizes.

Range Similar to a stovetop on a home oven; used to heat food in pots and pans.

Ratites (RAT-ites) Family of flightless birds, such as the ostrich, emu, and rhea. Their meat is a rich red color, lean, and low in fat.

Ravioli (rav-ee-OH-lee) Italian for "little wraps"; made by layering a filling between two sheets of pasta and then cutting out filled squares, rounds, or rectangles.

Raw bar Bar or counter at which raw shellfish is served.

Raw food cost Cost of all the ingredients that went into a single serving of the dish.

Reach-in Full-size refrigerator with a door that opens and shelves for storing food. May be a single unit or part of a bank of units.

Receptionist In formal restaurants, the person who assists the maître d' in greeting guests and taking telephone reservations. Referred to as the host or hostess in casual restaurants.

Recipe conversion factor (RCF) Amount you multiply a recipe's ingredients or yield to scale it up or down.

Recipe Written record of the ingredients and preparation steps needed to make a particular dish.

Recovery time Time it takes for a pan to heat up again after food is added.

Refined grains Grains that have been processed to remove some or all of the bran and germ; this process removes fiber, vitamins, and minerals from the grain.

Refined starch Starch (corn, rice, or potatoes) that has been processed enough to remove all but the starch itself; examples include cornstarch and arrowroot.

Refrigerated drawer Small undercounter refrigerator drawer within easy reach at a workstation.

Relish Sauce with a chunky texture, typically fruit- or vegetable-based and made with a sweet and sour flavoring. Served hot or cold.

Reservation policy Restaurant policy of accepting reservations.

Retail cuts Cuts made to subprimal pieces of meat to prepare smaller pieces, such as steaks, chops, roasts, stews, or ground meat.

Retarder Refrigerated cabinet used by bakers to slow down fermentation.

Ricer Device in which cooked food, typically potatoes, is pushed through a pierced container, resulting in rice-like pieces.

Rind (RYND) Surface of a cheese.

Ring-top range Cooking unit with thick concentric plates or rings of cast iron or steel set over the heat source. Removing one or more rings provides more intense direct heat.

Ripening Process when a fruit stops maturing and begins converting starches to sugar, changes color, and becomes ready to eat.

Risotto (rih-ZOT-toh) Creamy rice dish typically made with arborio rice, a short-grain rice.

Rivet Piece used to attach the handle of the knife to the blade; lies flush with the surface of the handle.

Roast station chef Chef responsible for all the roasted items cooked in a restaurant. Also called rôtisseur.

Roasting Dry heat method of cooking in which food is cooked by hot air trapped inside an oven. Roasting typically means you are preparing larger pieces of food than for baking.

Roasting pan Pan used for roasting and baking; has low sides and comes in various sizes. Roasting racks are placed inside the pan to hold foods as they cook so the bottom, sides, and top of the food all are cooked evenly.

Rock salt Salt that is less refined than table salt; not generally consumed.

Rolled oats Made by steaming oat groats and then rolling them into flat flakes; also called old-fashioned oats.

Rolled omelet Type of omelet made by stirring the egg mixture to produce curds and then rolling it out of the pan onto a plate.

Rolled-in yeast dough Yeast dough that has fat rolled and folded into it, creating alternating layers of fat and dough, adding flavor and flakiness to the finished product. Also called laminated yeast dough.

Rolling boil Description of liquid that is rapidly boiling.

Rondelles (rahn-DELLS) Round shapes produced by cutting through any cylindrical vegetable, such as a carrot or cucumber. Means "rounds" in French.

Room service Delivery of food to a hotel room; the food must be delivered quickly to keep it warm and fresh.

Rôtisseur (roh-tess-UHR) Chef responsible for all the roasted items cooked in a restaurant. Also called roast station chef.

Round fish Fish with eyes on both sides of their heads.

Roundsman Roving chef who fills in for absent chefs or assists chefs in other stations. Also called swing chef or tournant.

Roux (ROO) Cooked paste of wheat flour and a fat used to thicken simmering liquids, producing a sauce.

Rubbed-dough method In baking, process of cutting fat into chunks, chilling it, and then rubbing it into flour. This prevents the fat from fully combining with the flour and promotes flakiness.

Rubber spatula Scraping tool with a broad, flexible rubber or plastic tip. Used to scrape food from the inside of bowls and pans and also to mix in whipped cream or egg whites.

Russian service Serving style typically used for banquets in which completely prepared hot food is delivered from the kitchen in large platters to a table and then served to guests without plating at a side table. Also called platter service.

Rye berries Whole kernel of rye.

Rye flakes Rye kernels that have been cracked and rolled.

S

Sabayon Egg foam made by whipping egg yolks and sugar over heat.

Sachet d'épices Bag of fresh and dried herbs and spices tied up in a piece of cheesecloth; used to flavor a dish.

Saddle Half of an animal carcass, such as veal; it includes the right and left sides.

Safe foods Foods that won't make you sick or hurt you when you eat them.

Salad Combination of raw or cooked ingredients, served cold or warm, and coated with a salad dressing.

Salad dressing Use to flavor salads and sometimes to hold a salad together.

Salamander Small broiler used primarily to brown or melt foods.

Sales Money coming into a restaurant; also known as earnings or income.

Salsa Cold sauce or dip made from a combination of vegetables, typically tomatoes, onions, chilies, peppers, and other ingredients, and often seasoned with salt, pepper, and lime juice.

Saltwater fish Fish that live in oceans, seas, and the water of bays and gulfs.

Sanitizing Using either heat or chemicals to reduce the number of disease-causing organisms on a surface to a safe level.

Saturated fats Fats that come from animal sources (except for coconut oil and palm oil); they are usually solid at room temperature.

Saucepan Pan that has straight or slightly flared sides and a single long handle.

Saucepot Pot that is similar in shape to a stockpot but not as large. Has straight sides and two loop-style handles to ease lifting.

Saucier (saw-see-YAY) Chef responsible for sautéed dishes and accompanying sauces prepared in a restaurant. Also called sauté station chef.

Sauté pan Shallow, general-purpose pan.

Sauté station chef Chef responsible for sautéed dishes and accompanying sauces prepared in a restaurant. Also called saucier.

Sautéing Dry heat method of cooking in which food is cooked quickly, often uncovered, in a very small amount of fat in a pan over high heat.

Sauteuse (SAW-toose) Sauté pan that is wide and shallow with sloping sides and a single long handle.

Sautoir (SAW-twahr) Sauté pan that has straight sides and a long handle; often referred to as a skillet.

Savory (SAY-va-ree) Meaty or brothy flavor; the umami flavor.

Scale To change the amount of recipe ingredients to get the yield you need. Scale up to increase the yield or scale down to decrease it.

Scaling In baking, weighing liquid and solid ingredients to get precise measurements.

Scimitar Knife with a long curved blade, used for portioning raw meats.

Scone Rich biscuit that sometimes contains raisins and is served with butter, jam, or thick cream.

Scoring Slashes cut on the top of dough to release steam that builds up during baking.

Scotch barley Barley that retains most of its bran; also called pot barley.

Scrambled eggs Blended eggs that are stirred as they cook in a sauté pan or double boiler over low to medium heat.

Sea salt Salt made by evaporating sea water; not significantly refined.

Seams (in meat) Membranes that connect muscles, in cuts of meat.

Searing Dry heat method of cooking in which food is cooked, usually uncovered, in a small amount of fat just long enough to color the outside of the food.

Seasonings Ingredients that enhance, balance, or cut the richness in foods without changing the flavor of the food significantly.

Second chef Executive chef's principal assistant, responsible for scheduling personnel and temporarily replacing the executive chef or other chefs as needed. Also called sous-chef.

Seed Portion of a plant capable of producing a new plant.

Semi-soft cheeses Cheeses that are more solid than soft cheeses and retain their shape. They may be mild or strongly flavored as a result of the process used to make them.

Semisweet chocolate Dark chocolate that has more sugar than bittersweet chocolate.

Semolina flour (seh-muh-LEE-nuh) Pale yellow flour made from durum wheat. Semolina flour has a high protein content and makes an elastic dough; it is widely used for pasta.

Separate-course salad Salad that refreshes the appetite and provides a break before dessert.

Sequencing Arranging slices to overlap one another in the order they were cut.

Serrated edge Knife edge that has a row of teeth; works well for slicing foods with a crust or firm skin.

Service cart Cart used in a dining area to carry food and provide a work surface for carving, plating, assembling, and preparing dishes beside a table. Specialized carts are used to flame-finish (flambé) dishes, display pastry, warm food in a chafing dish, and prepare salads.

Service style How food and drink is delivered to a guest. Also called table service.

Serviceware Dishware and utensils used in the dining room, on or off the table.

Setting priorities In terms of mise en place, deciding which tasks are most important.

Seviche (seh-VEE-chee) Latin American dish of fish and seafood that is cooked in citrus juice and flavored with onions, chiles, and cilantro. A traditional cold appetizer or hors d'oeuvre. Also spelled ceviche.

Shallow poaching Moist cooking method in which food is cooked in just enough liquid to create some steam in the pan.

Sheet pan All-purpose baking pan; it is shallow and rectangular, with sides generally no higher than one inch.

Shellfish Aquatic animals protected by some type of shell. Can be one of two types: mollusks and crustaceans.

Shellfish stock Type of stock made by sautéing shellfish shells (lobster, shrimp, or crayfish) until bright red.

Sherbet Frozen desert that is similar to a sorbet but has meringue incorporated to make it lighter.

Shirred (SHURD) **eggs** Eggs topped with cream and baked in a small dish until they set.

Shortcake Dessert made with a foundation of biscuits topped with fruit and whipped cream, such as strawberry shortcake.

Shrimp cocktail Cold, steamed shrimp served with a spicy cocktail sauce; a traditional cold appetizer.

Shucked shellfish Seafood that has been removed from its shell.

Side salad Salad served on the plate to accompany the main dish.

Sides (of meat) Two halves of an animal carcass that are made by cutting down the length of the backbone.

Side-table service Service style in which dishes are prepared or finished off at the table on a mobile cart with a heat source.

Sil pad Flexible pan liner made of silicone that provides a nonstick, heat-resistant surface. Can be used repeatedly.

Silverskin Tough membrane that surrounds some cuts of meat; it is somewhat silver in color and is generally removed before cooking.

Simmering Moist heat method of cooking in which food is cooked at 170° F to 185° F.

Simple carbohydrates Carbohydrates that contain one sugar or two sugars. Found in fruit, milk, and refined sugars and are digested quickly.

Simple syrup Mixture of equal amounts of sugar and water, brought to a boil to dissolve the sugar crystals.

Single-rack dishwasher Dishwasher that processes small loads of dishes quickly.

Skewers (SKEW-ers) Long, thin, pointed rods made of wood or metal, used to cook meat, fish, poultry, or vegetables.

Skimmer Tool used to remove food from stocks and to skim fat from the tops of liquids; has a flat, perforated bowl and a long handle.

Slicer Knife with a long thin blade used to make smooth slices in a single stroke.

Smallware Hand tools, pots, and pans used for cooking.

Smoker Used for smoking and slow-cooking foods, which are placed on racks or hooks, allowing foods to smoke evenly. Some units can be operated at either cool or hot temperatures.

Smoothie Cold drink made by mixing fresh fruit (such as bananas and strawberries), juice, and ice in a blender until thick and smooth; can also contain milk or yogurt.

Smothering Dry heat method of cooking that is a variation of sweating; food, typically vegetables, is cooked covered over a low heat in a small amount of fat until food softens and releases moisture.

Sneeze guard See-through barrier that protects foods in a service station from cross-contamination caused by a sneeze. People can see and reach food under the guard.

Soda bread Quick bread leavened with baking soda and an acid ingredient, usually buttermilk, and shaped into a round loaf.

Sodium chloride Chemical name for salt.

Soft dough Lean dough with some sugar and fat added. Pullman loaves, the soft sliced bread used for sandwich making, are made from soft dough. Also called medium dough.

Soft, rind-ripened cheeses Soft cheeses that have been ripened by being exposed to a spray or dusting of "friendly" mold.

Soluble fiber Fiber that dissolves in water; foods that contain it help us feel full and help lower cholesterol levels in the blood.

Sommelier (suhm-uhl-YAY) Person responsible for buying and storing wines, maintaining proper wine inventory, counseling guests about wine choices, and serving wine properly at the table. Also called wine steward.

Sorbet Frozen dessert made from a flavored base that is frozen and aerated in an ice cream maker.

Soufflé (soo-FLAY) Light, puffed, baked egg dish.

Soufflé dish Baking dish that is round and straight-edged; comes in various sizes.

Soup and vegetables station chef Chef responsible for hot appetizers, pasta courses, and vegetable dishes. Also called entremetier.

Soup station chef Chef responsible for stocks and soups.

Sourdough In baking, a tangy, slightly sour dough starter made from wild yeast. Can be kept alive for a long time.

Sous-chef (SU-chef) Executive chef's principal assistant, responsible for scheduling personnel and temporarily replacing the executive chef or other chefs as needed. Also called second chef.

Spaetzle (SHPET-zuhl) Popular Austrian and German dumpling.

Spice blends Combination of spices (and in some cases, herbs) used to flavor a dish.

Spices Aromatic dried seeds, flowers, buds, bark, roots, or stems of various plants used to flavor food.

Spine Non-cutting edge of a knife blade.

Spiral mixer Mixer used for bread doughs in which the bowl turns instead of the mixing tool, which is a spiral-shaped hook.

Sponge In baking, a thick, batter-like mixture created when yeast is combined with water and some flour.

Springform pans Baking pan consisting of a hinged ring that clamps around a removable base; used for cakes that might otherwise be difficult to unmold.

Stabilizing In mixing, the step after which the mixed ingredients reach maximum volume and the mixer is slowed down to break the large air bubbles into smaller ones, providing better texture.

Stamped blade Knife blade made by cutting blade-shaped pieces from sheets of previously milled steel.

Standard breading Process of coating food prior to cooking it (typically by pan frying or deep frying). Involves dusting the food with flour, dipping it in beaten eggs, and then covering it in breadcrumbs.

Standard mirepoix Type of mirepoix consisting of 2 parts onion, 1 part carrot, and 1 part celery; used to flavor a dish.

Standardized ingredients Ingredients that have been processed, graded, or packaged according to established standards (eggs, shrimp, and butter, for instance).

Standardized recipe Recipe tailored to suit the needs of an individual kitchen.

Starch slurry Mixture of a refined starch and cold water used to thicken simmering liquids, producing a sauce.

Station chef Chef responsible for a particular type of food. Also called line chef.

Steam table Large freestanding unit that keeps food hot while it is being served. It holds several inserts or hotel pans, under which is steaming hot water. Large steam tables have a thermostat to control the heating elements that maintain the desired temperature.

Steamer Set of stacked pots or bamboo baskets with a tight-fitting lid. The upper pots or baskets have perforated bottoms so steam can gently cook or warm the contents of the pots or baskets. In a metal steamer, water is placed in the bottom pot and it is placed on the range. Bamboo steamers are generally placed over water in a wok.

Steaming Moist heat method of cooking in which food is in a closed pot or steamer and the steam trapped in the pot or steamer circulates around the food.

Steam-jacketed kettle Freestanding or tabletop cooking unit that circulates steam through the walls, providing even heat for cooking stocks, soups, and sauces. Units may tilt or have spigots or lids.

Steel Tool used to maintain a knife's edge between sharpenings; it is a rod made of textured steel or ceramic.

Stenciled cookies Delicate drop cookies made with batter that is spread very thin, sometimes onto sheet pans into stencils to make a perfect shape. They are often rolled or curled while still warm.

Stewing Combination cooking method in which food is first seared and then gently cooked in flavorful liquid. Stewing indicates that food is cut into smaller pieces and then cooked in enough liquid to completely cover the ingredients.

Stir frying Dry heat method of cooking that is a variation of sweating; food is typically cooked in a wok quickly and evenly while you constantly stir and toss it.

Stir-ins Savory or sweet ingredients that can be chopped up and added to muffins and quick breads; examples are vegetables, fruit, nuts, cheese, and chocolate.

Stock base Highly concentrated liquid or a dry powder that is mixed with water to make a stock.

Stock Flavorful liquid used primarily to prepare soups, sauces, stews, and braises. Made by simmering bones, shells, or vegetables, mirepoix, herbs, and spices in a liquid (typically water).

Stockpot Large pot that is taller than it is wide and has straight sides. Used to cook large quantities of liquid, such as stocks or soups. Some stockpots have a spigot at the base so the liquid can be drained off without lifting the heavy pot.

Stollen (STOH-len) Sweet, loaf shaped yeast bread filled with dried fruit and topped with icing and cherries; it is the traditional Christmas bread of Germany.

Stone Hard pit that covers a seed, such as the pit in a peach or apricot.

Straight dough-mixing method Method most commonly used for mixing yeast dough, in which all the ingredients are mixed together at the same time.

Strategies Skills and techniques you will use to get a job done.

Streusel (STRU-sel) Crumbly mixture of fat, sugar, and flour that may include spices and nuts; often applied to muffins or quick breads.

Sturgeon (STURH-jen) Large fish whose eggs are made into caviar.

Subprimal cuts Next level of cuts made to meat after cutting primals from saddles or quarters. Subprimal cuts can be trimmed, packed, and sold to restaurants or butcher shops.

Sugar glaze Thin liquid made by dissolving sugar in water, applied to baked goods.

Sugar syrup Concentrated solution of sugar and water.

Sunny-side-up egg Fried egg cooked without turning, keeping the yolk intact.

Superfine sugar Granulated sugar that is more finely ground than ordinary sugar so it dissolves more easily.

Supplier Person or business selling items to a restaurant. Also called vendor, purveyor, or producer.

Suprême Boneless, skinless breast of poultry with one wing joint still attached.

Sweating Dry heat method of cooking in which food, typically vegetables, is cooked uncovered over a low heat in a small amount of fat until food softens and releases moisture.

Sweet rich dough Lean dough that has added butter, oil, sugar, eggs, or milk. Also called enriched dough.

Swing chef Roving chef who fills in for absent chefs or assists chefs in other stations. Also called roundsman or tournant.

Swiss braiser Large shallow freestanding cooking unit, used to cook large quantities of meats or vegetables at one time. Most units have lids that allow the unit to function as a steamer.

Symmetrical In terms of plate presentation, equal numbers of items on either side of the plate.

T

Table d'hôte menu (TAH-buhl DOHT) Menu that offers a complete meal—from an appetizer to a dessert, and often including a beverage—for a set price.

Table salt Salt that is refined to remove other minerals.

Table service How food and drink is delivered to a guest. Also called service style.

Table tent menus Folded cards placed directly on restaurant tables to tell customers about specials.

Tableware Dishware and utensils used by customers.

Tabling method Method of cooling melted chocolate by moving it around on a marble slab.

Tagliatelle (tag-lee-ah-TEHL-ee) Thin, ribbon-shaped pasta.

Tahini (ta-HEE-nee) Sesame paste.

Take-out service Buying prepared food and taking it home or to work, as an alternative to cooking at home or eating in a restaurant.

Tang Section of knife blade that extends into the handle.

Tapenade (top-en-ODD) Vegetable-based dip made from black olives, capers, anchovies, garlic, herbs, lemon juice, and olive oil; originally from France's Provence region.

Taper ground edge Knife edge in which both sides of the blade taper smoothly to a narrow V-shape.

Tapioca Thickener made from the cassava root, a starchy tropical tuber; often used to thicken fruit pie fillings and to make pudding.

Tare weight Weight of the container holding food on a scale. To account for the tare, reset the scale to zero while weighing the empty container. If the scale cannot be reset, subtract the tare from the total weight.

Tart pans Baking pans made of tinned steel or ceramic, with short, often scalloped sides and usually a removable bottom. May be round, square, or rectangular.

Tasks In terms of mise en place, smaller jobs that lead to the completion of an assignment.

Taste One of the senses; the taste and aroma of a food.

Tea sandwich Simple, small sandwich usually made with firm, thinly sliced pullman loaves. Can be made both as closed sandwiches and as open-faced sandwiches. Also called finger sandwich.

Temperature danger zone Temperature range from 40° F to 140° F in which disease-causing organisms thrive.

Tempering Warming a liaison (a mixture of cream and egg yolks) so the yolks will not overcook when added to a simmering sauce or other liquid. Process of correctly crystallizing chocolate.

Termination Firing of an employee.

Terrine (teh-REEN). Mold for pâté. When pâté is served in its mold, it is call a terrine.

Terrine mold Oven cookware usually made of pottery but can also be metal, enameled cast iron, or ceramic. Produced in a wide range of sizes and shapes; some have lids.

Theme In a restaurant, the decorations, lighting, food, and prices, all of which tie the restaurant concept together.

Thermistor (therm-IS-tor) thermometer Thermometer that uses a resistor (electronic semiconductor) to measure temperature. Gives a fast reading, can measure temperature in thin and thick foods, and is not designed to stay in food while cooking.

Thermocouple thermometer Thermometer that uses two fine wires within the probe to measure temperature. Gives the fastest reading, can measure the temperature in thin and thick foods, and is not designed to stay in food while cooking.

Timeline In terms of the mise en place, a schedule that tells you when certain tasks have to be completed.

Time-temperature abused food Food that has been held in the temperature danger zone for more than two hours.

Tomato concassé (kon-kah-SAY) Tomatoes that have been peeled, seeded, and diced.

Tongs Tool used for picking up items such as meats, vegetables, or ice cubes. Can be spring-action or scissor-type.

Top crust Large piece of pastry dough rolled out and placed on top of a filled shell before baking.

Tortellini (tohr-te-LEEN-ee) Italian for "little twists"; made by cutting out circles or squares of fresh pasta, adding a filling, and then folding and twisting the dough to get a specific shape.

Tortilla (tohr-TEE-yuh) Mexico's unleavened bread; it is round, flat, made of corn or flour.

Tossed salad Salad in which all the ingredients are combined together with dressing.

Tournant (toor-NAHN) Roving chef who fills in for absent chefs or assists chefs in other stations. Also called swing chef or roundsman.

Tournée Paring knife with a curved blade. Also called a bird's beak knife.

Training Period of time during which new employees learn the job and practice it.

Trancheur (tran-SHUR) Person in charge of carving and serving meats or fish and their accompaniments. Also called carver.

Trans fats Also called trans fatty acids, a potentially harmful type of fat created from the process of hydrogenation; has been linked to heart disease.

Translucent (trans-LU-cent) Indicates that light will pass through an object.

Tray stand Used by serving staff to place a tray holding multiple dishes close to the table where they will be served.

Trueing Process of straightening a knife's edge.

Trussing Tying or securing poultry or other food so it maintains its shape while cooking.

Tube pans Baking pans with a center tube of metal that conducts heat through the center of the batter; they bake heavy batters evenly and quickly, without over-browning the outside of the cake. Typically made of thin metal with or without a nonstick coating.

Tuber Fleshy portion of certain plants; usually grows underground.

Turner Tool with a broad blade and a short handle that is bent to keep the user's hands off hot surfaces. Used to turn or lift hot foods from hot cookware, grills, broilers, and griddles. Also called an offset spatula or a flipper.

Turntable Used to decorate cakes; you turn the cake on the turntable with one hand while the other is free to use a palette knife, pastry bag, or cake comb.

Twice-baked cookies Cookies made of dough formed into a large log-shaped cookie and baked, and then cut into slices and baked a second time for a very crisp texture.

Two-stage cooling method Safely cooling foods to 70° F within two hours and to below 41° F within four hours, for a total cooling time of six hours to avoid foodborne illness.

Udon (oo-DOHN) Asian-style noodle, often purchased fresh.

Umami (OO-mam-ee) Meaty or brothy flavor; also called savory.

Undercounter dishwasher Dishwasher that holds portable dish or glass racks to allow for easy transfer of clean and dirty dishes.

Undercounter reach-in Refrigerator unit under a work-station counter used for storing a small amount of ingredients within easy reach. Also called a low boy.

Unsweetened chocolate Chocolate that has no sugar added. Also called baker's chocolate or chocolate liquor.

Utility knife Smaller, lighter version of a chef's knife, with a 5- to 7-inch blade.

Variable cost Business expense that can vary from one day, week, month, or year to the next, such as the cost of food.

Variety meat Organs and other portions of an animal, including the liver, heart, kidneys, and tongue. Also known as offal.

Veal Meat that comes from a young calf, generally two to three months old. It has delicate, tender flesh that is pale pink.

Vegan Person who eats no animal products whatsoever and consumes only plant-based foods.

Vegetable station chef Chef responsible for vegetables and starches.

Vegetable stock Type of stock made from a combination of vegetables.

Vegetarian Person who, for religious, ethical, economic, or nutritional reasons, does not eat meat, poultry, and fish.

Velouté (veh-loo-TAY) One of the grand sauces; a white sauce made by thickening a poultry, fish, or shellfish stock with a blond roux.

Vendor Person or business selling items to a restaurant. Also called supplier, purveyor, or producer.

Venison Meat from any member of the deer family, including antelope, caribou, elk, and moose.

Verbal feedback Form of feedback that is spoken.

Verbal warning When a manager tells an employee about the need for improvement in a particular area.

Vertical chopping machine (VCM) Machine to grind, whip, blend, or crush large amounts of foods. A motor at the base is permanently attached to a bowl with blades; the hinged lid must be locked in place before the unit will operate.

Vinaigrette (vin-eh-GRETT) Salad dressing made by combining oil and vinegar into an emulsion.

Viruses Biological hazards that can cause illness when they invade a cell and trick the cell into making more viruses.

Volume Measurement of the space occupied by a solid, liquid, or gas.

Walk-in Large refrigeration or freezing unit that usually has shelves arranged around the walls of the unit.

Walkout Customer who leaves the table without paying the bill.

Warewashing station Area for rinsing, washing, and holding tools, pots and pans, and dishes. Also includes trashcans, sinks, garbage disposals, and dishwashing equipment.

Water activity (A_w) Measurement of the amount of moisture available in a food; the scale runs from 0 to 1.0, with water at 1.0 and potentially hazardous foods at .85 or higher.

Water bath Method of baking in which a container of food is put into a pan of water in the oven to control the heat.

Water-soluble vitamins The B and C vitamins; they dissolve in water and are transported throughout the body in the bloodstream. They must be replenished often because they cannot be stored for very long in the body.

Well method Quick-bread mixing method in which liquid ingredients are added to a depression in the dry ingredients and mixed minimally to avoid overmixing.

Wet aging The process of storing meat in vacuum packaging under refrigeration to make it more tender and flavorful.

Wheat berries Whole kernels of wheat.

Whetstone Hard, fine-grained stone for honing tools; a general term for sharpening stones.

Whip Hand mixing tool similar to a whisk but narrower and with thicker wires; used to blend sauces or batters without adding too much air.

Whisk Hand mixing tool with thin wires in a sphere or an oval shape used to incorporate air for making foams. Very round whisks incorporate a large amount of air and are sometimes called balloon whisks.

White chocolate Chocolate made from cocoa butter, sugar, and milk powder; it contains no chocolate liquor.

White mirepoix Type of mirepoix consisting of onions, parsnips, celery, and, in some cases, leeks; used to flavor white stocks and soups.

White pepper Ripe berries from the pepper vine that have been allowed to dry and have had the husks removed; used as a seasoning.

White rice Rice with all its bran removed.

White stock Type of stock made from unroasted bones. The bones may be blanched before simmering.

Whole grains Grains that still have most of the nutrients found in the germ and bran, including fiber, vitamins, and minerals.

Wild rice Seed of an aquatic grass. Not related to other rice, but cooked like them.

Wine steward Person responsible for buying and storing wines, maintaining proper wine inventory, counseling guests about wine choices, and serving wine properly at the table. Also called sommelier.

Wok Cookware for fast stovetop cooking, such as stir-frying; has tall, sloped sides. Once one ingredient cooks, you can push it up the sides, leaving the hot center free for another ingredient.

Wontons (WAHN-tahns) Type of Chinese dumpling made with a fresh pasta wrapper; often served as an appetizer or with soup.

Work flow Planned movement of food and kitchen staff as food is prepared. (3.1) In terms of mise en place, putting ingredients, tools, and equipment in a logical order for accomplishing your task.

Work lines Geometric arrangements of workstation equipment and storage areas, designed to fit the available kitchen space and improve efficiency of staff. Examples include straight-line, L-shaped, U-shaped, back-to-back, and parallel.

Work sections Combination of workstations in a kitchen.

Work sequencing In terms of mise en place, doing the right thing at the right time.

Work simplification In terms of mise en place, getting things done in the fewest steps, the shortest time, and with the least amount of waste.

Worker's compensation Program run by each state that provides help for employees who are hurt or who become sick because of an accident on the job.

Workstation Work area containing equipment and tools for accomplishing a specific set of culinary tasks.

Wrap Sandwich that is rolled up, or otherwise enclosed in an edible wrapper, such as a tortilla.

Wrappers Type of pasta used in Asian cooking. Sold in squares, rounds, or rectangles; can be made from wheat or rice flour.

Written warning When a manager documents in writing to an employee that there is need for improvement in a particular area.

Y

Yeast hydration In baking, the soaking process that activates yeast.

Yield Measured output of a recipe, expressed in total weight, total volume, or total number of servings of a given portion.

Z

Zest Colored outer layer of citrus fruit peel.

Zester Tool that cuts away thin strips of citrus fruit peel, leaving the bitter pith.

Index

C

D

E

F

I

Q

R

Credits

Peter Arnold, Inc. 161

© Culinary Institute of America 172, 175, 185, 186-187, 189, 199, 203, 205, 261, 263-265, 267-268, 286, 300-306, 311, 319, 342, 352, 364, 397, 402-403, 406, 416, 420-422, 427, 439, 461, 463

© Dorling Kindersley 196, 203, 245, 370, 428

Andy Crawford © Dorling Kindersley 139, 173, 433

Neil Fletcher and Matthew Ward © Dorling Kindersley 200

Dave King © Dorling Kindersley 203, 246, 251, 363-364, 386, 437, 460

David Murray © Dorling Kindersley 200, 262, 383, 386

David Murray and Jules Selmes © Dorling Kindersley 151, 285, 300, 386, 420

Ian O'Leary © Dorling Kindersley 141, 200, 249, 319, 346

Gary Ombler © Dorling Kindersley 196

Roger Phillips © Dorling Kindersley 201, 320

Magnus Rew © Dorling Kindersley , Courtesy of Houston's, Winter Park. Orlando, Florida 224

Tim Ridley © Dorling Kindersley 203

Clive Streeter © Dorling Kindersley 257, 259, 278, 41

Jerry Young © Dorling Kindersley 324

Spencer Jones/Getty Images, Inc.—Photodisc 368

Ryan McVay/Getty Images, Inc.—Photodisc 186

Kim Steele/Getty Images, Inc.—Photodisc 188

Ken Whitmore/Getty Images, Inc.—Stone Allstock 129

Johnson & Wales University 462

© Masterfile 99, 105, 136

Ron Dahlquist/PacificStock.com 390

Tom Pantages 144

Pearson Education/PH College 295

Richard Embery/Pearson Education/PH College 154, 185, 196, 198, 261, 268, 282, 285, 297, 341-342, 393, 395, 406, 438-439, 441, 463-466

Mike Gallitelli/Pearson Education/PH College 168

Michal Heron/Pearson Education/PH College 120, 207-208

Les Lougheed/Pearson Education/PH College 207

Vincent P. Walter/Pearson Education/PH College 167, 170, 176, 178

Pearson Learning/Pearson Learning Photo Studio 141

Pearson Science 19

CNRI/Photo Researchers, Inc. 84

John Bavosi/Photo Researchers, Inc. 85

Simon Fraser/Photo Researchers, Inc. 115

John Radcliffe Hospital/Science Photo Library/Photo Researchers, Inc. 83

SINCLAIR STAMMERS/Photo Researchers, Inc. 162

Bill Aron/PhotoEdit Inc. 28

Amy Etra/PhotoEdit Inc. 190

Bonnie Kamin/PhotoEdit Inc. 67

Michael Newman/PhotoEdit Inc. 27

Bart's Medical Library/Phototake NYC 125

Yoav Levy/Phototake NYC 125

Martin Rotker/Phototake NYC 125

Russ Lappa/Prentice Hall School Division 378

Christopher LaMarca/Redux Pictures 127

Reuters/Corbis/Bettmann 98

© San Jamar 2009, www.sanjamar.com 175

Shutterstock.com 2-10, 13, 15-22, 24-27, 29-31, 33-34, 42, 45, 48, 50-51, 54-66, 68, 71-74, 76-82, 85-92, 94, 96-97, 102, 106-109, 111-114, 116-119, 123, 126, 130, 132-134, 137-138, 141, 145-146, 148, 153, 155-158, 160, 162, 165-166, 168-169, 171, 174, 176-177, 179, 180-182, 184-185, 187, 189, 191-192, 195, 197-198, 200-203, 205, 213-214, 216-217, 220, 225-237, 239-240, 243-244, 246-247, 250-251, 253-255, 258, 260, 265-267, 269-274, 276-277, 279-281, 283-284, 287-288, 290, 292, 294-296, 298-299, 304, 306-308, 310, 314-318, 320-323, 325-328, 330-332, 334-338, 340-344, 347-348, 350-354, 356-362, 365-366, 369, 372-377, 379-385, 387-388, 391-392, 394-402, 404-409, 411-417, 419, 422-425, 429, 430-432, 434-436, 438, 440-443, 445-450, 452-460, 464-470, 472-477

U.S. Department of Agriculture 36, 37